12/93

THE
NEW ENGLAND
SCHOOL OF ART
& DESIGN

The
ENCYCLOPEDIA
of
Wood

REVISED·EDITION

Sterling Publishing Co., Inc.   New York

Published in 1989 by Sterling Publishing Co., Inc.
387 Park Avenue South, New York, N.Y. 10016
Distributed in Canada by Oak Tree Press Ltd.
℅ Canadian Manda Group, P.O. Box 920, Station U
Toronto, Ontario, Canada M8Z 5P9
Reprint of the 1987 revised edition of "Wood Handbook:
Wood as an Engineering Material," issued by the
U.S. Government Printing Office
Manufactured in the United States of America
All rights reserved
Library of Congress Catalog Card No.: 88-39719
Sterling ISBN 0-8069-6994-6 Paper

Preface

Forests, distinct from all their other services and benefits, supply a basic raw material—wood—which from the earliest times has furnished mankind with necessities of existence and with comforts and conveniences beyond number.

One major use has always been in structures, particularly in housing. But despite wood's long service in structures, it has not always been used efficiently. In these days when the Nation is trying to utilize its resources more fully, better and more efficient use of the timber crop is vital.

Authorship

As an aid to more efficient use of wood as a material of construction, this handbook was prepared by the Forest Products Laboratory, a unit of the research organization of the Forest Service, U.S. Department of Agriculture. The Laboratory, established in 1910, is maintained at Madison, WI, in cooperation with the University of Wisconsin. It was the first institution in the world to conduct general research on wood and its utilization. The vast accumulation of information that has resulted from its engineering and allied investigations of wood and wood products over seven decades—along with knowledge of everyday construction practices and problems—is the chief basis for this handbook.

Purpose

This handbook provides engineers, architects, and others with a source of information on the physical and mechanical properties of wood and how these properties are affected by variations in the wood itself. Practical knowledge of wood has, over the years, resulted in strong and beautiful structures, even though exact engineering data were not always available. Continuing research and evaluation techniques promise to permit wider and more efficient utilization of wood and to encourage even more advanced industrial, structural, and decorative uses.

Organization

Individual chapters describe not only the wood itself, but wood-based products as well, together with the principles of how wood is dried, fastened, finished, and preserved from degradation in today's world. Each chapter is climaxed with a list of selected references to provide additional information. A glossary of terms is presented at the end of the handbook.

The problem of adequately presenting information for architects, engineers, and builders is complicated by the vast number of tree species they may encounter in the form of lumber, plywood, particleboard, fiberboard, hardboard, or other building material. To prevent confusion, the common and botanical names for different species mentioned in this volume conform to the official nomenclature of the Forest Service.

To reflect the increasing importance of imported species, information on selected foreign species is included.

English and metric systems of measurement are used for selected sections throughout this book. The system of measurement used in each case has been chosen to reflect the system most commonly used for the information presented in that chapter. This reflects user preference of measurement systems. A conversion table is given on page M1.

In some instances the abbreviations used are those commonly employed in design and construction, even though they are not standard U.S. Government usage: e.g. psi (for pounds-force per square inch) and pcf (for pounds-mass per cubic foot). In the same vein, some of the descriptions of how a particular species reacts to operations such as sawing, machining, polishing, or treating are included in the vernacular of the wood industry.

Acknowledgment

The Forest Products Laboratory appreciates the cooperation of other elements of the Forest Service, and representatives of many universities and segments of industry, in preparing this document. The number of such examples of assistance prevent the mentioning of individual contributions, but their assistance is gratefully acknowledged.

Max A. Davidson and Alan D. Freas, both retired from the Forest Products Laboratory where they were heavily involved in the 1974 revision of this book, volunteered to help in this latest revision in the coordinating, editing, and processing of the manuscript.

Revised June 1987

Contents

Chapter 1

Characteristics and Availability of Woods Commercially Important to the United States

Characteristics and Availability of Woods Commercially Important to the United States*

Through the ages the unique characteristics and comparative abundance of wood have made it a natural material for homes and other structures, furniture, tools, vehicles, and decorative objects. Today, for the same reasons, wood is prized for a multitude of uses.

All wood is composed of cellulose, lignin, ash-forming minerals, and extractives formed in a cellular structure. Variations in the characteristics and volume of the four components and differences in cellular structure result in some woods being heavy and some light, some stiff and some flexible, some hard and some soft. For a single species, the properties are relatively constant within limits; therefore, selection of wood by species alone may sometimes be adequate. However, to use wood to its best advantage and most effectively in engineering applications, the effect of specific characteristics or physical properties must be considered.

Historically, some woods have filled many purposes, while others that were not so readily available or so desirable qualitatively might serve only one or two needs. The tough, strong, and durable white oak, for example, was a highly prized wood for shipbuilding, bridges, cooperage, barn timbers, farm implements, railroad crossties, fenceposts, flooring, paneling, and other products. On the other hand, woods such as black walnut and cherry became primarily cabinet woods. Hickory was manufactured into tough, hard, resilient striking-tool handles. Black locust was prized for barn timbers and treenails. What the early builder or craftsman learned by trial and error became the basis for the decision as to which species to use for a given purpose, and what characteristics to look for in selecting a tree for a given use. It was commonly accepted that wood from trees grown in certain locations under certain conditions was stronger, more durable, and more easily worked with tools, or finer grained than wood from trees in some other locations. Modern wood quality research has substantiated that location and growth conditions do significantly affect wood properties.

The gradual utilization of the virgin forests in the United States has reduced the available supply of large clear logs for lumber and veneer. However, the importance of high-quality logs has diminished as new concepts of wood use have been introduced. Second-growth timber (fig. 1-1), the balance of the old-growth forests, and imports continue to fill the needs for wood in the quality required. Wood is as valuable an engineering material as it ever was, and in many cases technological advances have made it even more useful.

The inherent factors which keep wood in the forefront of raw materials are many and varied, but one of the chief attributes is its availability in many species, sizes, shapes, and conditions to suit almost every demand. It has a high ratio of strength to weight and a remarkable record for durability and performance as a structural material. Dry wood has good insulating properties against heat, sound, and electricity. It tends to absorb and dissipate vibrations under some conditions of use, yet is an incomparable material for such musical instruments as violins. Because of grain patterns and colors, wood is inherently an esthetically pleasing material, and its appearance may be easily enhanced by stains, varnishes, lacquers, and other finishes. It is easily shaped with tools and fastened with adhesives, nails, screws, bolts, and dowels. When wood is damaged, it is easily repaired, and wood structures are easily remodeled or altered. In addition, wood resists oxidation, acid, salt water, and other corrosive agents; has a high salvage value; has good shock resistance; takes treatments with preservatives and fire retardants; and combines with almost any other material for both functional and esthetic uses.

Timber Resources and Wood Uses

In the United States more than 100 woods are available to the prospective user, but it is very unlikely that all are available in any one locality. About 60 native woods are of major commercial importance. Another 30 woods are commonly imported in the form of logs, cants, lumber, and veneer for industrial uses, the building trades, and crafts.

A continuing program of timber inventory is in effect in the United States through cooperation of Federal agencies and the States. As new information regarding timber resources becomes available, it appears in State and Federal publications. One of the most valuable sourcebooks is "An Analysis of the Timber Situation in the United States 1952–2030," U.S. Department of Agriculture Forest Resource Report No. 23 (see Selected References at the end of this chapter for complete listing).

Current information on timber consumption, production, imports, and the demand and price situation is published periodically in a U.S. Department of Agriculture Miscellaneous Publication, "U.S. Timber Production, Trade, Consumption, and Price Statistics." These publications are available from the Superintendent of Documents, U.S. Government Printing Office, Washington, DC 20402.

Hardwoods and Softwoods

Trees are divided into two broad classes, usually referred to as "hardwoods" and "softwoods." Some softwoods, however, are actually harder than some of the hardwoods, and some hardwoods are softer than softwoods. For example, such softwoods as longleaf pine and Douglas-fir produce wood

---

* Revision by Regis B. Miller, Botanist.

Figure 1—1—Reforested area on the Kaniksu National Forest in Idaho. Foreground is stocked with western larch and Douglas-fir reproduced naturally. The central area, edged by mature timber, is a field-planted western white pine plantation.

(M513 614)

that is typically harder than the hardwoods basswood and aspen. Botanically, the softwoods are Gymnosperms or conifers which are plants with naked seeds, that is they are not enclosed in the ovary of the flower. Usually softwoods are cone-bearing plants with needlelike or scalelike evergreen leaves. Some conifers such as the larches and baldcypress, however, lose their needles during autumn and winter. Exam-

ples of softwoods are pines, spruces, redwoods, and junipers. Botanically, the hardwoods are Angiosperms whose seeds are enclosed in the ovary of the flower. Typically they are plants with broad leaves that, with few exceptions in the temperate region, lose their leaves in fall or during the winter. Most imported tropical woods are hardwoods.

Major resources of softwood species are spread across the

United States, except for the Great Plains where only small areas are forested. Species are often loosely grouped in three general producing areas:

Western softwoods

Alaska-cedar	Ponderosa pine
Incense-cedar	Sugar pine
Port-Orford-cedar	Western white pine
Douglas-fir	Western redcedar
White firs	Redwood
Western hemlock	Engelmann spruce
Western larch	Sitka spruce
Lodgepole pine	

Northern softwoods

Northern white-cedar	Eastern white pine
Balsam fir	Eastern redcedar
Eastern hemlock	Eastern spruces
Jack pine	Tamarack
Red pine	

Southern softwoods

Atlantic white-cedar	Southern pine
Baldcypress	Eastern redcedar

With some exceptions, most hardwoods occur east of the Great Plains area (fig. 1−2). The following classification is based on the principal producing region for each wood:

Southern hardwoods

Ash	Magnolia
Basswood	Soft maple
American beech	Red oak
Butternut	White oak
Cottonwood	Sassafras
Elm	Sweetgum
Hackberry	American sycamore
Pecan hickory	Tupelo
True hickory	Black walnut
Honeylocust	Black willow
Black locust	Yellow-poplar

Northern and Appalachian hardwoods

Ash	Hackberry
Aspen	True hickory
Basswood	Honeylocust
Buckeye	Black locust
Butternut	Hard maple
American beech	Soft maple
Birch	Red oak
Black cherry	White oak
American chestnut[1]	American sycamore
Cottonwood	Black walnut
Elm	Yellow-poplar

Western hardwoods

Red alder	Pacific madrone
Oregon ash	Bigleaf maple
Aspen	Paper birch
Black cottonwood	Tanoak
Golden chinquapin	California black oak
Oregon white oak	

Commercial Sources of Wood Products

Softwoods are available directly from the sawmill, wholesale and retail yards, or lumber brokers. Softwood lumber and plywood are used in construction for forms, scaffolding, framing, sheathing, flooring, ceiling, trim, paneling, cabinets, and many other building components. Softwoods may also appear in the form of shingles, sash, doors, and other millwork, in addition to some rough products such as round treated posts.

Hardwoods are used in construction for flooring, architectural woodwork, trim, and paneling. These items are usually available from lumberyards and building supply dealers. Most hardwood lumber and dimension stock are remanufactured into furniture, flooring, pallets, containers, dunnage, and blocking. Hardwood lumber and dimension stock are available directly from the manufacturer, through wholesalers and brokers, and in some retail yards.

Both softwood and hardwood forest products are distributed throughout the United States, although they tend to be more readily available in or near their area of origin. Local preferences and the availability of certain species may influence choice, but a wide selection of woods is generally available for building construction, industrial uses, remanufacturing, and home use.

Use Classes and Trends

Some of the many use classifications for wood are growing with the overall national economy, and others are holding about the same levels of production and consumption. The wood-based industries that are growing most vigorously convert wood to thin slices (veneer), particles (chips, flakes, etc.), or fiber pulps and reassemble the elements to produce plywood, numerous types of particleboard, paper, paperboard, and fiberboard products. Another growing wood industry specializes in producing laminated timbers. Annual production by the lumber industry has continued for a number of years at almost the same board footage. Forest products such as rail-

---

[1] American chestnut is no longer harvested as a living tree, but the lumber is still on the market as chestnut and prices are quoted in the Hardwood Market Report.

road crossties, cooperage, shingles, and shakes appear to be making modest increases in production.

Species of Commercial Importance in the United States

The following brief discussions of the principal localities of occurrence, characteristics, and uses of species, or groups of species, will aid in selecting woods for specific purposes. More detailed information on the properties of these and other species is given in various tables throughout this handbook.

Certain uses listed under the individual species are no longer important. They have been included to provide some information on the historical and traditional uses of the species.

The common and botanical names given for the different species follow the "Checklist of United States Trees," Agriculture Handbook No. 541.

Hardwoods

Alder, Red

Red alder (*Alnus rubra*) grows along the Pacific coast between Alaska and California. It is used commercially along the coasts of Oregon and Washington and is the most abundant commercial hardwood species in these States.

The wood of red alder varies from almost white to pale pinkish brown and has no visible boundary between heartwood and sapwood. It is moderately light in weight, intermediate in most strength properties, but low in shock resistance. Red alder has relatively low shrinkage.

The principal use of red alder is for furniture, but it is also used for sash, doors, panel stock, and millwork.

Ash

Important species of ash are white ash (*Fraxinus americana*), green ash (*F. pennsylvanica*), blue ash (*F. quadrangulata*), black ash (*F. nigra*), pumpkin ash (*F. profunda*), and Oregon ash (*F. latifolia*). The first five species grow in the eastern half of the United States. Oregon ash grows along the Pacific coast.

Commercial white ash is a group of species that consists mostly of white ash and green ash, although blue ash is also included. Heartwood of commercial white ash is brown; the sapwood is light colored or nearly white. Second-growth trees have a large proportion of sapwood. Old-growth trees, which characteristically have little sapwood, are scarce.

Second-growth commercial white ash is particularly sought because of the inherent qualities of this wood; it is heavy, strong, hard, stiff, and has high resistance to shock. Because of these qualities such tough ash is used principally for handles,

(M138 574)

Figure 1 – 2—Mixed northern hardwoods on Ottawa National Forest in Michigan.

oars, vehicle parts, baseball bats, and other sporting and athletic goods. Some handle specifications call for not less than five nor more than 17 growth rings per inch for handles of the best grade. The addition of a weight requirement of 43 or more pounds per cubic foot (pcf) at 12 percent moisture content will assure excellent material.

Oregon ash has somewhat lower strength properties than white ash, but it is used locally for the same purposes.

Black ash is important commercially in the Lake States. The wood of black ash and pumpkin ash runs considerably lighter in weight than that of commercial white ash. Ash trees growing in southern river bottoms, especially in areas that are frequently flooded for long periods, produce buttresses that contain relatively lightweight and weak wood. Such wood is sometimes separated from the heavier tougher ash when sold.

Ash wood of lighter weight, including black ash, is sold as cabinet ash, and is suitable for cooperage, furniture, and shipping containers. Some ash is cut into veneer for furniture, paneling, and wire-bound boxes.

Aspen

Aspen is a generally recognized name applied to bigtooth

aspen (*Populus grandidentata*) and to quaking aspen (*P. tremuloides*). Aspen does not include balsam poplar (*P. balsamifera*) and the other species of *Populus* that are included in the cottonwoods. In lumber statistics of the U.S. Bureau of the Census, however, the term ''cottonwood'' includes all of the preceding species. Also, the lumber of aspens and cottonwood may be mixed in trade and sold either as poplar (''popple'') or cottonwood. The name ''popple'' or ''poplar'' should not be confused with yellow-poplar (*Liriodendron tulipifera*), also known in the trade as ''poplar.''

Aspen lumber is produced principally in the Northeastern and Lake States. There is some production in the Rocky Mountain States.

The heartwood of aspen is grayish white to light grayish brown. The sapwood is lighter colored and generally merges gradually into heartwood without being clearly marked. Aspen wood is usually straight grained with a fine, uniform texture. It is easily worked. Well-seasoned aspen lumber does not impart odor or flavor to foodstuffs.

The wood of aspen is lightweight and soft. It is low in strength, moderately stiff, moderately low in resistance to shock, and has a moderately high shrinkage.

Aspen is cut for lumber (fig. 1−3), pallets, boxes and crating, pulpwood, particleboard, excelsior, matches, veneer, and miscellaneous turned articles.

Basswood

American basswood (*Tilia americana*) is the most important of the native basswood species; next in importance is white basswood (*T. heterophylla*) and no attempt is made to distinguish between them in lumber form. In commercial usage, ''white basswood'' is used to specify the white wood or sapwood of either species.

Basswood grows in the eastern half of the United States from the Canadian provinces southward. Most basswood lumber comes from the Lake, Middle Atlantic, and Central States.

The heartwood of basswood is pale yellowish brown with occasional darker streaks. Basswood has wide, creamy-white or pale brown sapwood that merges gradually into the heartwood. When dry, the wood is without odor or taste. It is soft and light in weight, has fine, even texture, is straight grained and easy to work with tools. Shrinkage in width and thickness during drying is rated as large; however, basswood seldom warps in use.

Basswood lumber is used mainly in venetian blinds, sash and door frames, molding, apiary supplies, woodenware, and boxes. Some basswood is cut for veneer, cooperage, excelsior, and pulpwood.

Beech, American

Only one species of beech, American beech (*Fagus grandifolia*), is native to the United States. It grows in the eastern one-third of the United States and adjacent Canadian pro-

vinces. Greatest production of beech lumber is in the Central and Middle Atlantic States.

Beech varies in color from nearly white sapwood to reddish-brown heartwood in some trees. Sometimes there is no clear line of demarcation between heartwood and sapwood. Sapwood may be 3 to 5 inches thick. The wood has little figure and is of close, uniform texture. It has no characteristic taste or odor.

The wood of beech is classed as heavy, hard, strong, high in resistance to shock, and highly suitable for steam bending. Beech shrinks substantially and therefore requires careful drying. It machines smoothly, is an excellent wood for turning, wears well, and is rather easily treated with preservatives.

Largest amounts of beech go into flooring, furniture, brush blocks, handles, veneer, woodenware, containers, cooperage, and laundry appliances. When treated, it is suitable for railway ties.

Birch

The important species of birch are yellow birch (*Betula alleghaniensis*), sweet birch (*B. lenta*), and paper birch (*B. papyrifera*). Other birches of some commercial importance are river birch (*B. nigra*), gray birch (*B. populifolia*), and western paper birch (*B. papyrifera* var. *commutata*).

Yellow birch, sweet birch, and paper birch grow principally in the Northeastern and Lake States. Yellow and sweet birch also grow along the Appalachian Mountains to northern Georgia. They are the source of most birch lumber and veneer.

Yellow birch has white sapwood and light reddish-brown heartwood. Sweet birch has light-colored sapwood and dark brown heartwood tinged with red. Wood of yellow birch and sweet birch is heavy, hard, strong, and has good shock-resisting ability. The wood is fine and uniform in texture. Paper birch is lower in weight, softer, and lower in strength than yellow and sweet birch. Birch shrinks considerably during drying.

Yellow and sweet birch lumber and veneer go principally into the manufacture of furniture, boxes, baskets, crates, woodenware, cooperage, interior finish, and doors. Birch veneer goes into plywood used for flush doors, furniture, paneling, radio and television cabinets, aircraft, and other specialty uses. Paper birch is used for turned products, including spools, bobbins, small handles, and toys.

Buckeye

Buckeye consists of two species, yellow buckeye (*Aesculus octandra*) and Ohio buckeye (*A. glabra*). They range from the Appalachians of Pennsylvania, Virginia, and North Carolina westward to Kansas, Oklahoma, and Texas. Buckeye is not customarily separated from other species when manufactured into lumber and can be utilized for the same purposes as aspen, basswood, and sap yellow-poplar.

The white sapwood of buckeye merges gradually into the

creamy or yellowish white heartwood. The wood is uniform in texture, generally straight-grained, light in weight, weak when used as a beam, soft, and low in shock resistance. It is rated low on machinability such as shaping, mortising, boring, and turning.

Buckeye is suitable for pulping for paper, and in lumber form has been used principally for furniture, boxes, and crates, food containers, woodenware, novelties, and planing mill products.

Butternut

Butternut (*Juglans cinerea*) is also called white walnut. It grows from southern New Brunswick and Maine west to Minnesota. Its southern range extends into northeastern Arkansas and eastward to western North Carolina.

The narrow sapwood is nearly white, and the heartwood is a light brown, frequently modified by pinkish tones or darker brown streaks. The wood is moderately light in weight—about the same as eastern white pine—rather coarse-textured, moderately weak in bending and endwise compression, relatively low in stiffness, moderately soft, and moderately high in shock resistance. Butternut machines easily and finishes well. In many ways it resembles black walnut especially when stained, but it does not have the strength or hardness. Principal uses are for lumber and veneer, which are further manufactured into furniture, cabinets, paneling, trim, and miscellaneous rough items.

Cherry, Black

Black cherry (*Prunus serotina*) is sometimes known as cherry, wild black cherry, or wild cherry. It is the only native species of the genus *Prunus* of commercial importance for lumber production. It is scattered from southeastern Canada throughout the eastern half of the United States. Production is centered chiefly in the Middle Atlantic States.

The heartwood of black cherry varies from light to dark reddish brown and has a distinctive luster. The sapwood is narrow in old trees and nearly white. The wood has a fairly uniform texture and very good machining properties. It is moderately heavy, strong, stiff, moderately hard, and has high shock resistance and moderately large shrinkage. After seasoning, it is very dimensionally stable.

Black cherry is used principally for furniture, fine veneer panels, architectural woodwork, and for backing blocks on which electrotype plates are mounted. Other uses include burial caskets, woodenware novelties, patterns, and paneling. It has proved satisfactory for gunstocks, but has a limited market for this purpose.

Chestnut, American

American chestnut (*Castanea dentata*) is known also as sweet chestnut. Before chestnut was attacked by a blight, it grew in commercial quantities from New England to northern Georgia. Practically all standing chestnut has been killed by

(M84 0471-3A.)

Figure 1 – 3—Aspen (*Populus tremuloides*) 2 × 4 studs made by the process called SDR (Saw, Dry, and Rip).

blight, and most supplies come from dead timber. There are still some quantities of standing dead chestnut in the Appalachian Mountains, which may be available because of the great natural resistance to decay.

The heartwood of chestnut is grayish brown or brown and becomes darker with age. The sapwood is very narrow and almost white. The wood is coarse in texture; the growth rings are made conspicuous by several rows of large, distinct pores at the beginning of each year's growth. Chestnut wood is moderately light in weight. It is moderately hard, moderately low in strength, moderately low in resistance to shock, and low in stiffness. It seasons well and is easy to work with tools.

Chestnut was used for poles, railroad crossties, furniture, caskets, boxes, crates, and core stock for veneer panels. It appears most frequently now as "wormy chestnut" for paneling, trim, and picture frames.

Cottonwood

Cottonwood includes several species of the genus *Populus*. Most important are eastern cottonwood (*P. deltoides* and varieties), also known as Carolina poplar and whitewood; swamp cottonwood (*P. heterophylla*), also known as cottonwood, river cottonwood, and swamp poplar; black cottonwood (*P. trichocarpa*); and balsam poplar (*P. balsamifera*).

Eastern cottonwood and swamp cottonwood grow throughout the eastern half of the United States. Greatest production of lumber is in the Southern and Central States. Black cottonwood grows in the West Coast States and in western Montana, northern Idaho, and western Nevada. Balsam poplar grows from Alaska across Canada, and in the northern Great Lakes States.

The heartwood of the three cottonwoods is grayish white to light brown. The sapwood is whitish and merges gradually with the heartwood. The wood is comparatively uniform in texture, and generally straight grained. It is odorless when well seasoned.

Eastern cottonwood is moderately low in bending and compressive strength, moderately stiff, moderately soft, and moderately low in ability to resist shock. Black cottonwood is slightly below eastern cottonwood in most strength properties. Both eastern and black cottonwood have moderately large shrinkage.

Some cottonwood is difficult to work with tools because of fuzzy surfaces. Tension wood is largely responsible for this characteristic.

Cottonwood is used principally for lumber, veneer, pulpwood, excelsior, and fuel. The lumber and veneer go largely into boxes, crates, baskets, and pallets.

Elm

Six species of elm grow in the eastern United States: American elm (*Ulmus americana*), slippery elm (*U. rubra*), rock elm (*U. thomasii*), winged elm (*U. alata*), cedar elm (*U. crassifolia*), and September elm (*U. serotina*). American elm is also known as white elm, water elm, and gray elm; slippery elm as red elm; rock elm as cork elm or hickory elm; winged elm as wahoo; cedar elm as red elm or basket elm; and September elm as red elm.

The supply of American elm is threatened by two diseases, Dutch Elm and phloem necrosis, which have killed hundreds of thousands of trees.

The sapwood of the elms is nearly white and the heartwood light brown, often tinged with red. The elms may be divided into two general classes, hard elm and soft elm, based on the weight and strength of the wood. Hard elm includes rock elm, winged elm, cedar elm, and September elm. American elm and slippery elm are the soft elms. Soft elm is moderately heavy, has high shock resistance, and is moderately hard and stiff. Hard elm species are somewhat heavier than soft elm. Elm has excellent bending qualities.

Production of elm lumber is chiefly in the Lake, Central, and Southern States.

Elm lumber is used principally in boxes, baskets, crates, and slack barrels; furniture; agricultural supplies and implements; caskets and burial boxes; and vehicles. The hard elms are preferred for some uses where more strength is required. Elm veneer is used for furniture, for fruit, vegetable, and cheese boxes, for baskets, and for decorative panels.

Hackberry

Hackberry (*Celtis occidentalis*) and sugarberry (*C. laevigata*) supply the lumber known in the trade as hackberry. Hackberry grows east of the Great Plains from Alabama, Georgia, Arkansas, and Oklahoma northward, except along the Canadian boundary. Sugarberry overlaps the southern part of the range of hackberry and grows throughout the Southern and South Atlantic States.

The sapwood of both species varies from pale yellow to greenish or grayish yellow. The heartwood is commonly darker. The wood resembles elm in structure.

Hackberry lumber is moderately heavy. It is moderately strong in bending, moderately weak in compression parallel to the grain, moderately hard to hard, high in shock resistance, but low in stiffness. It has moderately large to large shrinkage but keeps its shape well during seasoning.

Most hackberry is cut into lumber, with small amounts going into dimension stock and some into veneer. Most of it is used for furniture and some for containers.

Hickory, Pecan

Species of the pecan group include bitternut hickory (*Carya cordiformis*), pecan (*C. illinoensis*), water hickory (*C. aquatica*), and nutmeg hickory (*C. myristiciformis*). Bitternut hickory grows throughout the eastern half of the United States. Pecan hickory grows from central Texas and Louisiana to Missouri and Indiana. Water hickory grows from Texas to South Carolina. Nutmeg hickory occurs principally in Texas and Louisiana.

The wood of pecan hickory resembles that of true hickory. It has white or nearly white sapwood, which is relatively wide, and somewhat darker heartwood. The wood is heavy and sometimes has very large shrinkage.

Heavy pecan hickory finds use in tool and implement handles and flooring. The lower grades are used in pallets. Many higher grade logs are sliced to provide veneer for furniture and decorative paneling.

Hickory, True

True hickories are found throughout most of the eastern half of the United States. The species most important commercially are shagbark (*Carya ovata*), pignut (*C. glabra*), shellbark (*C. laciniosa*), and mockernut (*C. tomentosa*).

The greatest commercial production of the true hickories for all uses is in the Middle Atlantic and Central States. The Southern and South Atlantic States produce nearly half of all hickory lumber.

The sapwood of hickory is white and usually quite thick, except in old, slowly growing trees. The heartwood is reddish. From the standpoint of strength, no distinction should be made between sapwood and heartwood having the same weight.

The wood of true hickory is exceptionally tough, heavy, hard, strong, and shrinks considerably in drying. For some purposes, both rings per inch and weight are limiting factors where strength is important.

The major use for hickory is for tool handles, which require high shock resistance. It is also used for ladder rungs, athletic goods, agricultural implements, dowels, gymnasium apparatus, poles, and furniture.

A considerable quantity of lower grade hickory is not suitable for the special uses of high-quality hickory, because of knottiness or other growth features and low density. It is, however, useful for pallets and similar items. Hickory sawdust, chips, and some solid wood are used by the major packing companies to flavor meat by smoking.

Honeylocust

The wood of honeylocust (*Gleditsia triacanthos*) possesses many desirable qualities such as attractive figure and color, hardness, and strength, but is little used because of its scarcity. Although the natural range of honeylocust has been extended by planting, it is found most commonly in the eastern United States, except for New England and the South Atlantic and Gulf Coastal Plains.

The sapwood is generally wide and yellowish in contrast to the light red to reddish brown heartwood. The wood is very heavy, very hard, strong in bending, stiff, resistant to shock, and is durable when in contact with the ground. When available, it is restricted primarily to local uses, such as fenceposts and lumber for general construction. Occasionally it will show up with other species in lumber for pallets and crating.

Locust, Black

Black locust (*Robinia pseudoacacia*) is sometimes called yellow locust or post locust. This species grows from Pennsylvania along the Appalachian Mountains to northern Georgia and Alabama. It is also native to western Arkansas and south-

ern Missouri. The greatest production of black locust timber is in Tennessee, Kentucky, West Virginia, and Virginia.

Locust has narrow, creamy-white sapwood. The heartwood, when freshly cut, varies from greenish yellow to dark brown. Black locust is very heavy, very hard, very high in resistance to shock, and ranks very high in strength and stiffness. It has moderately small shrinkage. The heartwood has high decay resistance.

Black locust is used for round, hewed, or split mine timbers and for fenceposts, poles, railroad crossties, stakes, and fuel. Other uses are for rough construction, crating, and mine equipment. Important former products manufactured from black locust were insulator pins and ship treenails, a use for which the wood was well adapted because of its strength, decay resistance, and moderate shrinkage and swelling.

Magnolia

Commercial magnolia comprises three species—southern magnolia (*Magnolia grandiflora*), sweetbay (*M. virginiana*), and cucumbertree (*M. acuminata*). Other names for southern magnolia are evergreen magnolia, big laurel, bull bay, and laurel bay. Sweetbay is sometimes called swamp magnolia. The lumber produced by all three is simply called magnolia.

The natural range of sweetbay extends along the Atlantic and Gulf coasts from Long Island to Texas, and that of southern magnolia from North Carolina to Texas. Cucumbertree grows from the Appalachians to the Ozarks northward to Ohio. Louisiana leads in production of magnolia lumber.

The sapwood of southern magnolia is yellowish white, and the heartwood is light to dark brown with a tinge of yellow or green. The wood, which has close, uniform texture and is generally straight grained, closely resembles yellow-poplar. It is moderately heavy, moderately low in shrinkage, moderately low in bending and compressive strength, moderately hard and stiff, and moderately high in shock resistance. Sweetbay is much like southern magnolia. The wood of cucumbertree is similar to that of yellow-poplar, and cucumbertree growing in the yellow-poplar range is not separated from that species on the market.

Magnolia lumber is used principally in the manufacture of furniture, boxes, pallets, venetian blinds, sash, doors, veneer, and millwork.

Maple

Commercial species of maple in the United States include sugar maple (*Acer saccharum*), black maple (*A. nigrum*), silver maple (*A. saccharinum*), red maple (*A. rubrum*), boxelder (*A. negundo*), and bigleaf maple (*A. macrophyllum*). Sugar maple is also known as hard maple and rock maple; black maple as black sugar maple; silver maple as white maple, river maple, water maple, and swamp maple; red maple as soft maple, water maple, scarlet maple, white maple, and swamp maple; boxelder as ash-leaved maple, three-leaved maple, and cut-leaved maple; and bigleaf maple as Oregon maple.

Maple lumber comes principally from the Middle Atlantic and Lake States, which together account for about two-thirds of the production.

The wood of sugar maple and black maple is known as hard maple; that of silver maple, red maple, and boxelder as soft maple. The sapwood of the maples is commonly white with a slight reddish-brown tinge. It is from 3 to 5 or more inches thick. Heartwood is usually light reddish brown, but sometimes is considerably darker. Hard maple has a fine, uniform texture. It is heavy, strong, stiff, hard, resistant to shock, and has large shrinkage. Sugar maple is generally straight grained but the grain also occurs as "birdseye," "curly," and "fiddleback" grain. Soft maple is not so heavy as hard maple, but has been substituted for hard maple in the better grades, particularly for furniture.

Maple is used principally for lumber, veneer, railroad crossties, and pulpwood. A large proportion is manufactured into flooring, furniture, boxes, pallets and crates, shoe lasts, handles, woodenware, novelties, spools, and bobbins.

Oak (Red Oak Group)

Most red oak comes from the Southern States, the southern mountain regions, the Atlantic Coastal Plains, and the Central States. The principal species are: Northern red oak (*Quercus rubra*), scarlet oak (*Q. coccinea*), Shumard oak (*Q. shumardii*), pin oak (*Q. palustris*), Nuttall oak (*Q. nuttallii*), black oak (*Q. velutina*), southern red oak (*Q. falcata*), cherrybark oak (*Q. falcata* var. *pagodaefolia*), water oak (*Q. nigra*), laurel oak (*Q. laurifolia*), and willow oak (*Q. phellos*).

The sapwood is nearly white and usually 1 to 2 inches thick. The heartwood is brown with a tinge of red. Sawed lumber of red oak cannot be separated by species on the basis of the characteristics of the wood alone. Red oak lumber can be separated from white oak by the size and arrangement of pores in latewood and because, as a rule, it lacks tyloses in the pores. The open pores of the red oaks make these species unsuitable for tight cooperage, unless the barrels are lined with sealer or plastic. Quartersawed lumber of the oaks is distinguished by the broad and conspicuous rays, which add to its attractiveness.

Wood of the red oaks is heavy. Rapidly grown second-growth oak is generally harder and tougher than finer textured old-growth timber. The red oaks have fairly large shrinkage in drying.

The red oaks are largely cut into lumber, railroad crossties, mine timbers, fenceposts, veneer, pulpwood, and fuelwood. Ties, mine timbers, and fenceposts require preservative treatment for satisfactory service. Red oak lumber is remanufactured into flooring, furniture, general millwork, boxes, pallets and crates, agricultural implements, caskets, woodenware,

and handles. It is also used in railroad cars and boats.

Oak (White Oak Group)

White oak lumber comes chiefly from the South, South Atlantic, and Central States, including the southern Appalachian area. Principal species are: white oak (*Quercus alba*), chestnut oak (*Q. prinus*), post oak (*Q. stellata*), overcup oak (*Q. lyrata*), swamp chestnut oak (*Q. michauxii*), bur oak (*Q. macrocarpa*), chinkapin oak (*Q. muehlenbergii*), swamp white oak (*Q. bicolor*), and live oak (*Q. virginiana*).

The heartwood of the white oaks is generally grayish brown, and the sapwood, which is from 1 to 2 or more inches thick, is nearly white. The pores of the heartwood of white oaks are usually plugged with tyloses. These tend to make the wood impenetrable by liquids, and for this reason most white oaks are suitable for tight cooperage. Chestnut oak lacks tyloses in many of its pores.

The wood of white oak is heavy, averaging somewhat higher in weight than that of the red oaks. The heartwood has moderately good decay resistance.

White oaks are used for furniture lumber, railroad crossties, cooperage, mine timbers, fenceposts, veneer, fuelwood, and many other products. High-quality white oak is especially sought for tight cooperage. Live oak is considerably heavier and stronger than the other oaks, and was formerly used extensively for ship timbers. An important use of white oak is for planking and bent parts of ships and boats, heartwood often being specified because of its decay resistance. It is also used for flooring, pallets, agricultural implements, railroad cars, truck floors, furniture, doors, millwork, and many other items.

Sassafras

The range of sassafras (*Sassafras albidum*) covers most of the eastern half of the United States from southeastern Iowa and eastern Texas eastward.

The wood of sassafras is easily confused with black ash, which it resembles in color, grain, and texture. The sapwood is light yellow and the heartwood varies from dull grayish brown to dark brown, sometimes with a reddish tinge. The wood has an odor of sassafras on freshly cut surfaces.

Sassafras is moderately heavy, moderately hard, moderately weak in bending and endwise compression, quite high in shock resistance, and quite durable when exposed to conditions conducive to decay. It was highly prized by the Indians for dugout canoes, and some sassafras lumber is now used for small boats. Locally, it is used for fenceposts and rails and general millwork.

Sweetgum

Sweetgum (*Liquidambar styraciflua*) grows from southwestern Connecticut westward into Missouri and southward to the Gulf. Lumber production is almost entirely from the Southern and South Atlantic States.

The lumber from sweetgum is usually divided into two classes—sapgum, the light-colored wood from the sapwood, and redgum, the reddish-brown heartwood.

Sweetgum often has interlocked grain, a form of cross grain, and must be carefully dried. When quartersawn, the interlocked grain produces a ribbon stripe that is desirable for interior finish and furniture. The wood is rated as moderately heavy and hard. It is moderately strong, moderately stiff, and moderately high in shock resistance.

Sweetgum is used principally for lumber, veneer, plywood, slack cooperage, railroad crossties, fuel, and pulpwood. The lumber goes principally into boxes and crates, furniture, radio and phonograph cabinets, interior trim, and millwork. Sweetgum veneer and plywood are used for boxes, pallets, crates, baskets, and interior woodwork.

Sycamore, American

American sycamore (*Platanus occidentalis*) is also known as sycamore and sometimes as buttonwood, buttonball-tree, and planetree. Sycamore grows from Maine to Nebraska, southward to Texas, and eastward to Florida. In the production of sycamore lumber, the Central States rank first.

The heartwood of sycamore is reddish brown; sapwood is lighter in color and from 1-1/2 to 3 inches thick. The wood has a fine texture and interlocked grain. It shrinks moderately in drying. Sycamore wood is moderately heavy, moderately hard, moderately stiff, moderately strong, and has good resistance to shock.

Sycamore is used principally for lumber, veneer, railroad crossties, slack cooperage, fenceposts, and fuel. Sycamore lumber is used for furniture, boxes (particularly small food containers), pallets, flooring, handles, and butcher's blocks. Veneer is used for fruit and vegetable baskets, and some decorative panels and door skins.

Tanoak

In recent years tanoak (*Lithocarpus densiflorus*) has gained some importance commercially, primarily in California and Oregon. It is also known as tanbark-oak because at one time high-grade tannin in commercial quantities was obtained from the bark. This species is found in southwestern Oregon and south to Southern California, mostly near the coast but also in the Sierra Nevadas.

The sapwood of tanoak is light reddish brown when first cut and turns darker with age to become almost indistinguishable from the heartwood, which also ages to dark reddish brown. The wood is heavy, hard, and except for compression perpendicular to the grain, has roughly the same strength properties as eastern white oak. Volumetric shrinkage during drying is more than for white oak, and it has a tendency to collapse during drying. It is quite susceptible to decay, but the sapwood takes preservatives easily. It has straight grain, machines and glues well, and takes stains readily.

Because of tanoak's hardness and abrasion resistance, it is an excellent wood for flooring in homes or commercial buildings. It is also suitable for industrial applications such as truck flooring. Tanoak treated with preservative has been used for railroad crossties. The wood has been manufactured into baseball bats with good results. It is also suitable for veneer, both decorative and industrial, and for high-quality furniture.

Tupelo

Tupelo includes water tupelo (*Nyssa aquatica*), also known as tupelo gum, swamp tupelo, and sourgum; black tupelo (*N. sylvatica*), also known as blackgum and sourgum; swamp tupelo (*N. sylvatica* var. *biflora*), also known as swamp blackgum, blackgum, and sourgum; Ogeechee tupelo (*N. ogeche*), also known as sour tupelo, gopher plum, and Ogeechee plum.

All except black tupelo grow principally in the southeastern United States. Black tupelo grows in the eastern United States from Maine to Texas and Missouri. About two-thirds of the production of tupelo lumber is from the Southern States.

Wood of the different tupelos is quite similar in appearance and properties. Heartwood is light brownish gray and merges gradually into the lighter colored sapwood, which is generally several inches wide. The wood has fine, uniform texture and interlocked grain. Tupelo wood is rated as moderately heavy. It is moderately strong, moderately hard and stiff, and moderately high in shock resistance. Buttresses of trees growing in swamps or flooded areas contain wood that is much lighter in weight than that from upper portions of the same trees. For some uses, as in the case of buttressed ash trees, this wood should be separated from the heavier wood to assure material of uniform strength. Because of interlocked grain, tupelo lumber requires care in drying.

Tupelo is cut principally for lumber, veneer, pulpwood, and some railroad crossties and slack cooperage. Lumber goes into boxes, pallets, crates, baskets, and furniture.

Walnut, Black

Black walnut (*Juglans nigra*) is also known as American black walnut. Its natural range extends from Vermont to the Great Plains and southward into Louisiana and Texas. About three-quarters of the walnut timber is produced in the Central States.

The heartwood of black walnut varies from light to dark brown; the sapwood is nearly white and up to 3 inches wide in open-grown trees. Black walnut is normally straight grained, easily worked with tools, and stable in use. It is heavy, hard, strong, stiff, and has good resistance to shock. Black walnut is well suited for natural finishes.

Because of its properties and interesting grain pattern, black walnut is used for furniture, architectural woodwork, decorative panels, gunstocks, cabinets, and interior finish.

Willow, Black

Black willow (*Salix nigra*) is the most important of the many willows that grow in the United States. It is the only one to supply lumber to the market under its own name.

Most black willow is produced in the Mississippi Valley from Louisiana to southern Missouri and Illinois.

The heartwood of black willow is grayish brown or light reddish brown frequently containing darker streaks. The sapwood is whitish to creamy yellow. The wood of black willow is uniform in texture, with somewhat interlocked grain. The wood is light in weight. It has exceedingly low strength as a beam or post and is moderately soft and moderately high in shock resistance. It has moderately large shrinkage.

Willow is cut principally into lumber. Small amounts are used for slack cooperage, veneer, excelsior, charcoal, pulpwood, artificial limbs, and fenceposts. Black willow lumber is remanufactured principally into boxes, pallets, crates, caskets, and furniture. Willow lumber is suitable for roof and wall sheathing, subflooring, and studding.

Yellow-Poplar

Yellow-poplar (*Liriodendron tulipifera*) is also known as poplar, tulip-poplar, and tulipwood. Sapwood from yellow-poplar is sometimes called white poplar or whitewood.

Yellow-poplar grows from Connecticut and New York southward to Florida and westward to Missouri. The greatest commercial production of yellow-poplar lumber is in the South and Southeast.

Yellow-poplar sapwood is white and frequently several inches thick. The heartwood is yellowish brown, sometimes streaked with purple, green, black, blue, or red. These colorations do not affect the physical properties of the wood. The wood is generally straight grained and comparatively uniform in texture. Old-growth timber is moderately light in weight and moderately low in bending strength, moderately soft, and moderately low in shock resistance. It has moderately large shrinkage when dried from a green condition but is not difficult to season and stays in place well after seasoning.

Much of the second-growth yellow-poplar is heavier, harder, and stronger than old growth. Selected trees produce wood heavy and strong enough for gunstocks. Lumber goes mostly into furniture, interior finish, siding, radio cabinets, musical instruments and structural components. Boxes, pallets, and crates are made from lower grade stock. Yellow-poplar is also made into plywood which is used for paneling, furniture, piano cases, and various other special products. In the past, yellow-poplar was used for excelsior and slack-cooperage staves.

Lumber from the cucumbertree (*Magnolia acuminata*) sometimes is included in shipments of yellow-poplar because of its similarity.

Softwoods

Alaska-Cedar

Alaska-cedar (*Chamaecyparis nootkatensis*) grows in the Pacific coast region of North America from southeastern Alaska southward through Washington to southern Oregon.

The heartwood of Alaska-cedar is bright, clear yellow. The sapwood is narrow, white to yellowish, and hardly distinguishable from the heartwood. The wood is fine textured and generally straight grained. It is moderately heavy, moderately strong and stiff, moderately hard, and moderately high in resistance to shock. Alaska-cedar shrinks little in drying, is stable in use after seasoning, and the heartwood is very resistant to decay. The wood has a mild, unpleasant odor.

Alaska-cedar is used for interior finish, furniture, small boats, cabinetwork, and novelties.

Baldcypress

Baldcypress (*Taxodium distichum*) is commonly known as cypress, also as southern-cypress, red-cypress, yellow-cypress, and white-cypress. Commercially, the terms "tidewater red-cypress," "gulf-cypress," "red-cypress (coast type)," and "yellow-cypress (inland type)" are frequently used.

About one-half of the cypress lumber comes from the Southern States and one-fourth from the South Atlantic States (fig. 1–4). It is not as readily available as it was several decades ago.

The sapwood of baldcypress is narrow and nearly white. The color of the heartwood varies widely, ranging from light yellowish brown to dark brownish red, brown, or chocolate. The wood is moderately heavy, moderately strong, and moderately hard. The heartwood of old-growth timber is one of our most decay-resistant woods; but second-growth timber is only moderately decay resistant. Shrinkage is moderately small, but somewhat greater than that of the cedars and less than that of southern pine.

Frequently the wood of certain baldcypress trees contains pockets or localized areas that have been attacked by a fungus. Such wood is known as pecky cypress. The decay caused by this fungus is arrested when the wood is cut into lumber and dried. Pecky cypress, therefore, is durable and useful where water tightness is unnecessary, and appearance is not important or a novel effect is desired. Examples of such usage are as paneling in restaurants, stores, and other buildings.

Baldcypress has been used principally for building construction, especially where resistance to decay is required. It was used for beams, posts, and other members in docks, warehouses, factories, bridges, and heavy construction.

It is well suited for siding and porch construction. It is also used for caskets, burial boxes, sash, doors, blinds, and general millwork, including interior trim and paneling. Other uses are in tanks, vats, ship and boat building, refrigerators, railroad-car construction, greenhouse construction, cooling

Figure 1–4—Cypress-tupelo swamp near New Orleans, LA. Species include baldcypress, tupelo, ash, willow, and elm. Swollen buttresses and cypress knees are typically present in the cypress. (M84 0472-13A)

towers, and stadium seats. It is also used for railroad crossties, poles, piles, shingles, cooperage, and fenceposts.

Douglas-Fir

Douglas-fir (*Pseudotsuga menziesii*) is also known locally as red-fir, Douglas-spruce, and yellow-fir.

The range of Douglas-fir extends from the Rocky Mountains to the Pacific coast and from Mexico to central British Columbia. The Douglas-fir production comes from the Coast States of Oregon, Washington, and California and from the Rocky Mountain States.

Sapwood of Douglas-fir is narrow in old-growth trees but may be as much as 3 inches wide in second-growth trees of commercial size. Fairly young trees of moderate to rapid growth have reddish heartwood and are called red-fir. Very narrow-ringed wood of old trees may be yellowish brown and is known on the market as yellow-fir.

The wood of Douglas-fir varies widely in weight and strength. When lumber of high strength is needed for structural uses, selection can be improved by applying the density rule. This rule uses percentage of latewood and rate of growth as they affect density. The higher density generally indicates stronger wood.

Douglas-fir is used mostly for building and construction purposes in the form of lumber, timbers (fig. 1–5), piles, and plywood. Considerable quantities go into railroad crossties, cooperage stock, mine timbers, poles, and fencing. Douglas-fir lumber is used in the manufacture of various products, including sash, doors, laminated beams, general millwork, railroad-car construction, boxes, pallets, and crates. Small amounts are used for flooring, furniture, ship and boat construction, and tanks. Douglas-fir plywood has found ever-increasing usefulness in construction, furniture, cabinets, and many other products.

Firs, True (Eastern Species)

Balsam fir (*Abies balsamea*) grows principally in New England, New York, Pennsylvania, and the Lake States. Fraser fir (*A. fraseri*) grows in the Appalachian Mountains of Virginia, North Carolina, and Tennessee.

The wood of the true firs, eastern as well as western species, is creamy white to pale brown. Heartwood and sapwood are generally indistinguishable. The similarity of wood structure in the true firs makes it impossible to distinguish the species by an examination of the wood alone.

Balsam fir is rated as light in weight, low in bending and compressive strength, moderately limber, soft, and low in resistance to shock.

The eastern firs are used mainly for pulpwood, although there is some lumber produced from them, especially in New England and the Lake States.

Firs, True (Western Species)

Six commercial species make up the western true firs: Subalpine fir (*Abies lasiocarpa*), California red fir (*A. magnifica*), grand fir (*A. grandis*), noble fir (*A. procera*), Pacific silver fir (*A. amabilis*), and white fir (*A. concolor*).

The western firs are light in weight but, with the exception of subalpine fir, have somewhat higher strength properties than balsam fir. Shrinkage of the wood is rated from small to moderately large.

The western true firs are largely cut for lumber in Washington, Oregon, California, western Montana, and northern Idaho and marketed as white fir throughout the United States. Lumber of the western true firs goes principally into building construction, boxes and crates, planing-mill products, sash, doors, and general millwork. In house construction the lumber is used for framing, subflooring, and sheathing. Some western true fir lumber goes into boxes and crates. High-grade lumber from noble fir is used mainly for interior finish, moldings, siding, and sash and door stock. Some of the best material is suitable for aircraft construction. Other special and exacting uses of noble fir are for venetian blinds and ladder rails.

Hemlock, Eastern

Eastern hemlock (*Tsuga canadensis*) grows from New England to northern Alabama and Georgia, and in the Lake States. Other names are Canadian hemlock and hemlock-spruce.

The production of hemlock lumber is divided fairly evenly between the New England States, the Middle Atlantic States, and the Lake States.

Figure 1 – 5—Wood is favored for waterfront structures, particularly fendering, because of its shock-absorbing qualities. The fendering on this dock in Key West, FL, is made of creosote-treated Douglas-fir. Some tropical species are resistant to attack by decay fungi and marine borers and are used for marine construction without preservative treatment.

(M84 0472-14A)

The heartwood of eastern hemlock is pale brown with a reddish hue. The sapwood is not distinctly separated from the heartwood but may be lighter in color. The wood is coarse and uneven in texture (old trees tend to have considerable shake); it is moderately light in weight, moderately hard, moderately low in strength, moderately limber, and moderately low in shock resistance.

Eastern hemlock is used principally for lumber and pulpwood. The lumber is used largely in building construction for framing, sheathing, subflooring, and roof boards, and in the manufacture of boxes, pallets, and crates.

Hemlock, Western

Western hemlock (*Tsuga heterophylla*) is also known as west coast hemlock, Pacific hemlock, British Columbia hemlock, hemlock-spruce, and western hemlock-fir. It grows along the Pacific coast of Oregon and Washington and in the northern Rocky Mountains, north to Canada and Alaska.

A relative, mountain hemlock, *T. mertensiana*, inhabits mountainous country from central California to Alaska. It is treated as a separate species in assigning lumber properties.

The heartwood and sapwood of western hemlock are almost white with a purplish tinge. The sapwood, which is sometimes lighter in color, is generally not more than 1 inch thick. The wood often contains small, sound, black knots that are usually tight and stay in place. Dark streaks are often found in the lumber; these are caused by hemlock bark maggots and generally do not reduce strength.

Western hemlock is moderately light in weight and moderate in strength. It is moderate in its hardness, stiffness, and shock resistance. It has moderately large shrinkage, about the same as Douglas-fir. Green hemlock lumber contains considerably more water than Douglas-fir, and requires longer kiln drying time.

Mountain hemlock has approximately the same density as western hemlock but is somewhat lower in bending strength and stiffness.

Western hemlock is used principally for pulpwood, lumber, and plywood. The lumber goes largely into building material, such as sheathing, siding, subflooring, joists, studding, planking, and rafters. Considerable quantities are used in the manufacture of boxes, pallets, crates, and flooring, and smaller amounts for furniture and ladders.

Mountain hemlock serves some of the same uses as western hemlock although the quantity available is much lower.

Incense-Cedar

Incense-cedar (*Libocedrus decurrens*) grows in California and southwestern Oregon, and extreme western Nevada. Most incense-cedar lumber comes from the northern half of California and the remainder from southern Oregon.

Sapwood of incense-cedar is white or cream colored, and the heartwood is light brown, often tinged with red. The wood has a fine, uniform texture and a spicy odor. Incense-cedar is light in weight, moderately low in strength, soft, low in shock resistance, and low in stiffness. It has small shrinkage and is easy to season with little checking or warping.

Incense-cedar is used principally for lumber and fenceposts. Nearly all the high-grade lumber is used for pencils and venetian blinds. Some is used for chests and toys. Much of the incense-cedar lumber is more or less pecky; that is, it contains pockets or areas of disintegrated wood caused by advanced stages of localized decay in the living tree. There is no further development of peck once the lumber is seasoned. This lumber is used locally for rough construction where low cost and decay resistance are important. Because of its resistance to decay, incense-cedar is well suited for fenceposts. Other products are railroad crossties, poles, and split shingles.

Larch, Western

Western larch (*Larix occidentalis*) grows in western Montana, northern Idaho, northeastern Oregon, and on the eastern slope of the Cascade Mountains in Washington. About two-thirds of the lumber of this species is produced in Idaho and Montana and one-third in Oregon and Washington.

The heartwood of western larch is yellowish brown and the sapwood yellowish white. The sapwood is generally not more than 1 inch thick. The wood is stiff, moderately strong and hard, moderately high in shock resistance, and moderately heavy. It has moderately large shrinkage. The wood is usually straight grained, splits easily, and is subject to ring shake. Knots are common but generally small and tight.

Western larch is used mainly in building construction for rough dimension, small timbers, planks and boards, and for railroad crossties and mine timbers. It is used also for piles, poles, and posts. Some high-grade material is manufactured into interior finish, flooring, sash, and doors. The properties of western larch are similar to those of Douglas-fir and sometimes they are sold mixed.

Pine, Eastern White

Eastern white pine (*Pinus strobus*) grows from Maine to northern Georgia and in the Lake States. It is also known as white pine, northern white pine, Weymouth pine, and soft pine.

About one-half the production of eastern white pine lumber occurs in the New England States, about one-third in the Lake States, and most of the remainder in the Middle Atlantic and South Atlantic States.

The heartwood of eastern white pine is light brown, often with a reddish tinge. It turns considerably darker on exposure. The wood has comparatively uniform texture and is straight grained. It is easily kiln dried, has small shrinkage, and ranks high in stability. It is also easy to work and can be readily glued.

Eastern white pine is light in weight, moderately soft, moderately low in strength, and low in resistance to shock.

Practically all eastern white pine is converted into lumber, which is put to a great variety of uses. A large proportion, which is mostly second-growth knotty lumber of the lower grades, goes into container and packaging applications. High-grade lumber goes into patterns for castings. Other important uses are sash, doors, furniture, trim, knotty paneling, finish, caskets and burial boxes, shade and map rollers, toys, and dairy and poultry supplies.

Pine, Jack

Jack pine (*Pinus banksiana*), sometimes known as scrub pine, gray pine, or black pine in the United States, grows naturally in the Lake States and in a few scattered areas in New England and northern New York. In lumber, jack pine is sometimes not separated from the other pines with which it grows, including red pine and eastern white pine.

The sapwood of jack pine is nearly white; the heartwood is light brown to orange. The sapwood may make up one-half or more of the volume of a tree. The wood has a rather coarse texture and is somewhat resinous. It is moderately light in weight, moderately low in bending strength and compressive strength, moderately low in shock resistance, and low in stiffness. It also has moderately small shrinkage. Lumber from jack pine is generally knotty.

Jack pine is used for pulpwood, box lumber, pallets, and fuel. Less important uses include railroad crossties, mine timber, slack cooperage, poles, and posts.

Pine, Lodgepole

Lodgepole pine (*Pinus contorta*), also known as knotty pine, black pine, spruce pine, and jack pine, grows in the

Rocky Mountain and Pacific coast regions as far northward as Alaska. The cut of this species comes largely from the central Rocky Mountain States; other producing regions are Idaho, Montana, Oregon, and Washington.

The heartwood of lodgepole pine varies from light yellow to light yellow-brown. The sapwood is yellow or nearly white. The wood is generally straight grained with narrow growth rings.

The wood is moderately light in weight, fairly easy to work, and has moderately large shrinkage. Lodgepole pine rates as moderately low in strength, moderately soft, moderately stiff, and moderately low in shock resistance.

Lodgepole pine is used for lumber, mine timbers, railroad crossties, and poles. Less important uses include posts and fuel. It is being used in increasing amounts for framing, siding, finish, and flooring.

Pine, Pitch

Pitch pine (*Pinus rigida*) grows from Maine along the mountains to eastern Tennessee and northern Georgia. The heartwood is brownish red and resinous; the sapwood is thick and light yellow. The wood of pitch pine is moderately heavy to heavy, moderately strong, stiff, and hard, and moderately high in shock resistance. Its shrinkage is moderately small to moderately large. It is used for lumber, fuel, and pulpwood. Pitch pine lumber is classified as a "minor species" in southern pine grading rules.

Pine, Pond

Pond pine (*Pinus serotina*) grows in the coast region from New Jersey to Florida. It occurs in small groups or singly, mixed with other pines on low flats. The wood is heavy, coarse-grained, and resinous, with dark, orange-colored heartwood and thick, pale yellow sapwood. Shrinkage is moderately large. The wood is moderately strong, stiff, moderately hard, and moderately high in shock resistance. It is used for general construction, railway crossties, posts, and poles. The lumber of this species is also graded as a "minor species" in southern pine grading rules.

Pine, Ponderosa

Ponderosa pine (*Pinus ponderosa*) is known also as pondosa pine, western soft pine, western yellow pine, bull pine, and blackjack pine. Jeffrey pine (*P. jeffreyi*), which grows in close association with ponderosa pine in California and Oregon, is usually marketed with ponderosa pine and sold under that name.

Major producing areas are in Oregon, Washington, and California (fig. 1–6). Other important producing areas are in Idaho and Montana; lesser amounts come from the southern Rocky Mountain region and the Black Hills of South Dakota and Wyoming.

Botanically, ponderosa pine belongs to the yellow pine group rather than the white pine group. A considerable pro-

Figure 1–6—Ponderosa pine (*Pinus ponderosa*) (M84 0472-18A) growing in an open or "park-like" habitat.

portion of the wood, however, is somewhat similar to the white pines in appearance and properties. The heartwood is light reddish brown, and the wide sapwood is nearly white to pale yellow.

The wood of the outer portions of ponderosa pine of sawtimber size is generally moderately light in weight, moderately low in strength, moderately soft, moderately stiff, and moderately low in shock resistance. It is generally straight grained and has moderately small shrinkage. It is quite uniform in texture and has little tendency to warp and twist.

Ponderosa pine is used mainly for lumber and to a lesser extent for piles, poles, posts, mine timbers, veneer, and railroad crossties. The clear wood goes into sash, doors, blinds, moldings, paneling, mantels, trim, and built-in cases and cabinets. Lower grade lumber is used for boxes and crates. Much of the lumber of intermediate or lower grades goes into sheathing, subflooring, and roof boards. Knotty ponderosa pine is used for interior finish. A considerable amount now goes into particleboard and paper.

Pine, Red

Red pine (*Pinus resinosa*) is frequently called Norway pine. It is occasionally known as hard pine and pitch pine. This species grows in the New England States, New York, Pennsylvania, and the Lake States. In the past, lumber from red pine was often marketed with white pine without distinction as to species.

The heartwood of red pine varies from pale red to reddish brown. The sapwood is nearly white with a yellowish tinge, and is generally from 2 to 4 inches wide. The wood resembles the lighter weight wood of southern pine. Latewood is distinct in the growth rings.

Red pine is moderately heavy, moderately strong and stiff, moderately soft and moderately high in shock resistance. It is generally straight grained, not so uniform in texture as eastern white pine, and somewhat resinous. The wood has moderately large shrinkage but is not difficult to dry and stays in place well when seasoned.

Red pine is used principally for lumber and to a lesser extent for piles, poles, cabin logs, posts, pulpwood, and fuel. The wood is used for many of the purposes for which eastern white pine is used. It goes mostly into building construction, siding, flooring, sash, doors, blinds, general millwork, and boxes, pallets, and crates.

Pine, Southern

A number of species are included in the group marketed as southern pine lumber. The four major southern pines, and their growth range, are:

(1) Longleaf pine (*Pinus palustris*), which grows from eastern North Carolina southward into Florida and westward into eastern Texas. (2) Shortleaf pine (*P. echinata*), which grows from southeastern New York and New Jersey southward to northern Florida and westward into eastern Texas and Oklahoma. (3) Loblolly pine (*P. taeda*), which grows from Maryland southward through the Atlantic Coastal Plain and Piedmont Plateau into Florida and westward into eastern Texas. (4) Slash pine (*P. elliottii*), which grows in Florida and the southern parts of South Carolina, Georgia, Alabama, Mississippi, and Louisiana east of the Mississippi River.

Lumber from any one or from any mixture of two or more of these species is classified as southern pine by the grading standards of the industry. These standards provide also for lumber that is produced from trees of the longleaf and slash pine species to be classified as longleaf pine if conforming to the growth-ring and latewood requirements of such standards. The lumber that is classified as longleaf in the domestic trade is known also as pitch pine in the export trade.

Southern pine lumber comes principally from the Southern and South Atlantic States. States that lead in production are Georgia, Alabama, North Carolina, Arkansas, and Louisiana.

The wood of these southern pines is quite similar in appearance. The sapwood is yellowish white and heartwood reddish brown. The sapwood is usually wide in second-growth stands. Heartwood begins to form when the tree is about 20 years old. In old, slow-growth trees, sapwood may be only 1 to 2 inches in width.

Longleaf and slash pine are classed as heavy, strong, stiff, hard, and moderately high in shock resistance. Shortleaf and loblolly pine are usually somewhat lighter in weight than longleaf. All the southern pines have moderately large shrinkage but are stable when properly seasoned.

To obtain heavy, strong wood of the southern pines for structural purposes, a density rule has been written that specifies a certain percentage of latewood and growth rates for structural timbers.

The denser and higher strength southern pine is used extensively in construction of factories, warehouses, bridges, trestles, and docks in the form of stringers, and for roof trusses, beams, posts, joists, and piles. Lumber of lower density and strength finds many uses for building material, such as interior finish, sheathing, subflooring, and joists and for boxes, pallets, and crates. Southern pine is used also for tight and slack cooperage. When used for railroad crossties, piles, poles, and mine timbers, it is usually treated with preservatives. The manufacture of structural grade plywood from southern pine has become a major wood-using industry.

Pine, Spruce

Spruce pine (*Pinus glabra*), also known as cedar pine, poor pine, Walter pine, and bottom white pine, is classified as one of the minor southern pine species. It grows most commonly on low moist lands of the coastal regions of southeastern South Carolina, Georgia, Alabama, Mississippi, and Louisiana, and northern and northwestern Florida.

Heartwood is light brown, and the wide sapwood zone is nearly white. Spruce pine wood is lower in most strength values than the major southern pines. It compares favorably with white fir in important bending properties, in crushing strength perpendicular and parallel to the grain, and in hardness. It is similar to the denser species such as coast Douglas-fir and loblolly pine in shear parallel to the grain.

Until recent years the principal uses of spruce pine were locally for lumber, and for pulpwood and fuelwood. The lumber reportedly was used for sash, doors, and interior finish because of its lower specific gravity and less marked distinction between earlywood and latewood. In recent years it has qualified for use in plywood.

Pine, Sugar

Sugar pine (*Pinus lambertiana*), the world's largest species of pine, is sometimes called California sugar pine. Most of the sugar pine lumber is produced in California and the remainder in southwestern Oregon.

The heartwood of sugar pine is buff or light brown, sometimes tinged with red. The sapwood is creamy white. The wood is straight grained, fairly uniform in texture, and easy to work with tools. It has very small shrinkage, is readily seasoned without warping or checking, and stays in place well. This species is light in weight, moderately low in strength, moderately soft, low in shock resistance, and low in stiffness.

Sugar pine is used almost entirely for lumber products. The largest amounts are used in boxes and crates, sash, doors, frames, blinds, general millwork, building construction, and foundry patterns. Like eastern white pine, sugar pine is suitable for use in nearly every part of a house because of the ease with which it can be cut, its ability to stay in place, and its good nailing properties.

Pine, Virginia

Virginia pine (*Pinus virginiana*), known also as Jersey pine and scrub pine, grows from New Jersey and Virginia throughout the Appalachian region to Georgia and the Ohio Valley. It is another southern pine classified as a "minor species" in the grading rules. The heartwood is orange and the sapwood nearly white and relatively thick. The wood is rated as moderately heavy, moderately strong, moderately hard, and moderately stiff and has moderately large shrinkage and high shock resistance. It is used for lumber, railroad crossties, mine timbers, pulpwood, and fuel.

Pine, Western White

Western white pine (*Pinus monticola*) is also known as Idaho white pine or white pine. About four-fifths of the cut comes from Idaho with the remainder mostly from Washington; small amounts are cut in Montana and Oregon.

Heartwood of western white pine is cream colored to light reddish brown and darkens on exposure. The sapwood is yellowish white and generally from 1 to 3 inches wide. The wood is straight grained, easy to work, easily kiln dried, and stable after seasoning.

This species is moderately light in weight, moderately low in strength, moderately soft, moderately stiff, moderately low in shock resistance, and has moderately large shrinkage.

Practically all western white pine is sawed into lumber and used mainly for building construction, matches, boxes, patterns, and millwork products, such as sash, frames, doors, and blinds. In building construction, boards of the lower grades are used for sheathing, knotty paneling, subflooring, and roof strips. High-grade material is made into siding of various kinds, exterior and interior trim, and finish. It has practically the same uses as eastern white pine and sugar pine.

Port-Orford-Cedar

Port-Orford-cedar (*Chamaecyparis lawsoniana*) is sometimes known as Lawson-cypress, Oregon-cedar, and white-cedar. It grows along the Pacific coast from Coos Bay, OR, southward to California. It does not extend more than 40 miles inland.

The heartwood of Port-Orford-cedar is light yellow to pale brown in color. Sapwood is thin and hard to distinguish. The wood has fine texture, generally straight grain, and a pleasant spicy odor. It is moderately light in weight, stiff, moderately strong and hard, and moderately resistant to shock. Port-Orford-cedar heartwood is highly resistant to decay. The wood shrinks moderately, has little tendency to warp, and is stable after seasoning.

Some high-grade Port-Orford-cedar was once used in the manufacture of storage battery separators, match sticks, and specialty millwork. Today other uses are mothproof boxes, archery supplies, sash and door construction, stadium seats, flooring, interior finish, furniture, and boatbuilding.

Redcedar, Eastern

Eastern redcedar (*Juniperus virginiana*) grows throughout the eastern half of the United States, except in Maine, Florida, and a narrow strip along the Gulf coast, and at the higher elevations in the Appalachian Mountain Range. Commercial production is principally in the southern Appalachian and Cumberland Mountain regions. Another species, southern redcedar (*J. silicicola*), grows over a limited area in the South Atlantic and Gulf Coastal Plains.

The heartwood of redcedar is bright red or dull red, and the thin sapwood is nearly white. The wood is moderately heavy, moderately low in strength, hard, and high in shock resistance, but low in stiffness. It has very small shrinkage and stays in place well after seasoning. The texture is fine and uniform. Grain is usually straight, except where deflected by knots, which are numerous. Eastern redcedar heartwood is very resistant to decay.

The greatest quantity of eastern redcedar is used for fenceposts. Lumber is manufactured into chests, wardrobes, and closet lining. Other uses include flooring, novelties, pencils, scientific instruments, and small boats. Southern redcedar is used for the same purposes.

Redcedar, Western

Western redcedar (*Thuja plicata*) grows in the Pacific Northwest and along the Pacific coast to Alaska. Western redcedar is also called canoe-cedar, giant arborvitae, shinglewood, and Pacific redcedar. Western redcedar lumber is produced principally in Washington, followed by Oregon, Idaho, and Montana.

The heartwood of western redcedar is reddish or pinkish brown to dull brown and the sapwood nearly white. The sapwood is narrow, often not over 1 inch in width. The wood is generally straight grained and has a uniform but rather coarse texture. It has very small shrinkage. This species is light in weight, moderately soft, low in strength when used as a beam or posts, and low in shock resistance. Its heartwood is very resistant to decay.

Western redcedar is used principally for shingles, lumber, poles, posts, and piles. The lumber is used for exterior siding, interior finish, greenhouse construction, ship and boat building, boxes and crates, sash, doors, and millwork.

Redwood

Redwood (*Sequoia sempervirens*) is a very large tree growing on the coast of California. A closely related species, giant sequoia (*Sequoiadendron giganteum*), grows in a limited area in the Sierra Nevada of California, but is used in very limited quantities. Other names for redwood are coast redwood, California redwood, and sequoia. Production of redwood lumber is limited to California, but a nationwide market exists.

The heartwood of redwood varies from a light cherry to a dark mahogany. The narrow sapwood is almost white. Typical old-growth redwood is moderately light in weight, moderately strong and stiff, and moderately hard. The wood is easy to work, generally straight grained, and shrinks and swells comparatively little. The heartwood from old-growth trees has high decay resistance; but heartwood from second-growth trees generally ranges from resistant to moderately decay resistant.

Most redwood lumber is used for building. It is remanufactured extensively into siding, sash, doors, blinds, finish, casket stock, and containers. Because of its durability, it is useful for cooling towers, tanks, silos, wood-stave pipe, and outdoor furniture. It is used in agriculture for buildings and equipment. Its use as timbers and large dimension in bridges and trestles is relatively minor. The wood splits readily and the manufacture of split products, such as posts and fence material, is an important business in the redwood area. Some redwood veneer is manufactured for decorative plywood.

Spruce, Eastern

The term "eastern spruce" includes three species, red (*Picea rubens*), white (*P. glauca*), and black (*P. mariana*). White and black spruce grow principally in the Lake States and New England, and red spruce in New England and the Appalachian Mountains. All three species have about the same properties, and in commerce no distinction is made between them. The wood dries easily and is stable after drying, is moderately light in weight and easily worked, has moderate shrinkage, and is moderately strong, stiff, tough, and hard. The wood is light in color, and there is little difference between the heartwood and sapwood.

The largest use of eastern spruce is for pulpwood. It is also used for framing material, general millwork, boxes and crates, and piano sounding boards.

Spruce, Englemann

Engelmann spruce (*Picea engelmannii*) grows at high elevations in the Rocky Mountain region of the United States. This species is sometimes known by other names, such as white spruce, mountain spruce, Arizona spruce, silver spruce, and balsam. About two-thirds of the lumber is produced in the southern Rocky Mountain States. Most of the remainder comes from the northern Rocky Mountain States and Oregon.

The heartwood of Engelmann spruce is nearly white with a slight tinge of red. The sapwood varies from 3/4 inch to 2 inches in width and is often difficult to distinguish from heartwood. The wood has medium to fine texture and is without characteristic taste or odor. It is generally straight grained. Engelmann spruce is rated as light in weight. It is low in strength as a beam or post. It is limber, soft, low in shock resistance, and has moderately small shrinkage. The

lumber typically contains numerous small knots.

Engelmann spruce is used principally for lumber and for mine timbers, railroad crossties, and poles. It is used also in building construction in the form of dimension lumber, flooring, sheathing, and studding. It has excellent properties for pulp and papermaking.

Spruce, Sitka

Sitka spruce (*Picea sitchensis*) is a tree of large size growing along the northwestern coast of North America from California to Alaska. It is generally known as Sitka spruce, although other names may be applied locally, such as yellow spruce, tideland spruce, western spruce, silver spruce, and west coast spruce Much Sitka spruce timber is grown in Alaska, but most of the logs cut are sawn into cants for export to the Orient. Material for U.S. consumption is produced largely in Washington and Oregon.

The heartwood of Sitka spruce is a light pinkish brown. The sapwood is creamy white and shades gradually into the heartwood; it may be 3 to 6 inches wide or even wider in young trees. The wood has a comparatively fine, uniform texture, generally straight grain, and no distinct taste or odor. It is moderately light in weight, moderately low in bending and compressive strength, moderately stiff, moderately soft, and moderately low in resistance to shock. It has moderately small shrinkage. On the basis of weight, it rates high in strength properties and can be obtained in clear, straight-grained pieces.

Sitka spruce is used principally for lumber, pulpwood, and cooperage. Boxes and crates account for a considerable amount of the remanufactured lumber. Other important uses are furniture, planing-mill products, sash, doors, blinds, millwork, and boats. Sitka spruce has been by far the most important wood for aircraft construction. Other specialty uses are ladder rails and sounding boards for pianos.

Tamarack

Tamarack (*Larix laricina*) is a small- to medium-sized tree with a straight, round, slightly tapered trunk. In the United States it grows from Maine to Minnesota, with the bulk of the stand in the Lake States. It was formerly used in considerable quantity for lumber, but in recent years production for that purpose has been small.

The heartwood of tamarack is yellowish brown to russet brown. The sapwood is whitish, generally less than an inch wide. The wood is coarse in texture, without odor or taste, and the transition from earlywood to latewood is abrupt. The wood is intermediate in weight and in most mechanical properties.

Tamarack is used principally for pulpwood, lumber, railroad crossties, mine timbers, fuel, fenceposts, and poles. Lumber goes into framing material, tank construction, and boxes, pallets, and crates.

White-Cedar, Northern and Atlantic

Two species of white-cedar grow in the eastern part of the United States—northern white-cedar (*Thuja occidentalis*) and Atlantic white-cedar (*Chamaecyparis thyoides*). Northern white-cedar is also known as arborvitae, or simply cedar. Atlantic white-cedar is also known as southern white-cedar, swamp-cedar, and boat-cedar.

Northern white-cedar grows from Maine along the Appalachian Mountain Range and westward through the northern part of the Lake States. Atlantic white-cedar grows near the Atlantic coast from Maine to northern Florida and westward along the Gulf coast to Louisiana. It is strictly a swamp tree.

Production of northern white-cedar lumber is probably greatest in Maine and the Lake States. Commercial production of Atlantic white-cedar centers in North Carolina and along the Gulf coast.

The heartwood of white-cedar is light brown, and the sapwood is white or nearly so. The sapwood is usually thin. The wood is light in weight, rather soft and low in strength, and low in shock resistance. It shrinks little in drying. It is easily worked, holds paint well, and the heartwood is highly resistant to decay. The two species are used for similar purposes, mostly for poles, railroad crossties, lumber, posts, and decorative fencing. White-cedar lumber is used principally where a high degree of durability is needed, as in tanks and boats, and for woodenware.

Imported Woods

This section does not purport to discuss all of the woods that have been at one time or another imported into the United States. Only some species at present considered to be of commercial importance are included. The same species may be marketed in the United States under other common names. Because of the variation in common names, numerous cross-references are included.

Text information is necessarily brief, but when used in conjunction with the shrinkage and strength tables (chs. 3 and 4), a reasonably good picture may be obtained of a particular wood. The Selected References at the end of this chapter contain information on many species not described here.

Hardwoods

Afara (See Limba)

Afrormosia

Afrormosia or kokrodua (*Pericopsis elata*), a large West African tree, is sometimes used as a substitute for teak (*Tectona grandis*). The heartwood is fine textured, with straight to interlocked grain. The wood is brownish yellow with darker streaks, moderately hard and heavy, weighing about 44 pounds per cubic foot (pcf) at 15 percent moisture content. The wood strongly resembles teak in appearance but lacks the oily nature of teak and is finer textured.

The wood seasons readily with little degrade and has good dimensional stability. It is somewhat heavier and stronger than teak.

The heartwood is highly resistant to decay fungi and termite attack and should prove extremely durable under adverse conditions. The wood is often used for the same purposes as teak, such as boat construction, joinery, flooring, furniture, interior trim, and decorative veneer.

Albarco

Albarco or jequitiba, as it is known in Brazil, is the common name applied to species in the genus *Cariniana*. The 10 species in this genus are distributed from eastern Peru and northern Bolivia through central Brazil to Venezuela and Colombia.

The heartwood is reddish or purplish brown sometimes with dark streaks. It is usually not sharply demarcated from the pale brown sapwood. The texture is medium and the grain is straight to interlocked.

Albarco generally works satisfactorily with only a slight blunting on tool cutting edges due to the presence of silica. Veneers cut without difficulty. The wood is rather strong and moderately heavy weighing about 35 pcf at 12 percent moisture content. In general, the wood has about the same strength as American oaks.

The heartwood is durable, particularly the deeply colored material. It also has good resistance to dry-wood termite attack. The heartwood is extremely resistant to preservative treatment.

Albarco is a general construction and carpentry wood but is also used for furniture components, shipbuilding, flooring, veneer for plywood, and turnery.

Amaranth (See Purpleheart)

Anani (See Manni)

Anaura (See Marishballi)

Andiroba

Because of the widespread distribution of andiroba (*Carapa guianensis*) in tropical America, the wood is known under a variety of names that include cedro macho, carapa, crabwood, and tangare. These names are also applied to the related species *Carapa nicaraguensis*, whose properties are generally inferior to those of *C. guianensis*.

The heartwood color varies from reddish brown to dark reddish brown. The texture (size of pores) is like that of American mahogany (*Swietenia*) and the wood is sometimes substituted for mahogany. The grain is usually interlocked but is rated as easy to work, paint, and glue. The wood is rated as durable to very durable with respect to decay and insects. Andiroba is heavier than mahogany and accordingly is markedly superior in all static bending properties, compression parallel to the grain, hardness, shear, and toughness.

On the basis of its properties, andiroba appears to be suited for such uses as flooring, frame construction in the tropics, furniture and cabinetwork, millwork, and utility and decorative veneer and plywood.

Angelin (See Sucupira)

Angelique

Angelique (*Dicorynia guianensis*), comes from French Guiana and Surinam. Because of the variability in heartwood color between different trees, two forms are commonly recognized by producers. Heartwood that is russet colored when freshly cut, and becomes superficially dull brown with a purplish cast, is referred to as ''gris.'' Heartwood that is more distinctly reddish and frequently shows wide bands of purplish color is called angelique rouge.

The texture is somewhat coarser than that of black walnut. The grain is generally straight or slightly interlocked. In strength, angelique is superior to teak and white oak, when either green or air dry, in all properties except tension perpendicular to grain. Angelique is rated as highly resistant to decay, and resistant to marine borer attack. Machining properties vary and may be due to differences in density, moisture content, and silica content. After the wood is thoroughly air dried or kiln dried, it can be worked effectively only with carbide-tipped tools.

The strength and durability of angelique make it especially suitable for heavy construction, harbor installations, bridges, heavy planking for pier and platform decking, and railroad bridge ties. The wood is particularly suitable for ship decking, planking, boat frames, industrial flooring, and parquet blocks and strips.

Apa (See Wallaba)

Apamate (See Roble)

Apitong (See Keruing)

Avodire

Avodire (*Turraeanthus africanus*) has a rather extensive range from Sierra Leone westward to the Congo region and southward to Zaire and Angola. It is most common in the eastern region of the Ivory Coast and is scattered elsewhere. It is a medium-size tree of the rainforest in which it forms fairly dense but localized and discontinuous stands.

The wood is cream to pale yellow in color with a high natural luster and eventually darkens to a golden yellow. The grain is sometimes straight but more often is wavy or irregularly interlocked, which produces an unusual and attractive mottled figure when sliced or cut on the quarter.

Although its weight is only 85 percent that of oak, avodire has almost identical strength properties except that it is lower in shock resistance and in shear. The wood works fairly easily with hand and machine tools and finishes well in most operations. Figured material is usually converted into veneer for use in decorative work, and it is this kind of material that is chiefly imported into the United States. Other uses include furniture, fine joinery, cabinetwork, and paneling.

Azobe/Ekki

Ekki or azobe (*Lophira alata*) is found in West Africa and extends into the Congo basin. The heartwood is dark red, chocolate brown, or purple brown with conspicuous white deposits in the vessels. The texture is coarse, and the grain is usually interlocked.

The wood is strong and its density averages about 70 pcf at 12 percent moisture content. It is very difficult to work with hand and machine tools and severe blunting occurs if machined when dry. It can be dressed to a smooth finish, and the gluing properties are usually good. Drying is very difficult without excessive degrade.

The heartwood is rated as very durable but only moderately resistant to termite attack. It is very resistant to acids and has good weathering properties. It is also resistant to teredo attack. The heartwood is extremely resistant to preservative treatments.

Ekki is an excellent wood for heavy durable construction work, harbor work, heavy-duty flooring, and railroad crossties.

Bagtikan (See Lauan/Meranti Groupings)

Balata

Balata or bulletwood (*Manilkara bidentata*) is widely distributed throughout the West Indies, Central America, and northern South America. Its heartwood is light to dark reddish brown and not sharply demarcated from the pale brown sapwood. Texture and uniform, and the grain is straight to occasionally wavy or interlocked.

Balata is a strong and very heavy wood (air-dry density 66 pcf). It is generally difficult to air season, with a tendency to develop severe checking and warp. The wood is moderately easy to work despite its high density and is rated good to excellent in all operations.

Balata is very resistant to attack by decay fungi and highly resistant to subterranean termites but only moderately resistant to dry-wood termites. It has a high resistance to preservation treatments and absorption of moisture.

Balata is a hard, heavy wood suitable for heavy construction, textile and pulpmill equipment, furniture parts, turnery, tool handles, flooring, boat frames and other bentwork, railroad crossties, violin bows, billiard cues, and other specialty uses.

Balsa

Balsa (*Ochroma pyramidale*) is widely distributed throughout tropical America from southern Mexico to southern Brazil and Bolivia, but Ecuador has been the principal source of supply since the wood gained commercial importance. It is usually found at lower elevations, especially on bottom-land soils along streams and in clearings and cutover forests. Today it is often cultivated in plantations.

Balsa possesses several characteristics that make possible a wide variety of uses. It is the lightest and softest of all woods on the market. The lumber selected for use in the United States weighs, on the average, about 11 pcf when dry and often as little as 6 pfc. Because of its light weight and exceedingly porous composition, balsa is highly efficient in uses where buoyancy, insulation against heat and cold, or absorption of sound and vibration are important considerations.

The wood is readily recognized by its light weight, nearly white or oatmeal color often with a yellowish or pinkish hue, and its unique ''velvety'' feel.

The principal uses of balsa are in life-saving equipment, floats, rafts, core stock, insulation, cushioning, sound modifiers, models, and novelties.

Balau (See Lauan/Meranti Groupings)

Banak/Cuangare

Various species of banak (*Virola*) occur in tropical America, from Belize and Guatemala southward to Venezuela, the Guianas, the Amazon region of northern Brazil, southern Brazil, and on the Pacific Coast to Peru and Bolivia. The bulk of the timber known as banak, however, is supplied by *V. koschnyi* of Central America, and *V. surinamensis* and *V. sebifera* of northern South America. Cuangare (*Dialyanthera*) is botanically closely related to banak and the woods are so similar that they are generally mixed in the trade. The main commercial supply of cuangare comes from Colombia and Ecuador. Banak and cuangare are common in swamp and marsh forests and may occur in almost pure stands in some areas.

The heartwood of both banak and cuangare is usually pinkish brown or grayish brown in color and is generally not differentiated from the sapwood. The wood is straight grained and is of a medium to coarse texture (fig 1 − 7).

The various species are nonresistant to decay and insect attack but can be readily treated with preservatives. Their machining properties are very good, but, when zones of tension wood are present, machining may result in surface fuzziness and torn grain. The wood finishes readily and is easily glued. It is rated as a first-class veneer species. Its strength properties are similar to yellow-poplar. Banak is considered as a general utility wood in both lumber and plywood form and is now used as molding, millwork, and furniture components.

Benge/Ovangkol/Bubinga

Although benge (*Guibourtia arnoldiana*), ovangkol or ehie (*G. ehie*), and bubinga (*Guibourtia spp.*) belong to the same west African genus, they differ rather markedly with respect to their color and somewhat to their texture. The heartwood of benge is pale yellowish-brown to medium brown with gray to almost black striping. Ehie heartwood tends to be more golden brown to dark brown with gray to almost black stripes. The heartwood of bubinga is pink, vivid red, or red-brown with purple streaks, which upon exposure becomes yellow or medium brown with a reddish tint. The texture of ehie is moderately coarse, whereas benge and bubinga have a fine to moderately fine texture.

All three woods are moderately hard and heavy, but they can be worked well with hand and machine tools. They are listed as moderately durable and resistant to preservation treatments. Drying may be difficult, but with care the wood seasons well.

These woods are used in turnery, flooring, furniture components, cabinetwork, and decorative veneers.

Bubinga (See Benge)

Bulletwood (See Balata)

Carapa (See Andiroba)

Cativo

Cativo (*Prioria copaifera*) is one of the few tropical American species that occur in abundance and often in nearly pure stands. Commercial stands are found in Nicaragua, Costa Rica, Panama, and Colombia. The sapwood is usually thick, and in trees up to 30 inches in diameter the heartwood may be only 7 inches in diameter. The sapwood that is utilized commercially may be a very pale pinkish color or may be distinctly reddish. The grain is straight and the texture of the wood is uniform, comparable to that of American mahogany. Figure on flat-sawn surfaces is rather subdued and results from the exposure of the narrow bands of parenchyma tissue.

The wood can be seasoned rapidly and easily with very little degrade. The dimensional stability of the wood is very good; it is practically equal to that of American mahogany. Cativo is classed as a nondurable wood with respect to decay and insects. Cativo may contain appreciable quantities of gum, which may interfere with finishes. In wood that has been properly seasoned, however, the aromatics in the gum are removed and it presents no difficulties in finishing.

Considerable quantities of cativo are used for interior trim, and resin-stabilized veneer is an important pattern material. Cativo is widely used for furniture and cabinet parts, lumber core for plywood, picture frames, edge banding for doors, joinery, and millwork.

Cedro (See Spanish-Cedar)

Cedro Macho (See Andiroba)

Cedro-Rana (See Tornillo)

Ceiba

Ceiba (*Ceiba pentandra*) is a large tree growing to 200 feet in height with a straight cylindrical bole 40 to 60 feet long. Trunk diameters 6 feet and more are common. It grows in West Africa from the Ivory Coast and Sierra Leone to Liberia, Nigeria, and the Congo region. A related species is lupuna (*Ceiba samauma*) from South America.

The sapwood and heartwood are not clearly demarcated. The wood is whitish, pale brown, or pinkish brown, often

Figure 1–7—Highly perishable cuangare (*Dialyanthera spp.*) and banak (*Virola spp.*) logs harvested from coastal lowlands in southwest Colombia are ready for pond storage.

(M510 273-16)

with yellowish or grayish streaks. Texture is coarse, and the grain is interlocked or occasionally irregular.

Ceiba is very soft and light (airdry density 20 pcf). In strength the wood is comparable to basswood.

Ceiba seasons rapidly without marked deterioration. It is difficult to saw cleanly and dress smoothly due to a high percentage of tension wood. It peels to give good veneers and is easy to nail and glue.

Ceiba is very susceptible to attack by decay fungi and insects. It requires rapid harvest and conversion to prevent deterioration. Treatability, however, is rated as good.

The wood is available in large sizes and its low density combined with a rather high degree of dimensional stability make it ideally suited for pattern and core stock. Other uses include blockboard, boxes and crates, joinery, and furniture components.

Chewstick (See Manni)

Courbaril

The genus *Hymenaea* consists of about 25 species occurring in the West Indies and from southern Mexico, through Central America, into the Amazon basin of South America. The best known and most important species is *H. courbaril*, which occurs throughout the range of the genus.

The sapwood of courbaril, or jatoba, as it is often called in Brazil, is gray white and usually quite wide. The heartwood is sharply differentiated from the sapwood and is salmon red to orange brown when fresh and becoming russet or reddish brown when seasoned. Often the heartwood is marked with dark streaks. The texture is medium to rather coarse, and the grain is mostly interlocked. The wood is hard and heavy (about 50 pcf at 12 pct moisture content). The strength properties of courbaril are quite high and very similar to those of

1-23

shagbark hickory, a species of lower specific gravity.

Courbaril is rated as very resistant to resistant to attack by decay fungi and dry-wood termites. Heartwood is not treatable, but the sapwood is responsive.

Courbaril is moderately difficult to saw and machine because of its high density, but it can be machined to a smooth surface. Its turning, gluing, and finishing properties are satisfactory. Planing, however, is somewhat difficult due to interlocked grain. It compares favorably with white oak in steam bending behavior.

Courbaril is used for tool handles and other applications where good shock resistance is needed. It is also used for steam-bent parts, flooring, turnery, furniture and cabinetwork, veneer and plywood, railroad crossties, and other specialty items.

Crabwood (See Andiroba)

Cristobal (See Macawood)

Cuangare (See Banak)

Degame

Degame or lemonwood (*Calycophyllum candidissimum*) occurs in Cuba and ranges from southern Mexico through Central America to Colombia and Venezuela. It may grow in pure stands and is common on shaded hillsides and along waterways.

The heartwood of degame ranges from a light brown to oatmeal color and is sometimes grayish. The sapwood is lighter in color and merges gradually with the heartwood. The texture is fine and uniform. The grain is usually straight or infrequently shows a shallow interlocking which may produce a narrow and indistinct stripe on quartered faces. In strength, degame is above the average for woods of similar density (air-dry density 51 pcf). Tests show degame superior to persimmon (*Diospyros virginiana*) in all respects but hardness. Natural durability is low when the wood is used under conditions favorable to stain, decay, and insect attack. However, degame is reported to be highly resistant to marine borers.

Degame is moderately difficult to machine because of its density and hardness, although it produces no appreciable dulling effect on cutting tools. Machined surfaces are very smooth.

Degame is little used in the United States at the present time, but the characteristics of the wood should make it particularly adaptable for shuttles, picker sticks, and other textile industry items in which resilience and strength are required. It was prized in the manufacture of archery bows and fishing rods. It is also suitable for tool handles and turnery.

Determa

Determa (*Ocotea rubra*) is native to the Guianas, Trinidad, and the lower Amazon region of Brazil. The heartwood is light reddish brown with a golden sheen and is distinct from the dull-gray or pale yellowish-brown sapwood. Texture is rather coarse, and the grain is interlocked to straight.

Determa is a moderately strong and heavy (air-dry density 40 to 45 pcf) wood that is moderately difficult to air season. It can be worked readily with hand and machine tools with little dulling effect on cutters. It can be glued readily and polished fairly well.

Heartwood is durable to very durable in resistance to decay fungi and moderately resistant to dry-wood termites. Weathering characteristics are excellent, and the wood is highly resistant to moisture absorption.

Uses include furniture, general construction, boat planking, tanks and cooperage, heavy marine construction, turnery, and parquet flooring.

Ehie (See Benge)

Encino (See Oak)

Ekki (See Azobe)

Ekop

Ekop or gola (*Tetraberlinia tubmaniana*) is known only from Liberia. The heartwood is light reddish brown and is distinct from the lighter colored sapwood which may be up to 2 inches thick. The wood is medium-to coarse-textured, and the grain is interlocked, showing a narrow strip pattern on quartered surfaces.

The wood weighs about 46 pcf at 12 percent moisture content. It dries fairly well but with a marked tendency to end and surface check. Ekop works well with hand and machine tools and is an excellent wood for turnery. It also slices well into veneer and has good gluing properties. The heartwood is only moderately durable and is moderately resistant to impregnation with preservative treatments.

Gola (See Ekop)

Goncalo Alves

The major and early imports of goncalo alves (*Astronium graveolens* and *A. fraxinifolium*) have been from Brazil. These species range from southern Mexico through Central America into the Amazon basin.

When fresh, the heartwood is russet brown, orange brown, or reddish brown to red with narrow to wide irregular stripes of medium to very dark brown. After exposure it becomes brown, red, or dark reddish brown with nearly black stripes. The sapwood is grayish white and sharply demarcated from the heartwood. The texture is fine to medium and uniform. Grain is variable from straight to interlocked and wavy.

Goncalo alves turns readily, finishes very smoothly, and takes a high natural polish. The heartwood is highly resistant to moisture absorption, and the pigmented areas, because of their high density, may present some difficulties in gluing.

The heartwood is very durable, being resistant to both white-rot and brown-rot organisms. Treatment of the heartwood or sapwood is not very adequate.

The high air-dry density (63 pcf) of the wood is accompanied by equally high strength values, which are considerably higher in most respects than those of any well known U. S. species. Despite its strength, however, goncalo alves will be imported primarily for its beauty.

In the United States the greatest value of goncalo alves is in its use for specialty items such as archery bows, billiard cue butts, brushbacks, cutlery handles, and for fine and attractive products for turnery or carving.

Greenheart

Greenheart (*Ocotea rodiaei*) is essentially a Guyana tree although small stands also occur in Surinam. The heartwood varies in color from light to dark olive green or nearly black. The texture is fine and uniform and the grain is straight to wavy.

Greenheart is stronger and stiffer than white oak and generally more difficult to work with tools because of its high density (over 60 pcf air-dry density).

The heartwood is rated as very resistant to decay fungi and termites. It also is very resistant to marine borers in temperate waters but much less so in warm tropical waters (fig. 1-8).

Greenheart is used principally where strength and resistance to wear are required. Uses include ship and dock building, lock gates, wharves, piers, jetties, vats, piling, planking, industrial flooring, bridges, and some specialty items (fishing rods and billiard cue butts).

Guatambu (See Pau Marfim)
Guayacan (See Ipe)
Hura

Hura (*Hura crepitans*) grows throughout the West Indies and from Central America to northern Brazil and Bolivia. It is a large tree commonly reaching heights of 90 to 130 feet with clear boles of 40 to 75 feet. Often the diameters reach 3 to 5 feet and at times 6 to 9 feet.

The pale yellowish brown or pale olive gray heartwood is often indistinct from the yellowish white sapwood. Texture is fine to medium, and the grain is straight to interlocked.

Hura is a low-strength and low-density wood (air-dry density 15 to 28 pcf) that is moderately difficult to air dry. Warping is variable and sometimes severe. The wood usually machines easily, but green material is somewhat difficult to work due to tension wood which results in a fuzzy surface. The wood finishes well and is easy to glue and nail.

Hura is variable in resistance to attack by decay fungi but is highly susceptible to blue stain and very susceptible to dry-wood termites. However, the wood is easy to treat.

Hura is a light colored low-density wood that is often used in general carpentry, boxes and crates, and lower grade furniture. Other important uses are in veneer and plywood, fiberboard, and particleboard.

Figure 1–8—Hand hewing greenheart (*Ocotea rodiaei*) for heavy marine construction. Work is being done on a river landing in Guyana. (M150 272-14)

Ilomba

Ilomba (*Pycnanthus angolensis*) is a tree of the rainforest and ranges from Guinea and Sierra Leone through West tropical Africa to Uganda and Angola. Other common names include pycnanthus, walele, and otie.

The wood is grayish white to pinkish brown and in some trees may be a uniform light brown. There is generally no distinction between heartwood and sapwood. The texture is medium to coarse, and the grain is generally straight. This species is generally similar to banak (*Virola*) but is somewhat coarser textured.

The air-dry density is about 32 pcf and the wood is about as strong as yellow-poplar.

Ilomba seasons rapidly but is prone to collapse, warp, and splits. It is easily sawn and can be worked well with hand and machine tools. It is an excellent peeler and has good gluing and nailing characteristics.

The wood is perishable and subject to insect and fungal attack. Logs require rapid extraction and conversion to avoid degrade. Both sapwood and heartwood are permeable.

This species is used in the United States only in the form of plywood for general utility purposes. However, it is definitely suited for furniture components, interior joinery, and a general utility timber.

Ipe

The lapacho group of the genus *Tabebuia* consists of about 20 species of trees and occurs in practically every Latin American country except Chile. Other commonly used names are guayacan and lapacho.

The sapwood is relatively thick, yellowish gray or gray brown and sharply differentiated from the heartwood, which is a light to dark olive brown. The texture is fine to medium. Grain is closely and narrowly interlocked.

The wood is very heavy and averages about 64 pcf at 12 percent moisture content. Thoroughly air-dried specimens of heartwood generally sink in water.

Because of its high density and hardness, ipe is moderately difficult to machine, but glassy smooth surfaces can be produced.

Ipe is also very strong in all properties and in the air-dry condition is comparable to greenheart. Its hardness is two to three times that of oak or keruing.

The wood is highly resistant to decay and insects, including both subterranean and dry-wood termites. It is, however, susceptible to marine borer attack. The heartwood is impermeable, but the sapwood can be readily treated with preservatives.

Ipe is used almost exclusively for heavy duty and durable construction. Because of its hardness and good dimensional stability, it is particularly well suited for heavy duty flooring in trucks and boxcars. It is also used in railroad crossties, turnery, tool handles, decorative veneers, and some specialty items in textile mills.

Ipil (See Merbau)

Iroko

Iroko consists of two species (*Chlorophora excelsa* and *C. regia*). *C. excelsa* extends across the entire width of tropical Africa from the Ivory Coast southward to Angola and eastward across the continent to East Africa. *C. regia*, however, is limited to extreme West Africa from Gambia to Ghana and is less resistant to drought than *C. excelsa*.

The heartwood varies from a pale yellowish-brown to dark chocolate brown with lighter markings occurring most conspicuously on flat-sawn surfaces. Texture is medium to coarse, and the grain is typically interlocked.

Iroko can be worked easily with hand or machine tools but with some tearing of interlocked grain. Occasional deposits of calcium carbonate severely damage cutting edges. The wood dries rapidly with little or no degrade. The strength is similar to red maple, and the weight is about 43 pcf at 12 percent moisture content.

Heartwood is very resistant to decay fungi and is resistant to termite and marine borer attack.

Iroko has been suggested as a teak substitute, mostly because of its color. Due to its durability it is used in boat building, piles and marine work, railroad crossties, and other areas where durability is required. Other uses include joinery, domestic flooring, furniture, veneer, cabinetwork, and shop fittings.

Jacaranda (See Rosewood, Brazilian)

Jarrah

Jarrah (*Eucalyptus marginata*) is native to the coastal belt of southwestern Australia and one of the principal timbers of the country's sawmill industry.

The heartwood is a uniform pinkish to dark red, often a rich, dark red mahogany hue, turning to a deep brownish red with age and exposure. The sapwood is pale in color and usually very narrow in old trees. The texture is even and moderately coarse. The grain is frequently interlocked or wavy. The wood weighs about 54 pcf at 12 percent moisture content. The common defects of jarrah include gum veins or pockets which, in extreme instances, separate the log into concentric shells.

Jarrah is a heavy, hard timber possessing correspondingly high strength properties. It is resistant to attack by termites and rated as very durable with respect to fungus attack. The heartwood is rated as extremely resistant to preservative treatment.

The wood is difficult to work with hand and machine tools because of the high density and irregular grain.

Jarrah is used for decking and underframing of piers, jetties, and bridges, and also for piles and fenders in dock and harbor installations. As a flooring timber, it has a high resistance to wear but is inclined to splinter under heavy traffic. It is also used for railroad crossties and other heavy construction.

Jatoba (See Courbaril)

Jelutong

Jelutong (*Dyera costulata*) is an important species in Malaya where it is best known for its latex production in the manufacture of chewing gum rather than its timber.

The wood is white or straw-colored and there is no differentiation between heartwood and sapwood. The texture is moderately fine and even. The grain is straight, and luster is low. The wood weighs about 29 pcf at 12 percent moisture content.

The wood is very easy to season with little tendency to split or warp, but staining may cause trouble.

It is easy to work in all operations, finishes well, and glues satisfactorily.

The wood is rated as nondurable, but readily permeable to preservatives.

Jelutong would make an excellent core stock if it were economically feasible to fill the latex channels which radiate outward in the stem at the branch whorls. Because of its low density and ease of working, it is well suited for sculpture and pattern making, wooden shoes, picture frames, and drawing boards. Jelutong is essentially a "short-cutting" species, because the wood between the channels is remarkably free of other defects.

Jequitiba (See Albarco)

Kakaralli (See Manbarklak)

Kaneelhart

Kaneelhart or brown silverballi are names applied to the genus *Licaria*. Species of this genus are centered mostly in the Guianas and are found in association with greenheart on hilly terrain and also in wallaba forests on sandy soils.

The orange or brownish yellow, freshly cut heartwood darkens to yellowish brown or coffee brown on exposure. Sometimes there is a tinge of red or violet in the wood. Texture is fine to medium, and the grain is straight to slightly interlocked. The wood has a fragrant odor which is lost on drying.

Kaneelhart is a very strong and very heavy (air-dry density 52 to 72 pcf) wood that is difficult to work. It cuts smoothly, takes an excellent finish, but requires care in gluing.

Kaneelhart has excellent resistance to both brown-rot and white-rot fungi and is also rated very high in resistance to dry-wood termites.

Uses of kaneelhart include furniture, turnery, boatbuilding, heavy construction, and parquet flooring.

Kapur

The genus *Dryobalanops* comprises some nine species distributed over parts of Malaya, Sumatra, and Borneo, including North Borneo and Sarawak. For the export trade, however, the species are combined under the name kapur.

The heartwood is reddish brown, and clearly demarcated from the pale colored sapwood. The wood is fairly coarse textured but uniform. In general appearance the wood resembles that of keruing, but on the whole it is straighter grained and not quite so coarse in texture. The wood averages about 45 to 50 pcf at 12 percent moisture content.

Strength property values are similar to keruing of comparable specific gravity.

The heartwood is rated resistant to attack by decay fungi, but it is reported to be vulnerable to termites. It is extremely resistant to preservative treatment.

The wood works with moderate ease in most hand and machine operations, but blunting of cutters may be severe because of the silica content, particularly when machining dry wood. A good surface is obtainable from the various machining operations, but there is a tendency toward raised grain if dull cutters are used. Kapur takes nails and screws satisfactorily. Glue bonds are not durable in exterior plywood bonded with phenolic adhesives. However, the wood glues well with urea formaldehyde.

Kapur provides good and very durable construction timbers and is suitable for all purposes for which keruing is used

in the United States. In addition, kapur is extensively used in plywood either alone or in construction with species of *Shorea* (lauan/meranti).

Karri

Karri (*Eucalyptus diversicolor*) is a very large tree limited to southwestern Australia.

Karri resembles jarrah (*E. marginata*) in structure and general appearance. It is usually paler in color, and, on the average, slightly heavier (57 pcf at 12 pct moisture content).

Karri is a heavy hardwood possessing mechanical properties of a correspondingly high order, even somewhat higher than jarrah.

The heartwood is rated as moderately durable though less so than jarrah. It is extremely difficult to treat with preservatives.

The wood is fairly hard to work in machines and difficult to cut with hand tools. It is generally more resistant to cutting than jarrah and has slightly more dulling effect on tool edges.

It is inferior to jarrah for underground use and waterworks, but where flexural strength is required, such as in bridges (fig. 1−9), floors, rafters, and beams, it is an excellent timber. Karri is popular in the heavy construction field because of its strength and availability in large sizes and long lengths that are free of defects.

Kauta (See Marishballi)

Kempas

Kempas (*Koompassia malaccensis*) is distributed throughout the lowland forest in rather swampy areas of Malaysia and Indonesia.

The brick-red freshly cut heartwood darkens on exposure to an orange red or red brown with numerous yellow brown streaks due to soft tissue associated with the pores. The texture is rather coarse, and the grain is typically interlocked. Kempas is a hard, heavy wood (air-dry density 55 pcf) that is difficult to work with hand and machine tools. The timber dries well though with some tendency to warping and checking.

The heartwood is resistant to attack by decay fungi but vulnerable to termite activity. However, it treats readily with absorptions of preservative oils as high as 20 pcf.

Kempas is ideal for heavy construction work, railroad crossties, and flooring.

Keruing/Apitong

Keruing or apitong is widely scattered throughout the Indo-Malayan region. More than 70 species are in this group and most are marketed under the name keruing. Other important species of the genus *Dipterocarpus* are marketed as apitong in the Philippine Islands and yang in Thailand.

The heartwood varies from light-to-dark red brown or brown to dark brown, sometimes with a purple tint, which is usually well defined from the gray or buff sapwood. The texture is

Figure 1–9—Heavy timbers used in bridge construction. Species unknown, but this bridge in Pakistan is reportedly over 300 years old.

(M84 0472-16A)

moderately coarse and the grain is straight or shallowly interlocked. The wood is strong, hard, and heavy (air-dry density 45 to 50 pcf) and is characterized by the presence of resin ducts which occur singly or in short arcs as seen on end-grain surfaces. This resinous condition and the presence of silica can both be troublesome problems.

Although the heartwood is fairly resistant to decay and insect attack, the wood should be treated with preservatives when it is used in contact with the ground. Often the durability varies with species but generally the wood is classified as moderately durable.

Generally, keruing saws and machines well, particularly when green, but saws and cutters dull easily due to a high silica content in the wood. Resin adheres to machinery and tools and may be troublesome. Also, resin may cause gluing and finishing difficulties.

Keruing is used for general construction work, framework for boats, flooring, pallets, chemical-processing equipment, veneer and plywood, railroad crossties (if treated), truck floors, and boardwalks.

Khaya (See Mahogany, African)
Kokrodua (See Afrormosia)
Korina (See Limba)
Krabak (See Mersawa)
Kwila (See Merbau)
Lapacho (See Ipe)
Lauan/Meranti Groupings

The term lauan or "Philippine mahogany" is applied commercially to Philippine woods belonging to three genera—*Shorea*, *Parashorea*, and *Pentacme*. Meranti and seraya, on the other hand, are general southeast Asian names applied to woods of the genus *Shorea*, which contains almost 200 species. The various species of the three genera can be placed into five groups based on heartwood color and weight. The heartwood color and air-dry density of these groups are:

1-28

Name	Color	Air-dry density
Balau (also called selangan batu)	Light to deep red-brown	53 pcf
Dark red meranti (also called tanguile)	Dark brown, medium to deep red, sometimes with a purplish tinge	40+ pcf
Light red meranti (also called bagtikan or white lauan)	Variable from almost white to pale pink to dark red, or pale brown to deep brown	25−40 pcf, averaging 32
White meranti	Whitish when freshly cut, becoming light yellow-brown on exposure	30−42 pcf
Yellow meranti	Light yellow or yellow brown, sometimes with a greenish tinge, darkening on exposure	30−40 pcf

Lauan/meranti species as a whole have a coarser texture than American mahogany (*Swietenia*) or the "African mahoganies" (*Khaya*) and do not have dark-colored deposits in the pores. All lauan/meranti species have axial resin ducts aligned in long continuous tangential lines in the end surfaces of the wood. Sometimes these ducts contain white deposits that are visible to the naked eye, but the wood is not resinous like some keruing (*Dipterocarpus*) species that are similar in appearance to the lauans.

Lauan/meranti species are not durable to only slightly durable in exposed conditions or in ground contact. Generally, the heartwood is extremely resistant to moderately resistant to preservative treatments.

Due to its high density, the balau or selangan batu group is stronger than the other groups, and is rather difficult to machine. The strength and shrinkage properties of the four lower density groups compare favorably with oak. All four groups machine easily except white meranti, which dulls cutters due to a high silica content in the wood.

Principal uses for balau are heavy construction, boat frames, and flooring. Lauan/meranti species in the other four groups make up a large percentage of the total hardwood plywood imported into this country. Other uses include joinery, furniture and cabinetwork, vats, flooring, and general construction.

Lemonwood (See Degame)

Lignumvitae

For a great many years the only species of lignumvitae used on a large scale was *Guaiacum officinale*, which is native to the West Indies, northern Venezuela, northern Colombia, and Panama. With the near exhaustion of *G. officinale* the harvesters turned to *G. sanctum* which is now the principal species of commerce. The latter species occupies the same range as *G. officinale*, but is more extensive and includes the Pacific side of Central America as well as southern Mexico.

Lignumvitae is one of the heaviest and hardest woods on the market. The wood is characterized by its unique green color and oily or waxy feel. The wood has a fine, uniform texture and closely interlocked grain. Its resin content may constitute up to about one-fourth of the air-dry weight of the heartwood.

Lignumvitae wood is used chiefly for bearing or bushing blocks for the lining of stern tubes for steamship propeller shafts. The great strength and tenacity of lignumvitae, combined with the self-lubricating properties that are due to the high resin content, make it especially adaptable for underwater use. It is also used for such articles as mallets, pulley sheaves, caster wheels, stencil and chisel block, various turned articles, and brush backs.

Limba

Limba, ofram, or afara (*Terminalia superba*) is widely distributed from Sierra Leone to Angola and Zaire in the rain and savanna forest. It is also favored as a plantation species in West Africa.

The heartwood varies in color from a gray white to creamy or yellow brown and may contain dark streaks which are nearly black, producing an attractive figure which is valued for decorative veneer purposes. The light color of the wood is considered an important asset for the manufacture of blond furniture. The wood is generally straight grained and of uniform but coarse texture.

The wood is easy to season and the shrinkage is reported to be rather small. Limba is not resistant to decay, insects, or termites. It is easy to work with all types of tools and is veneered without difficulty. Principal uses include plywood, furniture, interior joinery, and sliced decorative veneers. Selected limba plywood is sold in the United States under the copyright name of "korina."

Lupuna (See Ceiba)

Macacauba (See Macawood)

Macawood/Trebol

Macawood and trebol are common names applied to species in the genus *Platymiscium*. Other common names include cristobal and macacauba. The distribution of this genus includes continental tropical America from southern Mexico to the Brazilian Amazon region and Trinidad.

The bright red to reddish purplish brown heartwood is more or less distinctly striped. Darker specimens look waxy, and the sapwood is sharply demarcated from the heartwood. The texture is medium to fine, and the grain is straight to roey or striped.

The wood is not very difficult to work, and it finishes

smoothly and takes on high polish. Generally it air dries slowly with a slight tendency to warp and check. The strength is quite high, and the air-dry density ranges from 55 to 73 pcf.

The heartwood is reported to be highly resistant to attack by decay fungi, insects, and dry-wood termites. Although the sapwood responds with good absorption, the heartwood is resistant to preservative treatments.

Macawood is a fine furniture and cabinet wood. It is also used in decorative veneers, musical instruments, turnery, joinery, and specialty items such as violin bows and billiard cues.

Machinmango (See Manbarklak)

Mahogany

The name mahogany is presently applied to several distinct kinds of commercial woods. The original mahogany wood, produced by the genus *Swietenia*, came from the American West Indies; in Europe during the 1600's it was the premier wood for fine furniture, cabinet work, and ship-building. Because the good reputation associated with the name mahogany is based on this wood, American mahogany is sometimes referred to as true mahogany. A related African wood, of the genus *Khaya*, has long been marketed as "African mahogany," and the similar properties and overall appearance allow it to be used for much the same purposes as American mahogany. A third kind of mahogany, and the one most commonly encountered in the market, is "Philippine mahogany." This name is applied to a group of Asian woods belonging to three distinct genera: *Shorea*, *Parashorea*, and *Pentacme*. In this work, information on the "Philippine mahoganies" is given under Lauan/Meranti groupings.

Mahogany, African

The bulk of "African Mahogany" shipped from west central Africa is *Khaya ivorensis*, which is the most widely distributed and most plentiful species of the genus found in the coastal belt of the so-called high forest. The closely allied species, *K. anthotheca*, has a more restricted range and is found farther inland in regions of lower rainfall but well within the area now being worked for the export trade.

The heartwood varies from a pale pink to a dark reddish brown. The grain is frequently interlocked, and the texture is medium to coarse which is comparable to that of American mahogany (*Swietenia*).

"African Mahogany" is very well known in the United States and large quantities are imported annually. The wood is easy to season, but machining properties are rather variable. Nailing and gluing properties are good, and an excellent finish is readily obtained. It is easy to slice and peel. In decay resistance, it is generally rated moderately durable, which is below American mahogany.

Principal uses include furniture and cabinetwork, interior finish, boat construction, and veneer.

Mahogany, American

True mahogany, Honduras mahogany, or American mahogany (*Swietenia macrophylla*) ranges from southern Mexico through Central America into South America as far south as Bolivia. Plantations have been established within its natural range and elsewhere throughout the tropics.

The heartwood varies from a pale pink or salmon color to a dark reddish brown. The grain is generally straighter than that of "African mahogany;" however, a wide variety of grain patterns are obtained from mahogany. The texture is rather fine to coarse.

Mahogany is easily air seasoned or kiln dried without appreciable warping or checking, and it has excellent dimensional stability properties. It is rated as durable in resistance to decay fungi and moderately resistant to dry-wood termites. Both heartwood and sapwood are resistant to impregnation with preservatives.

The wood is very easy to work with hand and machine tools and it slices and rotary cuts into fine veneer without difficulty. It also is easy to finish and takes an excellent polish.

The air-dry strength of mahogany is similar to American elm. The air-dry density varies from 30 to 52 pcf. The principal uses for mahogany are fine furniture and cabinet making, interior trim, pattern making, boat construction, fancy veneers, musical instruments, precision instruments, paneling, turnery, carving, and many other uses where an attractive and dimensionally stable wood is required.

Mahogany, Philippine (See Lauan/Meranti Groupings)

Manbarklak

Manbarklak is a common name applied to species in the genus *Eschweilera*. Other common names include kakaralli, machinmango, and mata-mata. About 80 species of this genus are distributed from eastern Brazil through the Amazon basin to the Guianas, Trinidad, and Costa Rica.

The heartwood of most species is light brown, grayish brown, reddish brown, or brownish buff. The texture is fine and uniform, and the grain is typically straight.

Manbarklak is a very hard and heavy wood (air-dry density ranges from 48 to 74 pcf) that is rated as fairly difficult to air season. Most of the species are difficult to work because of the high density and high silica content.

Most species are highly resistant to attack by decay fungi. Also most of the species have gained wide recognition for their high degree of resistance to marine borer attack. Resistance to dry-wood termite attack is variable depending on species.

Manbarklak is an ideal species for marine and other heavy

construction uses. It also is used for industrial flooring, pulpmill equipment, railroad crossties, piles, and turnery.

Manni

Manni (*Symphonia globulifera*) is native to two continents. In the New World it occurs in the West Indies, Central America, and North and South America. In the Old World it occurs in tropical West Africa. Other common names include ossol (Gabon), anani (Brazil), waika (Africa), and chewstick (Belize), a name acquired because of its use as a primitive toothbrush and flossing agent.

The heartwood is yellowish, grayish, or greenish brown or striped in these shades. The texture is coarse, and the grain is straight to irregular. The wood is very easy to work with both hand and machine tools, but surfaces tend to roughen in planing and shaping. Generally manni air seasons rapidly with only moderate warping and checking. Its strength is similar to hickory, and the air-dry density is 44 pcf.

The heartwood is durable in ground contact but only moderately resistant to dry-wood and subterranean termites. The wood is rated as resistant to impregnation.

Manni is a general purpose wood that is used for railroad crossties, general construction, cooperage, furniture components, flooring, utility plywood, and is suggested as an oak substitute.

Marishballi

Marishballi is one of many common names applied to species of *Licania*. Other common names include kauta and anaura. Species of *Licania* are widely distributed in tropical America but are most abundant in the Guianas and the lower Amazon region of Brazil.

The heartwood is generally a yellowish brown to brown or dark brown, sometimes with a reddish tinge. The texture is usually fine and close, and the grain is usually straight.

Marishballi is strong and very heavy (air-dry density 52 to 72 pcf). The woods are rated easy to moderately difficult to air season. Due to the high density and high silica content, the woods of *Licania* are difficult to work. Especially hardened cutters are suggested to obtain smooth surfaces.

Durability varies with species but all species are generally considered to have low to moderately low resistance to attack by decay fungi. However, all are known for their high resistance to attack by marine borers. Permeability also varies with species but generally the heartwood is moderately responsive to treatment.

Marishballi is ideal for underwater marine construction, heavy construction above ground, and railroad crossties (treated). Locally it is used for charcoal and fuel.

Mata-Mata (See Manbarklak)
Mayflower (See Roble)
Meranti (See Lauan/Meranti Groupings)

Merbau

Merbau (Malaya), ipil (Philippines), and kwila (New Guinea) are common names applied to species of the genus *Intsia*, most commonly *I. bijuga*. *Intsia* is distributed throughout the Indo-Malayan region, Indonesia, Philippines, and many of the western Pacific islands as well as Australia.

The yellowish to orange brown freshly cut heartwood turns brown or dark red brown on exposure. The texture is rather coarse, and the grain is straight to interlocked or wavy.

The air-dry strength of merbau is comparable to hickory but the density is somewhat lower (50 pcf at 12 pct moisture content). The wood seasons well with little degrade but stains black in the presence of iron and moisture. It is rather difficult to saw because it gums saw teeth and dulls cutting edges. However, it dresses smoothly in most operations and finishes well.

Merbau has good durability and high resistance to termite attack. The heartwood is impermeable, but the sapwood is treatable.

Merbau is used in furniture, fine joinery, turnery, cabinetmaking, flooring, musical instruments, and specialty items.

Mersawa

Mersawa is one of the *Anisoptera*, a genus of about 15 species distributed from the Philippine Islands and Malaysia to East Pakistan. Names applied to the timber vary with the source and three names are generally encountered in the lumber trade: krabak (Thailand), mersawa (Malaysia), and palosapis (Philippines).

Anisoptera species produce wood of light color and moderately coarse texture. The heartwood when freshly sawn is pale yellow or yellowish brown and darkens on exposure. Some timber may show a pinkish cast or pink streaks, but these eventually disappear on exposure. The wood weighs between 34 and 47 pcf in the seasoned condition at 12 percent moisture content and about 59 pcf when green.

The sapwood is susceptible to attack by powderpost beetles and the heartwood is not resistant to termites. With respect to fungus resistance, the heartwood is rated as moderately resistant and should not be used under conditions favoring decay. The heartwood does not absorb preservative solutions readily.

The wood machines readily, but because of the presence of silica, the dulling effect on the cutting edges of ordinary tools is severe and is very troublesome with saws.

It appears probable that the major volume of these timbers will be used in the form of plywood, because conversion in this form presents considerably less difficulty than does lumber production.

Mora

Mora (*Mora excelsa* and *M. gonggrijpii*) is widely distributed in the Guianas and less so in the Orinoco delta of

Venezuela. The yellowish red-brown, reddish brown, or dark red heartwood with paler streaks is distinct from the yellowish to pale brown sapwood. Texture is moderately fine to rather coarse. Grain is straight to commonly interlocked.

Mora is a strong and heavy (air-dry density 59 to 65 pcf) wood that is moderately difficult to work but yields smooth surfaces in sawing, planing, turning, or boring. Generally the wood is rated as moderately difficult to season.

The wood is rated durable to very durable in resistance to brown-rot and white-rot fungi. *M. gonggrijpii* is rated very resistant to drywood termites but *M. excelsa* is considerably less so. Sapwood responds readily to preservative treatments, but the heartwood resists impregnation.

Mora is used for industrial flooring, railroad crossties, shipbuilding, and heavy construction.

Oak

The oaks (*Quercus* spp.) are abundantly represented in Mexico and Central America with about 150 species, which are nearly equally divided between red and white oak groups. Mexico is represented with over 100 species and Guatemala with about 25; the numbers diminish southward to Colombia, which has two species. The usual Spanish name applied is encino or roble and both names are used interchangeably irrespective of species or use of the wood.

The wood of the various species is similar in heartwood color, texture, and grain characteristics to the oaks in the United States, especially the southern live oaks. In most cases the tropical oaks are heavier (44 to 62 pcf air dry) than the U.S. species.

Strength data are available for only four species, and the values obtained fall between those of white oak and live oak (*Quercus virginiana*) or are equal to those of the latter. The average specific gravity for these four species is 0.72 based on volume when green and weight ovendry, with an observed maximum average for one species from Guatemala of 0.86.

Heartwood is rated as very resistant to decay fungi and difficult to treat with preservatives.

Utilization of the tropical oaks is very limited at present due to difficulties encountered in the drying of the wood. The major volume is used in the form of charcoal, but the wood is used in flooring, railroad crossties, mine timbers, tight cooperage, boat and ship construction, and decorative veneers.

Obeche

Obeche (*Triplochiton scleroxylon*) trees of west central Africa reach heights of 150 feet or more and diameters of up to 5 feet. The trunk is usually free of branches for considerable heights so that clear lumber of considerable size is obtainable (fig. 1−10).

The wood is creamy white to pale yellow with little or no difference between the sapwood and heartwood. It is fairly soft, of uniform medium to coarse texture, and the grain is straight or more often interlocked. The wood weighs about 24 pcf in the air-dry condition.

The wood seasons readily with little degrade. It is not resistant to decay, and the sapwood blue stains readily unless appropriate precautions are taken after the trees are felled as well as after they have been converted into lumber.

The wood is easy to work and machine, veneers and glues well, and takes nails and screws without splitting.

The characteristics of this species make it especially suitable for veneer and core stock. Other uses include furniture components, millwork, blockboard, boxes and crates, particleboard and fiberboard, pattern making, and artificial limbs.

Ofram (See Limba)

Okoume

The natural distribution of okoume (*Aucoumea klaineana*) is rather restricted, being found only in west central Africa and Guinea. However, it is extensively planted throughout its natural range. This species has been popular in European markets for many years, but its extensive use in the United States is rather recent. When first introduced in volume for plywood and doors, its acceptance was phenomenal because it provided attractive appearance at moderate cost.

The wood has a salmon-pink color with a uniform texture and high luster. The texture is slightly coarser than that of birch. The nondurable heartwood dries readily with little degrade. Sawn lumber is somewhat difficult to machine because of the silica content, but okoume glues, nails, and peels into veneer easily. Okoume offers unusual flexibility in finishing because the color, which is of medium intensity, permits toning to either lighter or darker shades.

In the United States it is generally used for decorative plywood paneling, general utility plywood, and doors. Other uses include furniture components, joinery, and light construction.

Opepe

Opepe (*Nauclea diderrichii*) is widely distributed from Sierra Leone to the Congo region and eastward to Uganda and is often found in pure stands. The orange or golden yellow heartwood darkens on exposure and is clearly defined from the whitish or pale yellow sapwood. Texture is rather coarse, and the grain is usually interlocked or irregular.

The air-dry density (47 pcf) is about the same as for hickory, but the air-dry strength properties are somewhat lower. Quartersawn stock dries rapidly with little checking or warp, but flat-sawn lumber may develop considerable degrade. The timber works moderately well with hand and machine tools. It also glues and finishes satisfactorily.

The heartwood is rated as very resistant to decay and moderately resistant to termite attacks. The sapwood is permeable, but the heartwood is moderately resistant to impregnation.

Opepe is a general construction wood that is used in dock

and marine work, boatbuilding, railroad crossties, flooring, and furniture.

Ossol (See Manni)

Otie (See Ilomba)

Ovangkol (See Benge)

Palosapis (See Mersawa)

Para-Angelim (See Sucupira)

Pau Marfim

The growing range of pau marfim (*Balfourodendron riedelianum*) is rather limited, extending from the State of Sao Paulo, Brazil, into Paraguay and the provinces of Corrientes and Missiones of northern Argentina. In Brazil it is generally known as pau marfim and in Argentina and Paraguay as guatambu.

In color and general appearance the wood is very similar to birch or hard maple sapwood. Although growth rings are present, they do not show as distinctly as in birch and maple. The wood is straight grained, easy to work and finish but is not considered to be resistant to decay. There is no apparent difference in color between heartwood and sapwood.

The average basic specific gravity of pau marfim is about 0.73 based on the volume when green and weight when ovendry. On the basis of its specific gravity, its strength values would be above those of hard maple which has an average basic specific gravity of 0.56.

In its areas of growth pau marfim is used for much the same purposes as hard maple and birch are in the United States. Pau marfim was introduced to the U.S. market in the late 1960's and has been very well received and is especially esteemed for turned items.

Peroba, White (See Peroba de Campos)

Peroba de Campos

Peroba de campos or white peroba (*Paratecoma peroba*) grows in the coastal forests of eastern Brazil ranging from Bahia to Rio de Janeiro. It is the only species in the genus.

The heartwood is variable in color but generally is in shades of brown with tendencies toward casts of olive and reddish color. The sapwood is a yellowish gray, clearly defined from the heartwood. The texture is relatively fine and approximates that of birch. The grain is commonly interlocked with a narrow stripe or roey figure.

The wood machines easily but, when smooth surfaces are required, particular care must be taken in planing to prevent excessive grain tearing of quartered surfaces. There is some evidence that the fine dust from machining operations may produce allergic responses in certain individuals.

Peroba de campos averages about 50 to 55 pcf (air-dry densities). It is heavier than teak or white oak and is proportionately stronger than either of these species.

The heartwood is rated as very durable with respect to fungus attack and is rated as resistant to preservative treatment.

Figure 1–10—Obeche (*Triplochiton scleroxylon*) (M150 282-2) logs 5 feet and more in diameter yield lumber favored for joinery and millwork.

In Brazil the wood is used in the manufacture of fine furniture, flooring, and decorative paneling. The principal use in the United States is in shipbuilding, where it serves as an alternative to white oak for all purposes except bent members.

Peroba Rosa

Peroba rosa is the common name applied to a number of similar species in the genus *Aspidosperma*. They occur in southeastern Brazil and parts of Argentina.

The heartwood is a distinctive rose-red to yellowish, often variegated or streaked with purple or brown, becoming brownish yellow to dark brown upon exposure. The texture is fine and uniform and the grain is straight to irregular.

The wood is moderately heavy (47 pcf air dry) and compares closely in strength properties with oak. The wood dries with little checking or splitting. It works with moderate ease, and it glues and finishes satisfactorily.

The heartwood is resistant to decay fungi but is susceptible to dry-wood termite attack. Although the sapwood treats moderately well, the heartwood resists preservation treatments.

Peroba rosa is suited for general construction work and is favored for fine furniture and cabinetwork and decorative veneers. Other uses include flooring, interior trim, sash and doors, and turnery.

Pilon

The two main species of pilon or suradan are *Hyeronima alchorneoides* and *H. laxiflora*. These species range from southern Mexico to southern Brazil including the Guianas, Peru, and Colombia. They are also found throughout the West Indies.

The heartwood is a light reddish brown to chocolate brown or sometimes to dark red. The texture is moderately coarse, and the grain is interlocked.

The wood air seasons rapidly with only a moderate amount of warp and surface checking. It has good working properties in all operations except planing, which is rated poor due to the characteristic interlocked grain. The strength is comparable to hickory, and the air-dry density ranges from 46 to 53 pcf.

Pilon is rated moderately durable to very durable in ground contact and resistant to moderately resistant to subterranean and dry-wood termites. Both heartwood and sapwood are reported to treat moderately well using both open tank and pressure vacuum systems.

Pilon is especially suited for heavy construction, railway crossties, marinework, and flooring. It is also used for furniture, cabinetwork, decorative veneers, turnery, and joinery.

Piquia

Piquia is the common name generally applied to species in the genus *Caryocar*. This genus is distributed from Costa Rica southward into northern Colombia and the upland forest of the Amazon valley to eastern Brazil and the Guianas.

The yellowish to light grayish brown heartwood is hardly distinguishable from the sapwood. The texture is medium to rather coarse and the grain is generally interlocked.

The wood dries at a slow rate and warping and checking may develop, but only to a minor degree. It is reported easy to moderately difficult to saw, and with a rapid dulling of cutting edges.

The heartwood is very durable and resistant to decay fungi and dry-wood termites but only moderately resistant to marine borers.

Piquia is a wood that is recommended for general and marine construction, heavy flooring, railway crossties, boat parts, and furniture components. It is especially suitable where hardness and high wear resistance are needed.

Primavera

The natural distribution of primavera (*Cybistax donnell-smithii*) is restricted to southwestern Mexico, the Pacific Coast of Guatemala and El Salvador, and North Central Honduras.

Primavera is regarded as one of the primary light colored woods but its use was limited because of its rather restricted range and relative scarcity of wild trees within its natural growing area.

Plantations now coming into production have increased the availability of this species and provided a more constant source of supply. The quality of the plantation-grown wood is equal in all respects to the wood obtained from wild trees.

The heartwood is whitish to straw-yellow, and in some logs may be tinted with pale brown or pinkish streaks. Texture is medium to rather coarse, and the grain is straight to roey, which produces a wide variety of figure patterns. The wood also has a very high luster.

Shrinkage is rather low for primavera, and the wood shows a high degree of dimensional stability. Despite considerable grain variation, primavera machines remarkably well. The air-dry density is 29 pcf, and the wood is comparable in strength to tupelo.

Laboratory tests indicate a variable resistance to both brown-rot and white-rot fungi. Weathering characteristics are good.

The dimensional stability, ease of working, and pleasing appearance recommend primavera for solid furniture, paneling, interior trim, and special exterior uses.

Purpleheart

Purpleheart or amaranth are common names applied to species in the genus *Peltogyne*. The center of distribution is in the north middle part of the Brazilian Amazon region, but the combined range of all species is from Mexico through Central America and southward to southern Brazil.

When freshly cut, the heartwood is brown but turns a deep purple upon exposure. Texture is medium to fine, and the grain is usually straight. This strong and heavy wood (air-dry density 50 to 66 pcf) air dries easily to moderately difficult. It is moderately difficult to work with either hand or machine tools and dulls cutters rather quickly. Gummy resin exudes when heated by dull tools. For best cutting a slow feed rate and specially hardened cutters are suggested. The wood does turn smoothly, however, is easy to glue, and takes finishes well.

The heartwood is rated as highly durable in resistance to attack by decay fungi and very resistant to dry-wood termites. It is extremely resistant to impregnation with preservative oils.

Due to its unusual and unique color it is used in turnery, marquetry, cabinets, fine furniture, parquet flooring, and many specialty items such as billiard cue butts, and carvings. Other uses include heavy construction, shipbuilding, and chemical vats.

Pycnanthus (See Ilomba)

Ramin

Ramin (*Gonystylus bancanus*) is one of the very few moderately heavy woods that are classified as a "blond" wood. This species is native to southeast Asia from the Malay Peninsula to Sumatra and Borneo.

The wood is a uniform pale straw or yellowish to whitish in color. The grain is straight or shallowly interlocked. The texture is moderately fine, similar to that of mahogany (*Swietenia*), and even. The wood is without figure or luster.

Ramin is moderately hard and heavy, weighing about 42 pcf in the air-dry condition. The wood is easy to work, finishes well, and glues satisfactorily.

With respect to natural durability ramin is rated as perishable, but it is permeable with regard to preservative treatment.

Ramin is used for plywood, interior trim, furniture, turnery, joinery, moldings, flooring, dowels, nonstriking handles (brooms), and as a general utility wood.

Roble (See Oak)

Roble

Roble (species in the roble group of *Tabebuia*, generally *T. rosea*) ranges from southern Mexico through Central America to Venezuela and Ecuador. The name "roble" comes from the Spanish word for oak (*Quercus*). In addition, *T. rosea* is called roble because the wood superficially resembles oak. Other common names for *T. rosea* are mayflower and apamate.

The sapwood becomes a pale brown upon exposure. The heartwood varies through the browns, from a golden to a dark brown. Texture is medium, and grain is closely and narrowly interlocked. Heartwood is without distinctive odor or taste. The wood weighs about 38 pcf at 12 percent moisture content.

Roble has excellent working properties in all machine operations. It finishes attractively in natural color and takes finishes with good results. It averages lighter in weight than the average of the American white oaks, but is comparable with respect to bending and compression parallel to grain. The white oaks are superior with respect to side hardness and shear.

The heartwood of roble is generally rated as durable to very durable with respect to fungus attack; the darker colored and heavier wood is regarded as more resistant than the lighter forms.

Roble is used extensively for furniture, interior trim, doors, flooring, boat building, ax handles, and general construction. The wood veneers well and produces an attractive paneling. For some applications roble is suggested as a substitute for ash and oak.

Rosewood, Brazilian

Brazilian rosewood or jacaranda (*Dalbergia nigra*) occurs in the eastern forests from the State of Bahia to Rio de Janeiro. Having been exploited for a long period of time, Brazilian rosewood is nowhere abundant now.

The wood of commerce is very variable with respect to color ranging through shades of brown, red, and violet and is irregularly and conspicuously streaked with black. Many kinds are distinguished locally on the basis of prevailing color. The texture is coarse, and the grain is generally straight. Heartwood has an oily or waxy appearance and feel. The odor is fragrant and distinctive. The wood is hard and heavy; thoroughly air-dried wood will barely float in water (47 to 56 pcf).

The strength properties are high, and are more than adequate for the purposes for which Brazilian rosewood is utilized. For example, Brazilian rosewood is far harder than any U.S. native hardwood species used in the furniture and veneer field.

The wood machines and veneers well. It can be glued satisfactorily, providing the necessary precautions are taken to ensure good glue bonds, especially with oily specimens.

Brazilian rosewood has an excellent reputation for durability with respect to fungus and insect attack, including termites, although the wood is not used for purposes where durability would present a problem.

Brazilian rosewood is used primarily in the form of veneer for decorative plywood. Limited quantities are used in the solid form for specialty items such as cutlery handles, brush backs, billiard cue butts, and fancy articles of turnery.

Rosewood, Indian

Indian rosewood (*Dalbergia latifolia*) is native to most provinces of India except in the northwest.

The heartwood varies in color from golden brown to dark purplish brown with denser blackish streaks terminating the growth zones and giving rise to an attractive figure on flat-sawn surfaces. The average weight is about 53 pcf at 12 percent moisture content. The texture is uniform and moderately coarse. The wood of this species is quite similar in appearance to that of the Brazilian and Honduras rosewood. The timber is said to kiln dry well, but rather slowly, and the color is said to improve during drying.

Indian rosewood is a heavy timber with high strength properties and is particularly hard for its weight after being thoroughly seasoned.

The wood is moderately hard to work with handtools and offers a fair resistance in machine operations. Lumber containing calcareous deposits tends to blunt tools rapidly. The wood turns well and has high screw-holding properties. If a very smooth surface is required for certain purposes, pores may need to be filled.

Indian rosewood is essentially a decorative wood for high-class furniture and cabinetwork. In the United States it is used primarily in the form of veneer.

Sande

Practically all sande (*Brosimum* spp.—utile group) available commercially comes from Pacific Ecuador and Colombia. However, the range of sande is from the Atlantic Coast in Costa Rica southward to Colombia and Ecuador.

The sapwood and heartwood show no distinction, being a uniform yellowish white to yellowish brown or light brown. The texture is medium to moderately coarse and even. The grain is straight to widely and shallowly interlocked. Luster is high.

The wood is nondurable with respect to stain, decay, and insect attack, and care must be exercised to prevent degrade from these agents. The air-dry density ranges from 24 to 38 pcf, and the strength is comparable to oak. The lumber air seasons rapidly and easily with little or no degrade. However, material containing tension wood is subject to warp and the tension wood may cause fuzzy grain, as well as burning of saws due to pinching. The wood stains and finishes readily and presents no gluing problems.

Sande is used for plywood, particleboard, fiberboard, carpentry, light construction, furniture components, and moldings.

Santa Maria

Santa Maria (*Calophyllum brasiliense*) ranges from the West Indies to southern Mexico and southward through Central America into northern South America.

The heartwood is pinkish to brick red or rich reddish brown and marked by a fine and slightly darker striping on flat-sawn surfaces. The sapwood is lighter in color and generally distinct from the heartwood. Texture is medium and fairly uniform, and the grain is generally interlocked. Luster is medium. The heartwood is rather similar in appearance to the red lauan of the Philippines.

The wood is moderately easy to work and good surfaces can be obtained when attention is paid to machining operations. The wood averages about 38 pcf at 12 percent moisture content.

Santa Maria is in the density class of hard maple and its strength properties are generally similar, with the exception of hardness, in which property hard maple is superior to Santa Maria.

The heartwood is generally rated as moderately durable to very durable in contact with the ground, but apparently has little resistance against termites and marine borers.

The inherent natural durability, color, and figure on the quarter suggest utilization as face veneer for plywood in boat construction. It also offers possibilities for use in flooring, furniture, cabinetwork, millwork, and decorative plywood.

Sapele

Sapele (*Entandrophragma cylindricum*) is a large African rain forest tree, occurring from Sierra Leone to Angola and eastward through the Congo to Uganda.

The heartwood ranges in color from that of mahogany to a dark reddish or purplish brown. The lighter colored and distinct sapwood may be up to 4 inches thick. Texture is rather fine. Grain is interlocked and produces a narrow and uniform stripe pattern on quartered surfaces.

The wood averages about 47 pcf at 12 percent moisture content, and its mechanical properties are in general higher than those of white oak.

The wood works fairly easily with machine tools, although interlocked grain offers difficulties in planing and molding. Sapele finishes and glues well.

The heartwood is rated as moderately durable and as resistant to preservative treatment.

Sapele is used extensively, primarily in the form of veneer for decorative plywood, but also in the solid form for furniture and cabinetwork, joinery, and flooring.

Selangan Batu (See Lauan/Meranti Groupings)
Sepetir

The name sepetir applies to species in the genus *Sindora* and to *Pseudosindora palustris*. These species are distributed throughout Malaysia, Indochina, and the Philippines. The heartwood is brown with a pink or golden tinge which darkens on exposure. Dark brown or black streaks are sometimes present in species of *Sindora*. Texture is moderately fine and even, and the grain is straight or shallowly interlocked.

The strength of sepetir is similar to shellbark hickory, and the air-dry densities are also similar (40 to 45 pcf). The wood seasons well but dries rather slowly with a tendency to end-splitting. The wood is difficult to work with hand tools and has a rather rapid dulling effect on cutters. Gums tend to accumulate on the teeth of saws causing additional problems.

Sepetir is rated as nondurable in ground contact under Malayan exposure. The heartwood is extremely resistant to preservative treatments; however, the sapwood is only moderately resistant.

Sepetir is a general carpentry wood that is also used for furniture and cabinetwork, joinery, flooring, especially truck flooring, plywood, and decorative veneers.

Seraya (See Lauan/Meranti Groupings)
Silverballi, Brown (See Kaneelhart)
Spanish-Cedar

Spanish-cedar or cedro (*Cedrela* spp.) comprises a group of about seven species that are widely distributed in tropical America from southern Mexico to northern Argentina. Spanish-cedar and mahogany are the classic timbers of Latin America.

The wood is more or less distinctly ring porous, and the heartwood varies from light reddish brown to dark reddish brown. The texture is rather fine and uniform to coarse and uneven. The grain is usually straight. The heartwood is characterized by a distinctive cedarlike odor.

The wood seasons readily. It is not high in strength but is roughly rated to be similar to Central American mahogany in most properties except in hardness and compression perpendicular to the grain where mahogany is definitely superior. It is considered decay resistant and works and glues well.

Spanish-cedar is used locally for all purposes where an easily worked, light but straight grained, and durable wood is required. In the United States the wood is favored for millwork, cabinets, fine furniture, boatbuilding patterns, cigar wrappers and boxes, and decorative and utility plywoods.

Sucupira/Angelim/Para-Angelim

These three common names apply to species in four genera of legumes (Leguminosae) from South America. Sucupira refers to *Bowdichia nitida* from northern Brazil, *B. virgilioides* from Venezuela, the Guianas, and Brazil, and *Diplotropis purpurea* from the Guianas and southern Brazil.

Angelin (*Andira inermis*) is a widespread species occurring throughout the West Indies and from southern Mexico through Central America to northern South America and Brazil, whereas para-angelim (*Hymenolobium excelsum*) is generally restricted to Brazil.

The heartwood of sucupira is chocolate to chocolate reddish brown turning to a lighter brown especially in *Diplotropis purpurea*. Angelin heartwood is yellowish-brown to dark reddish brown, whereas para-angelim heartwood is pale brown upon exposure. The texture of these three woods is coarse and uneven, and the grain is often straight.

The air-dry density of these woods ranges from 45 to 60 pcf, which makes them generally heavier than hickory. Their strength properties are also higher than hickory.

The heartwood is rated very durable to durable in resistance to decay fungi but only moderately resistant to attack by dry-wood termites. Angelin is reported difficult to treat, but para-angelim and sucupira treat adequately.

Angelin saws and works fairly well except that it is difficult to plane to a smooth surface because of the alternating bands or hard and soft (parenchyma) tissue. Para-angelim works well in all operations. Sucupira is difficult to moderately difficult to work on account of its high density, interlocked and irregular grain, and coarse texture.

These woods are ideal for heavy construction, railroad crossties, and other uses not requiring much fabrication. Other suggested uses include flooring, boatbuilding, furniture, turnery, tool handles, and decorative veneers.

Suradan (See Pilon)

Tangare (See Andiroba)

Tanguile (See Lauan/Meranti Groupings)

Teak

Teak (*Tectona grandis*) occurs in commercial quantities in India, Burma, Thailand, Laos, Cambodia, North and South Vietnam, and the East Indies. Numerous plantations have been developed within its natural range and in tropical areas of Latin America and Africa, and many of these are now producing timber.

The heartwood varies from a yellow brown to dark golden brown and eventually turns a rich brown upon exposure. It has a coarse uneven texture (ring porous), is usually straight grained, and has a distinctly oily feel. The heartwood has excellent dimensional stability and possesses a very high degree of natural durability.

Although not generally used in the United States where strength is of prime importance, the values for teak are generally on a par with those of native American oaks.

Teak generally is worked with moderate ease with hand and machine tools. However, silica in the wood often dulls tools. Finishing and gluing are satisfactory although pretreatment may be necessary to ensure good bonding of finishes and glues.

Intrinsically, teak is one of the most valuable of all woods, but its use is limited by scarcity and high cost. Teak is unique in that it does not cause rust or corrosion when in contact with metal; hence, it is extremely useful in the shipbuilding industry, tanks and vats, and fixtures requiring high resistance to acids. Teak is currently used in the construction of expensive boats, furniture, flooring, decorative objects, and veneer for decorative plywood.

Tornillo

Tornillo or Cedro-rana (*Cedrelinga catenaeformis*) grows in the Loreton Huanuco provinces of Peru and in the humid ''terrafirma'' of the Brazilian Amazon region. Tornillo can be a large tree up to 160 feet tall with trunk diameters of 5 to 9 feet. Trees felled in Peru are often 4 feet in diameter with merchantable heights of 45 feet and more.

The heartwood is pale brown with a golden luster prominently marked with red vessel lines and gradually merging into the lighter colored sapwood. The texture is coarse. The air-dry density of material collected in Brazil averages 40 pcf and for Peruvian stock the average is about 30. The wood is comparable in strength to American elm.

Tornillo cuts easily and can be finished smoothly but areas of tension wood may result in woolly surfaces. The heartwood is fairly durable and reported to have a good weathering resistance.

Tornillo is a general construction wood that can be used for furniture components in lower grade furniture.

Trebol (See Macawood)

Virola (See Banak)

Waika (See Manni)

Walele (See Ilomba)

Wallaba

Wallaba is a common name applied to the species in the genus *Eperua*. Other common names include wapa and apa. The center of distribution is in the Guianas, but the distribution extends into Venezuela and the Amazon region of northern Brazil. Wallaba generally occurs in pure stands or as a dominant tree in the forest.

The heartwood is light to dark red to reddish or purplish brown with characteristic dark, gummy streaks. The texture is rather coarse, and the grain is typically straight.

Wallaba is a hard, heavy wood (air-dry density 58 pcf). The strength is higher than shagbark hickory.

The wood dries very slowly with a marked tendency to check, split, and warp. Though the wood has a high density it is easy to work with hand and machine tools. However, high gum content clogs saw teeth and cutters. Once the wood is kiln dried, gum exudates are not a serious problem in machining.

The heartwood is reported to be very durable and resistant to subterranean termites and fairly resistant to dry-wood termites.

Wallaba is well suited for heavy construction, railroad crossties, poles, industrial flooring, and tank staves. It is also highly favored for charcoal.

Wapa (See Wallaba)

Yang (See Keruing)

Softwoods

Cypress, Mexican

Mexican cypress (*Cupressus lusitanica*) is native to Mexico and probably Guatemala but is now widely planted at high elevations throughout the tropical world.

The heartwood is yellowish, pale brown, or pinkish with an occasional streaking or variegation. The texture is fine and uniform, and the grain is usually straight. The wood is fragrantly scented. The air-dry density is 32 pcf, and the strength is comparable to Alaska-cedar or western hemlock.

The wood is easy to work with hand and machine tools, and it nails, stains, and polishes well. The wood air dries very rapidly with little or no end or surface checking. Reports on durability are conflicting. The heartwood is not treatable by the open tank process and seems to have an irregular response to pressure-vacuum systems.

Mexican cypress is used mainly for posts and poles, furniture components, and general construction. Many more uses may be found for Mexican cypress as plantations of this wood mature.

Parana Pine

The wood that is commonly called "Parana pine" (*Araucaria angustifolia*) is not a true pine. It is a softwood that comes from southeastern Brazil and adjacent areas of Paraguay and Argentina.

"Parana pine" has many desirable characteristics. It is available in large-size clear boards with uniform texture. The small pinhead knots (leaf traces) that appear on flat-sawn surfaces and the light brown or reddish-brown heartwood, which is frequently streaked with red, provide desirable figured effects for matching in paneling and interior finishes. The growth rings are fairly distinct and more nearly like those of white pine (*Pinus strobus*) rather than those of the yellow pines. The wood has relatively straight grain, takes paint well, glues easily, and is free from resin ducts, pitch pockets, and pitch streaks.

The strength values of this species compare favorably with those softwood species of similar density found in the United States and, in some cases, approach the strength values of species with greater specific gravity. It is especially good in shearing strength, hardness, and nail-holding ability, but notably deficient in strength in compression across the grain.

Some tendency towards splitting of kiln-dried "Parana pine" and warping of seasoned and ripped lumber is caused by the presence of compression wood, an abnormal type of wood structure with intrinsically large shrinkage along the grain. Boards containing compression wood should be excluded from exacting uses. The principal uses of "Parana pine" include framing lumber, interior trim, sash and door stock, furniture, case goods, and veneer.

Pine, Caribbean

Caribbean pine (*Pinus caribaea*) occurs along the Caribbean side of Central America from Belize to northeastern Nicaragua. It is also native to the Bahamas and Cuba. This tree of the low elevations is widely introduced as a plantation species throughout the world tropics.

The heartwood is golden brown to red brown and distinct from the sapwood which is 1 to 2 inches in thickness and a light yellow. This softwood species has a strong resinous odor and a greasy feel. The weight varies considerably and may range from 26 to 51 pcf at 12 percent moisture content.

The lumber can be kiln dried satisfactorily.

Caribbean pine is easy to work in all machining operations, but the high resin content may necessitate occasional stoppages to permit removal of accumulated resin from the equipment.

Caribbean pine may be appreciably heavier than slash pine (*P. elliottii*) but the mechanical properties of these two species are rather similar. The durability and resistance to insect attack varies with resin content but generally the heartwood is rated moderately durable. The sapwood is highly permeable and is easily treated by open tank or pressure-vacuum systems. The heartwood is rated as moderately resistant and depends on the resin content.

Caribbean pine is used for the same purposes as the southern pines of the United States.

Pine, Ocote

Ocote pine (*Pinus oocarpa*) is a species of the higher elevations and occurs from northwestern Mexico southward through Guatemala into Nicaragua (fig. 1—11). The largest and most extensive stands occur in Guatemala, Nicaragua, and Honduras.

The sapwood is a pale yellowish brown and generally up to 3 inches in thickness. The heartwood is a light reddish brown.

Figure 1–11—In the highlands of El Salvador, ocote pine (*Pinus oocarpa*) is cut into boards by pit sawing. Finished lumber is sent down the mountainside on the backs of unattended burros.

(M150 272-15)

Grain is straight. Luster is medium. The wood has a resinous odor, and weighs about 41 pcf at 12 percent moisture content.

The strength properties of ocote pine are comparable in most respects with those of longleaf pine (*P. palustris*).

Decay resistance studies show ocote pine heartwood to be very durable with respect to attack by a white-rot fungus and moderately durable with respect to brown rot.

Ocote pine is comparable to the southern pines in workability and machining characteristics. Ocote pine is a general construction timber and is suited for the same uses as the southern pines.

Pine, Radiata

Radiata or Monterey pine (*Pinus radiata*) is planted extensively in the southern hemisphere, mainly in Chile, New Zealand, Australia, and South Africa. Plantation-grown trees may reach a height of 80 to 90 feet in 20 years.

Plantation-grown heartwood is light brown to pinkish brown and is distinct from the paler creamy sapwood. Growth rings are mostly wide and distinct. False rings may be common.

Texture is moderately even and fine, and the grain is usually straight.

Plantation-grown radiata pine averages about 30 pcf at 12 percent moisture content. The strength is comparable to red pine (*P. resinosa*) although there may be considerable variation due to location and growth rates.

The wood air or kiln dries rapidly with little degrade. The timber machines easily though the grain tends to tear around large knots. It nails and glues easily and takes paint and varnish well.

The sapwood is prone to attack by stain fungi and vulnerable to boring insects. However, plantation-grown stock is mostly sapwood, and the sapwood treats readily. The heartwood is rated as durable above ground and is moderately resistant to treatment.

Radiata pine can be used for the same purposes as other pines in the United States. These uses include veneers, plywood, pulp, fiberboard, particleboard, light construction, boxes, and millwork.

1-39

Selected References

Berni, C. A.; Bolza, E.; Christensen, F. J. South American timbers—the characteristics, properties, and uses of 190 species. Melbourne, Australia: Commonwealth Scientific and Industrial Research Organization, Division of Building Research; 1979.

Bolza, E.; Keating, W. G. African timbers—the properties, uses, and characteristics of 700 species. Melbourne, Australia: Commonwealth Scientific and Industrial Research Organization, Division of Building Research; 1972.

Building Research Establishment, Department of Environment. A handbook of softwoods. London: H. M. Stationery Office; 1977.

Building Research Establishment, Princes Risborough Laboratory; Farmer, R. H., rev. Handbook of hardwoods. 2nd ed. London: H. M. Stationery Office; 1972.

Chudnoff, Martin. Tropical timbers of the world. Agric. Handb. 607. Washington, DC: U.S. Department of Agriculture; 1984.

Hardwood Market Report: Lumber News Letter. Memphis, TN; (see current edition).

Keating, W. G.; Bolza, E. Characteristics, properties, and uses of timbers: Vol. 1. Southeast Asia, Northern Australia, and the Pacific. Melbourne, Australia: Inkata Press; 1982.

Kukachka, B. F. Properties of imported tropical woods. Res. Pap. FPL 125. Madison, WI: U.S. Department of Agriculture, Forest Service, Forest Products Laboratory; 1970.

Little E. L. Checklist of United States trees (native and naturalized). Agric. Handb. 541. Washington DC: U.S. Department of Agriculture; 1979.

Markwardt, L. J. Comparative strength properties of woods grown in the United States. Tech. Bull. 158. Washington, DC: U.S. Department of Agriculture; 1930.

Panshin, A. J.; deZeeuw, C. Textbook of wood technology. 4th ed. New York: McGraw-Hill; 1980.

Record, S. J.; Hess, R. W. Timbers of the new world. New Haven, CT: Yale University Press; 1949.

Ulrich, Alice H. U.S. timber production, trade, consumption, and price statistics, 1950−1980. Misc. Pub. 1408. Washington, DC: U.S. Department of Agriculture; 1981.

U.S. Department of Agriculture. An analysis of the timber situation in the United States, 1952−2030. For. Res. Rep. 23. Washington, DC: U.S. Department of Agriculture; 1982.

Chapter 2

Structure of Wood

Structure of Wood *

The fibrous nature of wood strongly influences how it is used. Specifically, wood is composed mostly of hollow, elongate, spindle-shaped cells that are arranged parallel to each other along the trunk of a tree. When lumber and other products are cut from the tree, the characteristics of these fibrous cells and their arrangement affect such properties as strength and shrinkage, as well as grain pattern of the wood.

A brief description of some elements of wood structure is given in this chapter.

Bark, Wood, and Pith

A cross section of a tree (fig. 2−1) shows the following well-defined features in succession from the outside to the center: (1) Bark, which may be divided into the outer, corky, dead part that varies greatly in thickness with different species and with age of trees, and the thin, inner living part; (2) wood, which in merchantable trees of most species is clearly differentiated into sapwood and heartwood; and (3) the pith, the small core of tissue located at the center of tree stems, branches, and twigs.

As the tree grows in height, branching is initiated by lateral bud development. The lateral branches are intergrown with the wood of the trunk as long as they are alive. These living branches constitute intergrown knots. After the branch dies, the trunk continues to increase in diameter and surrounds the portion of the branch which was projecting from the trunk at the time the branch died. Such enclosed portions of dead branches constitute the loose or encased knots. After the dead branches drop off, the dead stubs become overgrown and, subsequently, clear wood is formed.

Most growth in thickness of bark and wood arises by cell division at a layer of thin-walled living cells called the cambium. This layer of cells is located between the bark and wood and is invisible without a microscope. No growth in either diameter or length takes place in wood already formed; new growth is purely the addition of new cells, not the further development of old ones. New wood cells are formed on the inside of the cambium and new bark cells on the outside. Thus, new wood is laid down to the outside of old wood and the diameter of the woody trunk increases. The existing bark is pushed outward by the formation of new bark, and the outer bark layers become stretched, cracked, and ridged and are finally sloughed off.

Sapwood and Heartwood

Sapwood is located between the bark and the heartwood. The sapwood contains both living and dead cells and func-

tions primarily in the storage of food and the mechanical transport of water or sap.

The sapwood layer may vary in thickness and in the number of growth rings contained in it. Sapwood commonly ranges from 1-1/2 to 2 inches in radial thickness. In certain species, such as catalpa and black locust, the sapwood contains few growth rings and sometimes does not exceed 1/2 inch in thickness. The maples, hickories, ashes, some of the southern pines, ponderosa pine, and cativo (*Prioria copaifera*), ehie (*Guibourtia ehie*), and courbaril (*Hymenaea courbaril*) of tropical origin may have sapwood 3 to 6 inches or more in thickness, especially in second-growth trees. As a rule, the more vigorously growing trees have wider sapwood layers. Many second-growth trees of merchantable size consist mostly of sapwood.

In general, heartwood consists of inactive cells that do not function in either water conduction or food storage. The transition from sapwood to heartwood, however, is accompanied by an increase in the extractive content. Frequently these extractives darken the heartwood and give species such as black walnut and cherry their characteristic color. Species in which heartwood does not darken to a great extent include the spruces (except Sitka spruce), hemlock, the true firs, basswood, cottonwood, buckeye, ceiba, obeche, and ramin. Heartwood extractives in some species such as black locust, western redcedar, and redwood make the wood resistant to fungi or insect attack. All dark-colored heartwood is not resistant to decay and some nearly colorless heartwood provides decay resistance, as in northern white-cedar. Sapwood of all species, however, is not resistant to decay. Heartwood extractives may also affect wood in the following ways: (1) Reduce the permeability making the heartwood slower to dry and more difficult to impregnate with chemical preservatives, (2) increase the stability to changing moisture conditions, (3) increase the weight slightly, and (4) dull cutting tools. However, as sapwood becomes heartwood, no cells are added or taken away, nor do any cells change shape. The basic strength of the wood is essentially not affected by sapwood cells becoming heartwood cells.

In some species such as the ashes, hickories, and certain oaks, the pores (vessels) become plugged to a greater or lesser degree with ingrowths known as tyloses. Heartwood having pores tightly plugged by tyloses, as in white oak, is suitable for tight cooperage, since this prevents the passage of liquid through the pores. Tyloses also make impregnation with liquid preservatives difficult.

Growth Rings

With most species in temperate climates, there is sufficient difference between the wood formed early and that formed late in a growing season to produce well-marked annual growth

---

* Revision by Regis B. Miller, Botanist.

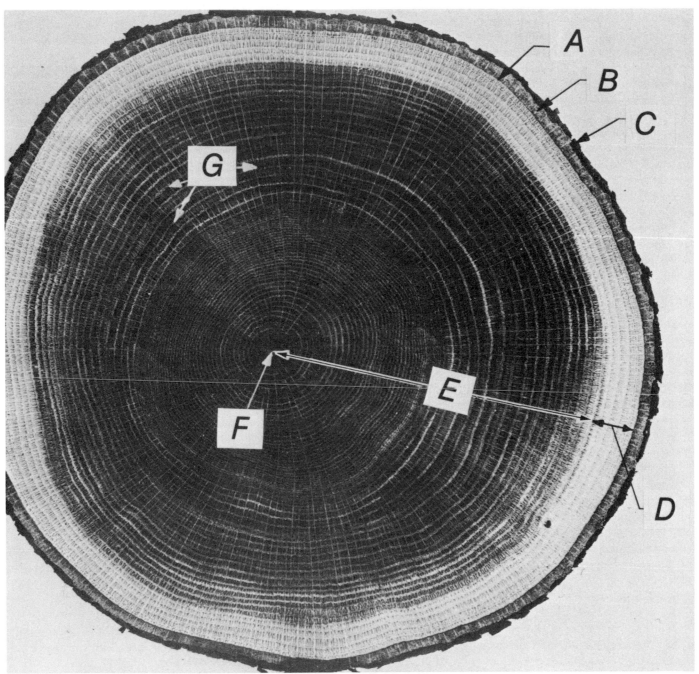

Figure 2–1—Cross section of a white oak tree trunk: *A,* Cambium layer (microscopic) is inside inner bark and forms wood and bark cells. *B,* Inner bark is moist, soft, and contains living tissue; carries prepared food from leaves to all growing parts of tree. *C,* Outer bark containing corky layers is composed of dry dead tissue. Gives general protection against external injuries. Inner and outer bark are separated by a bark (cork) cambium. *D,* Sapwood, which contains both living and dead tissues, is the light-colored wood beneath the bark. Carries sap from roots to leaves. *E,* Heartwood (inactive) is formed by a gradual change in the sapwood. *F,* Pith is the soft tissue about which the first wood growth takes place in the newly formed twigs. *G,* Wood rays connect the various layers from pith to bark for storage and transfer of food.

(M88 620)

rings. The age of a tree at the stump or the age at any cross section of the trunk may be determined by counting these rings (fig. 2—2). However, if the growth in diameter is interrupted by drought or defoliation by insects, more than one ring may be formed in the same season. In such an event, the inner rings usually do not have sharply defined boundaries and are termed false rings. Trees that have only very small crowns or that have accidentally lost most of their foliage may form only an incomplete growth layer, sometimes called a discontinuous ring.

The inner part of the growth ring formed first in the growing season is called earlywood and the outer part formed later in the growing season, latewood. Actual time of formation of these two parts of a ring may vary with environmental and weather conditions. Earlywood is characterized by cells having relatively large cavities and thin walls. Latewood cells have smaller cavities and thicker walls. The transition from earlywood to latewood may be gradual or abrupt, depending on the kind of wood and the growing conditions at the time it was formed.

Growth rings are most readily seen in species with sharp contrast between earlywood, last year's latewood and next year's earlywood, such as in the native ring-porous hardwoods ash and oak, and softwoods such as southern pine. In some other species, such as water tupelo, aspen, and sweetgum, differentiation of early and late growth is slight, and the annual growth rings are difficult to recognize. In many tropical regions, growth may be practically continuous throughout the year, and no well-defined annual rings are formed.

When growth rings are prominent, as in most softwoods and ring-porous hardwoods, earlywood differs markedly from latewood in physical properties. Earlywood is lighter in weight, softer, and weaker than latewood. Because of the greater density of latewood, the proportion of latewood is sometimes used to judge the strength of wood. This method is useful with such species as the southern pines, Douglas-fir, and the ring-porous hardwoods—ash, hickory, and oak.

Wood Cells

Wood cells that make up the structural elements of wood are of various sizes and shapes and are quite firmly cemented together. Dry wood cells may be empty or partly filled with deposits, such as gums and resins, or with tyloses. The majority of wood cells are considerably elongated and pointed at the ends; they are customarily called fibers or tracheids. The length of wood fibers is highly variable within a tree and among species. Hardwood fibers average about one twenty-fifth of an inch (1 mm) in length; softwood fibers (called tracheids) range from one-eighth to one-third of an inch (3 to 8 mm) in length.

In addition to their fibers, hardwoods have cells of relatively large diameter known as vessels (or pores). These form the main arteries in the movement of sap. Softwoods do not contain vessels for conducting sap longitudinally in the tree; this function is performed by the tracheids.

Both hardwoods and softwoods have cells (usually grouped into structures or tissues) that are oriented horizontally in the direction from the pith toward the bark. These groups of cells conduct sap radially across the grain and are called rays or wood rays. The rays are most easily seen on edge-grained or quartersawed surfaces. They vary greatly in size in different species. In oaks and sycamores, the rays are conspicuous and add to the decorative features of the wood.

Wood also has other cells, known as longitudinal or axial parenchyma cells, that function mainly for the storage of food.

Chemical Composition of Wood

Dry wood is made up chiefly of cellulose, lignin, hemicelluloses, and minor amounts (5-10 percent) of extraneous materials.

Cellulose, the major constituent, comprises approximately 50 percent of wood substance by weight. It is a high-molecular-weight linear polymer consisting of chains of bonded glucose monomers. During growth of the tree, the cellulose molecules are arranged into ordered strands called fibrils, which in turn are organized into the larger structural elements comprising the cell wall of wood fibers. Delignified wood fibers which are mostly cellulose have great commercial value when formed into paper. Moreover, they may be chemically altered to form synthetic textiles, films, lacquers, and explosives.

Lignin comprises 23 to 33 percent of softwoods, and 16 to 25 percent of hardwoods. It occurs in the wood throughout the cell wall, but is concentrated toward the outside of the cells and between cells. Lignin is a three-dimensional phenyl-propane polymer. Lignin structure and distribution in wood are still not fully understood. To remove lignin from wood on a commercial scale may require vigorous reagents or high temperatures or high pressures.

Theoretically, lignin might be converted to a variety of chemical products but, practically, a large percentage of the lignin removed from wood during pulping operations is a troublesome byproduct. It is often burned for heat and recovery of pulping chemicals. One sizable commercial use for lignin is in the formulation of oil-well drilling muds. It is also used in rubber compounding and in concrete mixes. Lesser amounts are processed to yield vanillin for flavoring purposes and to produce solvents.

The hemicelluloses are associated with cellulose and are polymers built from several different kinds of sugar monomers.

Figure 2–2—Cross section of a ponderosa pine log showing growth rings: Light bands are early-wood, dark bands are latewood. An annual (growth) ring is composed of the earlywood ring and the latewood ring outside it. (M60 729)

The relative amounts of these sugars vary markedly with species. The hemicelluloses play an important role in fiber-to-fiber bonding in the papermaking process. The component sugars of hemicellulose are of potential interest for conversion into chemical products.

Unlike the major constituents just discussed, the extraneous materials are not structural components of wood. They are both organic and inorganic. The organic component contributes to such properties of wood as color, odor, taste, decay resistance, density, hygroscopicity, and flammability. Extractives include tannins and other polyphenolics, coloring matters, essential oils, fats, resins, waxes, gums, starch, and simple metabolic intermediates. This component is termed extractive because it can be removed from wood by extraction with such solvents as water, alcohol, acetone, benzene, and ether. In quantity, the extractives may range from roughly 5 to 30 percent, depending on such factors as species, growth conditions, and time of year the tree is cut.

The inorganic component of the extraneous material generally comprises 0.2 to 1.0 percent of the wood substance, although higher values are occasionally reported. Calcium, potassium, and magnesium are the more abundant elemental constituents. Trace amounts (<100 p/m) of phosphorus, sodium, iron, silicon, manganese, copper, zinc, and perhaps a few others are also usually present.

A significant dollar value of nonfibrous products is produced from wood including naval stores, pulp byproducts, vanillin, ethyl alcohol, charcoal, extractives, and bark products.

Identification

Many species of wood have unique physical, mechanical, or chemical properties. Efficient utilization dictates that species should be matched to use requirements through an understanding of properties. This requires identification of the species in wood form, independent of bark, foliage, and other characteristics of the tree.

Field identification can often be made on the basis of readily visible characteristics such as color, odor, density, presence of pitch, or grain pattern. Where more positive identification is required, a laboratory investigation of the microscopic anatomy of the wood can be made. Detailed descriptions of identifying characteristics are given in texts such as "Textbook of Wood Technology" by Panshin and de Zeeuw and "Understanding Wood: A Craftsman's Guide to Wood Technology" by B. R. Hoadley.

Selected References

Bratt, L. C. Trends in the production of silvichemicals in the United States and abroad. Atlanta, GA: Tappi 48(7): 46A–49A; 1965.

Core, H. A.; Côté, W. A.; Day, A. C. Wood structure and identification. Syracuse, NY: Syracuse University Press; 1979.

Hamilton, J. K., Thompson, N. C. A comparison of the carbohydrates of hardwoods and softwoods. Atlanta, GA: Tappi 42: 752–760; 1959.

Hoadley, B. R. Understanding wood: a craftsmen's guide to wood technology. Newtown, CT: Taunton Press; 1980.

Kribs, D. A. Commercial woods on the American market. New York: Dover Publications; 1968.

Ott, E.; Spurlin, H. M.; Grafflin, M. W. Cellulose and cellulose derivatives. Volume V. Parts I, II, and III (1955) of High Polymers. New York: Interscience Publishers; 1954.

Panshin, A. J.; de Zeeuw, C. Textbook of wood technology. 4th ed. New York: McGraw-Hill; 1980.

Sarkanen, K. V.; Ludwig, C. H. (editors). Lignins: occurrence, formation, structure and reactions. New York: Wiley-Interscience; 1971.

Sjöström, E. Wood chemistry: fundamentals and applications. New York: Academic Press; 1981.

Wise, L. E.; Jahn, E. C. Wood chemistry, Volumes I and II. New York: Reinhold; 1952.

Chapter 3

Physical Properties of Wood

Physical Properties of Wood*

The versatility of wood is demonstrated by a wide variety of products. This variety is a result of a spectrum of desirable physical characteristics or properties among the many species of wood. Often more than one property of wood is important to an end product. For example, to select a species for a product, the value of appearance-type properties such as texture, grain pattern, or color may be evaluated against the influence of such characteristics as machinability, dimensional stability, or decay resistance. This chapter discusses the physical properties most often of interest in the design of wood products.

Some physical properties discussed and tabulated are influenced by species as well as variables like moisture content; other properties tend to be more independent of species. The thoroughness of sampling and the degree of variability influences the confidence with which species-dependent properties are known. In this chapter an effort is made to indicate either the general or specific nature of the properties tabulated.

Appearance

Grain and Texture

The terms "grain" and "texture" are commonly used rather loosely in connection with wood. Grain is often used in reference to annual rings as in fine grain and coarse grain, but it is also employed to indicate the direction of fibers, as in straight grain, spiral grain, and curly grain. Grain, as a synonym for fiber direction, is discussed in more detail relative to mechanical properties in chapter 4. Wood finishers refer to wood as open grained and close grained as terms reflecting the relative size of the pores, which determines whether the surface needs a filler. Texture is often used synonymously with grain, but usually it refers to the finer structure of wood rather than to annual rings. When the words "grain" or "texture" are used in connection with wood, the meaning intended should be made perfectly clear (see glossary).

Plainsawed and Quartersawed Lumber

Lumber can be cut from a log in two distinct ways: Tangent to the annual rings, producing "plainsawed" lumber in hardwoods and "flat-grained" or "slash-grained" lumber in softwoods; and radially to the rings or parallel to the rays, producing "quartersawed" lumber in hardwoods and "edge-

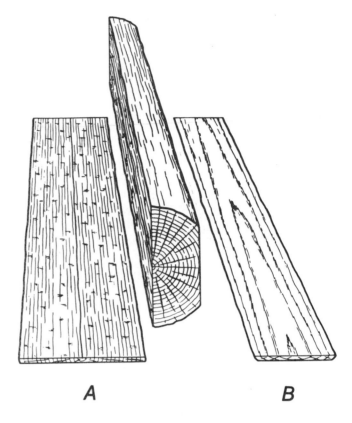

A   **B**

Figure 3—1—Quartersawed *A* and plainsawed *B* boards cut from a log.   (M554)

grained" or "vertical-grained" lumber in softwoods (fig. 3—1). Usually quartersawed or edge-grained lumber is not cut strictly parallel with the rays; and often in plainsawed boards the surfaces next to the edges are far from being tangent to the rings. In commercial practice, lumber with rings at angles of 45° to 90° with the wide surface is called quartersawed, and lumber with rings at angles of 0° to 45° with the wide surface is called plainsawed. Hardwood lumber in which annual rings make angles of 30° to 60° with the wide faces is sometimes called "bastard sawn."

For many purposes either plainsawed or quartersawed lumber is satisfactory. Each type has certain advantages, however, that may be important in a particular use. Some of the advantages are given in table 3—1.

Decorative Features of Common Woods

The decorative value of wood depends upon its color, figure, luster, and the way in which it bleaches or takes fillers, stains, and transparent finishes. Because of the combinations of color and the multiplicity of shades found in wood, it is impossible to give detailed descriptions of colors of the vari-

* Revision by William T. Simpson, Forest Products Technologist; William L. James, Physicist; William C. Feist, Chemist; Robert H. White, Forest Products Technologist; Regis B. Miller, Botanist; Robert R. Maeglin, Forest Products Technologist; Wallace E. Eslyn, Plant Pathologist; and Harold Stewart, Forest Products Technologist.

Table 3−1—*Some advantages of plainsawed and quartersawed lumber*

Plainsawed	Quartersawed
Figure patterns resulting from the annual rings and some other types of figure are brought out more conspicuously by plainsawing.	Quartersawed lumber shrinks and swells less in width.
	It twists and cups less. It surface-checks and splits less in seasoning and in use.
Round or oval knots that may occur in plainsawed boards affect the surface appearance less than spike knots that may occur in quartersawed boards. Also, a board with a round or oval knot is not as weak as a board with a spike knot.	Raised grain caused by separation in the annual rings does not become so pronounced.
	It wears more evenly.
Shakes and pitch pockets, when present, extend through fewer boards.	Types of figure due to pronounced rays, interlocked grain, and wavy grain are brought out more conspicuously.
It is less susceptible to collapse in drying.	
It shrinks and swells less in thickness.	It does not allow liquids to pass into or through it so readily in some species.
It may cost less because it is easier to obtain.	It holds paint better in some species.
	The sapwood appearing in boards is at the edges and its width is limited according to the width of the sapwood in the log.

ous kinds. Sapwood of most species, however, is light in color, and in some species it is practically white. White sapwood of certain species, such as maple, may be preferred to the heartwood for specific uses. In some species, such as hemlock, spruce, the true firs, basswood, cottonwood, and beech, there is typically little or no difference in color between sapwood and heartwood, but in most species heartwood is darker and fairly uniform in color. Table 3−2 describes in a general way the color of heartwood of the more common kinds of wood.

In plainsawed boards and rotary-cut veneer, the annual growth rings frequently form ellipses and parabolas that make striking figures, especially when the rings are irregular in width and outline on the cut surface. On quartersawed surfaces, these rings form stripes, which are not especially ornamental unless they are irregular in width and direction. The relatively large rays sometimes appear as flecks which can form a conspicuous figure in quartersawed oak and sycamore. With interlocked grain, which slopes in alternate directions in successive layers from the center of the tree outward, quarter-

Table 3–2—*Color and figure of common kinds of domestic wood*

Species	Color of dry heartwood[1]	Plainsawed lumber or rotary-cut veneer	Quartersawed lumber or quarter-sliced veneer
			Type of figure in—

HARDWOODS

Species	Color of dry heartwood[1]	Plainsawed lumber or rotary-cut veneer	Quartersawed lumber or quarter-sliced veneer
Alder, red	Pale pinkish brown	Faint growth ring	Scattered large flakes, sometimes entirely absent
Ash:			
Black	Moderately dark grayish brown	Conspicuous growth ring; occasional burl	Distinct, not conspicuous growth-ring stripe; occasional burl
Oregon	Grayish brown, sometimes with reddish tinge	do.	Do.
White	do.	do.	Do.
Aspen	Light brown	Faint growth ring	None
Basswood	Creamy white to creamy brown, sometimes reddish	do.	Do.
Beech, American	White with reddish tinge to reddish brown	do.	Numerous small flakes up to ⅛ inch in height
Birch:			
Paper	Light brown	do.	None
Sweet	Dark reddish brown	Distinct, not conspicuous growth ring; occasionally wavy	Occasionally wavy
Yellow	Reddish brown	do.	
Butternut	Light chestnut brown with occasional reddish tinge or streaks.	Faint growth ring	None
Cherry, black	Light to dark reddish brown	Faint growth ring; occasional burl	Occasional burl

[1] The sapwood of all species is light in color or virtually white unless discolored by fungus or chemical stains.

Table 3–2—Color and figure of common kinds of domestic wood—Continued

Species	Color of dry heartwood[1]	Type of figure in—	
		Plainsawed lumber or rotary-cut veneer	Quartersawed lumber or quarter-sliced veneer
HARDWOODS—continued			
Chestnut, American	Grayish brown	Conspicuous growth ring	Distinct, not conspicuous growth-ring stripe
Cottonwood	Grayish white to light grayish brown	Faint growth ring	None
Elm:			
American and rock	Light grayish brown, usually with reddish tinge	Distinct, not conspicuous with fine wavy pattern within each growth ring	Faint growth-ring stripe
Slippery	Dark brown with shades of red	Conspicuous growth ring with fine pattern within each growth ring	Distinct, not conspicuous growth-ring stripe
Hackberry	Light yellowish or greenish gray	Conspicuous growth ring	Do.
Hickory	Reddish brown	Distinct, not conspicuous growth ring	Faint growth-ring stripe
Honeylocust	Cherry red	Conspicuous growth ring	Distinct, not conspicuous growth-ring stripe
Locust, black	Golden brown, sometimes with tinge of green	do.	Do.
Magnolia	Light to dark yellowish brown with greenish or purplish tinge	Faint growth ring	None
Maple: Black, bigleaf, red, silver, and sugar	Light reddish brown	Faint growth ring, occasionally birds-eye, curly, and wavy	Occasionally curly and wavy

[1] The sapwood of all species is light in color or virtually white unless discolored by fungus or chemical stains.

Table 3 – 2—Color and figure of common kinds of domestic wood—Continued

Species	Color of dry heartwood[1]	Plainsawed lumber or rotary-cut veneer	Quartersawed lumber or quarter-sliced veneer
		Type of figure in—	
HARDWOODS—continued			
Oak:			
All red oaks	Light brown usually with pink or red tinge	Conspicuous growth ring	Pronounced flake; distinct, not conspicuous growth-ring stripe
All white oaks	Light to dark brown, rarely with reddish tinge	do.	Do.
Sweetgum	Reddish brown	Faint growth ring; occasional irregular streaks	Distinct, not pronounced ribbon; occasional streak
Sycamore	Light to dark or reddish brown	Faint growth ring	Numerous pronounced flakes up to ¼ inch in height
Tupelo:			
Black and water	Pale to moderately dark brownish gray	do.	Distinct, not pronounced ribbon
Walnut, black	Chocolate brown occasionally with darker, sometimes purplish streaks	Distinct, not conspicuous growth ring; occasionally wavy, curly, burl, and other types	Distinct, not conspicuous growth-ring stripe; occasionally wavy, curly, burl, crotch, and other types
Yellow-poplar	Light to dark yellowish brown with greenish or purplish tinge	Faint growth ring	None
SOFTWOODS			
Baldcypress	Light yellowish brown to reddish brown	Conspicuous irregular growth ring	Distinct, not conspicuous growth-ring stripe
Cedar:			
Alaska-	Yellow	Faint growth ring	None

[1] The sapwood of all species is light in color or virtually white unless discolored by fungus or chemical stains.

3-6

Table 3 – 2 — *Color and figure of common kinds of domestic wood*—Continued

Species	Color of dry heartwood[1]	Type of figure in—	
		Plainsawed lumber or rotary-cut veneer	Quartersawed lumber or quarter-sliced veneer
SOFTWOODS—continued			
Cedar (con.)			
Atlantic white-	Light brown with reddish tinge	Distinct, not conspicuous growth ring	Do.
Eastern redcedar	Brick red to deep reddish brown	Occasionally streaks of white sapwood alternating with heartwood	Occasionally streaks of white sapwood alternating with heartwood
Incense-	Reddish brown	Faint growth ring	Faint growth-ring stripe
Northern white-	Light to dark brown	do.	Do.
Port-Orford-	Light yellow to pale brown	do.	None
Western redcedar	Reddish brown	Distinct, not conspicuous growth ring	Faint growth-ring stripe
Douglas-fir	Orange red to red; sometimes yellow	Conspicuous growth ring	Distinct, not conspicuous growth-ring stripe
Fir:			
Balsam	Nearly white	Distinct, not conspicuous growth ring	Faint growth-ring stripe
White	Nearly white to pale reddish brown	Conspicuous growth ring	Distinct, not conspicuous growth-ring stripe
Hemlock:			
Eastern	Light reddish brown	Distinct, not conspicuous growth ring	Faint growth-ring stripe
Western	do.	do.	Do.
Larch, western	Russet to reddish brown	Conspicuous growth ring	Distinct, not conspicuous growth-ring stripe

[1] The sapwood of all species is light in color or virtually white unless discolored by fungus or chemical stains.

Table 3-2—*Color and figure of common kinds of domestic wood*—Continued

Species	Color of dry heartwood[1]	Type of figure in— Plainsawed lumber or rotary-cut veneer	Type of figure in— Quartersawed lumber or quarter-sliced veneer
		SOFTWOODS—continued	
Pine:			
Eastern white	Cream to light reddish brown	Faint growth ring	None
Lodgepole	Light reddish brown	Distinct, not conspicuous growth ring; faint "pocked" appearance	None
Ponderosa	Orange to reddish brown	Distinct, not conspicuous growth ring	Faint growth-ring stripe
Red	do.	do.	Do.
Southern: Longleaf, loblolly, shortleaf, and slash	do.	Conspicuous growth ring	Distinct, not conspicuous growth-ring stripe
Sugar	Light creamy brown	Faint growth ring	None
Western white	Cream to light reddish brown	do.	None
Redwood	Cherry to deep reddish brown	Distinct, not conspicuous growth ring; occasionally wavy and burl	Faint growth-ring stripe; occasionally wavy and burl
Spruce:			
Black, Engelmann, red, white	Nearly white	Faint growth ring	None
Sitka	Light reddish brown	Distinct, not conspicuous growth ring	Faint growth-ring stripe
Tamarack	Russet brown	Conspicuous growth ring	Distinct, not conspicuous growth-ring stripe

[1] The sapwood of all species is light in color or virtually white unless discolored by fungus or chemical stains.

sawed surfaces show a ribbon effect, either because of the difference in reflection of light from successive layers when the wood has a natural luster or because cross grain of varying degree absorbs stains unevenly. Much of this type of figure is lost in plainsawed lumber.

In open-grained hardwoods, the appearance of both plainsawed and quartersawed lumber can be varied greatly by the use of fillers of different colors. In softwoods, the annual growth layers can be made to stand out more by applying a stain.

Knots, pin wormholes, bird pecks, decay in isolated pockets, birdseye, mineral streaks, swirls in grain, and ingrown bark are decorative in some species when the wood is carefully selected for a particular architectural treatment.

Moisture Content

Moisture content of wood is defined as the weight of water in wood expressed as a fraction, usually as a percentage, of the weight of ovendry wood. Weight, shrinkage, strength, and other properties depend upon moisture content of wood.

In trees, moisture content may range from about 30 percent to more than 200 percent of the weight of wood substance. In softwoods the moisture content of sapwood is usually higher than heartwood. In hardwoods the difference in moisture content between heartwood and sapwood depends on the species. Table 3-3 gives some moisture content values for heartwood and sapwood of some domestic species. These values are considered typical, but there is considerable variation within and between trees. Variability of moisture content exists even within individual boards cut from the same tree. Information on heartwood and sapwood moisture content is not given for imported species.

Green Wood and Fiber Saturation Point

Moisture can exist in wood as water or water vapor in cell lumens (cavities) and as water "bound" chemically within cell walls. Green wood often is defined as wood in which the cell walls are completely saturated with water; however, green wood usually contains additional water in the lumens. The moisture content at which cell walls are completely saturated (all "bound" water) but no water exists in cell cavities is called the "fiber saturation point." The fiber saturation point of wood averages about 30 percent moisture content, but individual species and individual pieces of wood may vary by several percentage points from that value.

The fiber saturation point also is often considered as that moisture content below which the physical and mechanical properties of wood begin to change as a function of moisture content.

Equilibrium Moisture Content

The moisture content of wood below the fiber saturation point is a function of both relative humidity and temperature of the surrounding air. The equilibrium moisture content is defined as that moisture content at which the wood is neither gaining nor losing moisture; an equilibrium condition has been reached. The relationship between equilibrium moisture content, relative humidity, and temperature is shown in table 3-4. The values in this table illustrate that below the fiber saturation point wood will attain a moisture content in equilibrium with widely differing atmospheric conditions. For most practical purposes the values in table 3-4 may be applied to wood of any species.

The data in table 3-4 can also be represented in equation form.

$$M = \frac{1800}{W} \left[\frac{KH}{1 - KH} + \frac{K_1KH + 2K_1K_2K^2H^2}{1 + K_1KH + K_1K_2K^2H^2} \right]$$

$$(3-1)$$

where

$W = 330 + 0.452T + 0.00415T^2$
$K = 0.791 + 0.000463T - 0.000000844T^2$
$K_1 = 6.34 + 0.000775T - 0.0000935T^2$
$K_2 = 1.09 + 0.0284T - 0.0000904T^2$

and

T = temperature (°F)
H = relative humidity (pct)
M = moisture content (pct)

Wood in service usually is exposed to both long-term (seasonal) and short-term (such as daily) changes in the relative humidity and temperature of the surrounding air. Thus, wood virtually always is undergoing at least slight changes in moisture content. These changes usually are gradual, and short-term fluctuations tend to influence only the wood surface. Moisture content changes may be retarded, but not prevented, by protective coatings, such as varnish, lacquer, or paint. The practical objective of all wood seasoning, handling, and storing methods should be to minimize moisture content changes in wood in service. Favored procedures are those that bring the wood to a moisture content corresponding to the average atmospheric conditions to which it will be exposed (see chs. 14 and 16).

Table 3—3—_Average moisture content of green wood, by species_

Species	Moisture content[1]		Species	Moisture content[1]	
	Heartwood	Sapwood		Heartwood	Sapwood
	- - - - - - Percent - - - - - -			- - - - - - Percent - - - - - -	
HARDWOODS			**HARDWOODS—continued**		
Alder, red	—	97	Sycamore, American	114	130
Apple	81	74	Tupelo:		
Ash:			Black	87	115
Black	95	—	Swamp	101	108
Green	—	58	Water	150	116
White	46	44	Walnut, black	90	73
Aspen	95	113	Yellow-poplar	83	106
Basswood, American	81	133	**SOFTWOODS**		
Beech, American	55	72	Baldcypress	121	171
Birch:			Cedar:		
Paper	89	72	Alaska-	32	166
Sweet	75	70	Eastern redcedar	33	—
Yellow	74	72	Incense-	40	213
Cherry, black	58	—	Port-Orford-	50	98
Chestnut, American	120	—	Western redcedar	58	249
Cottonwood, black	162	146	Douglas-fir:		
Elm:			Coast type	37	115
American	95	92	Fir:		
Cedar	66	61	Grand	91	136
Rock	44	57	Noble	34	115
Hackberry	61	65	Pacific silver	55	164
Hickory, pecan:			White	98	160
Bitternut	80	54	Hemlock:		
Water	97	62	Eastern	97	119
Hickory, true:			Western	85	170
Mockernut	70	52	Larch, western	54	110
Pignut	71	49	Pine:		
Red	69	52	Loblolly	33	110
Sand	68	50	Lodgepole	41	120
Magnolia	80	104	Longleaf	31	106
Maple:			Ponderosa	40	148
Silver	58	97	Red	32	134
Sugar	65	72	Shortleaf	32	122
Oak:			Sugar	98	219
California Black	76	75	Western white	62	148
Northern red	80	69	Redwood, old-growth	86	210
Southern red	83	75	Spruce:		
Water	81	81	Eastern	34	128
White	64	78	Engelmann	51	173
Willow	82	74	Sitka	41	142
Sweetgum	79	137	Tamarack	49	—

[1] Based on weight when ovendry.

Table 3–4—Moisture content of wood in equilibrium with stated dry-bulb temperature and relative humidity[1]

Temperature (dry-bulb) °F	Relative humidity — Percent																			
	5	10	15	20	25	30	35	40	45	50	55	60	65	70	75	80	85	90	95	98
30	1.4	2.6	3.7	4.6	5.5	6.3	7.1	7.9	8.7	9.5	10.4	11.3	12.4	13.5	14.9	16.5	18.5	21.0	24.3	26.9
40	1.4	2.6	3.7	4.6	5.5	6.3	7.1	7.9	8.7	9.5	10.4	11.3	12.3	13.5	14.9	16.5	18.5	21.0	24.3	26.9
50	1.4	2.6	3.6	4.6	5.5	6.3	7.1	7.9	8.7	9.5	10.3	11.2	12.3	13.4	14.8	16.4	18.4	20.9	24.3	26.9
60	1.3	2.5	3.6	4.6	5.4	6.2	7.0	7.8	8.6	9.4	10.2	11.1	12.1	13.3	14.6	16.2	18.2	20.7	24.1	26.8
70	1.3	2.5	3.5	4.5	5.4	6.2	6.9	7.7	8.5	9.2	10.1	11.0	12.0	13.1	14.4	16.0	17.9	20.5	23.9	26.6
80	1.3	2.4	3.5	4.4	5.3	6.1	6.8	7.6	8.3	9.1	9.9	10.8	11.7	12.9	14.2	15.7	17.7	20.2	23.6	26.3
90	1.2	2.3	3.4	4.3	5.1	5.9	6.7	7.4	8.1	8.9	9.7	10.5	11.5	12.6	13.9	15.4	17.3	19.8	23.3	26.0
100	1.2	2.3	3.3	4.2	5.0	5.8	6.5	7.2	7.9	8.7	9.5	10.3	11.2	12.3	13.6	15.1	17.0	19.5	22.9	25.6
110	1.1	2.2	3.2	4.0	4.9	5.6	6.3	7.0	7.7	8.4	9.2	10.0	11.0	12.0	13.2	14.7	16.6	19.1	22.4	25.2
120	1.1	2.1	3.0	3.9	4.7	5.4	6.1	6.8	7.5	8.2	8.9	9.7	10.6	11.7	12.9	14.4	16.2	18.6	22.0	24.7
130	1.0	2.0	2.9	3.7	4.5	5.2	5.9	6.6	7.2	7.9	8.7	9.4	10.3	11.3	12.5	14.0	15.8	18.2	21.5	24.2
140	.9	1.9	2.8	3.6	4.3	5.0	5.7	6.3	7.0	7.7	8.4	9.1	10.0	11.0	12.1	13.6	15.3	17.7	21.0	23.7
150	.9	1.8	2.6	3.4	4.1	4.8	5.5	6.1	6.7	7.4	8.1	8.8	9.7	10.6	11.8	13.1	14.9	17.2	20.4	23.1
160	.8	1.6	2.4	3.2	3.9	4.6	5.2	5.8	6.4	7.1	7.8	8.5	9.3	10.3	11.4	12.7	14.4	16.7	19.9	22.5
170	.7	1.5	2.3	3.0	3.7	4.3	4.9	5.6	6.2	6.8	7.4	8.2	9.0	9.9	11.0	12.3	14.0	16.2	19.3	21.9
180	.7	1.4	2.1	2.8	3.5	4.1	4.7	5.3	5.9	6.5	7.1	7.8	8.6	9.5	10.5	11.8	13.5	15.7	18.7	21.3
190	.6	1.3	1.9	2.6	3.2	3.8	4.4	5.0	5.5	6.1	6.8	7.5	8.2	9.1	10.1	11.4	13.0	15.1	18.1	20.7
200	.5	1.1	1.7	2.4	3.0	3.5	4.1	4.6	5.2	5.8	6.4	7.1	7.8	8.7	9.7	10.9	12.5	14.6	17.5	20.0
210	.5	1.0	1.6	2.1	2.7	3.2	3.8	4.3	4.9	5.4	6.0	6.7	7.4	8.3	9.2	10.4	12.0	14.0	16.9	19.3
220	.4	.9	1.4	1.9	2.4	2.9	3.4	3.9	4.5	5.0	5.6	6.3	7.0	7.8	8.8	9.9	*	*	*	*
230	.3	.8	1.2	1.6	2.1	2.6	3.1	3.6	4.2	4.7	5.3	6.0	6.7	*	*	*	*	*	*	*
240	.3	.6	.9	1.3	1.7	2.1	2.6	3.1	3.5	4.1	4.6	*	*	*	*	*	*	*	*	*
250	.2	.4	.7	1.0	1.3	1.7	2.1	2.5	2.9	*	*	*	*	*	*	*	*	*	*	*
260	.2	.3	.5	.7	.9	1.1	1.4	*	*	*	*	*	*	*	*	*	*	*	*	*
270	.1	.1	.2	.3	.4	.4	*	*	*	*	*	*	*	*	*	*	*	*	*	*

[1] Asterisks indicate conditions not possible at atmospheric pressure.

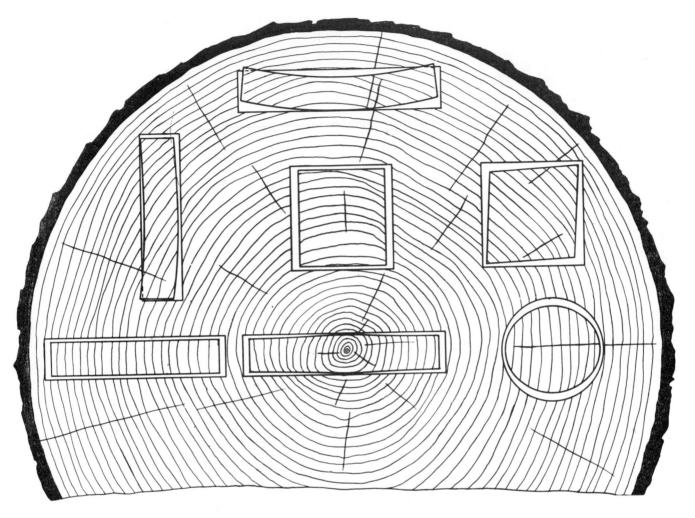

Figure 3–2—Characteristic shrinkage and distortion of flats, squares, and rounds as affected by the direction of the growth rings. Tangential shrinkage is about twice as great as radial. (M12 494)

Shrinkage

Wood is dimensionally stable when the moisture content is above the fiber saturation point. Wood changes dimension as it gains or loses moisture below that point. It shrinks when losing moisture from the cell walls and swells when gaining moisture in the cell walls. This shrinking and swelling may result in warping, checking, splitting, or performance problems that detract from its usefulness. It is therefore important that these phenomena be understood and considered when they may affect a product in which wood is used.

Wood is an anisotropic material with respect to shrinkage characteristics. It shrinks most in the direction of the annual growth rings (tangentially), about one-half as much across the rings (radially), and only slightly along the grain (longitudinally). The combined effects of radial and tangential shrinkage can distort the shape of wood pieces because of the difference in shrinkage and the curvature of annual rings. The major types of distortion due to these effects are illustrated in figure 3–2.

Transverse and Volumetric Shrinkage

Data have been collected to represent the average radial, tangential, and volumetric shrinkage of numerous domestic species by methods described in American Society for Testing and Materials (ASTM) D 143—Standard Method of Testing Small Clear Specimens of Timber. These shrinkage values, expressed as a percentage of the green dimension, are summarized in table 3–5. Shrinkage values collected from the world literature for selected imported species are summarized in table 3–6.

Table 3 – 5—*Shrinkage values of domestic woods*

Species	Shrinkage from green to ovendry moisture content[1]			Species	Shrinkage from green to ovendry moisture content[1]		
	Radial	Tangential	Volumetric		Radial	Tangential	Volumetric
	- - - - - - - Percent - - - - - - -				- - - - - - - Percent - - - - - - -		

HARDWOODS

Species	Radial	Tangential	Volumetric	Species	Radial	Tangential	Volumetric
Alder, red	4.4	7.3	12.6	Hickory, True:			
Ash:				Mockernut	7.7	11.0	17.8
Black	5.0	7.8	15.2	Pignut	7.2	11.5	17.9
Blue	3.9	6.5	11.7	Shagbark	7.0	10.5	16.7
Green	4.6	7.1	12.5	Shellbark	7.6	12.6	19.2
Oregon	4.1	8.1	13.2	Holly, American	4.8	9.9	16.9
Pumpkin	3.7	6.3	12.0	Honeylocust	4.2	6.6	10.8
White	4.9	7.8	13.3	Locust, black	4.6	7.2	10.2
Aspen:				Madrone, Pacific	5.6	12.4	18.1
Bigtooth	3.3	7.9	11.8	Magnolia:			
Quaking	3.5	6.7	11.5	Cucumbertree	5.2	8.8	13.6
Basswood,				Southern	5.4	6.6	12.3
American	6.6	9.3	15.8	Sweetbay	4.7	8.3	12.9
Beech, American	5.5	11.9	17.2	Maple:			
Birch:				Bigleaf	3.7	7.1	11.6
Alaska paper	6.5	9.9	16.7	Black	4.8	9.3	14.0
Gray	5.2	—	14.7	Red	4.0	8.2	12.6
Paper	6.3	8.6	16.2	Silver	3.0	7.2	12.0
River	4.7	9.2	13.5	Striped	3.2	8.6	12.3
Sweet	6.5	9.0	15.6	Sugar	4.8	9.9	14.7
Yellow	7.3	9.5	16.8	Oak, red:			
Buckeye, yellow	3.6	8.1	12.5	Black	4.4	11.1	15.1
Butternut	3.4	6.4	10.6	Laurel	4.0	9.9	19.0
Cherry, black	3.7	7.1	11.5	Northern red	4.0	8.6	13.7
Chestnut,				Pin	4.3	9.5	14.5
American	3.4	6.7	11.6	Scarlet	4.4	10.8	14.7
Cottonwood:				Southern red	4.7	11.3	16.1
Balsam poplar	3.0	7.1	10.5	Water	4.4	9.8	16.1
Black	3.6	8.6	12.4	Willow	5.0	9.6	18.9
Eastern	3.9	9.2	13.9	Oak, white:			
Elm:				Bur	4.4	8.8	12.7
American	4.2	9.5	14.6	Chestnut	5.3	10.8	16.4
Cedar	4.7	10.2	15.4	Live	6.6	9.5	14.7
Rock	4.8	8.1	14.9	Overcup	5.3	12.7	16.0
Slippery	4.9	8.9	13.8	Post	5.4	9.8	16.2
Winged	5.3	11.6	17.7	Swamp			
Hackberry	4.8	8.9	13.8	chestnut	5.2	10.8	16.4
Hickory, Pecan	4.9	8.9	13.6	White	5.6	10.5	16.3

Table 3 – 5—*Shrinkage values of domestic woods*—Continued

Species	Shrinkage from green to ovendry moisture content[1]			Species	Shrinkage from green to ovendry moisture content[1]		
	Radial	Tangential	Volumetric		Radial	Tangential	Volumetric
	- - - - - - - Percent - - - - - - -				- - - - - - - Percent - - - - - - -		

HARDWOODS—continued

Species	Radial	Tangential	Volumetric	Species	Radial	Tangential	Volumetric
Persimmon,				Tupelo:			
common	7.9	11.2	19.1	Black	5.1	8.7	14.4
Sassafras	4.0	6.2	10.3	Water	4.2	7.6	12.5
Sweetgum	5.3	10.2	15.8	Walnut, black	5.5	7.8	12.8
Sycamore,				Willow, black	3.3	8.7	13.9
American	5.0	8.4	14.1	Yellow-poplar	4.6	8.2	12.7
Tanoak	4.9	11.7	17.3				

SOFTWOODS

Species	Radial	Tangential	Volumetric	Species	Radial	Tangential	Volumetric
Baldcypress	3.8	6.2	10.5	Hemlock (con.):			
Cedar:				Western	4.2	7.8	12.4
Alaska-	2.8	6.0	9.2	Larch, western	4.5	9.1	14.0
Atlantic white-	2.9	5.4	8.8	Pine:			
Eastern				Eastern white	2.1	6.1	8.2
redcedar	3.1	4.7	7.8	Jack	3.7	6.6	10.3
Incense-	3.3	5.2	7.7	Loblolly	4.8	7.4	12.3
Northern				Lodgepole	4.3	6.7	11.1
white-	2.2	4.9	7.2	Longleaf	5.1	7.5	12.2
Port-Orford-	4.6	6.9	10.1	Pitch	4.0	7.1	10.9
Western				Pond	5.1	7.1	11.2
redcedar	2.4	5.0	6.8	Ponderosa	3.9	6.2	9.7
Douglas-fir:[2]				Red	3.8	7.2	11.3
Coast	4.8	7.6	12.4	Shortleaf	4.6	7.7	12.3
Interior north	3.8	6.9	10.7	Slash	5.4	7.6	12.1
Interior west	4.8	7.5	11.8	Sugar	2.9	5.6	7.9
Fir:				Virginia	4.2	7.2	11.9
Balsam	2.9	6.9	11.2	Western white	4.1	7.4	11.8
California red	4.5	7.9	11.4	Redwood:			
Grand	3.4	7.5	11.0	Old-growth	2.6	4.4	6.8
Noble	4.3	8.3	12.4	Young-growth	2.2	4.9	7.0
Pacific silver	4.4	9.2	13.0	Spruce:			
Subalpine	2.6	7.4	9.4	Black	4.1	6.8	11.3
White	3.3	7.0	9.8	Engelmann	3.8	7.1	11.0
Hemlock:				Red	3.8	7.8	11.8
Eastern	3.0	6.8	9.7	Sitka	4.3	7.5	11.5
Mountain	4.4	7.1	11.1	Tamarack	3.7	7.4	13.6

[1] Expressed as a percentage of the green dimension.

[2] Coast Douglas-fir is defined as Douglas-fir growing in the States of Oregon and Washington west of the summit of the Cascade Mountains. Interior West includes the State of California and all counties in Oregon and Washington east of but adjacent to the Cascade summit. Interior North includes the remainder of Oregon and Washington and the States of Idaho, Montana, and Wyoming.

Table 3—6—*Shrinkage for some woods imported into the United States*[1]

Species	Shrinkage from green to ovendry moisture content[2]		Species	Shrinkage from green to ovendry moisture content[2]	
	Radial	Tangential		Radial	Tangential
	- - - Percent - - -			- - - Percent - - -	
Afrormosia (*Pericopsis elata*)	3.0	6.4	Manni (*Symphonia globulifera*)	5.7	9.7
Albarco (*Cariniana* spp.)	2.8	5.4	Marishballi (*Licania* spp.)	7.5	11.7
Andiroba (*Carapa guianensis*)	3.1	7.6	Merbau (*Intsia bijuga* and		
Angelin (*Andira inermis*)	4.6	9.8	palembanica)	2.7	4.6
Angelique (*Dicorynia guianensis*)	5.2	8.8	Mersawa (*Anisoptera* spp.)	4.0	9.0
Apitong (*Dipterocarpus* spp.)	5.2	10.9	Mora (*Mora* spp.)	6.9	9.8
Azobe (*Lophira alata*)	8.4	11.0	Obeche (*Triplochiton scleroxylon*)	3.0	5.4
Balata (*Manilkara bidentata*)	6.3	9.4	Ocota pine (*Pinus oocarpa*)	4.6	7.5
Balsa (*Ochroma pyramidale*)	3.0	7.6	Okoume (*Aucoumea klaineana*)	4.1	6.1
Banak (*Virola* spp.)	4.6	8.8	Opepe (*Nauclea* spp.)	4.5	8.4
Benge (*Guibourtia arnoldiana*)	5.2	8.6	Parana pine (*Araucaria*		
Bubinga (*Guibourtia* spp.)	5.8	8.4	angustifolia)	4.0	7.9
Caribbean pine (*Pinus caribaea*)	6.3	7.8	Pau Marfim (*Balfourodendron*		
Cativo (*Prioria copaifera*)	2.4	5.3	riedelianum)	4.6	8.8
Courbaril (*Hymenaea courbaril*)	4.5	8.5	Peroba Rosa		
Cuangare (*Dialyanthera* spp.)	4.2	9.4	(*Aspidosperma* spp.)	3.8	6.4
Determa (*Ocotea rubra*)	3.7	7.6	Piquia (*Caryocar* spp.)	5.0	8.0
Ebony (*Diospyros* spp.)	5.5	6.5	Pilon (*Hyeronima* spp.)	5.4	11.7
Gmelina (*Gmelina arborea*)	2.4	4.9	Primavera (*Cybistax*		
Greenheart (*Ocotea rodiaei*)	8.8	9.6	donnell-smithii)	3.1	5.1
Hura (*Hura crepitans*)	2.7	4.5	Purpleheart (*Peltogyne* spp.)	3.2	6.1
Ipe (*Tabebuia* spp.)	6.6	8.0	Ramin (*Gonystylus* spp.)	4.3	8.7
Iroko (*Chlorophora excelsa*			Roble (*Quercus* spp.)	6.4	11.7
and *regia*)	2.8	3.8	Roble (*Tabebuia* spp.		
Jarrah (*Eucalyptus marginata*)	4.6	6.6	Roble group)	3.6	6.1
Kaneelhart (*Licaria* spp.)	5.4	7.9	Rosewood, Brazilian		
Kapur (*Dryobalanops* spp.)	4.6	10.2	(*Dalbergia nigra*)	2.9	4.6
Karri (*Eucalyptus diversicolor*)	7.2	10.7	Rosewood, Indian		
Kempas (*Koompassia*			(*Dalbergia latifolia*)	2.7	5.8
malaccensis)	6.0	7.4	Santa Maria (*Calophyllum*		
Keruing (*Dipterocarpus* spp.)	5.2	10.9	brasiliense)	4.6	8.0
Lauan (*Shorea* spp.)	3.8	8.0	Sapele (*Entandrophragma*		
Limba (*Terminalia superba*)	4.5	6.2	cylindricum)	4.6	7.4
Macawood (*Platymiscium* spp.)	2.7	3.5	Sepetir (*Pseudosindora* and		
Mahogany, African (*Khaya* spp.)	2.5	4.5	Sindora spp.)	3.7	7.0
Mahogany, true			Spanish-cedar (*Cedrela* spp.)	4.2	6.3
(*Swietenia macrophylla*)	3.0	4.1	Teak (*Tectona grandis*)	2.5	5.8
Manbarklak (*Eschweilera* spp.)	5.8	10.3	Wallaba (*Eperua* spp.)	3.6	6.9

[1] Shrinkage values in this table were obtained from world literature and may not represent a true species average.
[2] Expressed as a percentage of the green dimension.

The shrinkage of wood is affected by a number of variables. In general, greater shrinkage is associated with greater density. The size and shape of a piece of wood may also affect shrinkage, as may the temperature and rate of drying for some species. Transverse and volumetric shrinkage variability can be expressed by a coefficient of variation of approximately 15 percent, based on a study of 50 species.

Longitudinal Shrinkage

Longitudinal shrinkage of wood (shrinkage parallel to the grain) is generally quite small. Average values for shrinkage in going from green to ovendry are between 0.1 and 0.2 percent for most species of wood. Certain atypical types of wood, however, exhibit excessive longitudinal shrinkage, and these should be avoided in uses where longitudinal stability is important. Reaction wood, whether compression wood in softwoods or tension wood in hardwoods, tends to shrink excessively parallel to the grain. Wood from near the center of trees (juvenile wood) of some species also shrinks excessively lengthwise. Wood with cross grain exhibits increased shrinkage along the longitudinal axis of the piece.

Reaction wood exhibiting excessive longitudinal shrinkage may occur in the same board with normal wood. The presence of this type of wood, as well as cross grain, can cause serious warping such as bow, crook, or twist, and cross breaks may develop in the zones of high shrinkage.

Moisture-Shrinkage Relationship

The shrinkage of a small piece of wood normally begins at about the fiber saturation point and continues in a fairly linear manner until the wood is completely dry. However, in the normal drying of lumber or other large pieces the surface of the wood dries first. When the surface gets below the fiber saturation point, it begins to shrink. Meanwhile, the interior may still be quite wet. The exact form of the moisture content-shrinkage curve depends on several variables, principally size and shape of the piece, species of wood, and drying conditions employed.

Considerable variation in shrinkage occurs for any species. Figure 3−3 is a plot of shrinkage data for Douglas-fir boards, 7/8 by 5-1/2 inches in cross section; the material was grown in one locality and dried under mild conditions from green to near equilibrium at 65 °F and 30 percent relative humidity. The figure shows that it is impossible to predict accurately the shrinkage of an individual piece of wood; the average shrinkage of a quantity of pieces is more predictable.

If the shrinkage-moisture content relationship is not known for a particular product and drying condition, the data in tables 3−5 and 3−6 can be used to estimate shrinkage from the green condition to any moisture content:

$$S_m = S_o \left[\frac{30-m}{30} \right] \qquad (3-2)$$

where S_m is shrinkage (in percent) from the green condition to moisture content m (below 30 pct) and S_o is total shrinkage (in percent) from table 3−5 or 3−6. If the moisture content at which shrinkage from the green condition begins is known to be other than 30 percent for a species, the shrinkage estimate can be improved by replacing 30 in the equation with the appropriate moisture content.

S_o may be an appropriate value of radial, tangential, or volumetric shrinkage. Tangential values should be used for estimating width shrinkage of flat-sawed material; radial values for quartersawed material. For mixed or unknown ring orientations, the tangential values are suggested. Individual pieces will vary from predicted shrinkage values. As noted previously, shrinkage variability is characterized by a coefficient of variation of approximately 15 percent. This applies to pure tangential or radial ring orientation, and is probably somewhat higher in commercial lumber where ring orientation is seldom aligned perfectly parallel or perpendicular to board faces. Chapter 14 contains further discussion of shrinkage-moisture content relations.

Weight-Density-Specific Gravity

Two primary sources of variation affect the weight of wood products. One is the density of the basic wood structure; the other is the variable moisture content. A third source, minerals and extractable substances, has a marked effect only on a limited number of species. The density of wood, exclusive of water, varies greatly both within and between species. While the density of most species falls between about 20 and 45 pounds-mass per cubic foot, the range of densities actually extends from about 10 pounds-mass per cubic foot for balsa to over 65 pounds-mass per cubic foot for some other imported woods. A coefficient of variation of about 10 percent is considered suitable for describing the variability of density within common domestic species.

Wood is used in a wide range of conditions and thus has a wide range of moisture contents in use. Since moisture makes up part of the weight of each product in use, the density must reflect this fact. This has resulted in the density of wood often being determined and reported on a moisture content-in-use condition.

The calculated density of wood, including the water contained in it, is usually based on average species characteristics.

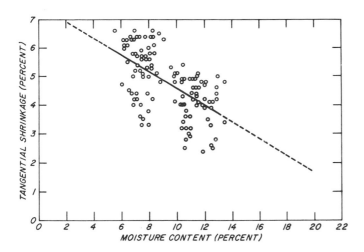

Figure 3-3—An illustration of variation in individual tangential shrinkage values of several boards of Douglas-fir from one locality, dried from green condition.

(M140 413)

This value should always be considered an approximation because of the natural variation in anatomy, moisture content, and the ratio of heartwood to sapwood that occurs. Nevertheless, this determination of density usually is sufficiently accurate to permit proper utilization of wood products where weight is important. Such applications range from estimation of structural loads to the calculation of approximate shipping weights.

To standardize comparisons of species or products and estimations of product weights, specific gravity is used as a standard reference basis, rather than density. The traditional definition of specific gravity is the ratio of the density of the wood to the density of water at a specified reference temperature (often 40 °F where the density of water is 1.0000 g/cm^3). To reduce any confusion introduced by the variable moisture content, the specific gravity of wood usually is based on the ovendry weight and the volume at some specified moisture content. A coefficient of variation of about 10 percent describes the variability inherent in many common domestic species.

In research activities specific gravity may be reported on the basis of both weight and volume ovendry. For engineering work, the basis commonly is ovendry weight and volume at the moisture content of test or use. Often the moisture content of use is taken as 12 percent. Some specific gravity data are reported in table 4—2, chapter 4, on this basis.

If the specific gravity of wood is known, based on ovendry weight and volume at a specified moisture content, the specific gravity at any other moisture content between 0 and 30 percent can be approximated from figure 3—4. This figure adjusts for average shrinkage and swelling that affects the volume of the wood. The specific gravity of wood based on

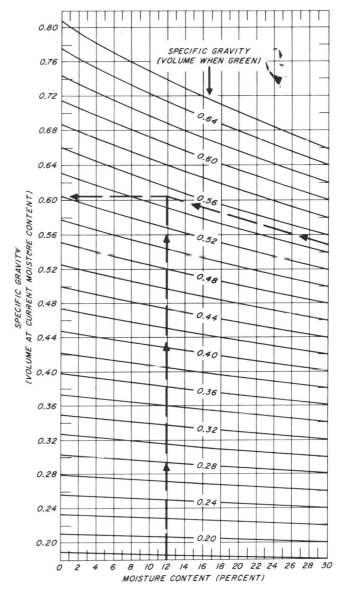

Figure 3—4—Relation of specific gravity and moisture content.

(M140 416)

ovendry weight does not change at moisture contents above approximately 30 percent (the approximate fiber saturation point). To use figure 3—4 locate the point corresponding to the known specific gravity (volume when green) on the vertical axis and the specified moisture content on the horizontal axis. From this point, move left or right parallel to the inclined lines until vertically above the target moisture content. Then read the new specific gravity corresponding to this point at the left-hand side of the graph.

3-17

For example, to estimate the density of white ash at 12 percent moisture content, consult table 4−2 in chapter 4. The average green specific gravity for the species is 0.55. Using figure 3−4, the 0.55 green specific gravity curve is found to intersect with the vertical 12 percent moisture content line at a point corresponding to a specific gravity of 0.605 based on ovendry weight and volume at 12 percent moisture content (see dashed lines in fig. 3−4). The density of wood including water at that moisture content can then be obtained from table 3−7 which converts the specific gravity of 0.605 to a density of 42 pounds-mass per cubic foot.

Working Qualities

The ease of working wood with hand tools generally varies directly with the specific gravity of the wood. The lower the specific gravity, the easier it is to cut the wood with a sharp tool. Tables 4−2 and 4−3 list the specific gravity values for various native and imported species. These specific gravity figures can be used as a general guide to the ease of working with hand tools.

A wood species that is easy to cut does not necessarily develop a smooth surface when it is machined. Consequently,

Table 3−7—*Density of wood as a function of specific gravity and moisture content*

Moisture content of wood	Density in pounds per cubic foot when the specific gravity[1] is—									
	0.30	0.32	0.34	0.36	0.38	0.40	0.42	0.44	0.46	0.48
Pct										
0	18.7	20.0	21.2	22.5	23.7	25.0	26.2	27.5	28.7	30.0
4	19.5	20.8	22.1	23.4	24.7	26.0	27.2	28.6	29.8	31.2
8	20.2	21.6	22.9	24.3	25.6	27.0	28.3	29.6	31.0	32.3
12	21.0	22.4	23.8	25.2	26.6	28.0	29.4	30.8	32.2	33.5
16	21.7	23.2	24.6	26.0	27.5	29.0	30.4	31.8	33.3	34.7
20	22.5	24.0	25.5	27.0	28.4	30.0	31.4	32.9	34.4	35.9
24	23.2	24.8	26.3	27.8	29.4	31.0	32.5	34.0	35.6	37.1
28	24.0	25.6	27.2	28.8	30.4	31.9	33.5	35.1	36.7	38.3
32	24.7	26.4	28.0	29.7	31.3	32.9	34.6	36.2	37.9	39.5
36	25.5	27.2	28.9	30.6	32.2	33.9	35.6	37.3	39.0	40.7
40	26.2	28.0	29.7	31.4	33.2	34.9	36.7	38.4	40.2	41.9
44	27.0	28.8	30.6	32.3	34.1	35.9	37.7	39.5	41.3	43.1
48	27.7	29.6	31.4	33.2	35.1	36.9	38.8	40.6	42.5	44.3
52	28.5	30.4	32.2	34.1	36.0	37.9	39.8	41.7	43.6	45.5
56	29.2	31.2	33.1	35.0	37.0	38.9	40.9	42.8	44.8	46.7
60	30.0	31.9	33.9	35.9	37.9	39.9	41.9	43.9	45.9	47.9
64	30.7	32.7	34.8	36.8	38.9	40.9	43.0	45.0	47.1	49.1
68	31.4	33.5	35.6	37.7	39.8	41.9	44.0	46.1	48.2	50.3
72	32.2	34.3	36.5	38.6	40.8	42.9	45.1	47.2	49.4	51.5
76	32.9	35.1	37.3	39.5	41.7	43.9	46.1	48.3	50.5	52.7
80	33.7	35.9	38.2	40.4	42.7	44.9	47.2	49.4	51.7	53.9
84	34.4	36.7	39.0	41.3	43.6	45.9	48.2	50.5	52.8	55.1
88	35.2	37.5	39.9	42.2	44.6	46.9	49.3	51.6	54.0	56.3
92	35.9	38.3	40.7	43.1	45.5	47.9	50.3	52.7	55.1	57.5
96	36.7	39.1	41.6	44.0	46.5	48.9	51.4	53.8	56.3	58.7
100	37.4	39.9	42.4	44.9	47.4	49.9	52.4	54.9	57.4	59.9
110	39.3	41.9	44.6	47.2	49.8	52.4	55.0	57.7	60.3	62.9
120	41.2	43.9	46.7	49.4	52.2	54.9	57.7	60.4	63.1	65.9
130	43.1	45.9	48.8	51.7	54.5	57.4	60.3	63.1	66.0	68.9
140	44.9	47.9	50.9	53.9	56.9	59.9	62.9	65.9	68.9	71.9
150	46.8	49.9	53.0	56.2	59.3	62.4	65.5	68.6	71.8	74.9

[1] Based on mass when ovendry and volume at tabulated moisture content.

tests have been made with many U.S. hardwoods to evaluate them for machining properties. Results of these evaluations are given in table 3–8.

Machining evaluations are not available for many imported woods. However, it is known that three major factors other than density may affect production of smooth surfaces during wood machining. These factors are: interlocked and variable grain; hard mineral deposits; and reaction wood, particularly tension wood in hardwoods. Interlocked grain is characteristic of many tropical species and presents difficulty in planing quartersawn boards unless attention is paid to feed rate,

cutting angles, and sharpness of knives. Hard deposits in the cells, such as calcium carbonate and silica, may have a pronounced dulling effect on all cutting edges. This dulling effect becomes more pronounced as the wood is dried to the usual inservice requirements. Tension wood may cause fibrous and fuzzy surfaces. It can be very troublesome in species of lower density. Reaction wood may also be responsible for the pinching effect on saws due to stress relief. The pinching may result in burning and dulling of the sawteeth. Table 3–9 lists some of the imported species that have irregular grain, hard deposits, or tension wood.

Density in pounds per cubic foot when the specific gravity[1] is—										
0.50	0.52	0.54	0.56	0.58	0.60	0.62	0.64	0.66	0.68	0.70
31.2	32.4	33.7	34.9	36.2	37.4	38.7	39.9	41.2	42.4	43.7
32.4	33.7	35.0	36.3	37.6	38.9	40.2	41.5	42.8	44.1	45.4
33.7	35.0	36.4	37.7	39.1	40.4	41.8	43.1	44.5	45.8	47.2
34.9	36.3	37.7	39.1	40.5	41.9	43.3	44.7	46.1	47.5	48.9
36.2	37.6	39.1	40.5	42.0	43.4	44.9	46.3	47.8	49.2	50.7
37.4	38.9	40.4	41.9	43.4	44.9	46.4	47.9	49.4	50.9	52.4
38.7	40.2	41.8	43.3	44.9	46.4	48.0	49.5	51.1	52.6	54.2
39.9	41.5	43.1	44.7	46.3	47.9	49.5	51.1	52.7	54.3	55.9
41.2	42.8	44.5	46.1	47.8	49.4	51.1	52.7	54.4	56.0	57.7
42.4	44.1	45.8	47.5	49.2	50.9	52.6	54.3	56.0	57.7	59.4
43.7	45.4	47.2	48.9	50.7	52.4	54.2	55.9	57.7	59.4	61.2
44.9	46.7	48.5	50.3	52.1	53.9	55.7	57.5	59.3	61.1	62.9
46.2	48.0	49.9	51.7	53.6	55.4	57.3	59.1	61.0	62.8	64.6
47.4	49.3	51.2	53.1	55.0	56.9	58.8	60.7	62.6	64.5	66.4
48.7	50.6	52.6	54.5	56.5	58.4	60.4	62.3	64.2	66.2	68.1
49.9	51.9	53.9	55.9	57.9	59.9	61.9	63.9	65.9	67.9	69.9
51.2	53.2	55.3	57.3	59.4	61.4	63.4	65.5	67.5	69.6	71.6
52.4	54.5	56.6	58.7	60.8	62.9	65.0	67.1	69.2	71.3	73.4
53.7	55.8	58.0	60.1	62.3	64.4	66.5	68.7	70.8	73.0	75.1
54.9	57.1	59.3	61.5	63.7	65.9	68.1	70.3	72.5	74.7	76.9
56.2	58.4	60.7	62.9	65.1	67.4	69.6	71.9	74.1	76.4	78.6
57.4	59.7	62.0	64.3	66.6	68.9	71.2	73.5	75.8	78.1	80.4
58.7	61.0	63.3	65.7	68.0	70.4	72.7	75.1	77.4	79.8	82.1
59.9	62.3	64.7	67.1	69.5	71.9	74.3	76.7	79.1	81.5	83.9
61.2	63.6	66.0	68.5	70.9	73.4	75.8	78.3	80.7	83.2	85.6
62.4	64.9	67.4	69.9	72.4	74.9	77.4	79.9	82.4	84.9	87.4
65.5	68.1	70.8	73.4	76.0	78.6	81.2	83.9	86.5	89.1	91.7
68.6	71.4	74.1	76.9	79.6	82.4	85.1	87.9	90.6	93.4	96.1
71.8	74.6	77.5	80.4	83.2	86.1	89.0	91.9	94.7	97.6	100.5
74.9	77.9	80.9	83.9	86.9	89.9	92.9	95.8	98.8	101.8	104.8
78.0	81.1	84.2	87.4	90.5	93.6	96.7	99.8	103.0	106.1	109.2

Table 3–8—Some machining and related properties of selected domestic hardwoods

Kind of wood[1]	Planing—perfect pieces	Shaping—good to excellent pieces	Turning—fair to excellent pieces	Boring—good to excellent pieces	Mortising—fair to excellent pieces	Sanding—good to excellent pieces	Steam bending—unbroken pieces	Nail splitting—pieces free from complete splits	Screw splitting—pieces free from complete splits
					Percent				
Alder, red	61	20	88	64	52	—	—	—	—
Ash	75	55	79	94	58	75	67	65	71
Aspen	26	7	65	78	60	—	—	—	—
Basswood	64	10	68	76	51	17	2	79	68
Beech	83	24	90	99	92	49	75	42	58
Birch	63	57	80	97	97	34	72	32	48
Birch, paper	47	22	—	—	—	—	—	—	—
Cherry, black	80	80	88	100	100	—	—	—	—
Chestnut	74	28	87	91	70	64	56	66	60
Cottonwood	21	3	70	70	52	19	44	82	78
Elm, soft	33	13	65	94	75	66	74	80	74
Hackberry	74	10	77	99	72	—	94	63	63
Hickory	76	20	84	100	98	80	76	35	63
Magnolia	65	27	79	71	32	37	85	73	76
Maple, bigleaf	52	56	80	100	80	—	—	—	52
Maple, hard	54	72	82	99	95	38	57	27	52
Maple, soft	41	25	76	80	34	37	59	58	61
Oak, red	91	28	84	99	95	81	86	66	78
Oak, white	87	35	85	95	99	83	91	69	74
Pecan	88	40	89	100	98	—	78	47	69
Sweetgum	51	28	86	92	58	23	67	69	69
Sycamore	22	12	85	98	96	21	29	79	74
Tanoak	80	39	81	100	100	—	—	—	—
Tupelo, water	55	52	79	62	33	34	46	64	63
Tupelo, black	48	32	75	82	24	21	42	65	63
Walnut, black	62	34	91	100	98	—	78	50	59
Willow	52	5	58	71	24	24	73	89	62
Yellow-poplar	70	13	81	87	63	19	58	77	67

[1] Commercial lumber nomenclature.

Table 3–9—_Some characteristics of imported woods that may affect machining_

Irregular and interlocked grain	Hard mineral deposits (silica or calcium carbonate)	Reaction wood (tension wood)
Apamate	Angelique	Andiroba
Apitong	Apitong	Banak
Avodire	Kapur	Cativo
Courbaril	Okoume	Khaya
Gola	Palosapis	Lupuna
Goncalo alves	Rosewood,	Mahogany
Ishpingo	Indian	Nogal
Jarrah	Teak	Sande
Kapur		Spanish-cedar
Karri		
Khaya		
Kokrodua		
Lapacho		
Laurel		
Lignumvitae		
Limba		
Meranti		
Obeche		
Okoume		
Palosapis		
Peroba de campos		
Primavera		
Rosewood, Indian		
Santa Maria		
Sapele		

Weathering

Without protective treatment, freshly cut wood of all species exposed outdoors to the weather changes materially in color. Other changes due to natural weathering include warping, loss of some surface fibers, and surface roughening and checking. The effects of weathering on wood may range from desirable to undesirable, depending on the requirements for the particular wood product. The time required to reach the fully weathered appearance depends on the severity of the exposure to sun and rain. Once weathered, and in the absence of decay, stain, and mildew, wood remains nearly unaltered in appearance (fig. 3–5).

The color of wood is affected very soon on exposure to weather. With continued exposure all woods turn gray; however, only the wood at or near the exposed surfaces is

Figure 3–5—Weathered surfaces of softwood after 15 years of exposure at Madison, WI. (M150 988-10)

noticeably affected. This very thin gray layer is composed chiefly of partially degraded cellulose fibers and microorganisms. Further weathering causes fibers to be lost from the surface (a process called erosion), but the process is so slow that on the average only about 1/4 inch of wood is lost in a century. This erosion rate is less for most hardwoods and greater for certain softwoods. Other factors such as growth rate, degree of exposure, grain orientation, and temperature are important in determining the rate of erosion.

In the weathering process, chemical degradation is influenced greatly by the wavelength of light. The most severe effects are produced by exposure to ultraviolet light. As cycles of wetting and drying take place, most woods develop physical changes such as checks or cracks that are easily visible. Moderate to low-density woods acquire fewer checks than do high-density woods. Vertical-grain boards check less than do flat-grain boards.

As a result of weathering, boards tend to warp (particularly cup) and pull out their fastenings. The cupping tendency varies with the density, width, and thickness of a board. The greater the density and the greater the width in proportion to the thickness, the greater is the tendency to cup. Warping also is more pronounced in flat-grain boards than in vertical-grain boards. For best cup resistance, the width of a board should not exceed eight times its thickness.

Biological attack of a wood surface by micro-organisms is recognized as a contributing factor to color changes. When weathered wood has an unsightly dark gray and blotchy appearance, it is due to dark-colored fungal spores and mycelia on the wood surface. The formation of a clean, light gray, silvery sheen on weathered wood occurs most frequently where micro-organism growth is inhibited by a hot, arid climate or a salt atmosphere in coastal regions. The micro-organisms primarily responsible for the gray discoloration of wood are commonly referred to as mildew.

The contact of fasteners and other metallic products with the weathering wood surface is also a source of color, often undesirable if a natural color is desired. Chapter 16 discusses this effect in more detail.

Additional details of weathering and treatments to preserve the natural color, retard biological attack, or impart additional color to the wood are covered in chapters 16, 17, and 18.

Decay Resistance

Wood kept constantly dry does not decay. Further, if it is kept continuously submerged in water even for long periods of time, it is not decayed significantly by the common decay fungi regardless of the wood species or the presence of sapwood. Bacteria and certain soft-rot fungi can attack submerged wood but the resulting deterioration is very slow. A large proportion of wood in use is kept so dry at all times that it lasts indefinitely. Moisture and temperature, which vary greatly with local conditions, are the principal factors affecting rate of decay. When exposed to conditions that favor decay, wood deteriorates more rapidly in warm, humid areas than in cool or dry areas. High altitudes, as a rule, are less favorable to decay than are low altitudes because the average temperatures are lower and the growing seasons for fungi, which cause decay, are shorter.

The heartwoods of common native species of wood have varying degrees of natural decay resistance. Untreated sapwood of substantially all species has low resistance to decay and usually has a short service life under decay-producing conditions. The decay resistance of heartwood is greatly affected by differences in the preservative qualities of the wood extractives, the attacking fungus, and the conditions of exposure. Considerable difference in service life may be obtained from pieces of wood cut from the same species, even from the same tree, and used under apparently similar conditions. There are further complications because, in a few species, such as the spruces and the true firs (not Douglas-fir), heartwood and sapwood are so similar in color that they cannot be easily distinguished.

Marketable sizes of some species such as the southern and eastern pines and baldcypress are becoming largely second growth and contain a high percentage of sapwood. Consequently, substantial quantities of heartwood lumber of these species are not available.

Precise ratings of decay resistance of heartwood of different species are not possible because of differences within species and the variety of service conditions to which wood is exposed. However, broad groupings of many of the native species, based on service records, laboratory tests, and general experience, are helpful in choosing heartwood for use under conditions favorable to decay. Table 3–10 shows such

Table 3 – 10—Grouping of some domestic woods according to approximate relative heartwood decay resistance

Resistant or very resistant	Moderately resistant	Slightly or nonresistant
Baldcypress (old growth)	Baldcypress (young growth)	Alder
Catalpa	Douglas-fir	Ashes
Cedars	Honeylocust	Aspens
Cherry, black	Larch, western	Basswood
Chestnut	Oak, swamp chestnut	Beech
Cypress, Arizona	Pine, eastern white	Birches
Junipers	Southern Pine:	Buckeye
Locust, black[1]	Longleaf	Butternut
Mesquite	Slash	Cottonwood
Mulberry, red[1]	Tamarack	Elms
Oak:		Hackberry
Bur		Hemlocks
Chestnut		Hickories
Gambel		Magnolia
Oregon white		Maples
Post		Oak (red and black species)
White		Pines (other than longleaf, slash, and eastern white)
Osage orange[1]		Poplars
Redwood		Spruces
Sassafras		Sweetgum
Walnut, black		True firs (western and eastern)
Yew, Pacific[1]		Willows
		Yellow-poplar

[1] These woods have exceptionally high decay resistance.

groupings for some domestic woods, according to their average heartwood decay resistance, and table 3−11 gives similar groupings for some imported woods. The extent of variations in decay resistance of individual trees or wood samples of a species is much greater for most of the more resistant species than for the slightly or nonresistant species.

Where decay hazards exist, heartwood of species in the resistant or very resistant category generally gives satisfactory service, but heartwood of species in the other two categories will usually require some form of preservative treatment. For mild decay conditions, a simple preservative treatment—such as a short soak in preservative after all cutting and boring operations are complete—will be adequate for wood low in decay resistance. For more severe decay hazards, pressure treatments are often required; even the very decay-resistant species may require preservative treatment for important structural or other uses where failure would endanger life or require expensive repairs. Preservative treatments and methods are discussed in chapter 18.

Chemical Resistance

Wood is highly resistant to many chemicals. In the chemical processing industry, it is the preferred material for numerous applications, such as various types of tanks and other containers, and for structures adjacent to or housing chemical equipment. Wood is widely used in cooling towers where the hot water to be cooled contains boiler conditioning chemicals as well as dissolved chlorine for algae suppression. It is also used in the fabrication of buildings for bulk chemical storage where the wood may be in direct contact with chemicals.

Wood owes its extensive use in chemical processing operations largely to its superiority over cast iron and ordinary steel in resistance to mild acids and solutions of acidic salts. While iron is superior to untreated wood in resistance to alkaline solutions, wood may be treated to greatly enhance its durability in this respect.

In general, heartwood is more resistant to chemical attack than sapwood, basically because heartwood is more resistant to penetration by liquids. The heartwoods of cypress, southern pine, Douglas-fir, and redwood are preferred for water tanks. Heartwoods of the first three of these species are preferred where resistance to chemical attack is an important factor. These four species combine moderate to high resistance to water penetration with moderate to high resistance to chemical attack and decay.

Chemical solutions may affect wood by two general types of action. The first is an almost completely reversible effect involving swelling of the wood structure. The second type of action is irreversible and involves permanent changes in the wood structure due to alteration of one or more of its chemical constituents.

In the first type, liquids such as water, alcohols, and some other organic liquids swell the wood with no degradation of the wood structure. Removal of the swelling liquid allows the wood to return to its original condition. Petroleum oils and creosote do not swell wood.

The second type of action causes permanent changes due to hydrolysis of cellulose and hemicelluloses by acids or acidic salts, oxidation of wood substance by oxidizing agents, or delignification and solution of hemicelluloses by alkalies or alkaline salt solutions. Experience and available data indicate species and conditions where wood is equal or superior to other materials in resisting degradative actions of chemicals. In general, heartwood of such species as cypress, Douglas-fir, southern pine, redwood, maple, and white oak is quite

Table 3−11—*Grouping of some woods imported into the United States according to approximate relative heartwood decay resistance*

Resistant or very resistant	Moderately resistant	Slightly or nonresistant
Angelique	Andiroba[1]	Balsa
Apamate	Apitong[1]	Banak
Brazilian rosewood	Avodire	Cativo
Caribbean pine	Capirona	Ceiba
Courbaril	European walnut	Jelutong
Encino	Gola	Limba
Goncalo alves	Khaya	Lupuna
Greenheart	Laurel	Mahogany, Philippine:
Guijo	Mahogany, Philippine:	Mayapis
Iroko	Almon	White lauan
Jarrah	Bagtikan	Obeche
Kapur	Red lauan	Parana pine
Karri	Tanguile	Ramin
Kokrodua (Afrormosia)	Ocote pine	Sande
Lapacho	Palosapis	Virola
Lignumvitae	Sapele	
Mahogany, American		
Meranti[1]		
Peroba de campos		
Primavera		
Santa Maria		
Spanish-cedar		
Teak		

[1] More than 1 species included, some of which may vary in resistance from that indicated.

resistant to attack by dilute mineral and organic acids. Oxidizing acids, such as nitric acid, have a greater degradative action than nonoxidizing acids. Alkaline solutions are more destructive than acidic solutions, and hardwoods are more susceptible to attack by both acids and alkalies than softwoods.

Highly acidic salts tend to hydrolyze wood when present in high concentrations. Even relatively low concentrations of such salts in railroad ties, for instance, have shown signs that the salt may migrate to the surface of ties that are occasionally wet and dried in a hot, arid region. This migration, combined with the high concentrations of salt relative to the small amount of water present, causes an acidic condition sufficient to make wood brittle.

Iron salts, which develop at points of contact with tie plates, bolts, and the like, have a degradative action on wood, especially in the presence of moisture. In addition, iron salts probably precipitate toxic extractives and thus lower the natural decay resistance of wood. The softening and discoloration of wood around corroded iron fastenings is a commonly observed phenomenon; it is especially pronounced in acidic woods, such as oak, and in woods such as redwood which contain considerable tannin and related compounds. The oxide layer formed on iron is transformed through reaction with wood acids into soluble iron salts which not only degrade the surrounding wood but probably catalyze the further corrosion of the metal. The action is accelerated by moisture; oxygen may also play an important role in the process. This effect is not encountered with well-dried wood used in dry locations. Under damp use conditions, it can be avoided or minimized by using corrosion-resistant fastenings.

Many substances have been employed as impregnants to enhance the natural resistance of wood to chemical degradation. One of the more economical treatments involves pressure impregnation with a viscous coke-oven coal tar to retard liquid penetration. Acid resistance of wood is increased by impregnation with phenolic resin solutions followed by appropriate drying and curing. Treatment with furfuryl alcohol has been used to increase resistance to alkaline solutions. A newer development involves massive impregnation with a monomeric resin, such as methyl methacrylate, followed by polymerization. Coatings and finishes, other chemical treatments, and preservation are described in chapters 16, 17, and 18.

Thermal Properties

Four important thermal properties of wood are (1) thermal conductivity, (2) specific heat, (3) thermal diffusivity, and (4) coefficient of thermal expansion. Thermal conductivity is a measure of the rate of heat flow through materials subjected to a temperature gradient. Specific heat of a material is the ratio of the heat capacity of the material to the heat capacity of water; the heat capacity of a material is the thermal energy required to produce one unit change of temperature in one unit mass. Thermal diffusivity is a measure of how quickly a material can absorb heat from its surroundings; it is the ratio of thermal conductivity to the product of density and specific heat. The coefficient of thermal expansion is a measure of the change of dimension caused by temperature change.

Thermal Conductivity

The thermal conductivity of common structural woods is a small fraction of the conductivity of metals with which it often is mated in construction. It is about two to four times that of common insulating material. For example, structural softwood lumber has a conductivity of about 0.75 British thermal units per inch per hour per square foot per degree Fahrenheit (Btu · in/hr · ft^2 · °F) compared with 1,500 for aluminum, 310 for steel, 6 for concrete, 7 for glass, 5 for plaster, and 0.25 for mineral wool.

The thermal conductivity of wood is affected by a number of basic factors: (1) density, (2) moisture content, (3) extractive content, (4) grain direction, (5) structural irregularities such as checks and knots, (6) fibril angle, and (7) temperature. It is nearly the same in the radial and tangential direction with respect to the growth rings but is 2.0 to 2.8 times greater parallel to the grain than in either the radial or tangential directions. It increases as the density, moisture content, temperature, or extractive content of the wood increases.

Figure 3−6 shows the average thermal conductivity perpendicular to the grain as related to wood density and moisture content up to approximately 30 percent moisture content. This chart is a plot of the empirical equation:

$$k = S(1.39 + 0.028M) + 0.165 \qquad (3-3)$$

where k is thermal conductivity in Btu · in./hr · ft^2 · °F, S is specific gravity based on volume at current moisture content and weight when ovendry, and M is the moisture content in percent of ovendry weight. For wood at a moisture content of 30 percent or greater, the following equation has been applied:

$$k = S(1.39 + 0.038M) + 0.165 \qquad (3-4)$$

The equations presented were derived by averaging the results of studies on a variety of species. Individual wood specimen conductivity will vary from these predicted values because of the seven variability sources noted previously. Appropriate specific gravity for different species is given in chapter 4.

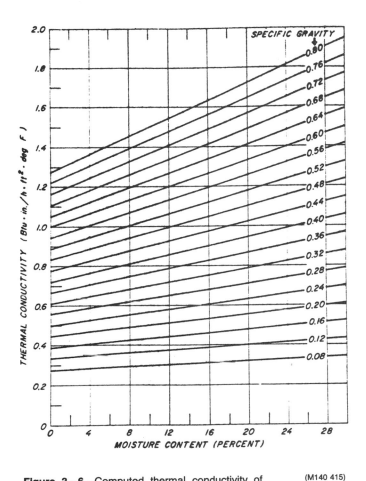

Figure 3-6—Computed thermal conductivity of wood perpendicular to grain as related to moisture content and specific gravity.

(M140 415)

Specific Heat

The specific heat of wood depends on the temperature and moisture content of the wood but is practically independent of density or species. Specific heat (cal/gm · °F) of dry wood is approximately related to temperature, t, in °F by

$$\text{Specific heat} = 0.25 + 0.0006t$$

When wood contains water, the specific heat is increased because the specific heat of water is larger than that of dry wood. The apparent specific heat of moist wood, however, is larger than would be expected from a simple sum of the separate effects of wood and water. The additional apparent specific heat is due to thermal energy absorbed by the wood-water bonds. As the temperature increases, the apparent specific heat increases because the energy of absorption of wood increases with temperature.

If the specific heat of water is considered to be unity, the specific heat of moist wood is given by:

$$\text{Specific heat} = \frac{m + c_o}{1 + m} + A \qquad (3-5)$$

where m is the fractional moisture content of the wood, c_o is the specific heat of dry wood, and A is the additional specific heat due to the wood-water bond energy. A increases with increasing temperature. For wood at 10 percent moisture content, A ranges from about 0.02 at 85 °F to about 0.04 at 140 °F. For wood at about 30 percent moisture content A ranges from about 0.04 at 85 °F to about 0.09 at 140 °F.

Thermal Diffusivity

Because of the small thermal conductivity and moderate density and specific heat of wood, the thermal diffusivity of wood is much smaller than that of other structural materials such as metals, brick, and stone. A typical value for wood is 0.00025 inch² per second compared to 0.02 inch² per second for steel, and 0.001 inch² per second for mineral wool. For this reason wood does not feel extremely hot or cold to the touch as do some other materials.

Few investigators have measured the diffusivity of wood directly. Since diffusivity is defined as the ratio of conductivity to the product of specific heat and density, conclusions regarding its variation with temperature and density often are based on calculating the effect of these variables on specific heat and conductivity.

All investigations illustrate that diffusivity is influenced slightly by both specific gravity and moisture content in an inverse fashion. The diffusivity increases approximately 0.0001 inch² per second over a decreasing specific gravity range of 0.65 to 0.30. Calculations suggest that the effect of moisture is to increase diffusivity by about 0.00004 inch² per second as the moisture content is reduced from 12 to 0 percent.

Coefficient of Thermal Expansion

The thermal expansion coefficients of completely dry wood are positive in all directions—that is, wood expands on heating and contracts on cooling. Only limited research has been carried out to explore the influence of wood property variability on thermal expansion. The linear expansion coefficient of ovendry wood parallel to the grain appears to be independent of specific gravity and species. In tests of both hardwoods and softwoods, the parallel-to-the-grain values have ranged from about 0.0000017 to 0.0000025 per °F.

The linear expansion coefficients across the grain (radial

and tangential) are proportional to wood density. These coefficients range from about 5 to over 10 times greater than the parallel-to-the-grain coefficients and thus are of more practical interest. The radial and tangential thermal expansion coefficients for ovendry wood, α_r and α_t, can be approximated by the following equations, over an ovendry specific gravity range of about 0.1. to 0.8:

$$\alpha_r = [(18) \text{ (specific gravity)} + 5.5] [10^{-6}] \text{ per } °F \quad (3-6)$$

$$\alpha_t = [(18) \text{ (specific gravity)} + 10.2] [10^{-6}] \text{ per } °F \quad (3-7)$$

Thermal expansion coefficients can be considered independent of temperature over the temperature range of -60 to $+130$ °F.

Wood that contains moisture reacts to varying temperature differently than does dry wood. When moist wood is heated, it tends to expand because of normal thermal expansion and to shrink because of loss in moisture content. Unless the wood is very dry initially (perhaps 3 or 4 percent MC or less), the shrinkage due to moisture loss on heating will be greater than the thermal expansion, so the net dimensional change on heating will be negative. Wood at intermediate moisture levels (about 8 to 20 percent) will expand when first heated, then gradually shrink to a volume smaller than the initial volume, as the wood gradually loses water while in the heated condition.

Even in the longitudinal (grain) direction, where dimensional change due to moisture change is very small, such changes will still predominate over corresponding dimensional changes due to thermal expansion unless the wood is very dry initially. For wood at usual moisture levels, net dimensional changes will generally be negative after prolonged heating.

Electrical Properties

The most important electrical properties of wood are conductivity, dielectric constant, and dielectric power factor.

The conductivity of a material determines the electric current that will flow when the material is placed under a given voltage gradient. The dielectric constant of a nonconducting material determines the amount of electric potential energy, in the form of induced polarization, that is stored in a given volume of the material when that material is placed in an electric field. The power factor of a nonconducting material determines the fraction of stored energy that is dissipated as heat when the material experiences a complete polarize-depolarize cycle.

Examples of industrial wood processes and applications in which electrical properties of wood are important include crossarms and poles for high-voltage powerlines, linemen's tools, and the heat-curing of adhesives in wood products by high-frequency electric fields. Moisture meters for wood utilize the relation between electrical properties and moisture content to estimate the moisture content.

Electrical Conductivity

The electrical conductivity of wood varies slightly with applied voltage and approximately doubles for each temperature increase of 10 °C. The electrical conductivity of wood or its reciprocal, resistivity, varies greatly with moisture content, especially below the fiber saturation point. As the moisture content of wood increases from near zero to fiber saturation, the electrical conductivity increases (resistivity decreases) by 10^{10} to 10^{13} times. The resistivity is about 10^{14} to 10^{16} ohm-meters for ovendry wood and 10^3 to 10^4 ohm-meters for wood at fiber saturation. As the moisture content increases from fiber saturation to complete saturation of the wood structure, the further increase in conductivity is smaller and erratic, generally amounting to less than a hundredfold.

Figure $3-7$ illustrates the change in resistance along the grain with moisture content, based on tests of many domestic species. Variability between test specimens is illustrated by the shaded area. Ninety percent of the experimental data points fall within this area. The resistance values were obtained using a standard moisture meter electrode at 80 °F. Conductivity is greater along the grain than across the grain and slightly greater in the radial direction than in the tangential direction. Relative conductivities in the longitudinal, radial, and tangential directions are in the approximate ratio of 1.0:0.55:0.50.

When wood contains abnormal quantities of water-soluble salts or other electrolytic substances, such as from preservative or fire-retardant treatment or prolonged contact with seawater, the electrical conductivity may be substantially increased. The increase is small when the moisture content of the wood is less than about 8 percent but becomes large rapidly as the moisture content exceeds 10 or 12 percent.

Dielectric Constant

The dielectric constant is the ratio of the dielectric permittivity of the material to that of free space; it is essentially a measure of the potential energy per unit volume stored in the material in the form of electric polarization when the material is in a given electric field. As measured by practical tests, the dielectric constant of a material is the ratio of the capacitance of a capacitor using the material as the dielectric, to the capacitance of the same capacitor using free space as the dielectric.

The dielectric constant of ovendry wood ranges from about

2 to 5 at room temperature, and decreases slowly but steadily with increasing frequency of the applied electric field. It increases as either temperature or moisture content increase, with a moderate positive interaction between temperature and moisture. There is an intense negative interaction between moisture and frequency: At 20 hertz the dielectric constant may range from about 4 for dry wood to near 1,000,000 for wet wood; at 1 kilohertz, from about 4 when dry to about 5,000 wet; and at 1 megahertz from about 3 when dry to about 100 wet. The dielectric constant is larger for polarization parallel to the grain than across the grain.

Dielectric Power Factor

When a nonconductor is placed in an electric field, it absorbs and stores potential energy. The amount of energy stored per unit volume depends upon the dielectric constant and the magnitude of the applied field. An ideal dielectric releases all of this energy to the external electric circuit when the field is removed, but practical dielectrics dissipate some of the energy as heat. The power factor is a measure of that portion of the stored energy converted to heat. Power factor values always fall between zero and unity. When the power factor does not exceed about 0.1, the fraction of the stored energy that is lost in one charge-discharge cycle is approximately equal to 2π times the power factor of the dielectric; for larger power factors, this fraction is approximated simply by the power factor itself.

The power factor of wood is large compared to inert plastic insulating materials, but some materials, for example some formulations of rubber, have equally large power factors. The power factor of wood varies from about 0.01 for dry low-density woods to as large as 0.95 for dense woods at high moisture levels. It is usually, but not always, greater for electric fields along the grain than across the grain.

The power factor of wood is affected by several factors, including frequency, moisture content, and temperature. These factors interact in complex ways to cause the power factor to have maximum and minimum values at various combinations of these factors.

Coefficient of Friction

The coefficient of friction depends on the moisture content of the wood and on surface roughness. It varies little with species except for those species that contain abundant oily or waxy extractives, such as lignumvitae.

Coefficients of friction for wood on most materials increase continuously as the moisture content of the wood increases from ovendry to fiber saturation, then remain roughly constant as the moisture content increases further until consider-

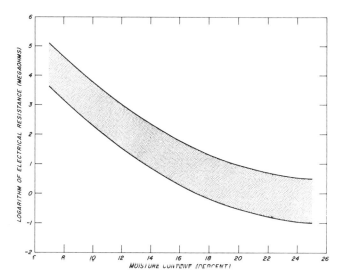

Figure 3–7—Change in electrical resistance of wood with varying moisture content levels for many U.S. species. Ninety percent of test values are represented by the shaded area. (M140 414)

able free water is present. When the surface is flooded with water, the coefficients of friction decrease again.

Static coefficients of friction are generally larger than for sliding, and the latter depend somewhat on the speed of sliding. Sliding coefficients of friction vary only slightly with speed when the wood moisture content is less than about 20 percent, but at high moisture content the coefficient of friction decreases substantially as the speed increases.

Coefficients of sliding friction for smooth, dry wood against hard, smooth surfaces commonly range from 0.3 to 0.5; at intermediate moisture content, 0.5 to 0.7; and near fiber saturation, 0.7 to 0.9.

Nuclear Radiation

Radiation passing through matter is reduced in intensity according to the relationship

$$I = I_o e^{-\mu x} \qquad (3-8)$$

where I is the reduced intensity of the beam at a depth x in the material, I_o is the incident intensity of a beam of radiation, and μ, the linear absorption coefficient of the material, is the fraction of energy removed from the beam per unit depth traversed. Where the density is a factor of interest in energy absorption, the linear absorption coefficient is divided by the density of the material to derive the mass absorption coefficient. The absorption coefficient of a material varies with type and energy of radiation.

The linear absorption coefficient for gamma (γ) radiation of wood is known to vary directly with moisture content and density, and inversely with the γ ray energy. As an example, the irradiation of ovendry yellow-poplar with 0.047 megaelectronvolt γ rays yielded linear absorption coefficients ranging from about 0.065 to about 0.11 per centimeter over the ovendry specific gravity range of about 0.33 to 0.62. An increase in the linear absorption coefficient of about 0.01 per centimeter occurs with an increase in moisture content from ovendry to fiber saturation. Absorption of γ rays in wood is of practical interest, in part, for measurement of the density of wood.

The interaction of wood with beta (β) radiation is similar in character to that with γ radiation, except that the absorption coefficients are larger. The linear absorption coefficient of wood with a specific gravity of 0.5 for a 0.5 megaelectronvolt β ray is about 3.0 per centimeter. The result of the larger coefficient is that even very thin wood products are virtually opaque to β rays.

The interaction of neutrons with wood is of interest because wood and the water it contains are compounds of hydrogen, and hydrogen has a relatively large probability of interaction with neutrons. High-energy-level neutrons lose energy much more quickly through interaction with hydrogen than with other elements found in wood. The lower energy neutrons that result from this interaction thus are a measure of the hydrogen density of the specimen. Measurement of the lower energy level neutrons can be related to the moisture content of the wood.

When neutrons interact with wood, an additional result is the production of radioactive isotopes of the elements present in the wood. The radioisotopes produced can be identified by the type, energy, and half-life of their emissions, and the specific activity of each indicates the amount of isotope present. This procedure, called neutron activation analysis, provides a sensitive nondestructive method of analysis for trace elements.

In the discussions above, moderate radiation levels that leave the wood physically unchanged have been assumed.

Very large doses of γ rays or neutrons can cause substantial degradation of wood. The effect of large radiation doses on the mechanical properties of wood is discussed in chapter 4.

Selected References

American Society of Heating, Refrigeration, and Air-Conditioning Engineers Handbook, 1981 Fundamentals. Atlanta, GA: ASHRAE; 1981.

American Society for Testing and Materials. Standard methods for testing small clear specimens of timber. ASTM D 143. Philadelphia, PA: ASTM; (see current edition).

Beall, F. C. Specific heat of wood—further research required to obtain meaningful data. Res. Note FPL−0184. Madison, WI: U.S. Department of Agriculture, Forest Service, Forest Products Laboratory; 1968.

James, W. L. Electric moisture meters for wood. Gen. Tech. Rep. FPL−6. Madison, WI: U.S. Department of Agriculture, Forest Service, Forest Products Laboratory; 1975.

Kleuters, W. Determining local density of wood by beta ray method. Madison, WI: Forest Products Journal. 14(9): 414; 1964.

Kollman, Franz F. P.; Côté, Wilfred A., Jr. Principles of wood science and technology I — solid wood. New York, Springer-Verlag New York, Inc; 1968.

Kubler, H.; Liang, L.; Chang, L. S. Thermal expansion of moist wood. Wood and Fiber. 5(3): 257−267; 1973.

Kukachka, B. F. Properties of imported tropical woods. Res. Pap. FPL 125. Madison, WI: U.S. Department of Agriculture, Forest Service, Forest Products Laboratory; 1970.

Lynn, R. Review of dielectric properties of wood and cellulose. Madison, WI: Forest Products Journal. 17(7): 61; 1967.

McKenzie, W. M.; Karpovich, H. Frictional behavior of wood. Munich: Wood Science and Technology. 2(2): 138; 1968.

Murase, Y. Frictional properties of wood at high sliding speed. Journal of the Japanese Wood Research Society. 26(2): 61−65; 1980.

Panshin, A. J.; deZeeuw, C. Textbook of wood technology. Vol 1, 3rd Ed. New York: McGraw-Hill; 1970.

Skaar, Christen. Water in wood. Syracuse, NY: Syracuse University Press; 1972.

Steinhagen, H. Peter. Thermal conductive properties of wood, green or dry, from -40° to +100° C: a literature review. Gen. Tech. Rep. FPL−9. Madison, WI: U.S. Department of Agriculture, Forest Service, Forest Products Laboratory; 1977.

Weatherwax, R. C.; Stamm, A. J. The coefficients of thermal expansion of wood and wood products. Transactions of American Society of Mechanical Engineers. 69(44): 421−432; 1947.

Chapter 4

Mechanical Properties of Wood

Mechanical Properties of Wood *

Mechanical properties discussed in this chapter have been obtained from tests of small pieces of wood termed "clear" and "straight grained" because they did not contain characteristics such as knots, cross grain, checks, and splits. These test pieces do contain wood structure characteristics such as growth rings that occur in consistent patterns within the piece. Clear wood specimens are usually considered "homogeneous" in wood mechanics.

Many of the mechanical properties of wood tabulated in this chapter were derived from extensive sampling and analysis procedures. These properties are represented as the average mechanical properties of the species and are used to derive allowable properties for design. Some properties, such as tension, and all properties for some imported species are based on a more limited number of specimens not subject to the same sampling and analysis procedures. The appropriateness of these latter properties to represent the average properties of a species is uncertain; nevertheless, they represent the best information available.

Variability, or variation in properties, is common to all materials. Because wood is a natural material and the tree is subject to numerous constantly changing influences (such as moisture, soil conditions, and growing space), wood properties vary considerably even in clear material. This chapter provides information where possible on the nature and magnitude of variability in properties.

Orthotropic Nature of Wood

Wood may be described as an orthotropic material; that is, it has unique and independent mechanical properties in the directions of three mutually perpendicular axes—longitudinal, radial, and tangential. The longitudinal axis (L) is parallel to the fiber (grain); the radial axis (R) is normal to the growth rings (perpendicular to the grain in the radial direction); and the tangential axis (T) is perpendicular to the grain but tangent to the growth rings. These axes are shown in figure 4−1.

Elastic Properties

Twelve constants (nine are independent) are needed to describe the elastic behavior of wood: Three moduli of elasticity, E, three moduli of rigidity, G, and six Poisson's

ratios, μ. The moduli of elasticity and Poisson's ratios are related by expressions of the form:

$$\frac{\mu_{ij}}{E_i} = \frac{\mu_{ij}}{E_j}, \quad i \neq j; i, j = L, R, T \quad (4-1)$$

General relations between stress and strain for a homogeneous, orthotropic material can be found in texts on anisotropic elasticity. Regression equations that may be used to predict some elastic parameters as a function of density have been developed.

Modulus of Elasticity

The three moduli of elasticity denoted by E_L, E_R, and E_T are, respectively, the elastic moduli along longitudinal, radial, and tangential axes of wood. These moduli are usually obtained from compression tests; however, data for E_R and E_T are not extensive. Average values of E_R and E_T for samples from a few species are presented in table 4−1 as ratios with E_L. The ratios, as well as the three elastic constants themselves, vary within and between species and with moisture content and specific gravity.

E_L determined from bending, rather than from an axial test, may be the only E available for a species. Average values of E_L obtained from bending tests are given in tables 4−2, 4−3, and 4−4. A representative coefficient of variation of E_L determined with bending tests for clear wood is reported in table 4−5. E_L as tabulated includes an effect of shear deflection. E_L from bending can be increased by 10 percent to approximately remove this effect. This adjusted bending E_L can be used to determine E_R and E_T based on ratios in table 4−1.

Modulus of Rigidity

The three moduli of rigidity denoted by G_{LR}, G_{LT}, G_{RT} are the elastic constants in the LR, LT, and RT planes, respectively. For example, G_{LR} is the modulus of rigidity based on shear strain in the LR plane and shear stresses in the LT and RT planes. Average values of shear moduli for samples of a few species expressed as ratios with E_L are given in table 4−1. As with moduli of elasticity, the moduli of rigidity vary within and between species and with moisture content and specific gravity.

Poisson's Ratio

The six Poisson's ratios are denoted by μ_{LR}, μ_{RL}, μ_{LT}, μ_{TL}, μ_{RT}, μ_{TR}. The first letter of the subscript refers to direction of applied stress and the second letter refers to direction

* Revision by B. Alan Bendtsen, Forest Products Technologist; William L. James, Physicist; Charles C. Gerhards, General Engineer; Jerrold E. Winandy, Forest Products Technologist; David W. Green, General Engineer; Martin Chudnoff, Forest Products Technologist; and Pamela J. Giese, Computer Programmer.

of lateral deformation. For example, μ_{LR} is the Poisson's ratio for deformation along the radial axis caused by stress along the longitudinal axis. Average values of Poisson's ratios for samples of a few species are given in table 4—1. Values for μ_{RL} and μ_{TL} are less precisely determined than are those for the other Poisson's ratios. Poisson's ratios vary within and between species and are affected slightly by moisture content.

Strength Properties

Common Properties

Strength values most commonly measured and represented as "strength properties" for design include the modulus of rupture in bending, the maximum stress in compression parallel to the grain, compressive strength perpendicular to the grain, and shear strength parallel to the grain. Additional measurements often made include work to maximum load in bending, impact bending strength, tensile strength perpendicular to the grain, and hardness. These properties, grouped according to the broad forest tree categories of hardwood and softwood (not correlated with hardness or softness), are given in tables 4—2, 4—3, and 4—4 for many of the commercially important species. Coefficients of variation for these properties from a limited sampling of specimens are reported in table 4—5.

Modulus of Rupture

Modulus of rupture in bending reflects the maximum load-carrying capacity of a member and is proportional to the maximum moment borne by a specimen. It is an accepted criterion of strength, although it is not a true stress because the formula by which it is computed is valid only to the proportional limit.

Work to Maximum Load in Bending

Work to maximum load in bending represents the ability to absorb shock with some permanent deformation and more or less injury to a specimen. It is a measure of the combined strength and toughness of wood under bending stresses.

Maximum Crushing Strength

Maximum crushing strength is the maximum stress sustained by a compression parallel-to-grain specimen having a ratio of length to least dimension of less than 11.

Compression Perpendicular to Grain

Strength in compression perpendicular to grain is reported as the stress at the proportional limit because there is no clearly defined ultimate stress for this property.

Shear Strength Parallel to Grain

Shear strength is a measure of ability to resist internal slipping of one part upon another along the grain. Values presented are the average of radial and tangential shear.

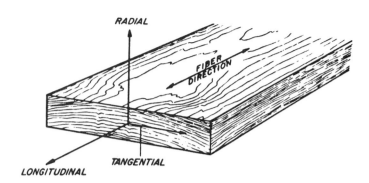

Figure 4—1—The three principal axes of wood with respect to grain direction and growth rings.　　(M140 728)

Impact Bending

In the impact bending test, a hammer of given weight is dropped upon a beam from successively increased heights until complete rupture occurs. The height of the maximum drop, or the drop that causes failure, is a comparative figure representing the ability of wood to absorb shocks that cause stresses beyond the proportional limit.

Tensile Strength Perpendicular to Grain

Tensile strength perpendicular to the grain is a measure of the resistance of wood to forces acting across the grain that tend to split a member. Values presented are the average of radial and tangential observations.

Hardness

Hardness represents the resistance of wood to wear and marring. It is measured by the load required to embed a 0.444-inch ball to one-half its diameter in the wood. Values presented are the average of radial and tangential penetrations.

Tension Parallel to Grain

Relatively few data are available on the tensile strength of various species parallel to grain. Table 4—6 lists average tensile strength values for a limited number of specimens of a few species. In the absence of sufficient tension test data, the modulus of rupture values are sometimes substituted for tensile strength of small, clear, straight-grained pieces of wood. The modulus of rupture is considered to be a low or conservative estimate of tensile strength for these specimens. Chapter 6 should be consulted for discussion of the tensile properties of structural members.

Less Common Properties

Strength properties less commonly measured in clear wood include torsion, toughness, and rolling shear. Other properties involving time under load include creep, creep-rupture or duration of load, and fatigue strength.

4-3

Torsion

For solid wood members, the torsional shear strength may be taken as the shear strength parallel to the grain. Two-thirds of this value may be used as the torsional shear stress at the proportional limit.

Toughness

Toughness represents the energy required to cause complete failure rapidly in a centrally loaded bending specimen. Table 4–7 gives average toughness values for samples of a few hardwood and softwood species. Table 4–5 records the average coefficient of variation for toughness as determined from approximately 50 species.

Creep and Creep-Rupture

Wood is known to creep under load. If the load is sufficiently high and acts long enough, failure (creep-rupture) will eventually occur. Duration of stress is commonly used as the term for time before rupture. Duration of stress is an important factor in setting design values for wood. Creep and duration of stress are described in later sections of this chapter.

Fatigue

Fatigue strength is that property which provides resistance to failure under specific combinations of cyclic loading conditions: frequency and number of cycles, the maximum stress, the ratio of maximum to minimum stress, and other factors of less importance. The main factors affecting fatigue in wood are discussed later in this chapter. Interpretation of fatigue data and a discussion of fatigue as a function of the service environment are also included.

Rolling Shear Strength

The term "rolling shear" describes the shear strength of wood where the shearing force is in any longitudinal plane and is acting perpendicular to the grain. Test procedures for rolling shear in solid wood are of recent origin; few test values have been reported. In limited tests, rolling shear strengths averaged 18 to 28 percent of the parallel-to-grain shear values. Rolling shear strength is about the same in the longitudinal-radial plane as in the longitudinal-tangential plane.

Vibration Properties

The vibration properties of primary interest in structural materials are the speed of sound and the damping capacity or internal friction.

Speed of Sound

The speed of sound in a structural material varies directly with the square root of modulus of elasticity and inversely with the square root of density. In wood, the speed of sound also varies strongly with grain direction because the transverse modulus of elasticity is much less than the longitudinal value (as little as 1/20); the speed of sound across the grain is about one-fifth to one-third of the longitudinal value. For example, a piece of wood with a longitudinal modulus of elasticity of 1,800,000 pounds per square inch (psi) and a density of 30 pounds per cubic foot would have a speed of sound in the longitudinal direction of about 150,000 inches per second. In the transverse direction, its modulus of elasticity would be about 100,000 psi and the speed of sound approximately 35,000 inches per second.

The speed of sound decreases with increasing temperature or moisture content in proportion to the influence of these variables on the modulus of elasticity and density. The speed of sound decreases slightly with increasing frequency and amplitude of vibration, although for most common applications this effect is too small to be significant. There is no recognized independent effect of species on the speed of sound. Variability in the speed of sound in wood is directly related to the variability of modulus of elasticity and density.

Internal Friction

When solid material is strained, some mechanical energy is dissipated as heat. Internal friction is the term used to denote the mechanism that causes this energy dissipation. The internal friction mechanism in wood is a complex function of temperature and moisture content. In general, there is a value of moisture content at which internal friction is minimum. On either side of this minimum, internal friction increases as moisture content varies down to zero or up to the fiber saturation point. The moisture content at which the minimum internal friction occurs varies with temperature. At room temperature (23 °C), the minimum occurs at about 6 percent moisture content; at −20 °C, it occurs at about 14 percent moisture content; and at 70 °C, at about 4 percent. At 90 °C the minimum is not well defined and occurs near zero moisture content.

Similarly, there are temperatures at which internal friction is minimum, and the temperatures of minimum internal friction vary with moisture content. The temperatures of minimum internal friction are higher as the moisture content is decreased. For temperatures above 0 °C, and moisture contents greater than about 10 percent, internal friction increases strongly as temperature increases with a strong positive interaction with moisture content. For very dry wood, there is a general tendency for internal friction to decrease as the temperature increases.

The value of internal friction, expressed by logarithmic decrement, ranges from about 0.1 for hot, moist wood to less than 0.02 for hot, dry wood. Cool wood, regardless of moisture content, would have an intermediate value.

Summary Tables on Mechanical Properties of Clear Straight-Grained Wood

The mechanical properties listed in tables 4−1 through 4−7 are based on a variety of sampling methods. Generally, the most extensive sampling is represented in tables 4−2, 4−3, and 4−4. The values in table 4−2 are averages derived for a number of species grown in the United States. The table value is an estimate of the average clear wood property of the species. Many of the values were obtained from test specimens taken at heights between 8 and 16 feet above the stump of the tree. Values reported in table 4−3 represent estimates of the average clear wood properties of species grown in Canada and commonly imported into the United States, while those in table 4−4 represent estimates of average properties of species imported from other countries.

Methods of data collection and analysis have changed over the years during which the data in tables 4−2 and 4−3 have been collected. In addition, the character of some forests changes with time. Because not all of the species have been reevaluated to reflect these changes, the appropriateness of the data should be reviewed when used for critical applications such as stress grades of lumber.

Values reported in table 4−4 were collected from the world literature; thus, the appropriateness of these properties to represent a species is not known. The properties reported in tables 4−1, 4−6, 4−7, and 4−13 may not necessarily represent average species characteristics because of inadequate sampling; they do suggest the relative influence of species and other specimen parameters on the mechanical behavior recorded.

Variability in properties can be important in both production and consumption of wood products. Often the fact that a piece may be stronger, harder, or stiffer than the average is of less concern to the user than if it is weaker; however, this may not be true if lightweight material is selected for a specific purpose or if harder or tougher material is hard to work. It is desirable, therefore, that some indication of the spread of property values be known. Average coefficients of variation for many mechanical properties are presented in table 4−5.

The mechanical properties reported in the tables are significantly affected by the moisture content of the specimens at the time of test. Some tables include properties evaluated at differing moisture levels; these moisture levels are reported. As indicated in the tables, many of the dry test data have been adjusted to a common moisture content base of 12 percent. The differences in properties displayed in the tables as a result of differing moisture levels are not necessarily consistent for larger wood pieces such as lumber. Guidelines for adjusting clear wood properties to arrive at allowable properties for lumber are discussed in chapter 6, "Lumber Stress Grades and Allowable Properties."

Specific gravity is reported in many of the tables because it is used as an index of clear wood mechanical properties. The specific gravity values given in tables 4−2 and 4−3 represent the estimated average clear wood specific gravity of the species. In the other tables, the specific gravity represents only the specimens tested. The variability of specific gravity, represented by the coefficient of variation derived from tests on 50 species, is included in table 4−5.

Mechanical and physical properties as measured and reported often reflect not only the characteristics of the wood but also the influence of the shape and size of test specimen and the mode of test. The methods of test used to establish properties in tables 4−2, 4−3, 4−6, and 4−7 are based on standard procedures, ASTM D 143. The methods of test for properties presented in other tables are referenced in the selected references at the end of this chapter.

Common names of species listed in the tables conform to standard nomenclature of the USDA, Forest Service. Other names may be used locally for a species. Also, one common name may be applied to groups of species for marketing.

Natural Characteristics Affecting Mechanical Properties

Clear straight-grained wood is used for determining fundamental mechanical properties; however, because of natural growth characteristics of trees, wood products vary in specific gravity, may contain cross grain, or have knots and localized slope of grain. Natural defects such as pitch pockets may occur due to biological or climatic elements acting on the living tree. These wood characteristics must be taken into account in assessing actual properties or estimating actual performance of wood products.

Specific Gravity

The substance of which wood is composed is actually heavier than water, its specific gravity being about 1.5 regardless of the species of wood. In spite of this, the dry wood of most species floats in water, and it is thus evident that part of the volume of a piece of wood is occupied by cell cavities and pores. Variations in the size of these openings and in the thickness of the cell walls cause some species to have more wood substance per unit volume than others and therefore to have higher specific gravity. Specific gravity thus is an excellent index of the amount of wood substance a piece of wood contains; it is a good index of mechanical properties so long as the wood is clear, straight grained, and free from defects. However, specific gravity values also reflect the presence of gums, resins, and extractives, which contribute little to mechanical properties.

(Text continues on page 4-27.)

Table 4—1—*Elastic ratios for various species*

Species	Approximate specific gravity[1]	Approximate moisture content	Modulus of elasticity ratios		Ratio of modulus of rigidity to modulus of elasticity			Poisson's ratios					
			E_T/E_L	E_R/E_L	G_{LR}/E_L	G_{LT}/E_L	G_{RT}/E_L	μ_{LR}	μ_{LT}	μ_{RT}	μ_{TR}	μ_{RL}	μ_{TL}
		Pct											
Balsa	0.13	9	0.015	0.046	0.054	0.037	0.005	0.23	0.49	0.67	0.23	0.02	0.01
Birch, yellow	.64	13	.050	.078	.074	.068	.017	.43	.45	.70	.43	.04	.02
Douglas-fir	.50	12	.050	.068	.064	.078	.007	.29	.45	.39	.37	.04	.03
Spruce, Sitka	.38	12	.043	.078	.064	.061	.003	.37	.47	.44	.24	.04	.02
Sweetgum	.53	11	.050	.115	.089	.061	.021	.32	.40	.68	.31	.04	.02
Walnut, black	.59	11	.056	.106	.085	.062	.021	.50	.63	.72	.38	.05	.04
Yellow-poplar	.38	11	.043	.092	.075	.069	.011	.32	.39	.70	.33	.03	.02

[1] Based on ovendry weight and volume at the moisture content shown.

Table 4–2—*Mechanical properties[1] of some commercially important woods grown in the United States*

Common names of species	Moisture condition	Specific gravity[2]	Static bending			Impact bending—height of drop causing complete failure[4]	Compression parallel to grain—maximum crushing strength	Compression perpendicular to grain—fiber stress at proportional limit	Shear parallel to grain—maximum shearing strength	Tension perpendicular to grain—maximum tensile strength	Side hardness—load perpendicular to grain
			Modulus of rupture	Modulus of elasticity[3]	Work to maximum load						
			Psi	Million psi	In–lb per in³	In	\-\-\- Psi \-\-\-				Lb
HARDWOODS											
Alder, red	Green	0.37	6,500	1.17	8.0	22	2,960	250	770	390	440
	Dry	.41	9,800	1.38	8.4	20	5,820	440	1,080	420	590
Ash:											
Black	Green	.45	6,000	1.04	12.1	33	2,300	350	860	490	520
	Dry	.49	12,600	1.60	14.9	35	5,970	760	1,570	700	850
Blue	Green	.53	9,600	1.24	14.7	—	4,180	810	1,540	—	—
	Dry	.58	13,800	1.40	14.4	—	6,980	1,420	2,030	—	—
Green	Green	.53	9,500	1.40	11.8	35	4,200	730	1,260	590	870
	Dry	.56	14,100	1.66	13.4	32	7,080	1,310	1,910	700	1,200
Oregon	Green	.50	7,600	1.13	12.2	39	3,510	530	1,190	590	790
	Dry	.55	12,700	1.36	14.4	33	6,040	1,250	1,790	720	1,160
White	Green	.55	9,500	1.44	15.7	38	3,990	670	1,350	590	960
	Dry	.60	15,000	1.74	16.6	43	7,410	1,160	1,910	940	1,320
Aspen:											
Bigtooth	Green	.36	5,400	1.12	5.7	—	2,500	210	730	—	—
	Dry	.39	9,100	1.43	7.7	—	5,300	450	1,080	—	—
Quaking	Green	.35	5,100	.86	6.4	22	2,140	180	660	230	300
	Dry	.38	8,400	1.18	7.6	21	4,250	370	850	260	350
Basswood, American	Green	.32	5,000	1.04	5.3	16	2,220	170	600	280	250
	Dry	.37	8,700	1.46	7.2	16	4,730	370	990	350	410
Beech, American	Green	.56	8,600	1.38	11.9	43	3,550	540	1,290	720	850
	Dry	.64	14,900	1.72	15.1	41	7,300	1,010	2,010	1,010	1,300
Birch:											
Paper	Green	.48	6,400	1.17	16.2	49	2,360	270	840	380	560
	Dry	.55	12,300	1.59	16.0	34	5,690	600	1,210	—	910
Sweet	Green	.60	9,400	1.65	15.7	48	3,740	470	1,240	430	970
	Dry	.65	16,900	2.17	18.0	47	8,540	1,080	2,240	950	1,470
Yellow	Green	.55	8,300	1.50	16.1	48	3,380	430	1,110	430	780
	Dry	.62	16,600	2.01	20.8	55	8,170	970	1,880	920	1,260

Table 4–2—*Mechanical properties[1] of some commercially important woods grown in the United States*—Continued

Common names of species	Moisture condition	Specific gravity[2]	Static bending Modulus of rupture (Psi)	Static bending Modulus of elasticity[3] (Million psi)	Static bending Work to maximum load (In-lb per in³)	Impact bending—height of drop causing complete failure[4] (In)	Compression parallel to grain—maximum crushing strength (Psi)	Compression perpendicular to grain—fiber stress at proportional limit (Psi)	Shear parallel to grain—maximum shearing strength (Psi)	Tension perpendicular to grain—maximum tensile strength (Psi)	Side hardness—load perpendicular to grain (Lb)
HARDWOODS—continued											
Butternut	Green	.36	5,400	.97	8.2	24	2,420	220	760	430	390
	Dry	.38	8,100	1.18	8.2	24	5,110	460	1,170	440	490
Cherry, black	Green	.47	8,000	1.31	12.8	33	3,540	360	1,130	570	660
	Dry	.50	12,300	1.49	11.4	29	7,110	690	1,700	560	950
Chestnut, American	Green	.40	5,600	.93	7.0	24	2,470	310	800	440	420
	Dry	.43	8,600	1.23	6.5	19	5,320	620	1,080	460	540
Cottonwood:											
Balsam poplar	Green	.31	3,900	.75	4.2	—	1,690	140	500	—	—
	Dry	.34	6,800	1.10	5.0	—	4,020	300	790	—	—
Black	Green	.31	4,900	1.08	5.0	20	2,200	160	610	270	250
	Dry	.35	8,500	1.27	6.7	22	4,500	300	1,040	330	350
Eastern	Green	.37	5,300	1.01	7.3	21	2,280	200	680	410	340
	Dry	.40	8,500	1.37	7.4	20	4,910	380	930	580	430
Elm:											
American	Green	.46	7,200	1.11	11.8	38	2,910	360	1,000	590	620
	Dry	.50	11,800	1.34	13.0	39	5,520	690	1,510	660	830
Rock	Green	.57	9,500	1.19	19.8	54	3,780	610	1,270	—	940
	Dry	.63	14,800	1.54	19.2	56	7,050	1,230	1,920	—	1,320
Slippery	Green	.48	8,000	1.23	15.4	47	3,320	420	1,110	640	660
	Dry	.53	13,000	1.49	16.9	45	6,360	820	1,630	530	860
Hackberry	Green	.49	6,500	.95	14.5	48	2,650	400	1,070	630	700
	Dry	.53	11,000	1.19	12.8	43	5,440	890	1,590	580	880
Hickory, pecan:											
Bitternut	Green	.60	10,300	1.40	20.0	66	4,570	800	1,240	—	—
	Dry	.66	17,100	1.79	18.2	66	9,040	1,680	—	—	—
Nutmeg	Green	.56	9,100	1.29	22.8	54	3,980	760	1,030	—	—
	Dry	.60	16,600	1.70	25.1	—	6,910	1,570	—	—	—
Pecan	Green	.60	9,800	1.37	14.6	53	3,990	780	1,480	680	1,310
	Dry	.66	13,700	1.73	13.8	44	7,850	1,720	2,080	—	1,820
Water	Green	.61	10,700	1.56	18.8	56	4,660	880	1,440	—	—
	Dry	.62	17,800	2.02	19.3	53	8,600	1,550	—	—	—

Table 4–2—*Mechanical properties[1] of some commercially important woods grown in the United States*—Continued

Common names of species	Moisture condition	Specific gravity[2]	Static bending — Modulus of rupture	Static bending — Modulus of elasticity[3]	Static bending — Work to maximum load	Impact bending—height of drop causing complete failure[4]	Compression parallel to grain—maximum crushing strength	Compression perpendicular to grain—fiber stress at proportional limit	Shear parallel to grain—maximum shearing strength	Tension perpendicular to grain—maximum tensile strength	Side hardness—load perpendicular to grain
			Psi	Million psi	In–lb per in³	In	--------- Psi ---------				Lb

HARDWOODS—continued

Common names of species	Moisture condition	Specific gravity[2]	Modulus of rupture (Psi)	Modulus of elasticity[3] (Million psi)	Work to maximum load (In–lb per in³)	Impact bending (In)	Compression parallel	Compression perpendicular	Shear parallel	Tension perpendicular	Side hardness (Lb)
Hickory, true:											
Mockernut	Green	.64	11,100	1.57	26.1	88	4,480	810	1,280	—	—
	Dry	.72	19,200	2.22	22.6	77	8,940	1,730	1,740	—	—
Pignut	Green	.66	11,700	1.65	31.7	89	4,810	920	1,370	—	—
	Dry	.75	20,100	2.26	30.4	74	9,190	1,980	2,150	—	—
Shagbark	Green	.64	11,000	1.57	23.7	74	4,580	840	1,520	—	—
	Dry	.72	20,200	2.16	25.8	67	9,210	1,760	2,430	—	—
Shellbark	Green	.62	10,500	1.34	29.9	104	3,920	810	1,190	—	—
	Dry	.69	18,100	1.89	23.6	88	8,000	1,800	2,110	—	—
Honeylocust	Green	.60	10,200	1.29	12.6	47	4,420	1,150	1,660	930	1,390
	Dry	—	14,700	1.63	13.3	47	7,500	1,840	2,250	900	1,580
Locust, black	Green	.66	13,800	1.85	15.4	44	6,800	1,160	1,760	770	1,570
	Dry	.69	19,400	2.05	18.4	57	10,180	1,830	2,480	640	1,700
Magnolia:											
Cucumbertree	Green	.44	7,400	1.56	10.0	30	3,140	330	990	440	520
	Dry	.48	12,300	1.82	12.2	35	6,310	570	1,340	660	700
Southern	Green	.46	6,800	1.11	15.4	54	2,700	460	1,040	610	740
	Dry	.50	11,200	1.40	12.8	29	5,460	860	1,530	740	1,020
Maple:											
Bigleaf	Green	.44	7,400	1.10	8.7	23	3,240	450	1,110	600	620
	Dry	.48	10,700	1.45	7.8	28	5,950	750	1,730	540	850
Black	Green	.52	7,900	1.33	12.8	48	3,270	600	1,130	720	840
	Dry	.57	13,300	1.62	12.5	40	6,680	1,020	1,820	670	1,180
Red	Green	.49	7,700	1.39	11.4	32	3,280	400	1,150	600	700
	Dry	.54	13,400	1.64	12.5	32	6,540	1,000	1,850	—	950
Silver	Green	.44	5,800	.94	11.0	29	2,490	370	1,050	560	590
	Dry	.47	8,900	1.14	8.3	25	5,220	740	1,480	500	700
Sugar	Green	.56	9,400	1.55	13.3	40	4,020	640	1,460	—	970
	Dry	.63	15,800	1.83	16.5	39	7,830	1,470	2,330	—	1,450

Table 4–2—Mechanical properties[1] of some commercially important woods grown in the United States—Continued

Common names of species	Moisture condition	Specific gravity[2]	Static bending			Impact bending—height of drop causing complete failure[4]	Compression parallel to grain—maximum crushing strength	Compression perpendicular to grain—fiber stress at proportional limit	Shear parallel to grain—maximum shearing strength	Tension perpendicular to grain—maximum tensile strength	Side hardness—load perpendicular to grain
			Modulus of rupture	Modulus of elasticity[3]	Work to maximum load						
			Psi	Million psi	In-lb per in³	In	Psi	Psi	Psi	Psi	Lb

HARDWOODS—continued

Common names of species	Moisture condition	Specific gravity[2]	Modulus of rupture (Psi)	Modulus of elasticity (Million psi)	Work to maximum load (In-lb per in³)	Impact bending (In)	Compression parallel (Psi)	Compression perpendicular (Psi)	Shear parallel (Psi)	Tension perpendicular (Psi)	Side hardness (Lb)
Oak, red:											
Black	Green	.56	8,200	1.18	12.2	40	3,470	710	1,220	—	1,060
	Dry	.61	13,900	1.64	13.7	41	6,520	930	1,910	—	1,210
Cherrybark	Green	.61	10,800	1.79	14.7	54	4,620	760	1,320	800	1,240
	Dry	.68	18,100	2.28	18.3	49	8,740	1,250	2,000	840	1,480
Laurel	Green	.56	7,900	1.39	11.2	39	3,170	570	1,180	770	1,000
	Dry	.63	12,600	1.69	11.8	39	6,980	1,060	1,830	790	1,210
Northern red	Green	.56	8,300	1.35	13.2	44	3,440	610	1,210	750	1,000
	Dry	.63	14,300	1.82	14.5	43	6,760	1,010	1,780	800	1,290
Pin	Green	.58	8,300	1.32	14.0	48	3,680	720	1,290	800	1,070
	Dry	.63	14,000	1.73	14.8	45	6,820	1,020	2,080	1,050	1,510
Scarlet	Green	.60	10,400	1.48	15.0	54	4,090	830	1,410	700	1,200
	Dry	.67	17,400	1.91	20.5	53	8,330	1,120	1,890	870	1,400
Southern red	Green	.52	6,900	1.14	8.0	29	3,030	550	930	480	860
	Dry	.59	10,900	1.49	9.4	26	6,090	870	1,390	510	1,060
Water	Green	.56	8,900	1.55	11.1	39	3,740	620	1,240	820	1,010
	Dry	.63	15,400	2.02	21.5	44	6,770	1,020	2,020	920	1,190
Willow	Green	.56	7,400	1.29	8.8	35	3,000	610	1,180	760	980
	Dry	.69	14,500	1.90	14.6	42	7,040	1,130	1,650	—	1,460
Oak, white:											
Bur	Green	.58	7,200	.88	10.7	44	3,290	680	1,350	800	1,110
	Dry	.64	10,300	1.03	9.8	29	6,060	1,200	1,820	680	1,370
Chestnut	Green	.57	8,000	1.37	9.4	35	3,520	530	1,210	690	890
	Dry	.66	13,300	1.59	11.0	40	6,830	840	1,490	—	1,130
Live	Green	.80	11,900	1.58	12.3	—	5,430	2,040	2,210	—	—
	Dry	.88	18,400	1.98	18.9	—	8,900	2,840	2,660	—	—
Overcup	Green	.57	8,000	1.15	12.6	44	3,370	540	1,320	730	960
	Dry	.63	12,600	1.42	15.7	38	6,200	810	2,000	940	1,190
Post	Green	.60	8,100	1.09	11.0	44	3,480	860	1,280	790	1,130
	Dry	.67	13,200	1.51	13.2	46	6,600	1,430	1,840	780	1,360

Table 4–2—*Mechanical properties¹ of some commercially important woods grown in the United States*—Continued

Common names of species	Moisture condition	Specific gravity[2]	Static bending — Modulus of rupture (Psi)	Static bending — Modulus of elasticity[3] (Million psi)	Static bending — Work to maximum load (In-lb per in³)	Impact bending—height of drop causing complete failure[4] (In)	Compression parallel to grain—maximum crushing strength (Psi)	Compression perpendicular to grain—fiber stress at proportional limit (Psi)	Shear parallel to grain—maximum shearing strength (Psi)	Tension perpendicular to grain—maximum tensile strength (Psi)	Side hardness—load perpendicular to grain (Lb)
HARDWOODS—continued											
Oak, white—con.											
Swamp chestnut	Green	.60	8,500	1.35	12.8	45	3,540	570	1,260	670	1,110
	Dry	.67	13,900	1.77	12.0	41	7,270	1,110	1,990	690	1,240
Swamp white	Green	.64	9,900	1.59	14.5	50	4,360	760	1,300	860	1,160
	Dry	.72	17,700	2.05	19.2	49	8,600	1,190	2,000	830	1,620
White	Green	.60	8,300	1.25	11.6	42	3,560	670	1,250	770	1,060
	Dry	.68	15,200	1.78	14.8	37	7,440	1,070	2,000	800	1,360
Sassafras	Green	.42	6,000	.91	7.1	—	2,730	370	950	—	—
	Dry	.46	9,000	1.12	8.7	—	4,760	850	1,240	—	—
Sweetgum	Green	.46	7,100	1.20	10.1	36	3,040	370	990	540	600
	Dry	.52	12,500	1.64	11.9	32	6,320	620	1,600	760	850
Sycamore, American	Green	.46	6,500	1.06	7.5	26	2,920	360	1,000	630	610
	Dry	.49	10,000	1.42	8.5	26	5,380	700	1,470	720	770
Tanoak	Green	.58	10,500	1.55	13.4	—	4,650	—	—	—	—
	—	—	—	—	—	—	—	—	—	—	—
Tupelo:											
Black	Green	.46	7,000	1.03	8.0	30	3,040	480	1,100	570	640
	Dry	.50	9,600	1.20	6.2	22	5,520	930	1,340	500	810
Water	Green	.46	7,300	1.05	8.3	30	3,370	480	1,190	600	710
	Dry	.50	9,600	1.26	6.9	23	5,920	870	1,590	700	880
Walnut, black	Green	.51	9,500	1.42	14.6	37	4,300	490	1,220	570	900
	Dry	.55	14,600	1.68	10.7	34	7,580	1,010	1,370	690	1,010
Willow, black	Green	.36	4,800	.79	11.0	—	2,040	180	680	—	—
	Dry	.39	7,800	1.01	8.8	—	4,100	430	1,250	—	—
Yellow-poplar	Green	.40	6,000	1.22	7.5	26	2,660	270	790	510	440
	Dry	.42	10,100	1.58	8.8	24	5,540	500	1,190	540	540

Table 4–2—*Mechanical properties[1] of some commercially important woods grown in the United States*—Continued

Common names of species	Moisture condition	Specific gravity[2]	Static bending			Impact bending—height of drop causing complete failure[4]	Compression parallel to grain—maximum crushing strength	Compression perpendicular to grain—fiber stress at proportional limit	Shear parallel to grain—maximum shearing strength	Tension perpendicular to grain—maximum tensile strength	Side hardness—load perpendicular to grain
			Modulus of rupture	Modulus of elasticity[3]	Work to maximum load						
			Psi	Million psi	In–lb per in³	In			Psi			Lb

SOFTWOODS

Common names of species	Moisture condition	Specific gravity[2]	Modulus of rupture (Psi)	Modulus of elasticity[3] (Million psi)	Work to maximum load (In–lb per in³)	Impact bending (In)	Compression parallel (Psi)	Compression perpendicular (Psi)	Shear parallel (Psi)	Tension perpendicular (Psi)	Side hardness (Lb)
Baldcypress	Green	.42	6,600	1.18	6.6	25	3,580	400	810	300	390
	Dry	.46	10,600	1.44	8.2	24	6,360	730	1,000	270	510
Cedar:											
Alaska-	Green	.42	6,400	1.14	9.2	27	3,050	350	840	330	440
	Dry	.44	11,100	1.42	10.4	29	6,310	620	1,130	360	580
Atlantic white-	Green	.31	4,700	.75	5.9	18	2,390	240	690	180	290
	Dry	.32	6,800	.93	4.1	13	4,700	410	800	220	350
Eastern redcedar	Green	.44	7,000	.65	15.0	35	3,570	700	1,010	330	650
	Dry	.47	8,800	.88	8.3	22	6,020	920	—	—	900
Incense-	Green	.35	6,200	.84	6.4	17	3,150	370	830	280	390
	Dry	.37	8,000	1.04	5.4	17	5,200	590	880	270	470
Northern white-	Green	.29	4,200	.64	5.7	15	1,990	230	620	240	230
	Dry	.31	6,500	.80	4.8	12	3,960	310	850	240	320
Port-Orford-	Green	.39	6,600	1.30	7.4	21	3,140	300	840	180	380
	Dry	.43	12,700	1.70	9.1	28	6,250	720	1,370	400	630
Western redcedar	Green	.31	5,200	.94	5.0	17	2,770	240	770	230	260
	Dry	.32	7,500	1.11	5.8	17	4,560	460	990	220	350
Douglas-fir.[5]											
Coast	Green	.45	7,700	1.56	7.6	26	3,780	380	900	300	500
	Dry	.48	12,400	1.95	9.9	31	7,230	800	1,130	340	710
Interior West	Green	.46	7,700	1.51	7.2	26	3,870	420	940	290	510
	Dry	.50	12,600	1.83	10.6	32	7,430	760	1,290	350	660
Interior North	Green	.45	7,400	1.41	8.1	22	3,470	360	950	340	420
	Dry	.48	13,100	1.79	10.5	26	6,900	770	1,400	390	600
Interior South	Green	.43	6,800	1.16	8.0	15	3,110	340	950	250	360
	Dry	.46	11,900	1.49	9.0	20	6,230	740	1,510	330	510

Table 4–2—*Mechanical properties[1] of some commercially important woods grown in the United States*—Continued

Common names of species	Moisture condition	Specific gravity[2]	Static bending			Impact bending—height of drop causing complete failure[4]	Compression parallel to grain—maximum crushing strength	Compression perpendicular to grain—fiber stress at proportional limit	Shear parallel to grain—maximum shearing strength	Tension perpendicular to grain—maximum tensile strength	Side hardness—load perpendicular to grain
			Modulus of rupture	Modulus of elasticity[3]	Work to maximum load						
			Psi	Million psi	In–lb per in³	In	Psi	Psi	Psi	Psi	Lb

SOFTWOODS—continued

Fir:											
Balsam	Green	.33	5,500	1.25	4.7	16	2,630	190	662	180	290
	Dry	.35	9,200	1.45	5.1	20	5,280	404	944	180	400
California red	Green	.36	5,800	1.17	6.4	21	2,760	330	770	380	360
	Dry	.38	10,500	1.50	8.9	24	5,460	610	1,040	390	500
Grand	Green	.35	5,800	1.25	5.6	22	2,940	270	740	240	360
	Dry	.37	8,900	1.57	7.5	28	5,290	500	900	240	490
Noble	Green	.37	6,200	1.38	6.0	19	3,010	270	800	230	290
	Dry	.39	10,700	1.72	8.8	23	6,100	520	1,050	220	410
Pacific silver	Green	.40	6,400	1.42	6.0	21	3,140	220	750	240	310
	Dry	.43	11,000	1.76	9.3	24	6,410	450	1,220	—	430
Subalpine	Green	.31	4,900	1.05	—	—	2,300	190	700	—	260
	Dry	.32	8,600	1.29	—	—	4,860	390	1,070	—	350
White	Green	.37	5,900	1.16	5.6	22	2,900	280	760	300	340
	Dry	.39	9,800	1.50	7.2	20	5,800	530	1,100	300	480
Hemlock:											
Eastern	Green	.38	6,400	1.07	6.7	21	3,080	360	850	230	400
	Dry	.40	8,900	1.20	6.8	21	5,410	650	1,060	—	500
Mountain	Green	.42	6,300	1.04	11.0	32	2,880	370	930	330	470
	Dry	.45	11,500	1.33	10.4	32	6,440	860	1,540	—	680
Western	Green	.42	6,600	1.31	6.9	22	3,360	280	860	290	410
	Dry	.45	11,300	1.63	8.3	23	7,200	550	1,290	340	540
Larch, western	Green	.48	7,700	1.46	10.3	29	3,760	400	870	330	510
	Dry	.52	13,000	1.87	12.6	35	7,620	930	1,360	430	830

Table 4 – 2—Mechanical properties[1] of some commercially important woods grown in the United States—Continued

Common names of species	Moisture condition	Specific gravity[2]	Static bending			Impact bending—height of drop causing complete failure[4]	Compression parallel to grain—maximum crushing strength	Compression perpendicular to grain—fiber stress at proportional limit	Shear parallel to grain—maximum shearing strength	Tension perpendicular to grain—maximum tensile strength	Side hardness—load perpendicular to grain
			Modulus of rupture	Modulus of elasticity[3]	Work to maximum load						
			Psi	Million psi	In–lb per in³	In			Psi			Lb

SOFTWOODS—continued

Common names of species	Moisture condition	Specific gravity[2]	Modulus of rupture (Psi)	Modulus of elasticity[3] (Million psi)	Work to maximum load (In–lb per in³)	Impact bending (In)	Compression parallel to grain (Psi)	Compression perpendicular to grain (Psi)	Shear parallel to grain (Psi)	Tension perpendicular to grain (Psi)	Side hardness (Lb)
Pine:											
Eastern white	Green	.34	4,900	.99	5.2	17	2,440	220	680	250	290
	Dry	.35	8,600	1.24	6.8	18	4,800	440	900	310	380
Jack	Green	.40	6,000	1.07	7.2	26	2,950	300	750	360	400
	Dry	.43	9,900	1.35	8.3	27	5,660	580	1,170	420	570
Loblolly	Green	.47	7,300	1.40	8.2	30	3,510	390	860	260	450
	Dry	.51	12,800	1.79	10.4	30	7,130	790	1,390	470	690
Lodgepole	Green	.38	5,500	1.08	5.6	20	2,610	250	680	220	330
	Dry	.41	9,400	1.34	6.8	20	5,370	610	880	290	480
Longleaf	Green	.54	8,500	1.59	8.9	35	4,320	480	1,040	330	590
	Dry	.59	14,500	1.98	11.8	34	8,470	960	1,510	470	870
Pitch	Green	.47	6,800	1.20	9.2	—	2,950	360	860	—	—
	Dry	.52	10,800	1.43	9.2	—	5,940	820	1,360	—	—
Pond	Green	.51	7,400	1.28	7.5	—	3,660	440	940	—	—
	Dry	.56	11,600	1.75	8.6	—	7,540	910	1,380	—	—
Ponderosa	Green	.38	5,100	1.00	5.2	21	2,450	280	700	310	320
	Dry	.40	9,400	1.29	7.1	19	5,320	580	1,130	420	460
Red	Green	.41	5,800	1.28	6.1	26	2,730	260	690	300	340
	Dry	.46	11,000	1.63	9.9	26	6,070	600	1,210	460	560
Sand	Green	.46	7,500	1.02	9.6	—	3,440	450	1,140	—	—
	Dry	.48	11,600	1.41	9.6	—	6,920	836	—	—	—
Shortleaf	Green	.47	7,400	1.39	8.2	30	3,530	350	910	320	440
	Dry	.51	13,100	1.75	11.0	33	7,270	820	1,390	470	690
Slash	Green	.54	8,700	1.53	9.6	—	3,820	530	960	—	—
	Dry	.59	16,300	1.98	13.2	—	8,140	1,020	1,680	—	—
Spruce	Green	.41	6,000	1.00	—	—	2,840	280	900	—	450
	Dry	.44	10,400	1.23		—	5,650	730	1,490	—	660
Sugar	Green	.34	4,900	1.03	5.4	17	2,460	210	720	270	270
	Dry	.36	8,200	1.19	5.5	18	4,460	500	1,130	350	380
Virginia	Green	.45	7,300	1.22	10.9	34	3,420	390	890	400	540
	Dry	.48	13,000	1.52	13.7	32	6,710	910	1,350	380	740

Table 4–2—Mechanical properties[1] of some commercially important woods grown in the United States—Continued

Common names of species	Moisture condition	Specific gravity[2]	Static bending			Impact bending—height of drop causing complete failure[4]	Compression parallel to grain—maximum crushing strength	Compression perpendicular to grain—fiber stress at proportional limit	Shear parallel to grain—maximum shearing strength	Tension perpendicular to grain—maximum tensile strength	Side hardness—load perpendicular to grain
			Modulus of rupture	Modulus of elasticity[3]	Work to maximum load						
			Psi	Million psi	In–lb per in³	In	In ---------- Psi ----------				Lb
SOFTWOODS—continued											
Pine—con.											
Western white	Green	.35	4,700	1.19	5.0	19	2,430	190	680	260	260
	Dry	.38	9,700	1.46	8.8	23	5,040	470	1,040	—	420
Redwood:											
Old-growth	Green	.38	7,500	1.18	7.4	21	4,200	420	800	260	410
	Dry	.40	10,000	1.34	6.9	19	6,150	700	940	240	480
Young-growth	Green	.34	5,900	.96	5.7	16	3,110	270	890	300	350
	Dry	.35	7,900	1.10	5.2	15	5,220	520	1,110	250	420
Spruce:											
Black	Green	.38	6,100	1.38	7.4	24	2,840	240	739	100	370
	Dry	.42	10,800	1.61	10.5	23	5,960	550	1,230	—	520
Engelmann	Green	.33	4,700	1.03	5.1	16	2,180	200	640	240	260
	Dry	.35	9,300	1.30	6.4	18	4,480	410	1,200	350	390
Red	Green	.37	6,000	1.33	6.9	18	2,720	260	750	220	350
	Dry	.40	10,800	1.61	8.4	25	5,540	550	1,290	350	490
Sitka	Green	.37	5,700	1.23	6.3	24	2,670	280	760	250	350
	Dry	.40	10,200	1.57	9.4	25	5,610	580	1,150	370	510
White	Green	.33	5,000	1.14	6.0	22	2,350	210	640	220	320
	Dry	.36	9,400	1.43	7.7	20	5,180	430	970	360	480
Tamarack	Green	.49	7,200	1.24	7.2	28	3,480	390	860	260	380
	Dry	.53	11,600	1.64	7.1	23	7,160	800	1,280	400	590

[1] Results of tests on small, clear, straight-grained specimens. Values in the first line for each species are from tests of green material; those in the second line are from tests of seasoned material adjusted to a moisture content of 12 percent.

[2] Specific gravity based on weight ovendry and volume at moisture content indicated.

[3] Modulus of elasticity measured from a simply supported, center-loaded beam, on a span-depth ratio of 14/1. The modulus can be corrected for the effect of shear deflection by increasing it 10 percent.

[4] 50-pound hammer.

[5] Coast Douglas-fir is defined as Douglas-fir growing in the States of Oregon and Washington west of the summit of the Cascade Mountains. Interior West includes the State of California and all counties in Oregon and Washington east of but adjacent to the Cascade summit. Interior North includes the remainder of Oregon and Washington and the States of Idaho, Montana, and Wyoming. Interior South is made up of Utah, Colorado, Arizona, and New Mexico.

Table 4–3—_Mechanical properties of some commercially important woods grown in Canada and imported into the United States[1,2]_

Common names of species	Specific gravity	Static bending		Compression parallel to grain— maximum crushing strength	Compression perpendicular to grain— fiber stress at proportional limit	Shear parallel to grain— maximum shearing strength
		Modulus of rupture	Modulus of elasticity[3]			
		Psi	Million Psi	- - - - - - - - - - - - - Psi - - - - - - - - - - - - -		

HARDWOODS

Aspen:						
Quaking	0.37	5,500	1.31	2,350	200	720
		9,800	1.63	5,260	510	980
Big-toothed	.39	5,300	1.08	2,390	210	790
		9,500	1.26	4,760	470	1,100
Cottonwood:						
Balsam, poplar	.37	5,000	1.15	2,110	180	670
		10,100	1.67	5,020	420	890
Black	.30	4,100	.97	1,860	100	560
		7,100	1.28	4,020	260	860
Eastern	.35	4,700	.87	1,970	210	770
		7,500	1.13	3,840	470	1,160

SOFTWOODS

Cedar:						
Alaska-	.42	6,600	1.34	3,240	350	880
		11,600	1.59	6,640	690	1,340
Northern white-	.30	3,900	.52	1,890	200	660
		6,100	.63	3,590	390	1,000
Western redcedar	.31	5,300	1.05	2,780	280	700
		7,800	1.19	4,290	500	810
Douglas-fir	.45	7,500	1.61	3,610	460	920
		12,800	1.97	7,260	870	1,380
Fir:						
Subalpine	.33	5,200	1.26	2,500	260	680
		8,200	1.48	5,280	540	980
Pacific silver	.36	5,500	1.35	2,770	230	710
		10,000	1.64	5,930	520	1,190
Balsam	.34	5,300	1.13	2,440	240	680
		8,500	1.40	4,980	460	910
Hemlock:						
Eastern	.40	6,800	1.27	3,430	400	910
		9,700	1.41	5,970	630	1,260
Western	.41	7,000	1.48	3,580	370	750
		11,800	1.79	6,770	660	940

Common names of species	Specific gravity	Static bending		Compression parallel to grain— maximum crushing strength	Compression perpendicular to grain— fiber stress at proportional limit	Shear parallel to grain— maximum shearing strength
		Modulus of rupture	Modulus of elasticity[3]			
		Psi	*Million Psi*	- - - - - - - - - - - *Psi* - - - - - - - - - - - - -		

SOFTWOODS—continued

Common names of species	Specific gravity	Modulus of rupture	Modulus of elasticity[3]	Compression parallel maximum crushing strength	Compression perpendicular stress at proportional limit	Shear parallel maximum shearing strength
Larch, western	.55	8,700	1.65	4,420	520	920
		15,500	2.08	8,840	1,060	1,340
Pine:						
Eastern white	.36	5,100	1.18	2,590	240	640
		9,500	1.36	5,230	490	880
Jack	.42	6,300	1.17	2,950	340	820
		11,300	1.48	5,870	830	1,190
Lodgepole	.40	5,600	1.27	2,860	280	720
		11,000	1.58	6,260	530	1,240
Ponderosa	.44	5,700	1.13	2,840	350	720
		10,600	1.38	6,130	760	1,020
Red	.39	5,000	1.07	2,370	280	710
		10,100	1.38	5,500	720	1,090
Western white	.36	4,800	1.19	2,520	240	650
		9,300	1.46	5,240	470	920
Spruce:						
Black	.41	5,900	1.32	2,760	300	800
		11,400	1.52	6,040	620	1,250
Engelmann	.38	5,700	1.25	2,810	270	700
		10,100	1.55	6,150	540	1,100
Red	.38	5,900	1.32	2,810	270	810
		10,300	1.60	5,590	550	1,330
Sitka	.35	5,400	1.37	2,560	290	630
		10,100	1.63	5,480	590	980
White	.35	5,100	1.15	2,470	240	670
		9,100	1.45	5,360	500	980
Tamarack	.48	6,800	1.24	3,130	410	920
		11,000	1.36	6,510	900	1,300

[1]Results of tests on small, clear, straight-grained specimens. Property values based on American Society for Testing and Materials Standard D 2555−70, "Standard methods for establishing clear wood values." Information on additional properties can be obtained from Department of Forestry, Canada, Publication No. 1104.

[2]The values in the first line for each species are from tests of green material; those in the second line are adjusted from the green condition to 12 percent moisture content using dry to green clear wood property ratios as reported in ASTM D 2555−70. Specific gravity is based on weight when ovendry and volume when green.

[3]Modulus of elasticity measured from a simply supported, center-loaded beam, on a span-depth ratio of 14/1. The modulus can be corrected for the effect of shear deflection by increasing it 10 percent.

Table 4–4—Mechanical properties[1],[2] of some woods imported into the United States

Common and botanical names of species	Moisture content	Specific gravity[3]	Static bending			Compression parallel to grain—maximum crushing strength	Shear parallel to grain—maximum shearing strength	Side hardness—load perpendicular to grain	Sample origin[5]
			Modulus of rupture	Modulus of elasticity[4]	Work to maximum load				
	Pct		Psi	Million Psi	In–lb per in³	- - - - Psi - - - -		Lb	
Afrormosia (Pericopsis elata)	Green	0.61	14,800	1.77	19.5	7,490	1,670	1,600	AF
	12		18,400	1.94	18.4	9,940	2,090	1,560	
Albarco (Cariniana spp.)	Green	.48	—	—	—	—	—	—	AM
	12		14,500	1.50	13.8	6,820	2,310	1,020	
Andiroba (Carapa guianensis)	Green	.54	10,300	1.69	9.8	4,780	1,220	880	AM
	12	—	15,500	2.00	14.0	8,120	1,510	1,130	
Angelin (Andira spp.)	Green	.65	—	—	—	—	—	—	AM
	12		18,000	2.49	—	9,200	1,840	1,750	
Angelique (Dicorynia guianensis)	Green	.60	11,400	1.84	12.0	5,590	1,340	1,100	AM
	12		17,400	2.19	15.2	8,770	1,660	1,290	
Avodire (Turraeanthus africanus)	Green	.48	—	—	—	—	—	—	AF
	12		12,700	1.49	9.4	7,150	2,030	1,080	
Azobe (Lophira alata)	Green	.87	16,900	2.16	12.0	9,520	2,040	2,890	AF
	12		24,500	2.47	—	12,600	2,960	3,350	
Balsa (Ochroma pyramidale)	Green	.16	—	—	—	—	—	—	AM
	12		3,140	.49	2.1	2,160	300	—	
Banak (Virola spp.)	Green	.42	5,600	1.64	4.1	2,390	720	320	AM
	12	—	10,900	2.04	10.0	5,140	980	510	
Benge (Guibourtia arnoldiana)	Green	.65	—	—	—	—	—	—	AF
	12		21,400	2.04	—	11,400	2,090	1,750	
Bubinga (Guibourtia spp.)	Green	.71	—	—	—	—	—	—	AF
	12		22,600	2.48	—	10,500	3,110	2,690	
Bulletwood (Manilkara bidentata)	Green	.85	17,300	2.70	13.6	8,690	1,900	2,230	AM
	12		27,300	3.45	28.5	11,640	2,500	3,190	

Table 4—4—Mechanical properties[1], [2] of some woods imported into the United States—Continued

Common and botanical names of species	Moisture content	Specific gravity[3]	Static bending			Compression parallel to grain—maximum crushing strength	Shear parallel to grain—maximum shearing strength	Side hardness—load perpendicular to grain	Sample origin[5]
			Modulus of rupture	Modulus of elasticity[4]	Work to maximum load				
	Pct		Psi	Million Psi	In—lb per in³	- - - - Psi - - - -		Lb	
Cativo (*Prioria copaifera*)	Green	.40	5,900	.94	5.4	2,460	860	440	AM
	12	—	8,600	1.11	7.2	4,290	1,060	630	
Ceiba (*Ceiba pentandra*)	Green	.25	2,200	.41	1.2	1,060	350	220	AM
	12	—	4,300	.54	2.8	2,380	550	240	
Courbaril (*Hymenaea courbaril*)	Green	.71	12,900	1.84	14.6	5,800	1,770	1,970	AM
	12	—	19,400	2.16	17.6	9,510	2,470	2,350	
Cuangare (*Dialyanthera* spp.)	Green	.31	4,000	1.01	—	2,080	590	230	AM
	12		7,300	1.52	—	4,760	830	380	
Cypress, Mexican (*Cupressus lusitanica*)	Green	.39	6,200	.92	—	2,880	950	340	AF[6]
	12		10,300	1.02	—	5,380	1,580	460	
Degame (*Calycophyllum candidissimum*)	Green	.67	14,300	1.93	18.6	6,200	1,660	1,630	AM
	12		22,300	2.27	27.0	9,670	2,120	1,940	
Determa (*Ocotea rubra*)	Green	.52	7,800	1.46	4.8	3,760	860	520	AM
	12		10,500	1.82	6.4	5,800	980	660	
Ekop (*Tetraberlinia tubmaniana*)	Green	.60	—	—	—	—	—	—	AF
	12		16,700	2.21	—	9,010	—	—	
Goncalo alves (*Astronium graveolens*)	Green	.84	12,100	1.94	6.7	6,580	1,760	1,910	AM
	12	—	16,600	2.23	10.4	10,320	1,960	2,160	
Greenheart (*Ocotea rodiaei*)	Green	.80	19,300	2.47	10.5	9,380	1,930	1,880	AM
	12		24,900	3.25	25.3	12,510	2,620	2,350	
Hura (*Hura crepitans*)	Green	.38	6,300	1.04	5.9	2,790	830	440	AM
	12		8,700	1.17	6.7	4,800	1,080	550	
Ilomba (*Pycnanthus angolensis*)	Green	.40	5,500	1.14	—	2,900	840	470	AF
	12		9,900	1.59	—	5,550	1,290	610	

Table 4 – 4—*Mechanical properties[1], [2] of some woods imported into the United States*—Continued

Common and botanical names of species	Moisture content	Specific gravity[3]	Static bending			Compression parallel to grain—maximum crushing strength	Shear parallel to grain—maximum shearing strength	Side hardness—load perpendicular to grain	Sample origin[5]
			Modulus of rupture	Modulus of elasticity[4]	Work to maximum load				
	Pct		Psi	Million Psi	In—lb per in³	- - - - Psi - - - -		Lb	
Ipe (*Tabebuia* spp. —Lapacho group)	Green	.92	22,600	2.92	27.6	10,350	2,120	3,060	AM
	12		25,400	3.14	22.0	13,010	2,060	3,680	
Iroko (*Chlorophora* spp.)	Green	.54	10,200	1.29	10.5	4,910	1,310	1,080	AF
	12		12,400	1.46	9.0	7,590	1,800	1,260	
Jarrah (*Eucalyptus marginata*)	Green	.67	9,900	1.48	—	5,190	1,320	1,290	AS
	12	—	16,200	1.88	—	8,870	2,130	1,910	
Jelutong (*Dyera costulata*)	Green	.36	5,600	1.16	5.6	3,050	760	330	AS
	15		7,300	1.18	6.4	3,920	840	390	
Kaneelhart (*Licaria* spp.)	Green	.96	22,300	3.82	13.6	13,390	1,680	2,210	AM
	12		29,900	4.06	17.5	17,400	1,970	2,900	
Kapur (*Dryobalanops* spp.)	Green	.64	12,800	1.60	15.7	6,220	1,170	980	AS
	12		18,300	1.88	18.8	10,090	1,990	1,230	
Karri (*Eucalyptus diversicolor*)	Green	.82	11,200	1.94	11.6	5,450	1,510	1,360	AS
	12		20,160	2.60	25.4	10,800	2,420	2,040	
Kempas (*Koompassia malaccenis*)	Green	.71	14,500	2.41	12.2	7,930	1,460	1,480	AS
	12		17,700	2.69	15.3	9,520	1,790	1,710	
Keruing (*Dipterocarpus* spp.)	Green	.69	11,900	1.71	13.9	5,680	1,170	1,060	AS
	12		19,900	2.07	23.5	10,500	2,070	1,270	
Lignumvitae (*Guaiacum* spp.)	Green	1.05	—	—	—	—	—	—	AM
	12	—	—	—	—	11,400	—	4,500	
Limba (*Terminalia superba*)	Green	.38	6,000	.77	7.7	2,780	880	400	AF
	12		8,800	1.01	8.9	4,730	1,410	490	
Macawood (*Platymiscium* spp.)	Green	.94	22,300	3.02	—	10,540	1,840	3,320	AM
	12		27,600	3.20	—	16,100	2,540	3,150	

Table 4-4—*Mechanical properties[1, 2] of some woods imported into the United States*—Continued

Common and botanical names of species	Moisture content	Specific gravity[3]	Static bending		Work to maximum load	Compression parallel to grain—maximum crushing strength	Shear parallel to grain—maximum shearing strength	Side hardness—load perpendicular to grain	Sample origin[5]
			Modulus of rupture	Modulus of elasticity[4]					
	Pct		Psi	Million Psi	In—lb per in³	- - - - Psi - - - -		Lb	
Mahogany, African (*Khaya* spp.)	Green	.42	7,400	1.15	7.1	3,730	931	640	AF
	12		10,700	1.40	8.3	6,460	1,500	830	
Mahogany, Honduras (*Swietenia macrophylla*)	Green	.45	9,000	1.34	9.1	4,340	1,240	740	AM
	12	—	11,500	1.50	7.5	6,780	1,230	800	
Manbarklak (*Eschweilera* spp.)	Green	.87	17,100	2.70	17.4	7,340	1,630	2,280	AM
	12		26,500	3.14	33.3	11,210	2,070	3,480	
Manni (*Symphonia globulifera*)	Green	.58	11,200	1.96	11.2	5,160	1,140	940	AM
	12		16,900	2.46	16.5	8,820	1,420	1,120	
Marishballi (*Lincania* spp.)	Green	.88	17,100	2.93	13.4	7,580	1,620	2,250	AM
	12		27,700	3.34	14.2	13,390	1,750	3,570	
Merbau (*Intsia* spp.)	Green	.64	12,900	2.02	12.8	6,770	1,560	1,380	AS
	15	—	16,800	2.23	14.8	8,440	1,810	1,500	
Mersawa (*Anisoptera* spp.)	Green	.52	8,000	1.77	—	3,960	740	880	AS
	12		13,800	2.28	—	7,370	890	1,290	
Mora (*Mora* spp.)	Green	.78	12,600	2.33	13.5	6,400	1,400	1,450	AM
	12		22,100	2.96	18.5	11,840	1,900	2,300	
Oak (*Quercus* spp.)	Green	.76	—	—	—	—	—	—	AM
	12		23,000	3.02	16.5	—	—	2,500	
Obeche (*Triplochiton scleroxylon*)	Green	.30	5,100	.72	6.2	2,570	660	420	AF
	12		7,400	.86	6.9	3,930	990	430	
Okoume (*Aucoumea klaineana*)	Green	.33	—	—	—	—	—	—	AF
	12		7,400	1.14	—	3,970	970	380	
Opepe (*Nauclea diderrichii*)	Green	.63	13,600	1.73	12.2	7,480	1,900	1,520	AF
	12		17,400	1.94	14.4	10,400	2,480	1,630	

4-21

Table 4–4—*Mechanical properties[1], [2] of some woods imported into the United States—Continued*

Common and botanical names of species	Moisture content	Specific gravity[3]	Static bending			Compression parallel to grain—maximum crushing strength	Shear parallel to grain—maximum shearing strength	Side hardness—load perpendicular to grain	Sample origin[5]
			Modulus of rupture	Modulus of elasticity[4]	Work to maximum load				
	Pct		Psi	Million Psi	In–lb per in³	- - - - Psi - - - -		Lb	
Ovangkol (*Guibourtia ehie*)	Green	.67	—	—	—	—	—	—	AF
	12		16,900	2.56	—	8,300	—	—	
Para-angelium (*Hymenolobium excelsum*)	Green	.63	14,600	1.95	12.8	7,460	1,600	1,720	AM
	12		17,600	2.05	15.9	8,990	2,010	1,720	
Parana-pine (*Araucaria augustifolia*)	Green	.46	7,200	1.35	9.7	4,010	970	560	AM
	12	—	13,500	1.61	12.2	7,660	1,730	780	
Pau marfim (*Balfourodendron riedelianum*)	Green	.73	14,400	1.66	—	6,070	—	—	AM
	15		18,900	—	—	8,190	—	—	
Peroba de campos (*Paratecoma peroba*)	Green	.62	—	—	—	—	—	—	AM
	12		15,400	1.77	10.1	8,880	2,130	1,600	
Peroba rosa (*Aspidosperma* spp.—Peroba group)	Green	.66	10,900	1.29	10.5	5,540	1,880	1,580	AM
	12		12,100	1.53	9.2	7,920	2,490	1,730	
Pilon (*Hyeronima* spp.)	Green	.65	10,700	1.88	8.3	4,960	1,200	1,220	AM
	12		18,200	2.27	12.1	9,620	1,720	1,700	
Pine, Caribbean (*Pinus caribaea*)	Green	.68	11,200	1.88	10.7	4,900	1,170	980	AM
	12	—	16,700	2.24	17.3	8,540	2,090	1,240	
Pine, ocote (*Pinus oocarpa*)	Green	.55	8,000	1.74	6.9	3,690	1,040	580	AM
	12	—	14,900	2.25	10.9	7,680	1,720	910	
Pine, radiata (*Pinus radiata*)	Green	.42	6,100	1.18	—	2,790	750	480	AS[6]
	12	—	11,700	1.48	—	6,080	1,600	750	
Piquia (*Caryocar* spp.)	Green	.72	12,400	1.82	8.4	6,290	1,640	1,720	AM
	12		17,000	2.16	15.8	8,410	1,990	1,720	
Primavera (*Cybistax donnell-smithii*)	Green	.40	7,200	.99	7.2	3,510	1,030	700	AM
	12		9,500	1.04	6.4	5,600	1,390	660	

Table 4 – 4—*Mechanical properties*[1], [2] *of some woods imported into the United States*—Continued

Common and botanical names of species	Moisture content	Specific gravity[3]	Static bending			Compression parallel to grain—maximum crushing strength	Shear parallel to grain—maximum shearing strength	Side hardness—load perpendicular to grain	Sample origin[5]
			Modulus of rupture	Modulus of elasticity[4]	Work to maximum load				
	Pct		Psi	Million Psi	In—lb per in³	- - - - Psi - - - -		Lb	
Purpleheart (*Peltogyne* spp.)	Green	.67	13,700	2.00	14.8	7,020	1,640	1,810	AM
	12		19,200	2.27	17.6	10,320	2,220	1,860	
Ramin (*Gonystylus* spp.)	Green	.52	9,800	1.57	9.0	5,390	990	640	AS
	12	—	18,500	2.17	17.0	10,080	1,520	1,300	
Robe (*Tabebuia* spp.—Roble group)	Green	.52	10,800	1.45	11.7	4,910	1,250	910	AM
	12		13,800	1.60	12.5	7,340	1,450	960	
Rosewood, Brazilian (*Dalbergia nigra*)	Green	.80	14,100	1.84	13.2	5,510	2,360	2,440	AM
	12	—	19,000	1.88	—	9,600	2,110	2,720	
Rosewood, Indian (*Dalbergia latifolia*)	Green	.75	9,200	1.19	11.6	4,530	1,400	1,560	AS
	12		16,900	1.78	13.1	9,220	2,090	3,170	
Sande (*Brosimum* spp.—Utile group)	Green	.49	8,500	1.94	—	4,490	1,040	600	AM
	12		14,300	2.39	—	8,220	1,290	900	
Santa Maria (*Calophyllum brasiliense*)	Green	.52	10,500	1.59	12.7	4,560	1,260	890	AM
	12	—	14,600	1.83	16.1	6,910	2,080	1,150	
Sapele (*Entandrophragma cylindricum*)	Green	.55	10,200	1.49	10.5	5,010	1,250	1,020	AF
	12	—	15,300	1.82	15.7	8,160	2,280	1,510	
Sepetir (*Pseudosindora palustris*)	Green	.56	11,200	1.57	13.3	5,460	1,310	950	AS
	12		17,200	1.97	13.3	8,880	2,030	1,410	
Shorea (*Shorea* spp.—Baulau group)	Green	.68	11,700	2.10	—	5,380	1,440	1,350	AS
	12		18,800	2.61	—	10,180	2,190	1,780	
Shorea-Lauan-Meranti group, Dark red (*Shorea* spp.)	Green	.46	9,400	1.50	8.6	4,720	1,110	700	AS
	12		12,700	1.77	13.8	7,360	1,450	780	
Shorea-Lauan-Meranti group, Light red (*Shorea* spp.)	Green	.34	6,600	1.04	6.2	3,330	710	440	AS
	12		9,500	1.23	8.6	5,920	970	460	

Table 4 – 4—*Mechanical properties[1], [2] of some woods imported into the United States*—Continued

Common and botanical names of species	Moisture content	Specific gravity[3]	Static bending			Compression parallel to grain—maximum crushing strength	Shear parallel to grain—maximum shearing strength	Side hardness—load perpendicular to grain	Sample origin[5]
			Modulus of rupture	Modulus of elasticity[4]	Work to maximum load				
	Pct		Psi	Million Psi	In–lb per in³	- - - - Psi - - - -		Lb	
Shorea-White Meranti group (*Shorea* spp.)	Green	.55	9,800	1.30	8.3	5,490	1,320	1,000	AS
	15		12,400	1.49	11.4	6,350	1,540	1,140	AS
Shorea-Yellow Meranti group (*Shorea* spp.)	Green	.46	8,000	1.30	8.1	3,880	1,030	750	AS
	12		11,400	1.55	10.1	5,900	1,520	770	AS
Spanish-cedar (*Cedrela* spp.)	Green	.41	7,500	1.31	7.1	3,370	990	550	AM
	12	—	11,500	1.44	9.4	6,210	1,100	600	AM
Sucupira (*Bowdichia* spp.)	Green	.74	17,200	2.27	—	9,730	—	—	AM
	15		19,400	—	—	11,100	—	—	AM
Sucupira (*Diplotropis purpurea*)	Green	.78	17,400	2.68	13.0	8,020	1,800	1,980	AM
	12		20,600	2.87	14.8	12,140	1,960	2,140	AM
Teak (*Tectona grandis*)	Green	.55	11,600	1.37	13.4	5,960	1,290	930	AS
	12		14,600	1.55	12.0	8,410	1,890	1,000	AS
Tornillo (*Cedrelinga catenaeformis*)	Green	.45	8,400	—	—	4,100	1,170	870	AM
	12	—	—	—	—	—	—	—	AM
Wallaba (*Eperua* spp.)	Green	.78	14,300	2.33	—	8,040	—	1,540	AM
	12	—	19,100	2.28	—	10,760	—	2,040	AM

[1] Results of tests on small, clear, straight-grained specimens. Property values were taken from world literature (not obtained from experiments conducted at the Forest Products Laboratory). Other species may be reported in the world literature, as well as additional data on many of these species.
[2] Some property values have been adjusted to 12 percent moisture content; others are based on moisture content at time of test.
[3] Specific gravity based on weight ovendry and volume green.
[4] Modulus of elasticity measured from a simply supported, center-loaded beam, on a span-depth ratio of 14/1. The modulus can be corrected for the effect of shear deflection by increasing it 10 percent.
[5] Key to code letters: AF, Africa; AM, Tropical America; AS, Asia.
[6] Plantation grown.

Table 4—5—Average coefficient of variation for some mechanical properties of clear wood

Property	Coefficient of variation[1]
	Pct
Static bending:	
Fiber stress at proportional limit	22
Modulus of rupture	16
Modulus of elasticity	22
Work to maximum load	34
Impact bending, height of drop causing complete failure	25
Compression parallel to grain:	
Fiber stress at proportional limit	24
Maximum crushing strength	18
Compression perpendicular to grian, fiber stress at proportional limit	28
Shear parallel to grain, maximum shearing strength	14
Tension perpendicular to grain, maximum tensile strength	25
Hardness:	
Perpendicular to grain	20
Toughness	34
Specific gravity	10

[1]Values given are based on results of tests of green wood from approximately 50 species. Values for wood adjusted to 12 percent moisture content may be assumed to be approximately of the same magnitude.

Table 4—6—Average parallel-to-grain tensile strength of some species of wood[1]

Species	Tensile strength
	Psi
HARDWOODS	
Beech, American	12,500
Elm:	
Cedar	17,500
Maple, sugar	15,700
Oak:	
Overcup	11,300
Pin	16,300
Poplar, balsam	7,400
Sweetgum	13,600
Willow, black	10,600
Yellow-poplar	15,900
SOFTWOODS	
Baldcypress	8,500
Cedar:	
Port-Orford	11,400
Western redcedar	6,600
Douglas-fir, Interior North	15,600
Fir:	
California red	11,300
Pacific silver	13,800
Hemlock, western	13,000
Larch, western	16,200
Pine:	
Eastern white	10,600
Loblolly	11,600
Ponderosa	8,400
Virginia	13,700
Redwood:	
Virgin	9,400
Young-growth	9,100
Spruce:	
Engelmann	12,300
Sitka	8,600

[1]Results of tests on small, clear, straight-grained specimens tested in the green moisture condition. For the hardwood species, specimens tested at 12 percent moisture content averaged about 32 percent higher; softwoods, about 13 percent.

Species	Moisture content	Specific gravity[2]	Toughness[3]	
			Radial	Tangential
	Pct		In–lb	In–lb
HARDWOODS				
Birch, yellow	12	0.65	500	620
Hickory:				
(Mockernut, pignut, sand)	Green	.64	700	720
	12	.71	620	660
Maple, sugar	14	.64	370	360
Oak, red:				
Pin	12	.64	430	430
Scarlet	11	.66	510	440
Oak, white:				
Overcup	Green	.56	730	680
	13	.62	340	310
Sweetgum	Green	.48	340	330
	13	.51	260	260
Willow, black	Green	.38	310	360
	11	.40	210	230
Yellow-poplar	Green	.43	320	300
	12	.45	220	210
SOFTWOODS				
Cedar:				
Alaska-	10	.48	210	230
Western redcedar	9	.33	90	130
Douglas-fir:				
Coast	Green	.44	210	360
	12	.47	200	360
Interior West	Green	.48	200	300
	13	.51	210	340
Interior North	Green	.43	170	240
	14	.46	160	250
Interior South	Green	.38	130	180
	14	.40	120	180
Fir:				
California red	Green	.36	130	180
	12	.39	120	170
Noble	Green	.36	—	240
	12	.39	—	220
Pacific silver	Green	.37	150	230
	13	.40	170	260
White	Green	.36	140	220
	13	.38	130	200

Species	Moisture content	Specific gravity[2]	Toughness[3]	
			Radial	Tangential
	Pct		In–lb	In–lb
SOFTWOODS—continued				
Hemlock:				
Mountain	Green	.41	250	280
	14	.44	140	170
Western	Green	.38	150	170
	12	.41	140	210
Larch, western	Green	.51	270	400
	12	.55	210	340
Pine:				
Eastern white	Green	.33	120	160
	12	.34	110	120
Jack	Green	.41	200	380
	12	.42	140	240
Loblolly	Green	.48	310	380
	12	.51	160	260
Lodgepole	Green	.38	160	210
Ponderosa	Green	.38	190	270
	11	.43	150	190
Red	Green	.40	210	350
	12	.43	160	290
Shortleaf	Green	.47	290	400
	13	.50	150	230
Slash	Green	.55	350	450
	12	.59	210	320
Virginia	Green	.45	340	470
	12	.49	170	250
Redwood:				
Old-growth	Green	.39	110	200
	11	.39	90	140
Young-growth	Green	.33	110	140
	12	.34	90	110
Spruce, Engelmann	Green	.34	150	190
	12	.35	110	180

[1] Results of tests on small, clear, straight-grained specimens.
[2] Based on ovendry weight and volume at moisture content of test.
[3] Properties based on specimen size of 2 centimeters square by 28 centimeters long; radial indicates load applied to radial face and tangential indicates load applied to tangential face of specimens.

(This discussion began on page 4-5.)

Approximate relationships between various mechanical properties and specific gravity for clear straight-grained wood of hardwoods and softwoods are given in table 4—8 as power functions. Those relationships are based on average values for the 43 softwood and 66 hardwood species presented in table 4—2. It is recognized that the average data vary around the relationships, so that individual average species values or an individual specimen value are not accurately predicted by the relationships. In fact, mechanical properties within a species tend to be linearly related with specific gravity rather than curvilinearly, and where data are available for individual species, linear analysis is suggested.

Knots

A knot is that portion of a branch that has become incorporated in the bole of a tree. The influence of a knot on mechanical properties of a wood member is due to the interruption of continuity and change in direction of wood fibers associated with the knot. The influence of knots depends on their size, location, shape, soundness, attendant local slope of grain, and the type of stress to which a member is subjected.

The shape (form) of a knot appearing on a sawed surface depends upon the direction of the exposing cut. If, when sawing lumber from a log, a branch is sawed through at right angles to its length as in flat sawn boards, a nearly round knot results; when cut diagonally (as in a bastard-sawn board), an oval knot; and when sawed lengthwise (as in a quarter-sawn board), a spike knot.

Knots are further classified as intergrown or encased (fig. 4—2). As long as a limb remains alive, there is continuous growth at the junction of the limb and the bole of the tree, and the resulting knot is called intergrown. After the branch has died, additional growth on the trunk encloses the dead limb, and an encased knot results; fibers of the bole are not continuous with fibers of encased knots. Encased knots and knotholes tend to be accompanied by less cross grain than are intergrown knots and are, therefore, generally less serious with regard to most mechanical properties.

Most mechanical properties are lower at sections containing knots than in clear straight-grained wood because (1) the clear wood is displaced by the knot, (2) the fibers around the knot are distorted, causing cross grain, (3) the discontinuity of wood fiber leads to stress concentrations, and (4) checking often occurs around knots in drying. Hardness and strength in compression perpendicular to the grain are exceptions, where the knots may be objectionable only in that they cause nonuniform wear or nonuniform stress distributions at contact surfaces.

Knots have a much greater effect on strength in axial tension than in axial short-column compression, and the effects

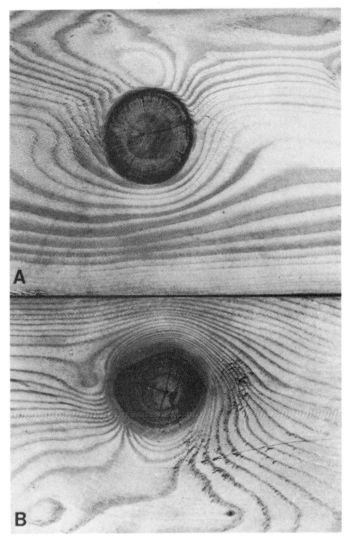

Figure 4—2—*A*, Encased knot. *B*, Intergrown knot.   (M84 0261)

on bending are somewhat less than those in axial tension. For this reason, in a simply supported beam a knot on the lower side (subjected to tensile stresses) has a greater effect on the load the beam will support than when the knot is on the upper side (subjected to compression stresses).

In long columns, knots are important in that they affect stiffness. In short or intermediate columns, the reduction in strength caused by knots is approximately proportional to the size of the knot; however, large knots have a somewhat greater relative effect than do small knots.

Knots in round timbers, such as poles and piles, have less effect on strength than knots in sawed timbers. Although the grain is irregular around knots in both forms of timber, its angle with the surface is less in naturally round than in sawed timber. Further, in round timbers there is no discontinuity in

Table 4–8—Functions relating mechanical properties to specific gravity of clear, straight-grained wood

Property		Specific gravity-strength relation[1]					
		Green wood			Wood at 12 percent moisture content		
		Softwoods	Hardwoods	All species[2]	Softwoods	Hardwoods	All species[2]
Static bending:							
Fiber stress at proportional limit	psi	$8,420G^{0.92}$	$8,480G^{1.04}$		$14,200G^{0.91}$	$12,200G^{0.80}$	
Modulus of elasticity	million psi	$2.44G^{0.81}$	$1.91G^{0.64}$		$3.13G^{0.90}$	$2.33G^{0.65}$	
Modulus of rupture	psi	$16,230G^{1.04}$	$16,700G^{1.12}$		$25,600G^{1.05}$	$24,400G^{1.10}$	
Work to maximum load	in-lb per in^3	$20.7G^{1.16}$	$31.4G^{1.48}$		$24.2G^{1.24}$	$29.7G^{1.47}$	
Total work	in-lb per in^3			$103G^{2.00}$			$72.7G^{1.75}$
Impact bending, height of drop causing complete failure	in.			$114G^{1.75}$			$94.6G^{1.75}$
Compression parallel to grain:							
Fiber stress at proportional limit	psi	$5,400G^{0.90}$	$4,930G^{0.96}$		$10,100G^{1.02}$	$6,210G^{0.57}$	
Modulus of elasticity	million psi	$3.24G^{0.92}$	$2.03G^{0.55}$		$3.72G^{0.91}$	$2.70G^{0.63}$	
Maximum crushing strength	psi	$7,740G^{1.02}$	$6,630G^{1.02}$		$14,600G^{1.04}$	$10,600G^{0.83}$	
Shear parallel to grain	psi	$1,560G^{0.72}$	$2,510G^{1.20}$		$2,430G^{0.86}$	$3,200G^{1.15}$	
Compression perpendicular to grain:							
Fiber stress at porportional limit	psi	$1,360G^{1.60}$	$2,380G^{2.32}$		$2,540G^{1.65}$	$2,920G^{2.03}$	
Hardness:							
End				$3,740G^{2.25}$			$4,800G^{2.25}$
Side				$3,420G^{2.25}$			$3,770G^{2.25}$

[1] The properties and values should be read as equations; for example: modulus of rupture for green wood of softwoods = $16,230G^{1.04}$, where G represents the specific gravity of ovendry wood, based on the volume at the moisture condition indicated.

[2] As reported in USDA Bulletin No. 676, "The relation of the shrinkage and strength properties of wood to its specific gravity" by J. A. Newlin and T.R.C. Wilson, 1919.

wood fibers caused by sawing through both local and general slope of grain.

The effects of knots in structural lumber are discussed in chapter 6.

Slope of Grain

In some wood product applications, the directions of important stresses may not coincide with the natural axes of fiber orientation in the wood. This may occur by choice in design, by the way the wood was removed from the log, or because of grain irregularities that occurred during growth.

Elastic properties in directions other than along the natural axes can be obtained from elastic theory. Strength properties in directions ranging from parallel to perpendicular to the fibers can be approximated using a Hankinson-type formula:

$$N = \frac{PQ}{P \sin^n\theta + Q \cos^n\theta} \qquad (4-2)$$

in which N represents the strength property at an angle θ from the fiber direction, Q is the strength perpendicular to the grain, P is the strength parallel to the grain, and n is an empirically determined constant. The formula has been used for modulus of elasticity as well as strength properties. Values of n and associated ratios of Q/P have been tabulated below from available literature:

Property	n	Q/P
Tensile strength	1.5−2	0.04−0.07
Compressive strength	2−2.5	0.03−0.4
Bending strength	1.5−2	0.04−0.1
Modulus of elasticity	2	0.04−0.12
Toughness	1.5−2	0.06−0.1

The Hankinson-type formula can be graphically depicted as a function of Q/P and n. Figure 4−3 shows the strength in any direction expressed as a fraction of the strength parallel to the fiber direction, plotted against angle to the fiber direction θ. The plot is for a range of values of Q/P and n.

The term "slope of grain" relates the fiber direction to the edges of a piece. Slope of grain is usually expressed by the ratio between a 1-inch deviation of the grain from the edge or long axis of the piece and the distance in inches within which this deviation occurs (tan θ). Table 4−9 gives the effect of grain slope on some properties of wood, as determined from tests. The values in table 4−9 for modulus of rupture fall very close to the curve in figure 4−3 for $Q/P = 0.1$ and $n = 1.5$. Similarly, the impact bending values fall close to the curve for $Q/P = 0.05$ and $n = 1.5$; and for compression, $Q/P = 0.1$, $n = 2.5$.

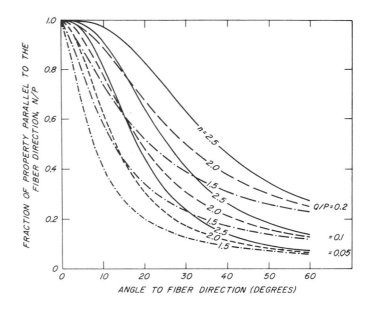

Figure 4−3—Effect of grain angle on mechanical property of clear wood according to a Hankinson-type formula. Q/P is the ratio of the mechanical property across the grain (Q) to that parallel to the grain (P); n is an empirically determined constant. (M140 730)

The term "cross grain" indicates the condition measured by slope of grain. Two important forms of cross grain are spiral grain and diagonal grain (fig. 4−4). Other types are wavy, dipped, interlocked, and curly grain. Some of the mechanical property values in table 4−4 are based on specimens with interlocked grain, because that is characteristic for some of the species.

Spiral grain in a tree is caused by fibers growing in a winding or spiral course about the bole of the tree instead of in a vertical course. In sawn products, spiral grain can be defined as fibers lying in the tangential plane of the growth rings, not parallel to the longitudinal axis of the product (see fig. 4−4B for a simple case). Spiral grain often goes undetected by ordinary visual inspection in sawn products. The best test for spiral grain is to split a sample section from the piece in the radial direction. A nondestructive method of determining the presence of spiral grain is to note the alignment of pores, rays, and resin ducts on a flat-sawn face. Drying checks on a flat-sawn surface follow the fibers and indicate the fiber slope.

Diagonal grain describes cross grain caused by growth rings not parallel to one or both surfaces of the sawn piece. Diagonal grain is produced by sawing parallel to the axis (pith) of the tree in a log having pronounced taper. It also occurs in lumber sawn from crooked or swelled logs.

Cross grain can be quite localized as a result of the distur-

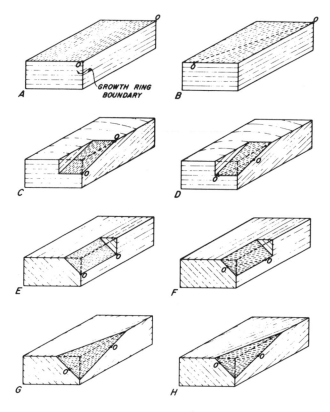

Figure 4–4—Schematic views of wood specimens containing straight grain and cross grain to illustrate the relationship of fiber orientation (0–0) to the axes of the piece. Specimens *A* through *D* have radial and tangential surfaces; *E* through *H* do not. *A* and *E* contain no cross grain. *B, D, F,* and *H* have spiral grain. *C, D, G,* and *H* have diagonal grain.

(M139 385)

bance of growth patterns by a branch. This condition, termed "local slope of grain," may be present even though the branch (knot) may have been removed in a sawing operation. Often the degree of local cross grain may be difficult to determine.

Any form of cross grain can have a serious effect on mechanical properties or machining characteristics.

Spiral and diagonal grain can combine to produce a more complex cross grain. To determine net cross grain, regardless of origin, fiber slopes on contiguous surfaces of a piece must be measured and combined. The combined slope of grain is determined by taking the square root of the sum of the squares of the two slopes. For example, assume the spiral grain slope on the flat-grain surface of figure 4–4*D* is 1 in 12 and the diagonal-grain slope is 1 in 18. The combined slope is

$$\sqrt{\left(\frac{1}{18}\right)^2 + \left(\frac{1}{12}\right)^2} = \frac{1}{10} \quad \text{or slope of 1 in 10}$$

Annual Ring Orientation

Stresses perpendicular to the fiber (grain) direction may be at any angle from 0° (T) to 90° (R) to the growth rings (fig. 4–5). Perpendicular-to-grain properties depend somewhat upon orientation of annual rings with respect to the direction of stress. Compression perpendicular-to-grain values in table 4–2 are derived from tests in which the load is applied parallel to the growth rings (T-direction); shear parallel-to-grain and tension perpendicular-to-grain values are averages of equal numbers of specimens with 0° and 90° growth ring

Table 4–9—*Strength of wood members with various grain slopes compared to strength of a straight-grained member, expressed as percentages*

Maximum slope of grain in member	Modulus of rupture	Impact bending—height of drop causing complete failure (50-lb hammer)	Compression parallel to grain—maximum crushing strength
	- *Percent* -		
Straight-grained	100	100	100
1 in 25	96	95	100
1 in 20	93	90	100
1 in 15	89	81	100
1 in 10	81	62	99
1 in 5	55	36	93

orientations. In some species, there is no difference in 0° and 90° orientation properties. Other species exhibit slightly higher shear parallel or tension perpendicular properties for the 0° orientation than for the 90° orientation; the converse is true for about an equal number of species.

The effects at intermediate annual ring orientations have been studied in a limited way. Modulus of elasticity, compression perpendicular-to-grain stress at the proportional limit, and tensile strength perpendicular to the grain tend to be about the same at 45° and 0°, but for some species the 45° orientation is 40 to 60 percent lower. For those species with lower properties at 45° ring orientation, properties tend to be about equal at 0° and 90° orientations. For species with about equal properties at 0° and 45° orientations, properties tend to be higher at 90° orientation.

Reaction Wood

Abnormal woody tissue is frequently associated with leaning boles and crooked limbs of both conifers and hardwoods. It is generally believed that it is formed as a natural response of the tree to return its limbs or bole to a more normal position, hence the term "reaction wood." In softwoods, the abnormal tissue is called "compression wood." It is common to all softwood species and is found on the lower side of the limb or inclined bole. In hardwoods, the abnormal tissue is known as "tension wood;" it is located on the upper side of the inclined member, although in some instances it is distributed irregularly around the cross section. Reaction wood is more prevalent in some species than in others.

Many of the anatomical, chemical, physical, and mechanical properties of reaction wood differ distinctly from those of normal wood. Perhaps most evident is the increase in the density over that of normal wood. The specific gravity of compression wood is commonly 30 to 40 percent greater than normal wood, while tension wood commonly ranges between 5 and 10 percent greater but may be as much as 30 percent greater than normal wood.

Compression and tension wood undergo extensive longitudinal shrinkage when subjected to moisture loss reaching below the fiber saturation point. Longitudinal shrinkage in compression wood ranges to 10 times that for normal wood and in tension wood perhaps 5 times that for normal wood. When reaction wood is present in the same board with normal wood, unequal longitudinal shrinkage causes internal stresses that result in warping. This warp sometimes occurs in rough lumber but more often in planed, ripped, or resawed lumber (fig. 4–6C). In extreme cases, the unequal longitudinal shrinkage results in axial tension failure over a portion of the cross section of the lumber (fig. 4–6B).

Reaction wood, particularly compression wood in the green

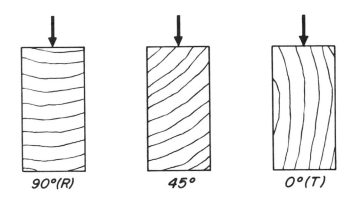

Figure 4–5—The direction of load in relation to the direction of the annual growth rings: 90° or perpendicular (R); 45°; 0° or parallel (T).

(M140 729)

condition, may be somewhat stronger than normal wood. However, when compared to normal wood of comparable specific gravity, the reaction wood is definitely weaker. Possible exceptions to this are compression parallel-to-grain properties of compression wood and impact bending properties of tension wood.

Because of its abnormal properties, it may be desirable to eliminate reaction wood from raw material. In logs, compression wood is characterized by eccentric growth about the pith and by the large proportion of summerwood at the point of greatest eccentricity (fig. 4–6A). Fortunately, pronounced compression wood in lumber can be detected by ordinary visual examination. It is usually somewhat darker than normal wood because of the greater proportion of summerwood and it frequently has a relatively lifeless appearance, especially in woods which normally have an abrupt transition from earlywood to latewood (fig. 4–6). Because it is more opaque than normal wood, intermediate stages of compression wood can be detected by transmitting light through thin cross sections, but borderline forms of compression wood that merge with normal wood are commonly detected only by microscopic examination.

Tension wood is more difficult to detect than compression wood. However, eccentric growth as seen on the transverse section suggests its presence. Also, the tough tension wood fibers resist being cut cleanly and this results in a woolly condition on the surfaces of sawn boards, especially when surfaced in the green condition (fig. 4–7). In some species, tension wood may show up on a smooth surface as areas of contrasting colors. Examples of this are the silvery appearance of tension wood in sugar maple and the darker color of tension wood in mahogany.

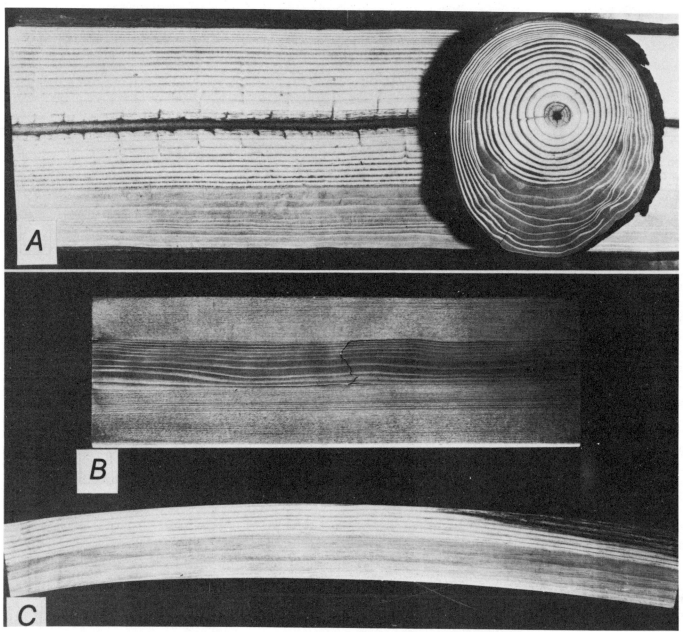

Figure 4−6—*A*, Eccentric growth about the pith in a cross section containing compression wood. The dark area in the lower third of the cross section is compression wood. *B*, Axial tension break caused by excessive longitudinal shrinkage of compression wood. *C*, Warp caused by excessive longitudinal shrinkage of compression wood.

(M41 434)

Compression Failures

Excessive bending of standing trees from wind or snow, felling trees across boulders, logs, or irregularities in the ground, or the rough handling of logs or lumber may produce excessive compression stresses along the grain which cause minute compression failures. In some instances, such failures are visible on the surface of a board as minute lines or zones formed by the crumpling or buckling of the cells (fig 4−8*A*), although usually they appear only as white lines or may even be invisible to the naked eye. Their presence may be indicated by fiber breakage on end grain (fig. 4−8*B*). Compression failures should not be confused with compression wood.

Products containing visible compression failures may have

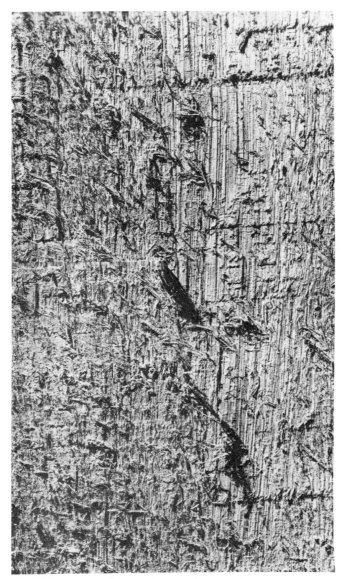

Figure 4 – 7—Projecting tension wood fibers on the sawn surface of a mahogany board. (M81 915)

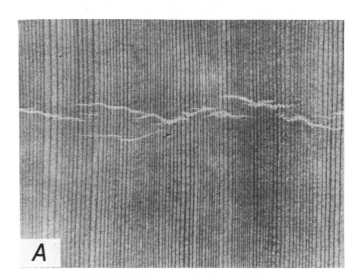

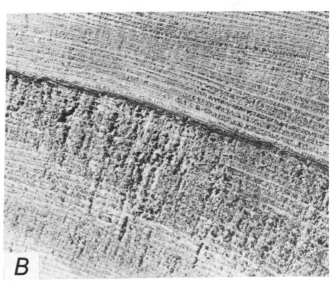

Figure 4 – 8—*A*, Compression failure is shown by the irregular lines across the grain. *B*, End-grain surfaces of spruce lumber show fiber breakage caused by compression failures below the dark line. (M45 594, M81 195)

low strength properties, especially in tensile strength and shock resistance. Tensile strength of wood containing compression failures may be as low as one-third of the strength of matched clear wood. Even slight compression failures, visible only under the microscope, may seriously reduce strength and cause brittle fracture. Because of the low strength associated with compression failures, many safety codes require certain structural members, such as ladder rails and scaffold planks, to be entirely free of them.

Compression failures are often difficult to detect with the unaided eye, and special efforts including optimum lighting are required to aid detection.

Pitch Pockets

A pitch pocket is a well-defined opening that contains free resin. It extends parallel to the annual rings and is almost flat on the pith side and curved on the bark side. Pitch pockets are confined to such species as the pines, spruces, Douglas-fir, tamarack, and western larch.

The effect of pitch pockets on strength depends upon their number, size, and location in the piece. A large number of pitch pockets indicates a lack of bond between annual growth layers, and a piece containing them should be inspected for shake or separations along the grain.

Bird Peck

Maple, hickory, white ash, and a number of other species are often damaged by small holes made by woodpeckers. These bird pecks are often in horizontal rows, sometimes encircling the tree, and a brown or black discoloration known as a mineral streak originates from each hole. Holes for tapping maple trees are also a source of mineral streaks. The streaks are caused by oxidation and other chemical changes in the wood.

Bird pecks and mineral streaks are not generally important in regard to strength, although they do impair the appearance of the wood.

Extractives

Many species of wood contain extraneous materials or extractives that can be removed by solvents that do not degrade the cellulosic/lignin structure of the wood. These extractives are especially abundant in species such as larch, redwood, western redcedar, and black locust.

A small decrease in modulus of rupture and strength in compression parallel to grain has been measured for some species after removal of extractives. The extent to which the extractives influence the strength is apparently a function of the amount of extractives, the moisture content of the piece, and the mechanical property under consideration.

Timber From Live Versus Dead Trees

Timber from trees killed by insects, blight, wind, or fire may be as good for any structural purpose as that from live trees, provided further insect attack, staining, decay, or seasoning degrade has not occurred. In a living tree, the heartwood is entirely dead, and in the sapwood only a comparatively few cells are living. Therefore, most wood is dead when cut, regardless of whether the tree itself is living or not. However, if a tree stands on the stump too long after its death, the sapwood is likely to decay or to be attacked severely by wood-boring insects, and in time the heartwood will be similarly affected. Such deterioration occurs also in logs that have been cut from live trees and improperly cared for afterwards. Because of variations in climatic and local weather conditions and in other factors that affect deterioration, the time during which dead timber may stand or lie in the forest without serious deterioration varies.

Tests on wood from trees that had stood as long as 15 years after being killed by fire demonstrated that this wood was as sound and as strong as wood from live trees. Also, logs of some of the more durable species have had thoroughly sound heartwood after lying on the ground in the forest for many years.

On the other hand, decay may cause great loss of strength within a very brief time, both in trees standing dead on the stump and in logs cut from live trees and allowed to lie on the ground. The important consideration is not whether the trees from which timber products are cut are alive or dead, but whether the products themselves are free from decay or other degrading factors that would render them unsuitable for use.

Effect of Manufacturing and Service Environment on Mechanical Properties

Moisture Content—Drying

Many mechanical properties are affected by changes in moisture content below the fiber saturation point. Most properties reported in tables 4–2, 4–3, and 4–4 increase with decrease in moisture content. The relation that describes these clear wood property changes at about 70 °F is:

$$P = P_{12} \left(\frac{P_{12}}{P_g} \right)^{\left(\frac{12-M}{M_p-12} \right)} \qquad (4-3)$$

where P is the property and M the moisture content in percent. M_p is the moisture content at the intersection of a horizontal line representing the strength of green wood and an inclined line representing the logarithm of strength-moisture content relationship for dry wood. This moisture content is slightly less than the fiber saturation point. Table 4–10 gives values of M_p for a few species; for other species, $M_p = 25$ may be assumed.

P_{12} is the property value at 12 percent moisture content, and P_g (green condition) is the property value for all moisture contents greater than M_p. Average property values of P_{12} and P_g are given for many species in tables 4–2, 4–3, and 4–4. The formula for moisture content adjustment is not recommended for work to maximum load, impact bending, and tension perpendicular. These properties are known to be erratic in their response to moisture content change.

The formula can be used to estimate a property at any moisture content below M_p from the species data given. For example, suppose the modulus of rupture of white ash at 8 percent moisture content is wanted. Using information from tables 4–2 and 4–10:

$$P_8 = (15,400) \left(\frac{15,400}{9,600} \right)^{\left(\frac{4}{12} \right)}$$

$$P_8 = 18,030 \text{ psi}$$

Table 4−10—*Intersection moisture content values for selected species*[1]

Species	M_p
	Pct
Ash, white	24
Birch, yellow	27
Chestnut, American	24
Douglas-fir	24
Hemlock, western	28
Larch, western	28
Pine, loblolly	21
Pine, longleaf	21
Pine, red	24
Redwood	21
Spruce, red	27
Spruce, Sitka	27
Tamarack	24

[1] Intersection moisture content is the point at which mechanical properties begin to change when drying from the green condition.

Table 4−11 tabulates approximate increases in property values at 6 percent moisture content and approximate decreases at 20 percent moisture content relative to those at 12 percent moisture content. The middle trend values are based on results of many reported studies. The values should be used with caution (i.e., treated only as trends) because of the variation in results from different studies.

The increase in mechanical properties discussed above assumes small, clear specimens in a drying process in which no deterioration of the product (degrade) occurs. The property changes applied to large wood specimens such as lumber are discussed in chapter 6.

Drying degrade can take several forms. Perhaps the most common degrade is surface and end checking. Checks most often limit mechanical properties.

Although visual signs of degrade may not be present, some loss of strength may occur in some species dried at high temperatures (110 °C and higher). Losses in excess of 10 percent have been observed in modulus of rupture and in shear of small clear wood specimens and in modulus of rupture and axial tensile strength of lumber.

Further information on moisture content is included in chapter 14.

Table 4−11—*Approximate middle trend effects of moisture content on mechanical properties of clear wood at about 20 °C*

Property	Relative change in property from 12 percent moisture content	
	At 6 percent moisture content	At 20 percent moisture content
	------------- Percent -------------	
Modulus of elasticity parallel to the grain	+9	−13
Modulus of elasticity perpendicular to the grain	+20	−23
Shear modulus	+20	−20
Bending strength	+30	−25
Tensile strength parallel to the grain	+8	−15
Compressive strength parallel to the grain	+35	−35
Shear strength parallel to the grain	+18	−18
Tensile strength perpendicular to the grain	+12	−20
Compressive strength perpendicular to the grain at the proportional limit	+30	−30

Temperature

Reversible Effects

In general, the mechanical properties of wood decrease when heated and increase when cooled. At a constant moisture content and below about 150 °C, mechanical properties are approximately linearly related to temperature. The change in properties that occurs when wood is quickly heated or cooled and then tested at that condition is termed an "immediate effect." At temperatures below 100 °C, the immediate effect is essentially reversible; that is, the property will return to the value at the original temperature if the temperature change is rapid.

Figure 4−9 illustrates the immediate effect of temperature on modulus of elasticity parallel to the grain, relative to values at 20 °C, based on a composite of results. Figure 4−10 gives similar information for modulus of rupture and figure 4−11 for compression parallel to grain. Figures 4−9 through 4−11 represent an interpretation of data from several investigators. The width of the band illustrates variability between and within reported trends.

Table 4−12 lists percentage changes in properties at −50 °C and +50 °C relative to those at 20 °C for a number of moisture conditions. The large changes at −50 °C for wet wood (at the fiber saturation point or wetter) reflect the presence of ice in the wood cell cavities.

Irreversible Effects

In addition to the reversible effect of temperature on wood, there is an irreversible effect at elevated temperature. This permanent effect is one of degradation of wood substance, which results in loss of weight and strength. The loss depends on factors which include moisture content, heating medium, temperature, exposure period, and, to some extent, species and size of piece involved.

The permanent decrease of modulus of rupture due to heating in steam and in water is shown as a function of temperature and heating time in figure 4−12, based on tests of Douglas-fir and Sitka spruce. From the same studies, work to maximum load was affected more than modulus of rupture by heating in water (fig. 4−13). The effect of oven heating (wood at 0 pct moisture content) on the modulus of rupture and modulus of elasticity is shown in figures 4−14 and 4−15, respectively, as derived from tests on four softwoods and two hardwoods. Note that the permanent property losses discussed above are based on tests conducted after the specimens have been cooled to room temperature and conditioned to the range of 7 to 12 percent moisture content. If specimens are tested hot, the percentage reductions due to permanent effects are based on values already reduced by the immediate effects.

Repeated exposure to elevated temperature has a cumula-

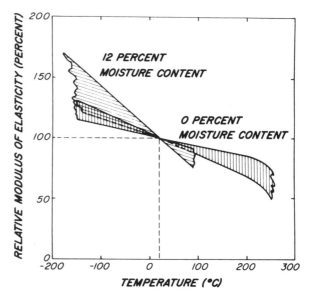

Figure 4−9—The immediate effect of temperature on modulus of elasticity parallel to the grain at two moisture contents relative to value at 20 °C. The plot is a composite of results from several studies. Variability in reported trends is illustrated by the width of bands. (ML84 5719)

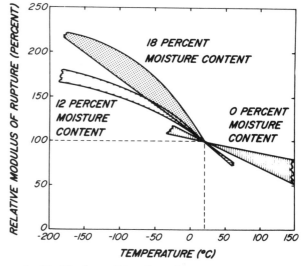

Figure 4−10—The immediate effect of temperature on modulus of rupture in bending at three moisture contents relative to value at 20 °C. The plot is a composite of results from several studies. Variability in reported trends is illustrated by the width. (ML84 5720)

tive effect on wood properties. For example, at a given temperature the property loss will be about the same after six exposure periods of 1 month each as it would after a single 6-month exposure period.

The shape and size of wood pieces are important in analyz-

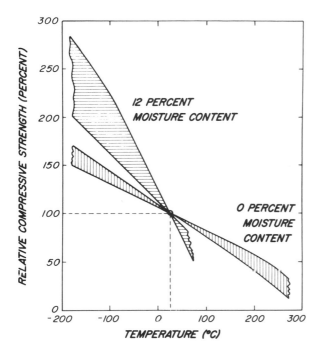

Figure 4–11—The immediate effect of temperature on compressive strength parallel to the grain at two moisture contents relative to value at 20 °C. The plot is a composite of results from several studies. Variability of reported trends is illustrated by the width of bands.

(ML84 5721)

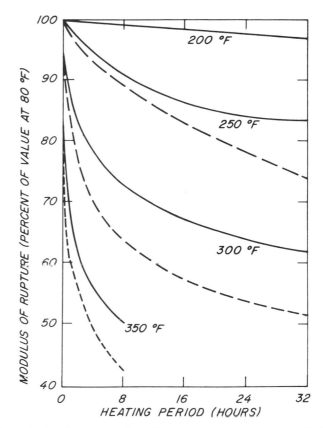

Figure 4–12—Permanent effect of heating in water (solid line) and in steam (dashed line) on the modulus of rupture. All data based on tests of Douglas-fir and Sitka spruce tested at room temperature.

(M140 731)

ing the influence of temperature. If the exposure is for only a short time, so that the inner parts of a large piece do not reach the temperature of the surrounding medium, the immediate effect on strength of inner parts will be less than for outer parts. The type of loading must be considered however. If the member is to be stressed in bending, the outer fibers of a piece are subjected to the greatest stress and will ordinarily govern the ultimate strength of the piece; hence, under this loading condition, the fact that the inner part is at a lower temperature may be of little significance.

For extended, noncyclic exposures, it can be assumed that the entire piece reaches the temperature of the heating medium and will, therefore, be subject to permanent strength losses throughout the volume of the piece, regardless of size and mode of stress application. However, wood often will not reach the daily extremes in temperature of the air around it in ordinary construction; thus, long-term effects should be based on the accumulated temperature experience of critical structural parts.

Time Under Load

Rate of Loading

Mechanical property values given in tables 4–2, 4–3, and 4–4 are usually referred to as static strength values.

Static strength tests are typically conducted at a rate of loading or rate of deformation to attain maximum load in about 5 minutes. Higher values of strength are obtained for wood loaded at more rapid rates and lower values are obtained at slower rates. For example, the load required to produce failure in a wood member in one second is approximately 10 percent higher than that obtained in a standard strength test. Over several orders of magnitude of rate of loading, strength is approximately an exponential function of rate.

Figure 4–16 illustrates how strength decreases with time to maximum load. The variability in the trend shown is based on results from several studies pertaining to bending, compression, and shear.

Creep/Relaxation

When first loaded, a wood member deforms elastically. If the load is maintained, additional time-dependent deformation occurs. This is called creep. Even at very low stresses, creep takes place and can continue over a period of years. For sufficiently high loads, failure will eventually occur. This failure phenomenon, termed "duration of load," is discussed in the next section.

Table 4 – 12—*Approximate middle trend effects of temperature on mechanical properties of clear wood at various moisture conditions*

Property	Moisture condition	Relative change in mechanical property from 20 °C	
		At -50 °C	At +50 °C
		-------------------- Percent --------------------	
Modulus of elasticity parallel to the grain	0	+11	−6
	12	+17	−7
	>FSP[1]	+50	—
Modulus of elasticity perpendicular to the grain	6	—	−20
	12	—	−35
	≧20	—	−38
Shear modulus	>FSP[1]	—	−25
Bending strength	≦4	+18	−10
	11−15	+35	−20
	18−20	+60	−25
	>FSP[1]	+110	−25
Tensile strength parallel to the grain	0−12	—	−4
Compressive strength parallel to the grain	0	+20	−10
	12−45	+50	−25
Shear strength parallel to the grain	>FSP[1]	—	−25
Tensile strength perpendicular to the grain	4−6	—	−10
	11−16	—	−20
	≧18	—	−30
Compressive strength perpendicular to the grain at the proportional limit	0−6	—	−20
	≧10	—	−35

[1] Moisture content higher than the fiber saturation point.

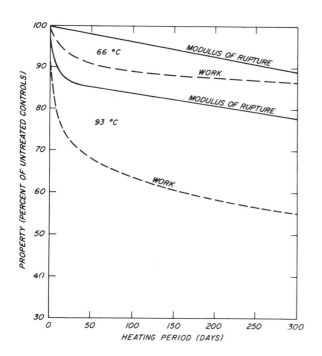

Figure 4 – 13—Permanent effect of heating in water (M140 732) on work to maximum load and on modulus of rupture. All data based on tests of Douglas-fir and Sitka spruce tested at room temperature.

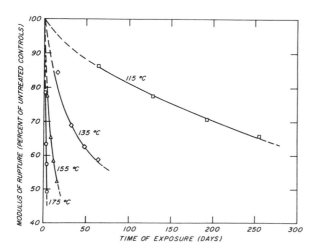

Figure 4 – 14—Permanent effect of oven heating at (M140 726) four temperatures on the modulus of rupture, based on four softwood and two hardwood species. All tests conducted at room temperature.

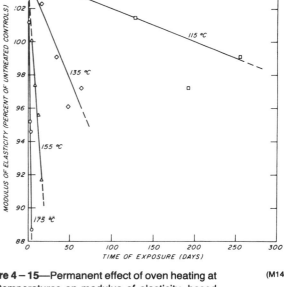

Figure 4 – 15—Permanent effect of oven heating at (M140 727) four temperatures on modulus of elasticity, based on four softwood and two hardwood species. All tests conducted at room temperature.

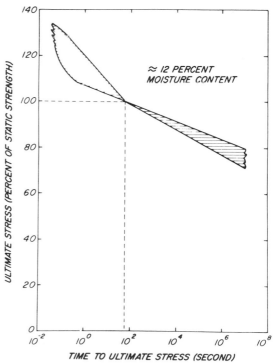

Figure 4 – 16—Relationship of ultimate stress at (ML84 5722) short time loading to that at 5-minute loading, based on a composite of results from rate of loading studies on bending, compression, and shear parallel to the grain. Variability in reported trends is indicated by width of band.

At typical design levels and use environments the additional deformation due to creep after several years may approximately equal the initial, instantaneous elastic deformation. For illustration, a creep curve based on creep as a function of initial deflection (relative creep) at several stress levels is shown in figure 4–17; creep is greater under higher stresses than lower ones.

Ordinary climatic variations in temperature and humidity will cause creep to increase. An increase of about 50 °F in temperature can cause a twofold to threefold increase in creep. Green wood may creep four to six times the initial deformation as it dries under load.

Unloading a member results in an immediate and complete recovery of the original elastic deformation and, after time, a recovery of approximately one-half of the creep deformation as well. Fluctuations in temperature and humidity increase the magnitude of the recovered deformation.

Relative creep at low stress levels is similar in bending, tension, or compression parallel to grain although it may be somewhat less in tension than in bending or compression under varying moisture conditions. Relative creep across the grain is qualitatively similar to, but likely to be greater than, creep parallel to the grain. The creep behavior of all species studied is approximately the same.

If, instead of controlling load or stress, a constant deformation is imposed and maintained on a wood member, the initial stress relaxes at a decreasing rate to about 60 to 70 percent of its original value within a few months. This reduction of stress with time is commonly termed "relaxation."

In limited bending tests carried out between approximately 18 °C and 49 °C over 2 to 3 months, the curve of stress vs. time that expresses relaxation is approximately the mirror image of the creep curve (deformation vs. time). These tests were carried out at initial stresses up to about 50 percent of the bending strength of the wood. As with creep, relaxation is markedly affected by fluctuations in temperature and humidity.

Duration of Stress

The duration of stress, or the time during which a load acts on a wood member either continuously or intermittently, is an important factor in determining the load that a member can safely carry. The duration of stress may be affected by changes in temperature and relative humidity.

The constant stress a wood member can sustain is approximately an exponential function of time to failure as illustrated in figure 4-18. The relationship is a composite of results of studies on small, clear wood specimens, conducted at constant temperature and relative humidity.

For a member that continuously carries a load for a long period of time, the load required to produce failure is much less than that determined from the strength properties in tables 4–2, 4–3, and 4–4. Based on figure 4–18, a wood mem-

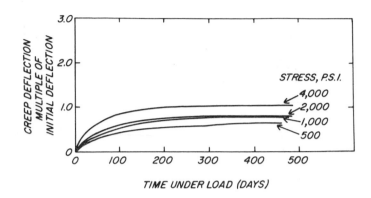

Figure 4–17—An illustration of creep as influenced by four levels of stress. (Adapted from Kingston.)   (M140 725)

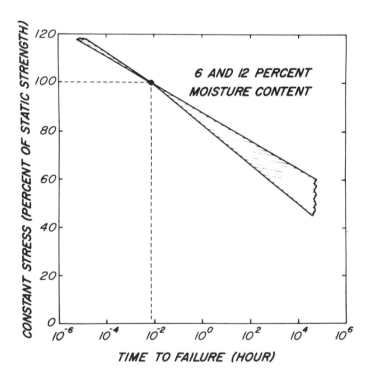

Figure 4–18—Relationship between stress due to constant load and time to failure for small clear wood specimens, based on 28 seconds' duration at 100 percent stress level. The figure is a composite of trends from several studies, mostly dealing with bending but with some on compression parallel to the grain and bending perpendicular to the grain. Variability in reported trends is indicated by the width of band.   (ML84 5723)

ber under the continuous action of bending stress for 10 years may carry only 60 percent (or perhaps less) of the load required to produce failure in the same specimen loaded in a standard bending strength test of only a few minutes' duration. Conversely, if the duration of stress is very short, the load-carrying capacity may be higher than that determined from strength properties given in the tables.

Time under intermittent loading has a cumulative effect. In tests where a constant load was periodically placed on a beam and then removed, the cumulative time the load was actually applied to the beam before failure was essentially equal to the time to failure for a similar beam under the same load continuously applied.

The time to failure under continuous or intermittent loading is looked upon as a creep-rupture process; a member has to undergo substantial deformation before failure. Deformation at failure is approximately the same for duration of load tests as for standard strength tests.

Changes in climatic conditions increase the rate of creep and shorten the duration during which a member can support a given load. This effect can be substantial for very small specimens of wood under large cyclic changes in temperature and relative humidity. Fortunately, changes in temperature and relative humidity in the typical service environment for wood are moderate.

Fatigue

The term "fatigue" in engineering is defined as the progressive damage that occurs in a material subjected to cyclic loading. This loading may be repeated (stresses of the same sign, i.e. always compression or always tension) or reversed (stresses of alternating sign). When sufficiently high and repetitious enough, cyclic loading stresses can result in fatigue failure.

Fatigue life is a term used to define the number of cycles that are sustained before failure. Fatigue strength, the maximum stress attained in the stress cycle used to determine fatigue life, is approximately exponentially related to fatigue life; that is, fatigue strength decreases approximately linearly as the logarithm of number of cycles increases. Fatigue strength and fatigue life also depend on several other factors: frequency of cycling; whether the loading is repeated or reversed; range factor (ratio of minimum to maximum stress per cycle); and other factors such as temperature, moisture content and specimen size. Positive range ratios imply repeated loading while negative ratios imply reversed loading.

A summary of results from several fatigue studies on wood is given in table 4–13. Most of the results are for repeated loading with a range ratio of 0.1, meaning that the minimum stress per cycle is 10 percent of the maximum stress. The maximum stress per cycle, expressed as a percent of estimated static strength, is associated with the fatigue life given

in millions of cycles. The first three lines of data, which list the same cyclic frequency, demonstrate the effect of range ratio on fatigue strength (maximum fatigue stress that can be maintained for a given fatigue life); fatigue bending strength decreases as range ratio decreases. Third-point bending results show the effect of small knots or slope of grain on fatigue strength at a range ratio of 0.1 and frequency of 8-1/3 hertz. Fatigue strength is lower for wood containing small knots or a 1 in 12 slope of grain than for clear straight-grained wood and even lower when wood contains a combination of both small knots and a 1 in 12 slope of grain. Fatigue strength is the same for a scarf joint in tension as for tension parallel to the grain but is a little lower for a finger joint in tension. Fatigue strength is slightly lower in shear than in tension parallel to the grain. Other comparisons do not have much meaning because range ratios or cyclic frequency differ; however, fatigue strength is high in compression parallel to the grain compared to other properties. Little is known about other factors that may affect fatigue strength in wood.

Creep, temperature rise, and loss of moisture content occur in testing wood for fatigue strength. At stresses that cause failure in about 10^6 cycles at 40 hertz, temperature rises of 15 °C have been reported for compression parallel fatigue (range ratio slightly greater than zero), for tension parallel fatigue (range ratio = 0), and for reversed bending fatigue (range ratio = −1). The rate of temperature rise is high initially, but it then diminishes to a moderate rate which is maintained more or less constant during a large percentage of fatigue life. During the latter stages of fatigue life, the rate of temperature rise increases until failure occurs. Smaller rises in temperature would be expected for slower cyclic loading or lower stresses. Decreases in moisture content are probably related to temperature rise.

Age

In relatively dry and moderate temperature conditions where wood is protected from deteriorating influences such as decay, the mechanical properties of wood show little change with time. Test results for very old timbers suggest that significant losses in strength occur only after several centuries of normal aging conditions. The soundness of centuries-old wood in some standing trees (redwood, for example) also attests to the durability of wood.

Chemicals

The effect of chemical solutions on mechanical properties depends on the specific type of chemical. Nonswelling liquids, such as petroleum oils and creosote, have no appreciable effect on properties. Properties are lowered in the presence of water, alcohol, or other wood-swelling organic liquids even though these liquids do not chemically degrade the wood substance. The loss in properties depends largely on amount of swelling, and this loss is regained upon removal of the swelling liquid. Liquid ammonia markedly reduces the strength and stiffness of wood, but most of the reduction is regained upon removal of the ammonia.

Chemical solutions that decompose wood substance have a permanent effect on strength. The following generalizations summarize the effect of chemicals: (1) Some species are quite resistant to attack by dilute mineral and organic acids, (2) oxidizing acids such as nitric acid degrade wood more than nonoxidizing acids, (3) alkaline solutions are more destructive than acidic solutions, and (4) hardwoods are more susceptible to attack by both acids and alkalies than are softwoods. Because both species and application are extremely important, reference to industrial sources with a specific his-

tory of use is recommended where possible. For example, large cypress tanks have survived long continuous use where exposure conditions involved mixed acids at the boiling point.

A general discussion of the resistance of wood to chemical degradation is given in chapter 3.

Wood Treatment With Chemicals

Wood products sometimes are treated with preservative or fire-retarding salts, usually in water solution, to impart resistance to decay or fire. Such products generally are kiln dried after treatment. At levels of preservative treatments required for underground or ground-contact service, mechanical properties are essentially unchanged except that work to maximum load, height of drop in impact bending, and toughness are reduced somewhat. Heavy salt treatments required for protection in marine environments may reduce bending

Table 4–13—A summary of reported results on cyclic fatigue[1]

Property	Range ratio	Cyclic frequency	Maximum stress per cycle, percentage of estimated static strength	Approximate fatigue life, 10^6 cycles
		Hz		
Bending, clear, straight grain				
Cantilever	0.45	30	45	30
Cantilever	0	30	40	30
Cantilever	−1.0	30	30	30
Center-point	−1.0	40	30	4
Rotational	−1.0	—	28	30
Third-point	.1	8-1/3	60	2
Bending, third-point				
Small knots	.1	8-1/3	50	2
Clear, 1:12 slope of grain	.1	8-1/3	50	2
Small knots,				
1:12 slope of grain	.1	8-1/3	40	2
Tension parallel to grain				
Clear, straight grain	.1	15	50	30
Clear, straight grain	0	40	60	3.5
Scarf joint	.1	15	50	30
Finger joint	.1	15	40	30
Compression parallel to grain				
Clear, straight grain	.1	40	75	3.5
Shear parallel to grain				
Glue laminated	.1	15	45	30

[1] Starting moisture contents about 12 to 15 percent.

strength by 10 percent or more and work properties by up to 50 percent. Further strength reduction may be observed if temperatures and pressures involved in treating and subsequent drying are not controlled within acceptable limits. See chapter 18 for details on treating conditions.

Strength properties are also affected to some extent by the combined effects of fire-retardant chemicals, treatment methods, and kiln drying. A variety of fire-retardant treatments have been studied. Collectively the studies indicate that modulus of rupture, work to maximum load, and toughness are reduced by varying amounts depending on species and type of fire retardant. Work to maximum load and toughness are most affected, with reductions of as much as 45 percent. A reduction in modulus of rupture of 20 percent has been observed; a design reduction of 10 percent is frequently used. Stiffness is not appreciably affected.

Wood is also sometimes impregnated with monomers, such as methyl methacrylate, which are subsequently polymerized. Many of the mechanical properties of the resulting wood-plastic composite are higher than those of the original wood, generally as a result of filling the void spaces in the wood structure with plastic. The polymerization process and both the chemical nature and quantity of monomers are variables that influence composite properties.

Nuclear Radiation

There are occasions when wood is subjected to nuclear radiation. Examples are wooden structures closely associated with nuclear reactors, or when nuclear radiation is used for polymerizing plastic impregnants in wood, or for nondestructive estimation of wood density and moisture content. Very large doses of gamma rays or neutrons can cause substantial degradation of wood. In general, irradiation with gamma rays in doses up to about 1 megarad has little effect on the strength properties of wood. As dosage increases above 1 megarad, tensile strength parallel to grain and toughness decrease. At a dosage of 300 megarads, tensile strength is reduced about 90 percent. Gamma rays also affect compressive strength parallel to grain above 1 megarad, but strength losses with further dosage are less than for tensile strength. Only about one-third of the compressive strength is lost when the total dose is 300 megarads. Effects of gamma rays on bending and shear strength are intermediate between the effects on tensile and compressive strength.

Molding and Staining Fungi

Molding and staining fungi do not seriously affect most mechanical properties of wood because they feed upon substances within the cell cavity or attached to the cell wall rather than on the structural wall itself. The duration of infection and the species of fungi involved are important factors in determining the extent of weakening.

Though little loss in strength is encountered at the lower levels of biological staining, intense staining may reduce specific gravity by 1 to 2 percent, surface hardness by 2 to 10 percent, bending and crushing strength by 1 to 5 percent, and toughness or shock resistance properties by 15 to 30 percent.

Although molds and stains usually do not have major effects on strength, conditions that favor these organisms are also ideal for the development of wood-destroying or decay fungi and the soft-rot fungi (see ch. 17). Pieces containing mold and stain should be examined closely for decay if they are used for structural purposes.

Decay

Unlike molding and staining fungi, the wood-destroying (decay) fungi seriously reduce strength. These decay fungi metabolize the cellulose fraction of wood which gives wood its strength.

Early stages of decay are virtually impossible to detect. For example, brown-rot fungi may reduce mechanical properties in excess of 10 percent before a measurable weight loss is observed and before there are visible signs of decay. When weight loss reaches 5−10 percent, mechanical properties are reduced from 20−80 percent. Toughness, impact bending, and work to maximum load in bending are reduced most, shear and hardness the least, while other properties show an intermediate effect. Thus, when strength is important, adequate measures should be taken to (1) prevent decay before it occurs, (2) control incipient decay by remedial measures (see ch. 17), or (3) replace any wood member in which decay is evident or believed to exist in a critical section. Decay can be prevented from starting or progressing if wood is kept dry (below 20 percent moisture content).

No method is known for estimating the amount of reduction in strength from the appearance of decayed wood. Therefore, when strength is an important consideration, the safe procedure is to discard every piece that contains even a small amount of decay. An exception may be pieces in which decay occurs in a knot but does not extend into the surrounding wood.

Insect Damage

Insect damage may occur in standing trees, logs, and unseasoned or seasoned lumber. Damage in the standing tree is difficult to control, but otherwise insect damage can be largely eliminated by proper control methods.

Insect holes are generally classified as pinholes, grub holes, and powderpost holes. The powderpost larvae, by their irregular burrows, may destroy most of the interior of a piece, while the surface shows only small holes, and the strength of the piece may be reduced virtually to zero.

No method is known for estimating the reduction in strength from the appearance of insect-damaged wood, and, when strength is an important consideration, the safe procedure is to eliminate pieces containing insect holes.

Selected References

American Society for Testing and Materials. Standard methods for testing small clear specimens of timber. ASTM D 143. Philadelphia, PA: ASTM; (see current edition.)

Bendtsen, B. A. Rolling shear characteristics of nine structural softwoods. Madison, WI: Forest Products Journal. 26(11): 51−56; 1976.

Bendtsen, B. A.; Freese, Frank; Ethington, R. L. Methods for sampling clear, straight-grained wood from the forest. Madison, WI: Forest Products Journal. 20(11): 38−47; 1970.

Bodig, J.; Goodman, J. R. Prediction of elastic parameters for wood. Madison, WI: Wood Science. 5(4): 249−264; 1973.

Bodig, J.; Jayne, B. A. Mechanics of wood and wood composites. New York: Van Nostrand Reinhold Co.; 1982.

Boller, K. H. Wood at low temperatures. Modern Packaging. 28(1): 153−157; 1954.

Chudnoff, Martin. Tropical timbers of the world. Agric. Handb. 607. Washington DC: U.S. Department of Agriculture, Forest Service; 1984.

Coffey, D. J. Effects of knots and holes on the fatigue strength of quarter-scale timber bridge stringers. M.S. Thesis. Madison, WI: University of Wisconsin, Department of Civil Engineering; 1962.

Gerhards, C. C. Effects of type of testing equipment and specimen size on toughness of wood. Res. Pap. FPL 97. Madison, WI: U.S. Department of Agriculture, Forest Service, Forest Products Laboratory; 1968.

Gerhards, C. C. Effect of duration and rate of loading on strength of wood and wood-based materials. Res. Pap. FPL 283. Madison, WI: U.S. Department of Agriculture, Forest Service, Forest Products Laboratory; 1977.

Gerhards, C. C. Effect of high-temperature drying on tensile strength of Douglas-fir 2 by 4's. Madison, WI: Forest Products Journal. 29(3): 39−46.;. 1979.

Gerhards, C. C. Effect of moisture content and temperature on the mechanical properties of wood: an analysis of immediate effects. Wood and Fiber. 14(1): 4−36; 1982.

Hearmon, R.F.S. The elasticity of wood and plywood. Special Rep. 7. London, England: Department of Scientific and Industrial Research, Forest Products Research; 1948.

Hearmon, R.F.S. An introduction to applied anisotropic elasticity. London, England: Oxford University Press; 1961.

Kingston, R.S.T. Creep, relaxation, and failure of wood. Research Applied in Industry. 15(4); 1962.

Kollmann, Franz F. P.; Côté, Wilfred A., Jr. Principles of wood science and technology. New York: Springer Verlag; 1968.

Koslik, C. J. Effect of kiln conditions on the strength of Douglas-fir and western hemlock. Rep. D−9. Corvallis, OR: Oregon State University, School of Forestry, Forestry Research Laboratory; 1967.

Little, E. L., Jr. Checklist of United States trees (native and naturalized). Agric. Handb. 541. Washington, DC: U.S. Department of Agriculture; 1979.

MacLean, J. D. Effect of steaming on the strength of wood. American Wood-Preservers' Association. 49: 88−112; 1953.

MacLean J. D. Effect of heating in water on the strength properties of wood. American Wood-Preservers' Association. 50: 253−281; 1954.

Millett, M. A.; Gerhards, C. C. Accelerated aging: Residual weight and flexural properties of wood heated in air at 115° to 175° C. Madison, WI: Wood Science. 4(4); 1972.

Nicholas, D. D. Wood deterioration and its prevention by preservative treatments. Vol. I. Degradation and protection of Wood. Syracuse, NY: Syracuse University Press; 1973.

Pillow, M. Y. Studies of compression failures and their detection in ladder rails. Rep. D 1733. Madison, WI: U.S. Department of Agriculture, Forest Service, Forest Products Laboratory; 1949.

U.S. Department of Defense. Design of wood aircraft structures. ANC−18 Bulletin. (Issued by Subcommittee on Air Force-Navy-Civil Aircraft, Design Criteria Aircraft Comm.) 2nd ed. Munitions Board Aircraft Committee; 1951.

Wangaard, F. F. Resistance of wood to chemical degradation. Madison, WI: Forest Products Journal. 16(2): 53−64; 1966.

Wilcox, W. Wayne. Review of literature on the effects of early stages of decay on wood strength. Wood and Fiber. 9(4): 252−257; 1978.

Wilson, T.R.C. Strength-moisture relations for wood. Tech. Bull. 282. Washington, DC: U.S. Department of Agriculture; 1932.

Chapter 5

Commercial Lumber

Commercial Lumber *

Commercial lumber in a broad sense is any lumber that is bought or sold in the normal channels of commerce. It may be found in a variety of forms, species, and types, and in various commercial establishments both wholesale and retail. Most commercial lumber is graded by standardized rules that make purchasing more or less uniform throughout the country.

A log when sawed yields lumber of varying quality. To enable users to buy the quality that best suits their purposes, lumber is graded into use categories, each having an appropriate range in quality.

Generally, the grade of a piece of lumber is based on the number, character, and location of features that may lower the strength, durability, or utility value of the lumber. Among the more common visual features are knots, checks, pitch pockets, shake, and stain, some of which are a natural part of the tree. Some grades are free or practically free from these features. Other grades, comprising the great bulk of lumber, contain fairly numerous knots and other features. With proper grading, lumber containing these features is entirely satisfactory for many uses.

The grading operation for most lumber takes place at the sawmill. Establishment of grading procedures is largely the responsibility of manufacturers' associations. Because of the wide variety of wood species, industrial practices, and customer needs, different lumber grading practices coexist. The grading practices of most interest are considered in the sections that follow, under the major categories of Hardwood Lumber and Softwood Lumber.

Hardwood Lumber

The principal use of hardwood lumber is for remanufacture into furniture, cabinetwork, and pallets, or directly into flooring, paneling, moulding, and millwork. It is mainly graded and marketed in three categories: factory lumber, dimension parts, and finished market products. Recently, several hardwood species are being graded under the American Softwood Lumber Standard and sold as structural lumber although the amounts are very small. Also, specially graded hardwood lumber can be used for structural glue-laminated lumber.

Before 1898, grading of hardwoods was done on an individual mill basis for local markets. In 1898 manufacturers and users formed the National Hardwood Lumber Association to standardize grading for hardwood lumber. Between 1898 and 1932 grading was based on the number and size of defects. In 1932, the basis for grading was changed to standard clear cutting sizes.

* Revision by B. Alan Bendtsen, Forest Products Technologist; Robert R. Maeglin, Forest Products Technologist; Kent A. McDonald, Forest Products Technologist; and Catherine M. Marx, General Engineer.

Both factory lumber and dimension parts are intended to serve the industrial customer; the important difference is that for factory lumber, the grades reflect the proportion of a piece that can be cut into useful smaller pieces while the dimension grades are based on use of the entire piece. Finished market products are graded for their unique end use with little or no remanufacture. Examples of finished products include moulding, stair treads, and hardwood flooring.

Factory Lumber

The rules adopted by the National Hardwood Lumber Association are considered standard in grading hardwood lumber intended for cutting into smaller pieces to make furniture or other fabricated products. In these rules the grade of a piece of hardwood lumber is determined by the proportion of a piece that can be cut into a certain number of smaller pieces of material, commonly called cuttings, generally clear on one side, the reverse face sound, and not smaller than a specified size.

The best grade in the factory lumber category is termed Firsts and the next grade Seconds. Firsts and Seconds are nearly always combined in one grade and referred to as FAS. A third grade is termed Selects followed by No. 1 Common, No. 2 Common, Sound Wormy, No. 3A Common, and No. 3B Common. A description of the standard hardwood lumber grades is given in table 5-1. This table illustrates, for example, that Firsts call for pieces that will allow at least 91-2/3 percent of their surface measure to be cut into clear face material. Thus not more than 8-1/3 percent of each piece can be unused in making the required cuttings. Except for Sound Wormy, the minimum acceptable length, width, surface measure, and percentage of piece that must work into a cutting decreases with decreasing grade. Figure 5-1 is an example of grading for cuttings.

This brief summary of grades for factory lumber should not be regarded as a complete set of grading rules because numerous details, exceptions, and special rules for certain species are not included. The complete official rules of the National Hardwood Lumber Association should be followed as the only full description of existing grades. The address is given in table 5-2. Table 5-3 lists names of commercial domestic hardwood species that are graded by NHLA rules.

Standard Lengths

Standard lengths of hardwood lumber are 4, 5, 6, 7, 8, 9, 10, 11, 12, 13, 14, 15, and 16 feet, but not more than 50 percent of odd-numbered lengths are allowed in any single shipment.

Standard Thickness

Standard thicknesses for hardwood lumber, rough and surfaced two sides (S2S)[1], are given in table 5-4. The thickness of S1S lumber is subject to contract agreement.

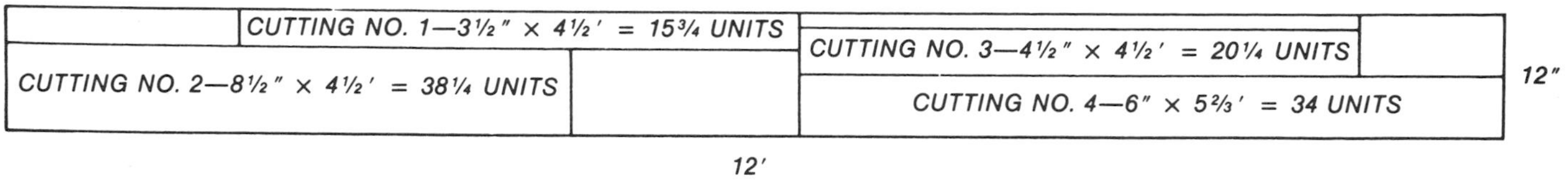

		CUTTING NO. 1—3½" × 4½' = 15¾ UNITS			
CUTTING NO. 2—8½" × 4½' = 38¼ UNITS			CUTTING NO. 3—4½" × 4½' = 20¼ UNITS		12"
			CUTTING NO. 4—6" × 5⅔' = 34 UNITS		

12'

1. Determine Surface Measure (S.M.) using lumber scale stick or from formula:

$$\frac{Width\ in\ inches \times length\ in\ feet}{12} = \frac{12" \times 12'}{12}$$

$$= 12\ sq.\ ft.\ S.M.$$

2. No. 1 Common is assumed grade of board. Percent of clear-cutting area required for No. 1 Common—66⅔% or ⁸⁄₁₂.

3. Determine maximum number of cuttings permitted.

For No. 1 Common grade (S.M. + 1) ÷ 3

$$= \frac{(12 + 1)}{3} = \frac{13}{3} = 4\ cuttings.$$

4. Determine minimum size of cuttings.

For No. 1 Common grade 4" × 2' or 3" × 3'.

5. Determine clear-face cutting units needed.

For No. 1 Common grade S.M. × 8 = 12 × 8 = 96 units.

6. Determine total area of permitted clear-face cutting in units.

Width in inches and fractions of inches × length in feet and fractions of feet

Cutting #1—3½" × 4½' = 15¾ units
Cutting #2—8½" × 4½' = 38   units
Cutting #3—4½" × 4½' = 20¼ units
Cutting #4—6" × 5⅔' = 34   units

Total Units   108

Units required for No. 1 Common—96.

7. Conclusion: Board meets requirements for No. 1 Common grade.

Figure 5–1—An example of hardwood grading for cuttings using a No. 1 Common lumber grade.

(ML84 5825)

Standard Widths

Hardwood lumber is usually manufactured to random width. The hardwood lumber grades do not specify standard widths; however, the grades do specify minimum widths for each grade as follows:

Firsts . . . . 6 inches
Seconds . . . . 6 inches
Selects . . . . 4 inches
Nos. 1, 2, 3A, 3B Common . . . . 3 inches

If width is specified by purchase agreement, S1E or S2E lumber is 3/8 inch scant of nominal size in lumber less than 8 inches wide and 1/2 inch scant in lumber 8 inches and wider.

Dimension Parts

Dimension parts for hardwoods signifies stock that is processed in specific thicknesses, widths and lengths, or multiples thereof. This stock is sometimes referred to as hardwood dimension stock or hardwood dimension lumber. This should not be confused with "dimension," a term used in the softwood structural lumber market to mean lumber nominally two to less than 5 inches thick. The user should be particularly alert to this conflict as some hardwoods are now being manufactured into "dimension" for use in structural application.

Dimension parts are normally kiln dried and generally graded under the rules of the National Dimension Manufacturers Association (NDMA). These rules encompass three classes of material: (1) hardwood dimension parts (flat stock), (2) solid kiln dried squares (rough), and (3) solid kiln dried squares (surfaced). Hardwood dimension parts (flat stock) has five grades: Clear Two Faces, Clear One Face, Paint, Core, and Sound. Squares have three grades if rough (Clear, Select, and Sound) and four if surfaced (Clear, Select, Paint, and Second). Each class may be further defined as rough, surfaced and semifabricated, or completely fabricated. The rough dimension parts are blank sawn and ripped to size. Surfaced and semifabricated parts have been through one or more manufacturing stages. Completely fabricated parts have been completely processed for their end use.

---

[1] See commonly used lumber abbreviations at the end of this chapter.

Table 5 – 1—*Standard hardwood lumber grades[1]*

Grade and lengths allowed	Widths allowed-	Allowable surface measure of pieces	Minimum amount of each piece in clear-face cuttings	Maximum cuttings allowed	Minimum size of cuttings allowed
Ft	*In*	*Ft²*	*Pct*	*Number*	
Firsts:[2]					
8 to 16 (will admit 30 percent of 8- to 11-foot, 1/2 of which may be 8- and 9-foot)	6+	4 to 9	91-2/3	1	4 inches by 5 feet, or 3 inches by 7 feet
		10 to 14	91-2/3	2	
		15 +	91-2/3	3	
Seconds:[2]					
8 to 16 (will admit 30 percent of 8- to 11-foot, 1/2 of which may be 8- and 9-foot)	6+	4 to 7	83-1/3	1	Do.
		6 and 7	91-2/3	2	
		8 to 11	83-1/3	2	
		8 to 11	91-2/3	3	
		12 to 15	83-1/2	3	
		12 to 15	91-2/3	4	
		16 +	83-1/3	4	
Selects:					
6 to 16 (will admit 30 percent of 6- to 11-foot, 1/6 of which may be 6- and 7-foot)	4+	2 and 3	91-2/3	1	Do.
		4 +	(³)	1	
No. 1 Common:					
4 to 16 (will admit 10 percent of 4- to 7-foot, 1/2 of which may be 4- and 5-foot)	3+	1	100	0	4 inches by 2 feet, or 3 inches by 3 feet
		2	75	1	
		3 and 4	66-2/3	1	
		3 and 4	75	2	
		5 to 7	66-2/3	2	
		5 to 7	75	3	
		8 to 10	66-2/3	3	
		11 to 13	66-2/3	4	
		14 +	66-2/3	5	

Table 5–1—Standard hardwood lumber grades[1]—Continued

Grade and lengths allowed (Ft)	Widths allowed (In)	Allowable surface measure of pieces (Ft²)	Minimum amount of each piece in clear-face cuttings (Pct)	Maximum cuttings allowed (Number)	Minimum size of cuttings allowed
No. 2 Common: 4 to 16 (will admit 30 percent of 4- to 7-foot, 1/3 of which may be 4- and 5-foot)	3 +	1	66-2/3	1	3 inches by 2 feet
		2 and 3	50	1	
		2 and 3	66-2/3	2	
		4 and 5	50	2	
		4 and 5	66-2/3	3	
		6 and 7	50	3	
		6 and 7	66-2/3	4	
		8 and 9	50	4	
		10 and 11	50	5	
		12 and 13	50	6	
		14 +	50	7	
Sound Wormy:[4]					
No. 3A Common: 4 to 16 (will admit 50 percent of 4- to 7-foot, 1/2 of which may be 4- and 5-foot)	3 +	1 +	[5]33-1/3	([6])	Do.
No. 3B Common: 4 to 16 (will admit 50 percent of 4- to 7-foot, 1/2 of which may be 4- and 5-foot)	3 +	1 +	[7]25	([6])	1-1/2 inches by 2 feet

[1] Inspection to be made on the poorer side of the piece, except in Selects.

[2] Firsts and Seconds are combined as 1 grade (FAS). The percentage of Firsts required in the combined grade varies from 20 to 40 percent, depending on the species.

[3] Same as Seconds with reverse side of board not below No. 1 Common or reverse side of cuttings sound.

[4] The grade of hardwood lumber called "Sound Wormy" has the same requirements as No. 1 Common and Better except that wormholes and limited sound knots and other imperfections are allowed in the cuttings.

[5] This grade also admits pieces that grade not below No. 2 Common on the good face and have the reverse side of cutting sound.

[6] Unlimited.

[7] The cuttings must be sound; clear face not required.

5-5

Finished Market Products

Some hardwood lumber products are graded in relatively finished form, with little or no further processing anticipated. Flooring is probably the highest volume finished market product. Other examples are lath, siding, ties, planks, carstock, construction boards, timbers, trim, moulding, stair treads, and risers. Grading rules promulgated for flooring anticipate final consumer use and are summarized in this section. Details on grades of other finished products are found in appropriate association grading rules.

Hardwood flooring generally is graded under the rules of the Maple Flooring Manufacturers Association and the rules of the National Oak Flooring Manufacturers Association. Tongued-and-grooved and end-matched hardwood flooring is commonly furnished. Square edge and square end strip flooring is also available as well as parquet flooring suitable for laying with mastic.

The Maple Flooring Manufacturers Association grading rules cover flooring manufactured from hard maple, beech, and birch. Each species is graded into four categories—First, Second, Third, and Fourth Grades. Combination grades of Second and Better and Third and Better are sometimes specified. There are also special grades based on color and species.

First-grade flooring must have one face practically free from all imperfections. Variations in the natural color of the wood are allowed.

Second-grade flooring admits tight, sound knots and other slight imperfections but must lay without waste.

Third-grade flooring has few restrictions as to imperfections permitted but must permit proper laying and provide a good, serviceable floor.

The standard thickness of maple, beech, and birch flooring is 25/32 inch. Face widths are 1-1/2, 2, 2-1/4, and 3-1/4 inches. Standard lengths are 2 feet and longer in First- and Second-grade flooring and 1-1/4 feet and longer in Third-grade flooring.

The grading rules of the National Oak Flooring Manufacturers Association mainly cover quartersawed and plainsawed oak flooring. Quartersawed flooring has two grades—Clear and Select. Plainsawed flooring has four grades—Clear, Select, No. 1 Common, and No. 2 Common. The Clear grade in both plainsawed and quartersawed flooring must have the face free from surface imperfections except that up to 3/8 inch of bright sap is allowed. Color is not considered in the Clear grade. Select flooring (plainsawed or quartersawed) may contain sap and will admit a few features such as pin wormholes and small tight knots. No. 1 Common plainsawed flooring must contain material that will make a sound floor

Table 5—2—*Hardwood grading associations in United States*[1]

Name and address	Species covered by grading rules
National Hardwood Lumber Association P.O. Box 34518 Memphis, TN 38184-0518	Hardwoods (furniture cuttings, construction lumber, siding, panels)
National Dimension Manufacturers Association 101 Village Pkwy., Suite 202 Marietta, GA 30067	Hardwoods (hardwood furniture dimension, squares, laminated stock, interior trim, stair treads and risers)
Maple Flooring Manufacturers Association 8600 Bryn Mawr Ave., Suite 720S Chicago, IL 60631	Maple, beech, birch (flooring)
National Oak Flooring Manufacturers Association 804 Sterick Building Memphis, TN 38103	Oak, pecan, beech, birch, and hard maple (flooring)

[1] Grading associations that include hardwood species in structural grades are listed in table 5—5.

Table 5 – 3—*Nomenclature of commercial hardwood lumber*

Commercial name for lumber	Official common tree name	Botanical name
Alder, red	Red alder	*Alnus rubra*
Ash:		
Black	Black ash	*Fraxinus nigra*
Oregon	Oregon ash	*F. latifolia*
	⎰ Blue ash	*F. quadrangulata*
White	⎨ Green ash	*F. pennsylvanica*
	⎱ White ash	*F. americana*
Aspen (popple)	⎰ Bigtooth aspen	*Populus grandidentata*
	⎱ Quaking aspen	*P. tremuloides*
Basswood	⎰ American basswood	*Tilia americana*
	⎱ White basswood	*T. heterophylla*
Beech	American beech	*Fagus grandifolia*
	⎧ Gray birch	*Betula populifolia*
	⎪ Paper birch	*B. papyrifera*
Birch	⎨ River birch	*B. nigra*
	⎪ Sweet birch	*B. lenta*
	⎩ Yellow birch	*B. alleghaniensis*
Box elder	Boxelder	*Acer negundo*
Buckeye	⎰ Ohio buckeye	*Aesculus glabra*
	⎱ Yellow buckeye	*A. octandra*
Butternut	Butternut	*Juglans cinerea*
Cherry	Black cherry	*Prunus serotina*
Chestnut	American chestnut	*Castanea dentata*
	⎧ Balsam poplar	*Populus balsamifera*
Cottonwood	⎨ Eastern cottonwood	*P. deltoides*
	⎩ Black cottonwood	*P. trichocarpa*
Cucumber	Cucumbertree	*Magnolia acuminata*
Dogwood	⎰ Flowering dogwood	*Cornus florida*
	⎱ Pacific dogwood	*C. nuttallii*
Elm:		
	⎧ Cedar elm	*Ulmus crassifolia*
Rock	⎨ Rock elm	*U. thomasii*
	⎪ September elm	*U. serotina*
	⎩ Winged elm	*U. alata*
Soft	⎰ American elm	*U. americana*
	⎱ Slippery elm	*U. rubra*
Gum	Sweetgum	*Liquidambar styraciflua*
Hackberry	⎰ Hackberry	*Celtis occidentalis*
	⎱ Sugarberry	*C. laevigata*

without cutting. No. 2 Common may contain grain and surface imperfections of all kinds but must provide a serviceable floor.

Standard thicknesses of oak flooring are 25/32, 1/2, and 3/8 inch. Standard face widths are 1-1/2, 2, 2-1/4, and 3-1/4 inches. Lengths in upper grades are 2 feet and up with a required average of 4-1/2 feet in a shipment. In the lower grades, lengths are 1-1/4 feet and up with a required average of 2-1/2 or 3 feet per shipment.

The rules of the National Oak Flooring Manufacturers Association also include specifications for flooring of pecan, hard maple, beech, and birch. The grades of pecan flooring are: First grade, practically clear but unselected for color; First grade red, practically clear with an all-heartwood face; First

Table 5 – 3—*Nomenclature of commercial hardwood lumber*—Continued

Commercial name for lumber	Official common tree name	Botanical name
Hickory	Mockernut hickory	Carya tomentosa
	Pignut hickory	C. glabra
	Shagbark hickory	C. ovata
	Shellbark hickory	C. laciniosa
Holly	American holly	Ilex opaca
Ironwood	Eastern hophornbeam	Ostrya virginiana
Locust	Black locust	Robinia pseudoacacia
	Honeylocust	Gleditsia triacanthos
Madrone	Pacific madrone	Arbutus menziesii
Magnolia	Southern magnolia	Magnolia grandiflora
	Sweetbay	M. virginiana
Maple:		
Hard	Black maple	Acer nigrum
	Sugar maple	A. saccharum
Oregon	Big leaf maple	A. macrophyllum
Soft	Red maple	A. rubrum
	Silver maple	A. saccharinum
Oak:		
Red	Black oak	Quercus velutina
	Blackjack oak	Q. marilandica
	California black oak	Q. kelloggi
	Cherrybark oak	Q. falcata var. pagodaefolia
	Laurel oak	Q. laurifolia
	Northern pin oak	Q. ellipsoidalis
	Northern red oak	Q. rubra
	Nuttall oak	Q. nuttallii
	Pin oak	Q. palustris
	Scarlet oak	Q. coccinea
	Shumard oak	Q. shumardii
	Southern red oak	Q. falcata
	Turkey oak	Q. laevis
	Willow oak	Q. phellos

grade white, practically clear with an all-bright sapwood face; Second grade, admits sound tight knots, pin wormholes, streak, and slight machining imperfections; Second grade red, similar to Second grade but must have a heartwood face; Third grade, must make a sound floor without cutting; and Fourth grade, must provide a serviceable floor. The standard sizes for pecan flooring are the same as those for oak flooring.

The National Oak Flooring Manufacturers Association rules for hard maple, beech, and birch flooring are the same as those of the Maple Flooring Manufacturers Association.

Hardwood Lumber Species

The names used by the trade to describe commercial lumber in the United States are not always the same as the names

Table 5 – 3—Nomenclature of commercial hardwood lumber—Continued

Commercial name for lumber	Official common tree name	Botanical name
Oak (con.)	Arizona white oak	Q. arizonica
	Blue oak	Q. douglasii
	Bur oak	Q. macrocarpa
	Valley oak	Q. lobata
	Chestnut oak	Q. primus
	Chinkapin oak	Q. muehlenbergii
	Emory oak	Q. emoryi
White	Gambel oak	Q. gambelii
	Mexican blue oak	Q. oblongifolia
	Live oak	Q. virginiana
	Oregon white oak	Q. garryana
	Overcup oak	Q. lyrata
	Post oak	Q. stellata
	Swamp chestnut oak	Q. michauxii
	Swamp white oak	Q. bicolor
	White oak	Q. alba
Oregon myrtle	California-laurel	Umbellularia californica
Osage orange	Osage-orange	Maclura pomifera
	Bitternut hickory	Carya cordiformis
Pecan	Nutmeg hickory	C. myristiciformis
	Water hickory	C. aquatica
	Pecan	C. illinoensis
Persimmon	Common persimmon	Diospyros virginiana
Poplar	Yellow-poplar	Liriodendron tulipifera
Sassafras	Sassafras	Sassafras albidum
Sycamore	Sycamore	Platanus occidentalis
Tanoak	Tanoak	Lithocarpus densiflorus
	Black tupelo; blackgum	Nyssa sylvatica
Tupelo	Ogeechee tupelo	N. ogeche
	Water tupelo	N. aquatica
Walnut	Black walnut	Juglans nigra
Willow	Black willow	Salix nigra
	Peachleaf willow	S. amygdaloides

Table 5-4—*Standard thicknesses for rough and surfaced (S2S) hardwood lumber*

Rough	Surfaced	Rough	Surfaced
- - - - - - *Inch* - - - - - -		- - - - - - *Inch* - - - - - -	
3/8	3/16	2-1/2	2-1/4
1/2	5/16	3	2-3/4
5/8	7/16	3-1/2	3-1/4
3/4	9/16	4	3-3/4
1	13/16	4-1/2	([1])
1-1/4	1-1/16	5	([1])
1-1/2	1-5/16	5-1/2	([1])
1-3/4	1-1/2	6	([1])
2	1-3/4		

[1] Finished size not specified in rules. Thickness subject to special contract.

of trees adopted as official by the USDA Forest Service. Table 5-3 shows the common trade name, the USDA Forest Service tree name, and the botanical name. United States agencies and associations that prepare rules for and supervise grading of hardwoods are given in table 5-2.

Softwood Lumber

Softwood lumber for many years has demonstrated the versatility of wood by serving as a primary raw material for construction and manufacture. In this role it has been produced in a wide variety of products from many different species. The first industry-sponsored grading rules (product descriptions) for softwoods were established before 1900 and were comparatively simple because the sawmills marketed their lumber locally and grades had only local significance. As new timber sources were developed and lumber was transported to distant points, each producing region continued to establish its own grading rules, so lumber from various regions differed in size, grade name, and allowable grade characteristics. When different species were graded under different rules and competed in the chief consuming areas, confusion and dissatisfaction were inevitable.

To eliminate unnecessary differences in the grading rules of softwood lumber and to improve and simplify these rules, a number of conferences were organized from 1919 to 1925 by the U.S. Department of Commerce. These were attended by representatives of lumber manufacturers, distributors, wholesalers, retailers, engineers, architects, and contractors. The result was a relative standardization of sizes, definitions,

and procedures for deriving allowable design properties, formulated as a voluntary American Lumber Standard. This standard has been modified several times since. The current edition of the standard is the American Softwood Lumber Standard PS 20-70 (ALS). Lumber cannot be graded as American Standard lumber unless the grade rules have been approved by the American Lumber Standards Committee.

Softwood lumber is classified for market use by form of manufacture, species, and grade. For many products the American Softwood Lumber Standard serves as a basic reference. For specific information on other products, reference must be made to industry marketing aids, trade journals, and grade rules.

Softwood Lumber Grades

Softwood lumber grades can be considered in the context of two major categories of use—construction and remanufacture. The term construction relates principally to lumber expected to function as graded and sized after primary processing (sawing and planing). The term remanufacture refers to lumber that will undergo a number of further manufacturing steps and reach the consumer in a significantly different form.

Lumber for Construction

The grading requirements of construction lumber are related specifically to the major construction uses intended and little or no further grading occurs once the piece leaves the sawmill. Construction lumber can be placed in three general categories—stress-graded, nonstress-graded, and appearance lumber. Stress-graded and nonstress-graded lumber are employed where the structural integrity of the piece is the primary requirement. Appearance lumber, as categorized here, encompasses those lumber products in which appearance is of primary importance; structural integrity, while sometimes important, is a secondary feature.

Stress-Graded Lumber—Almost all softwood lumber nominally 2 to 4 inches thick (dimension lumber) is stress graded and assigned allowable properties under the National Grading Rule, a part of the American Softwood Lumber Standard. Dimension should not be confused with hardwood dimension lumber, a name applied to hardwood stock intended for remanufacture into pieces of specified size called dimension parts. For lumber of this kind there is a single set of grade names and descriptions used throughout the United States although the allowable properties vary with species. Other stress-graded products include timbers, posts, stringers, beams, decking, and some boards. Stress-graded lumber may be graded visually or mechanically. The grade designation for mechanically or machine stress rated (MSR) lumber includes an E rating (modulus of elasticity).

Stress grades and the National Grading Rule are discussed in chapter 6.

Nonstress-Graded Lumber—In nonstress-graded structural lumber, the section properties (shape, size) of the pieces combine with the visual grade requirements to provide the degree of structural integrity intended.

Boards are the most important nonstress-graded product. The usual grades of boards are suitable for construction and general utility purposes. They are separated into three to five different grades depending upon the species and lumber manufacturing association involved. Grades may be described by number (No. 1, No. 2, No. 3) or by descriptive terms (Select Merchantable, Construction, Standard).

Because there are differences in the inherent properties of the various species and in corresponding names, the grades for different species are not always interchangeable in use. The top-grade boards (No. 1, Select Merchantable) are usually graded primarily for serviceability, but appearance is also considered. This grade is used for such purposes as siding, cornice, shelving, and paneling. Features such as knots and knotholes are permitted to be larger and more frequent as the grade level becomes lower. Intermediate-grade boards are often used for such purposes as subfloors, roof and wall sheathing, and rough concrete work. The lower grade boards are not selected for appearance but for adequate strength. They are used for roof and wall sheathing, subfloor, and rough concrete form work (fig. 5–2).

Grading provisions for other nonstress-graded products vary by species, product, and grading association. For detailed descriptions it is necessary to consult the appropriate grade rule for these products. Addresses are given in table 5–5.

Appearance Lumber—Appearance lumber often is nonstress-graded but forms a separate category because of the distinct importance of appearance in the grading process. This category of construction lumber includes most lumber that is machined to a pattern and lumber that is S4S. Secondary manufacture on these items is usually restricted to onsite fitting such as cutting to length and mitering. The appearance category includes trim, siding, flooring, ceiling, paneling, casing, base, stepping, and finish boards. This category of lumber is graded as Finish or Selects.

Most Finish lumber grades are described by letters and combinations of letters (B&BTR, C&BTR, D) or names (Superior, Prime) depending upon the grading agency. (See Commonly Used Lumber Abbreviations at the end of this chapter for definitions of letter grades.)

Select grades are described by numbers, letters, and names or combinations of them. Three grades are listed: B&BTR—1 and 2 Clear, C Select, and D Select. Special names are given to western white pine grades that correspond to the above categories: Supreme–IWP, Choice–IWP, and Quality–IWP.

The specification FG (flat grain), VG (vertical grain), or MG (mixed grain) is offered as a purchase option for some appearance lumber products.

In cedar and redwood, there is a pronounced difference in color between heartwood and sapwood. Heartwood also has high natural resistance to decay, so some grades are denoted as "heart." Since appearance grades emphasize the quality of one face, the reverse side may be lower in quality. Appearance grades are not uniform across species and products and official grade rules must be used for detailed reference.

Lumber for Remanufacture

A wide variety of species, grades, and sizes of softwood lumber is supplied to industrial accounts for cutting to specific smaller sizes which become integral parts of other products. In the secondary manufacturing process, grade descriptions, sizes, and often the entire appearance of the wood piece are changed. Thus the role of the grading process for these remanufacture items is to reflect as accurately as possible the yield to be obtained in the subsequent cutting operation. Typical of lumber for secondary manufacture are the factory grades, industrial clears, box lumber, moulding stock, and ladder stock. The variety of species available for these purposes has led to a variety of grade names and grade definitions. The following sections briefly outline some of the more common classifications. For details, reference must be made to industry sources. Availability and grade designation often vary by region and species.

Factory (Shop) Grades—Traditionally softwood lumber used for cuttings has been termed Factory or Shop. This lumber forms the basic raw material for many secondary manufacturing operations. Some grading associations refer to these grades as Factory while others refer to Shop. All impose a somewhat similar nomenclature in the grade structure. Shop lumber is graded on the basis of characteristics affecting its use for general cut-up purposes, or on the basis of size of cutting, such as for sash and doors. Factory Select and Select Shop are typical high grades, followed by No. 1, No. 2, and No. 3 Shop.

Grade characteristics of boards are influenced by the width, length, and thickness of the basic piece and are based on the amount of high-quality material that can be removed by cutting. Typically, a Select Shop would be required to contain either (a) 70 percent of cuttings of specified size, clear on both sides or (b) 70 percent cuttings of different size equal to a B&BTR grade on one side. No. 1 Shop would be required to have 50 percent; No. 2 Shop would be required to have 33-1/3 percent. Because of different characteristics assigned to grades with similar nomenclature, grades labeled Factory or Shop must be referenced to the appropriate industry source.

Industrial Clears—These grades are used for trim, cabinet stock, garage door stock, and other product components

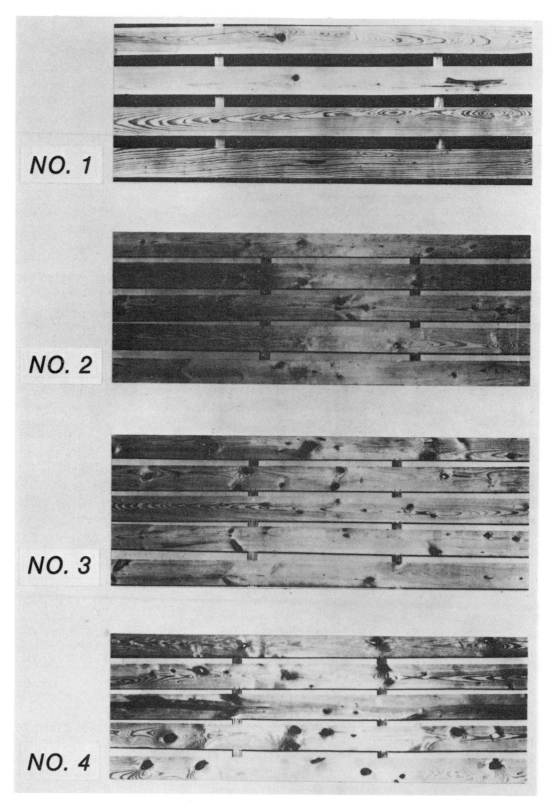

Figure 5—2—Typical examples of softwood boards
in grades No. 1, No. 2, No. 3, and No. 4.

(M85 0012)

5-12

Table 5 – 5—*Organizations promulgating softwood grades*

Name and address	Species covered by grading rules
National Hardwood Lumber Association P.O. Box 34518 Memphis, TN 38184-0518	Baldcypress, eastern redcedar
Northeastern Lumber Manufacturers Association, Inc. 4 Fundy Road Falmouth, ME 04105	Balsam fir, eastern white pine, red pine, eastern hemlock, black spruce, white spruce, red spruce, pitch pine, tamarack, jack pine, northern white cedar
Northern Hardwood and Pine Manufacturers Association, Inc. 8600 W. Bryn Mawr Ave., Suite 720-S Chicago, IL 60631	Aspen, cottonwood, eastern white pine, red pine, jack pine, black spruce, white spruce, red spruce, balsam fir, eastern hemlock, tamarack, yellow-poplar
Red Cedar Shingle & Handsplit Shake Bureau 515 116th Avenue NE., Suite 275 Bellevue, WA 98004	Western redcedar (shingles and shakes)
Redwood Inspection Service 591 Redwood Highway, Suite 3100 Mill Valley, CA 94941	Redwood
Southern Cypress Manufacturers Association 805 Sterick Building Memphis, TN 38103	Baldcypress
Southern Pine Inspection Bureau 4709 Scenic Highway Pensacola, FL 32504	Longleaf pine, slash pine, shortleaf pine, loblolly pine, Virginia pine, pond pine, pitch pine
West Coast Lumber Inspection Bureau Box 23145 6980 SW. Varns Road Portland, OR 97223	Douglas-fir, western hemlock, western redcedar, incense-cedar, Port-Orford-cedar, Alaska-cedar, western true firs, mountain hemlock, Sitka spruce
Western Wood Products Association 1500 Yeon Building Portland, OR 97204	Ponderosa pine, western white pine, Douglas-fir, sugar pine, western true firs, western larch, Engelmann spruce, incense-cedar, western hemlock, lodgepole pine, western redcedar, mountain hemlock, red alder, aspen

where excellent appearance, mechanical and physical properties, and finishing characteristics are important. The principal grades are B&BTR, C, and D Industrial. Grading is based primarily on the best face, although the influence of edge characteristics is important and varies depending upon piece width and thickness. In redwood the Industrial Clear All Heart grade includes an "all-heart" requirement for decay resistance in manufacture of cooling towers, tanks, pipe, and similar products.

Moulding, Ladder, Pole, Tank and Pencil Stock—Within producing regions, grading rules delineate the requirements for a variety of lumber classes oriented to specific consumer products. Custom and the characteristics of the wood supply lead to different grade descriptions and terminology. For example, in West Coast species, the ladder industry can choose from one "ladder and pole stock" grade plus two ladder rail grades and one ladder rail stock grade. In southern pine, ladder stock is available as Select and Industrial. Moulding stock, tank stock, pole stock, stave stock, stadium seat stock, box lumber, and pencil stock are other typical classes oriented to the final product. Some product classes have only one grade level; a few offer two or three levels. Special features of these grades may include a restriction on sapwood related to desired decay resistance, specific requirements for slope of grain and growth ring orientation for high-stress use such as ladders, and particular cutting requirements as in pencil stock. All references to these grades should be made directly to current lumber association grading rules.

Structural Laminations—Structural laminating grades describe the characteristics used to segregate lumber to be used in structural glued-laminated (glulam) timbers. Generally, allowable properties are not assigned separately to laminating grades, rather, the grades permitted are based on the expected effect of that grade of lamination on the combined glulam timber.

There are two kinds of graded material: visually graded and E-rated material. Visually graded material is graded according to one of three sets of rules. The first set is based on the rules of the American Softwood Lumber Standard with additional requirements for laminating. The second involves the laminating grades typically used for visually graded western species and includes the three basic categories called L1, L2, and L3. The third set includes special requirements for tension members and outer tension laminations on bending members. The visual grades have provisions for dense, close-grain, medium-grain, or coarse-grain lumber.

The E-rated grades are categorized by a combination of visual grading criteria and lumber stiffness. These grades are expressed in terms of the size of maximum edge characteristic permitted (as a fraction of the width) along with a specified long span modulus of elasticity (e.g., 1/6 − 2.2E).

Softwood Lumber Manufacture

Size

Lumber length is recorded in actual dimensions while width and thickness are traditionally recorded in "nominal" dimensions—the actual dimension being somewhat less.

Softwood lumber is manufactured in length multiples of 1 foot as specified in various grading rules. In practice, 2-foot multiples (in even numbers) are the rule for most construction lumber. Width of softwood lumber varies, commonly from 2 to 16 inches nominal. The thickness of lumber can be generally categorized as follows:

Boards.—lumber less than 2 inches in nominal thickness.

Dimension.—lumber from 2 inches to, but not including, 5 inches in nominal thickness.

Timbers.—lumber 5 or more inches in nominal thickness in the least dimension.

To standardize and clarify nominal-actual sizes the American Softwood Lumber Standard (ALS) specifies thickness and width for lumber that falls under the standard.

The standard sizes for stress-graded and nonstress-graded construction lumber are given in table 5−6. Timbers are usually surfaced while green and only green sizes are given. Dimension and boards may be surfaced green or dry at the prerogative of the manufacturer; therefore, both green and dry standard sizes are given. The sizes are such that a piece of green lumber, surfaced to the standard green size, will shrink to approximately the standard dry size as it dries down to about 15 percent moisture content. The ALS definition of dry is a moisture content of 19 percent or less with an average of 15 percent. Lumber may also be designated KD (kiln dried), having a maximum moisture content of 15 percent and an average of 12 percent. Many types of lumber are dried before surfacing and only dry sizes for these products are given in the standard.

Lumber for remanufacture is offered in specified sizes to fit end product requirements. Factory (Shop) grades for general cuttings are offered in nominal thicknesses from less than 1 inch to 4 inches. Nominal thicknesses of door cuttings start at 1-3/8 inches. Cuttings are of various lengths and widths. Laminating stock sometimes is offered oversize, compared to standard dimension sizes, to permit resurfacing prior to laminating. Industrial Clears can be offered rough or surfaced in a variety of sizes starting from less than 2 inches thick and as narrow as 3 inches. Sizes for special product grades such as moulding stock and ladder stock are specified in appropriate grading rules or handled by purchase agreements.

Surfacing

Lumber can be produced either rough or surfaced (dressed). Rough lumber has surface imperfections caused by the primary sawing operations. It may be greater than target size by variable amounts in both thicknesses and width, depending upon the type of sawmill equipment. Rough lumber serves as a raw material for further manufacture and also for some decorative purposes. A rough-sawn surface is common in post and timber products. Because of surface roughness, grading of rough lumber generally is difficult.

Surfaced lumber has been planed or sanded on one side (S1S), two sides (S2S), one edge (S1E), two edges (S2E), or combinations of sides and edges (S1S1E, S2S1E, S1S2E, or S4S). Surfacing may be done to attain smoothness or uniformity of size or both.

Imperfections or blemishes defined in the grading rules and caused by machining are classified as "manufacturing imperfections" or "mismanufacture." For example, chipped and torn grain are irregularities of the surface where particles of the surface have been torn out by the surfacing operation. Chipped grain is a "barely perceptible" characteristic, while torn grain is classified by depth. Raised grain, skip, machine burn and gouge, chip marks, and wavy dressing are other manufacturing imperfections. Manufacturing imperfections (mismanufacture) are defined in the American Lumber Standard and further detailed in the grading rules. Classifications

Table 5 – 6—American Standard lumber sizes for stress-graded and nonstress-graded lumber for construction[1]

Item	Thickness			Face width		
	Nominal	Minimum dressed		Nominal	Minimum dressed	
		Dry	Green		Dry	Green
	- - - - - - - - - - Inch - - - - - - - - - -			- - - - - - - - - - - Inch - - - - - - - - - - -		
Boards	1	3/4	25/32	2	1-1/2	1-9/16
	1-1/4	1	1-1/32	3	2-1/2	2-9/16
	1-1/2	1-1/4	1-9/32	4	3-1/2	3-9/16
				5	4-1/2	4-5/8
				6	5-1/2	5-5/8
				7	6-1/2	6-5/8
				8	7-1/4	7-1/2
				9	8-1/4	8-1/2
				10	9-1/4	9-1/2
				11	10-1/4	10-1/2
				12	11-1/4	11-1/2
				14	13-1/4	13-1/2
				16	15-1/4	15-1/2
Dimension	2	1-1/2	1-9/16	2	1-1/2	1-9/16
	2-1/2	2	2-1/16	3	2-1/2	2-9/16
	3	2-1/2	2-9/16	4	3-1/2	3-9/16
	3-1/2	3	3-1/16	5	4-1/2	4-5/8
	4	3-1/2	3-9/16	6	5-1/2	5-5/8
	4-1/2	4	4-1/16	8	7-1/4	7-1/2
				10	9-1/4	9-1/2
				12	11-1/4	11-1/2
				14	13-1/4	13-1/2
				16	15-1/4	15-1/2
Timbers	5 and greater		1/2 less than nominal	5 and greater		1/2 less than nominal

[1] Nominal sizes in the table are used for convenience. No inference should be drawn that they represent actual sizes.

of manufacturing imperfections (combinations of the imperfections allowed in the rules) are established in the rules as Standard "A", Standard "B", etc. For example, Standard "A" admits very light torn grain, occasional slight chip marks, and very slight knife marks. These classifications are used as part of the grade description of some lumber products to specify the allowable surfacing quality.

Patterns

Lumber which, in addition to being surfaced, has been matched, shiplapped, or otherwise patterned is often classed as "worked lumber." Figure 5−3 shows typical patterns.

Softwood Lumber Species

The names of lumber adopted by the trade as American Standard may vary from the names of trees adopted as official by the USDA Forest Service. Table 5−7 shows the ALS commercial name for lumber, the USDA Forest Service tree name, and the botanical name. Some softwood species are marketed primarily in combinations. Designations such as Southern Pine and Hem-Fir represent typical combinations. The grading organizations listed in table 5−5 should be contacted for questions regarding combination names and species not listed in table 5−7. Further discussion of species grouping is contained in chapter 6.

Softwood Lumber Grading

Most lumber is graded under the supervision of inspection bureaus and grading agencies. These organizations supervise lumber mill grading, and provide reinspection services to resolve disputes concerning lumber shipments. Some of the agencies also author grading rules that reflect the species and products in the geographic regions they represent.[2] Many of the grading rules and procedures follow the American Softwood Lumber Standard. This is important because it provides for recognized uniform grading procedures. Names and addresses of rules-writing organizations in the United States, and the species with which they are concerned, are given in table 5−5. Canadian softwood lumber imported into the United States is graded by inspection agencies in Canada, also by the procedures of the American Softwood Lumber Standard. Names and addresses of Canadian grading agencies may be obtained from the Canadian Lumber Standards Accreditation Board, Suite 4175, 1055 West Hastings Street, Vancouver, BC, Canada V6E 2E9.

---

[2] A limited number of hardwoods are also being graded under the provisions of the standards used for grading softwoods. These hardwoods include aspen, red alder, cottonwood, and yellow-poplar.

Purchasing Lumber

After primary manufacture, most lumber products are marketed through wholesalers to remanufacturing plants or to retail outlets. Because of the extremely wide variety of lumber products, wholesaling is very specialized with some organizations dealing only with a limited number of species or products. Where the primary manufacturer can readily identify the customers, direct sales may be made. Primary manufacturers often sell directly to large retail chains, contractors, manufacturers of mobile and modular housing, and truss fabricators.

Some primary manufacturers and wholesale organizations set up distribution yards in lumber-consuming areas to more effectively distribute both hardwood and softwood products. Retail yards draw inventory from distribution yards and, in wood-producing areas, from local lumber producers. The wide range of grades and species covered in the grade rules are not readily available in most retail outlets.

Transportation is a vital factor in lumber distribution. Often the lumber shipped by water is green because weight is not a major factor in this type of shipping. On the other hand, lumber reaching the East Coast from the Pacific Coast by rail is largely kiln dried because rail shipping rates are based on weight. A shorter rail haul places southern and northeastern species in a favorable shipping cost position in this same market.

Changing transportation costs have influenced shifts in market distribution of species and products. Trucks have become a major factor in lumber transport for regional remanufacture plants, for retail supply from distribution yards, and for much construction lumber distribution.

The development of foreign hardwood and softwood manufacturing and the availability of water transport has brought foreign lumber products to the United States market, particularly in coastal areas.

Retail Yard Inventory

The small retail yards throughout the United States carry softwoods required for ordinary construction purposes and often small stocks of one or two hardwoods in grades suitable for finishing or cabinetwork. Special orders must be made for other hardwoods. Trim items such as moulding in either softwood or hardwood are available cut to size and standard pattern. Cabinets are usually made by millwork plants ready for installation and many common styles and sizes are carried or cataloged by the modern retail yard. Hardwood flooring is available to the buyer only in standard patterns. Some retail yards may carry specialty stress grades of lumber such as structural light framing for truss fabrication.

The assortment of species in general construction items carried by retail yards depends largely upon geographic location, and both transportation costs and tradition are important factors. Retail yards within, or close to, a major lumber-producing region commonly emphasize the local timber. For example, a local retail yard on the coast in the Pacific Northwest may stock only green Douglas-fir and cedar in dimension grades, dry pine and hemlock in boards and moulding, plus assorted specialty items such as redwood posts, cedar shingles and shakes, and rough cedar siding. The only hardwoods carried may be walnut and "Philippine mahogany."[3] Retail yards farther from a major softwood supply, such as in the Midwest, may draw from several species-growing areas and for example may stock spruce and southern pine. Being located in a major hardwood production area, these yards will stock, or have available to them, a different and wider variety of hardwoods.

Geography has less influence where consumer demands are more specific. For example, where long construction lumber (20 to 26 ft) is required, West Coast species often are marketed because the size of the trees in several of the species makes long lengths a practical market item. Ease of treatability makes treated southern pine construction lumber available in a wide geographic area.

Stress-Graded Lumber for Construction

Dimension is the principal stress-graded lumber item available in a retail yard. It is primarily framing lumber for joists, rafters, and studs. Strength, stiffness, and uniformity of size are essential requirements. Dimension is stocked in all yards, frequently in only one or two of the general purpose construction woods such as pine, fir, hemlock, or spruce. Two by four and wider dimension is found in grades of Select Structural, No. 1, No. 2, and No. 3; often in combinations of No. 2&BTR or possibly No. 3&BTR. More often in 2 by 4, the grades available would normally be Construction and Standard, sold as Standard and Better (STD&BTR), Utility and Better (UTIL&BTR), or STUD (10 ft and shorter).

Dimension is often found in nominal 2-, 4-, 6-, 8-, 10-, or 12-inch widths and 8- to 18-foot lengths in multiples of 2 feet. Dimension formed by structural end-jointing procedures may be found. Dimension thicker than 2 inches and longer than 18 feet is not commonly available in large quantity in many retail yards.

Other stress-graded products generally present are posts and timbers, with some beams and stringers also possibly in stock. Typical stress grades in these products are Select Structural and No.1 Structural in Douglas-fir and No. 1SR and No. 2SR in southern pine.

---

[3] Common market name encompassing many species including tanguile, red lauan, and white lauan.

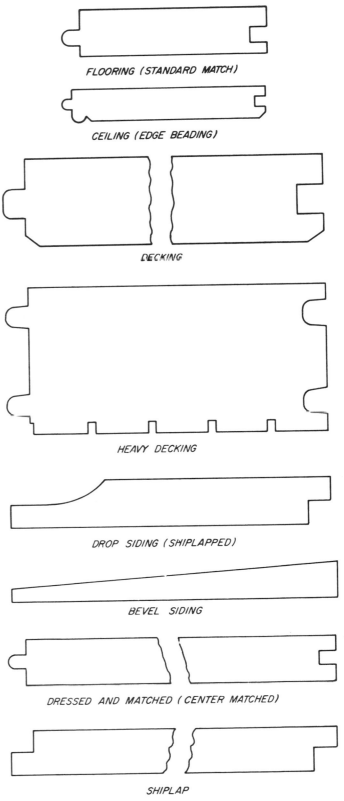

FLOORING (STANDARD MATCH)

CEILING (EDGE BEADING)

DECKING

HEAVY DECKING

DROP SIDING (SHIPLAPPED)

BEVEL SIDING

DRESSED AND MATCHED (CENTER MATCHED)

SHIPLAP

Figure 5–3—Typical patterns of worked lumber.   (M139 411)

Nonstress-Graded Lumber for Construction

Boards are the most common nonstress-graded general purpose construction lumber in the retail yard. Boards are stocked in one or more species, usually in nominal 1-inch thickness. Standard nominal widths are 2, 3, 4, 6, 8, 10, and 12 inches. Grades most generally available in retail yards are No. 1, No. 2, and No. 3 (or CONST, STD, and UTIL). These will often be combined in grade groups. Boards are sold square edged, dressed and matched (tongued and grooved), or with a shiplapped joint. Boards formed by end-jointing of shorter sections may form an appreciable portion of the inventory.

Appearance Lumber

Completion of a construction project usually depends on a variety of lumber items available in finished or semifinished form. The following items often may be stocked in only a few species, finishes, or in limited sizes depending on the yards.

Finish—Finish boards usually are available in a local yard in one or two species principally in grade C&BTR. Redwood and cedar have different grade designations. Grades such as

Table 5–7—*Nomenclature of commercial softwood lumber*

Standard lumber name under American Softwood Lumber Standards	Official Forest Service tree name used in this handbook	Botanical name
Cedar:		
Alaska	Alaska-cedar	*Chamaecyparis nootkatensis*
Eastern red	Eastern redcedar	*Juniperus virginiana*
Incense	Incense-cedar	*Libocedrus decurrens*
Northern white	Northern white-cedar	*Thuja occidentalis*
Port Orford	Port-Orford-cedar	*Chamaecyparis lawsoniana*
Southern white	Atlantic white-cedar	*C. thyoides*
Western red	Western redcedar	*Thuja plicata*
Cypress, red (coast type), yellow (inland type), white (inland type)	Baldcypress	*Taxodium distichum*
Douglas-fir	Douglas-fir	*Pseudotsuga menziesii*
Fir:		
Balsam	Balsam fir	*Abies balsamea*
	Fraser fir	*A. fraseri*
Noble	Noble fir	*A. procera*
	California red fir	*A. magnifica*
	Grand fir	*A. grandis*
White	Pacific silver fir	*A. amabilis*
	Subalpine fir	*A. lasiocarpa*
	White fir	*A. concolor*
Hemlock:		
Eastern	Eastern hemlock	*Tsuga canadensis*
Mountain	Mountain hemlock	*T. mertensiana*
West Coast	Western hemlock	*T. heterophylla*
	Alligator juniper	*Juniperus deppeana*
Juniper, western	Rocky Mountain juniper	*J. scopulorum*
	Utah juniper	*J. osteosperma*
	Western juniper	*J. occidentalis*
Larch, western	Western larch	*Larix occidentalis*

Clear Heart, A, or B are used in cedar; Clear All Heart, Clear, and Select are typical redwood grades. Finish boards are usually a nominal 1 inch thick, dressed two sides to 3/4 inch. The widths usually stocked are nominal 2 to 12 inches in even-numbered inches.

Siding—Siding as the name implies, is intended specifically to cover exterior walls. Beveled siding is ordinarily stocked only in white pine, ponderosa pine, western redcedar, cypress, or redwood. Drop siding, also known as rustic siding or barn siding, is usually stocked in the same species as beveled siding. Siding may be stocked as B&BTR or C&BTR except in cedar where Clear, A, and B may be available and redwood where Clear All Heart and Clear will be found. Vertical grain (VG) is sometimes a part of the grade designation. Drop siding sometimes is stocked also in sound knotted C and D grades of southern pine, Douglas-fir, and hemlock. Drop siding may be dressed and matched, or shiplapped.

Flooring—Flooring is made chiefly from hardwoods such as oak and maple, and the harder softwood species, such as

Table 5 – 7—*Nomenclature of commercial softwood lumber*—Continued

Standard lumber name under American Softwood Lumber Standards	Official Forest Service tree name used in this handbook	Botanical name
Pine:		
Idaho white	Western white pine	*Pinus monticola*
Jack	Jack pine	*P. banksiana*
Lodgepole	Lodgepole pine	*P. contorta*
Longleaf yellow[1]	Longleaf pine	*P. palustris*
	Slash pine	*P. elliottii*
Northern white	Eastern white pine	*P. strobus*
Norway	Red pine	*P. resinosa*
Ponderosa	Ponderosa pine	*P. ponderosa*
Southern Major	Longleaf pine	*P. palustris*
	Shortleaf pine	*P. echinata*
	Loblolly pine	*P. taeda*
	Slash pine	*P. elliottii*
Minor	Pitch pine	*P. rigida*
	Pond pine	*P. serotina*
	Sand pine	*P. clausa*
	Table mountain pine	*P. pungens*
	Virginia pine	*P. virginiana*
Sugar	Sugar pine	*P. lambertiana*
Redwood	Redwood	*Sequoia sempervirens*
Spruce:		
Eastern	Black spruce	*Picea mariana*
	Red spruce	*P. rubens*
	White spruce	*P. glauca*
Engelmann	Blue spruce	*P. pungens*
	Engelmann spruce	*P. engelmannii*
Sitka	Sitka spruce	*P. sitchensis*
Tamarack	Tamarack	*Larix laricina*
Yew, Pacific	Pacific yew	*Taxus brevifolia*

[1] The commercial requirements for longleaf pine lumber are that not only must it be produced from the species *Pinus elliottii* and *P. palustris*, but each piece must average either on 1 end or the other not less than 6 annual rings per inch and not less than 1/3 summerwood. Longleaf pine lumber is sometimes designated as pitch pine in the export trade.

Douglas-fir, western larch, and southern pine. Often at least one softwood and one hardwood are stocked. Flooring is usually nominal 1 inch thick. Thicker flooring is available for heavy-duty floors. Thinner flooring is available especially for re-covering old floors. Vertical and flat grain (also called quartersawed and plainsawed) flooring is manufactured from both softwoods and hardwoods. Vertical-grained flooring shrinks and swells less than flat-grained flooring, is more uniform in texture, wears more uniformly, and the joints do not open as much.

Softwood flooring is usually available in B&BTR grade, C Select, or D Select. The chief grades in maple are Clear No. 1 and No. 2. The grades in quartersawed oak are Clear and Select, and in plainsawed Clear, Select, and No. 1 Common. Quartersawed hardwood flooring has the same advantages as vertical-grained softwood flooring. In addition, the silver or flaked grain of quartersawed flooring is frequently preferred to the figure of plainsawed flooring.

Casing and Base—Casing and base are standard items in the more important softwoods and are stocked by most yards in at least one species. The chief grade, B&BTR, is designed to meet the requirements of interior trim for dwellings. Many casing and base patterns are dressed to 11/16 by 2-1/4; other sizes used include 9/16 by 3, 3-1/4, and 3-1/2. Hardwoods for the same purposes, such as oak and birch, may be carried in stock in the retail yard or may be obtained on special order.

Shingles and Shakes—Shingles usually available are sawn from western redcedar, northern white-cedar, and redwood. The shingle grades are: Western redcedar, No. 1, No. 2, No. 3; northern white-cedar, Extra, Clear, 2nd Clear, Clearwall, Utility; redwood, No. 1, No. 2 VG, and No. 2 MG.

Shingles that are all heartwood give greater resistance to decay than do shingles that contain sapwood. Edge-grained shingles are less likely to warp and split than flat-grained shingles; thick-butted shingles less likely than thin-butted shingles; and narrow shingles less likely than wide shingles. The standard thicknesses of thin butted shingles are described as 4/2, 5/2-1/4, and 5/2 (four shingles to 2 in. of butt thickness, five shingles to 2-1/4 in. of butt thickness, and five shingles to 2 in. of butt thickness). Lengths may be 16, 18, or 24 inches. Random widths and specified widths ("dimension" shingles) are available in western redcedar, redwood, and cypress.

Shingles are usually packed four bundles to the square. A square of shingles will cover 100 square feet of roof area when the shingles are applied at standard weather exposures.

Shakes are handsplit, or handsplit and resawn from western redcedar. Shakes are of a single grade and must be 100 percent clear, graded from the split face in the case of handsplit and resawn material. Handsplit shakes are graded from the best face. Shakes must be 100 percent heartwood. The standard thickness of shakes ranges from 3/8 to 1-1/4 inches. Lengths are 18 and 24 inches, and a 15-inch "Starter-Finish Course" length.

Important Purchase Considerations

The following outline lists some of the points to consider when ordering lumber or timbers.

1. Quantity—Lineal feet, board measure, number of pieces of definite size and length. Consider that the board measure depends on the thickness and width nomenclature used and that the interpretation of this must be clearly delineated. In other words, nominal or actual, pattern size, etc., must be considered.

2. Size—Thickness in inches—nominal and actual if surfaced on faces. Width in inches—nominal and also actual if surfaced on edges. Length in feet—may be nominal average length, limiting length, or a single uniform length. Often a trade designation, "random" length, is used to denote a nonspecified assortment of lengths. Note that such an assortment should contain critical lengths as well as a range. The limits allowed in making the assortment "random" can be established at the time of purchase.

3. Grade—As indicated in grading rules of lumber manufacturing associations. Some grade combinations (B&BTR) are official grades and others (STD&BTR) are unofficial combinations of grades and subject to purchase agreement. A typical assortment of the unofficial combinations labeled STD&BTR is 75 percent CONST and 25 percent STD. In ALS-graded softwood, each piece of such unofficial combinations of lumber is stamped with its grade, a name or number identifying the producing mill, the dryness at the time of surfacing, and a symbol identifying the inspection agency supervising the grading inspection. The grade designation stamped on a piece indicates the quality at the time the piece was graded. Subsequent exposure to unfavorable storage conditions, improper drying, or careless handling may cause the material to fall below its original grade.

Note that working or rerunning a graded product to a pattern may change or invalidate the original grade. The purchase specification should be clear regarding regrading or acceptance of worked lumber. In softwood lumber, grades for dry lumber generally are determined after kiln drying and surfacing. This practice is not general for hardwood factory lumber, however, where the grade is generally based on quality and size prior to kiln drying. To be certain the product grade is correct, refer to the grading rule by number and paragraph.

4. Species or groupings of wood—Douglas-fir, cypress, Hem-Fir, etc. Some species have been grouped for marketing

convenience; others are traded under a variety of names. Be sure the species or species group is correctly and clearly depicted on the purchase specification.

5. Product—Flooring, siding, timbers, boards, etc. Nomenclature varies by species, region, and grading association. To be certain the nomenclature is correct for the product, refer to the grading rule by number and paragraph.

6. Condition of seasoning—Air dry, kiln dry, etc. Softwood lumber dried to 19 percent moisture content or less is defined as dry by the American Lumber Standard. Associations designate other degrees of dryness such as air dry (AD), partially air dry (PAD), green (S-GRN), and 15 percent maximum (MC-15 or KD for southern pine). If the moisture requirement is critical, the level of moisture content and how it will be determined must be specified.

7. Surfacing and working—Rough (unplaned), dressed (surfaced), or patterned stock. Specify condition. If surfaced, indicate S4S, S1S1E, etc. If patterned, list pattern number with reference to the appropriate grade rules.

8. Grading rules—Official grading agency name and name of official rules under which a product is graded, product identification, paragraph and page number of rules, date of rules or official rule edition.

9. Manufacturer—Name of manufacturer or trade name of specific product or both. Most lumber products are sold without reference to a specific manufacturer. If proprietary names or quality features of a manufacturer are required, this must be stipulated clearly on the purchase agreement.

10. Reinspection—Procedures for resolution of purchase disputes. The American Softwood Lumber Standard provides for procedures to be followed in resolution of manufacturer-wholesaler-consumer conflicts over quality or quantity of American Standard grades of lumber. The dispute may be resolved by reinspecting the shipment. Time limits, liability, costs, and complaint procedures are outlined in the grade rules of both softwood and hardwood agencies under which the disputed shipment was graded and purchased.

Commonly Used Lumber Abbreviations

The following standard lumber abbreviations are commonly used in contracts and other documents for purchase and sale of lumber.

AAR	Association of American Railroads
AD	air dried
ADF	after deducting freight
AF	alpine fir
ALS	American Lumber Standard
AST	antistain treated; at ship tackle (western softwoods)

AV or avg	average
AW&L	all widths and lengths
B1S	see EB1S, CB1S, and E&CB1S
B2S	see EB2S, CB2S, and E&CB2S
B&B B&BTR	B and Better
B&S	beams and stringers
BD	board
BD FT	board feet
BDL	bundle
BEV	bevel or beveled
BH	boxed heart
B/L, BL	bill of lading
BM	board measure
BSND	bright sapwood no defect
BTR	better
CB	center beaded
CB1S	center bead on one side
CB2S	center bead on two sides
CC	cubical content
cft or cu. ft.	cubic foot or feet
CF	cost and freight
CIF	cost, insurance, and freight
CIFE	cost, insurance, freight, and exchange
CG2E	center groove on two edges
C/L	carload
CLG	ceiling
CLR	clear
CM	center matched
Com	common
CONST	construction
CS	calking seam
CSG	casing
CV	center V
CV1S	center V on one side
CV2S	center V on two sides
DB Clg	double beaded ceiling (E&CB1S)
DB Part	double beaded partition (E&CB2S)
DET	double end trimmed
DF	Douglas-fir
DF-L	Douglas-fir, larch
DIM	dimension
DKG	decking
D/S, DS, D/Sdg	drop siding
D1S, D2S	see S1S and S2S
D&M	dressed and matched
D&CM	dressed and center matched
D&SM	dressed and standard matched
D2S&CM	dressed two sides and center matched
D2S&SM	dressed two sides and standard matched

E	edge	H or M	hit or miss
EB1S	edge bead one side	IC	incense cedar
EB2S, SB2S	edge bead on two sides	IN, in.	inch or inches
		Ind	industrial
EE	eased edges	IWP	Idaho white pine
EG	edge (vertical or rift) grain	J&P	joists and planks
EM	end matched	JTD	jointed
EV1S, SV1S	edge V one side	KD	kiln dried
		L	western larch
EV2S, SV2S	edge V two sides	LBR, Lbr	lumber
		LCL	less than carload
E&CB1S	edge and center bead one side	LGR	longer
E&CB2S, DB2S, BC&2S	edge and center bead two sides	LGTH	length
		Lft, Lf	lineal foot or feet
		LIN, Lin	lineal
E&CV1S, DV1S, V&CV1S	edge and center V one side	LL	longleaf
		LNG, Lng	lining
		LP	lodgepole pine
E&CV2S, DV2S, V&CV2S	edge and center V two sides	M	thousand
		MBM, MBF, M.BM	thousand (feet) board measure
ES	Engelmann spruce	MC, M.C.	moisture content
F_b, F_t, F_c, F_v, F_{cx}	allowable stress (psi) in bending; in tension, compression and shear parallel to grain; and in compression perpendicular to grain, respectively	MERCH, Merch	merchantable
		MG	medium grain or mixed grain
		MH	mountain hemlock
FA	facial area	MLDG, Mldg	moulding
Fac	factory		
FAS	free alongside (vessel)	Mft	thousand feet
FAS	Firsts and Seconds	M-S	mixed species
FAS1F	Firsts and Seconds one face	MSR	machine stress rated
FBM, Ft. BM	feet board measure	N	nosed
		NBM	net board measure
FG	flat or slash grain	No.	number
FJ	finger joint. End-jointed lumber using a finger-joint configuration	N1E or N2E	nosed one or two edges
FLG, Flg	flooring	Ord	order
FOB	free on board (named point)	PAD	partially air dry
FOHC	free of heart center	PAR, Par	paragraph
FOK	free of knots	PART, Part	partition
FRT, Frt	freight	PAT, Pat	pattern
FT, ft	foot or feet	Pcs.	pieces
FT. SM	feet surface measure	PE	plain end
G	girth	PET	precision end trimmed
GM	grade marked	PP	ponderosa pine
G/R	grooved roofing	P&T	posts and timbers
HB, H.B.	hollow back	P1S, P2S	see S1S and S2S
HEM	hemlock	RDM	random
H-F	hem-fir	REG, Reg	regular
Hrt	heart	Rfg.	roofing
H&M	hit and miss	RGH, Rgh	rough

R/L, RL	random lengths		S2E	surfaced two edges
R/W, RW	random widths		S1S	surfaced one side
RES	resawn		S2S	surfaced two sides
SB1S	single bead one side		S4S	surfaced four sides
SDG, Sdg	siding		S1S&CM	surfaced one side and center matched
S-DRY	surfaced dry; lumber 19 percent moisture content or less per American Lumber Standard for softwood		S2S&CM	surfaced two sides and center matched
			S4S&CS	surfaced four sides and calking seam
			S1S1E	surfaced one side, one edge
SE	square edge		S1S2E	surfaced one side, two edges
SEL, Sel	select or select grade		S2S1E	surfaced two sides, one edge
SE&S	square edge and sound		S2S&SL	surfaced two sides and shiplapped
SG	slash or flat grain		S2S&SM	surfaced two sides and standard matched
S-GRN	surfaced green; lumber unseasoned, in excess of 19 percent moisture content per American Lumber Standard for softwood		TBR	timber
			T&G	tongued and grooved
			UTIL	utility
SGSSND	Sapwood, gum spots and streaks, no defect		VG	vertical (edge) grain
SIT. SPR	Sitka spruce		V1S	see EV1S, CV1S, and E&CV1S
S/L, SL, S/Lap	shiplap		V2S	see EV2S, CV2S, and E&CV2S
			WC	western cedar
SM	surface measure		WCH	West Coast hemlock
Specs	specifications		WCW	West Coast woods
SP	sugar pine		WDR, wdr	wider
SQ	square		WF	white fir
SQRS	squares		WHAD	worm holes a defect
SR	stress rated		WHND	worm holes no defect
STD, Std	standard		WT	weight
Std. lgths.	standard lengths		WTH	width
STD. M	standard matched		WRC	western redcedar
SS	Sitka spruce		WW	white woods (Engelmann spruce, any true firs, any hemlocks, any pines)
SSND	sap stain no defect (stained)			
STK	stock		YP	yellow pine
STPG	stepping			
STR, STRUCT	structural			
SYP	southern pine			
S&E	side and edge (surfaced on)			
S1E	surfaced one edge			

Selected References

U.S. Department of Commerce. Strip oak flooring. Comm. Stand. CS 56. Washington, DC: USDC; (see current edition).

U.S. Department of Commerce. American softwood lumber standard. Prod. Stand. PS 20. Washington, DC: USDC; (see current edition).

Chapter 6

Lumber Stress Grades and Allowable Properties

Lumber Stress Grades and Allowable Properties[*]

Lumber of any species and size, as it is sawed from the log, is quite variable in its mechanical properties. Pieces may differ in strength by several hundred percent. For simplicity and economy in use, pieces of lumber of similar mechanical properties are placed in classes called stress grades.

Stress grades are characterized by: (1) One or more sorting criteria. (2) A set of allowable properties for engineering design. (3) A unique grade name.

This chapter discusses sorting criteria for two stress-grading methods, along with the philosophy of how allowable properties for engineering design are derived. The allowable properties depend upon the particular sorting criteria and on additional factors that are independent of the sorting criteria. Allowable properties are usually much lower than the properties of clear, straight-grained wood tabulated in chapter 4.

From one to six allowable properties are associated with a stress grade—modulus of elasticity and stresses in tension and compression parallel to the grain, in compression perpendicular to the grain, in shear parallel to the grain, and in extreme fiber in bending. As with any structural material, the allowable engineering design properties must be either inferred or measured nondestructively. In wood, the properties are inferred through visual grading criteria or a nondestructive measurement such as flatwise bending stiffness. This nondestructive test provides both a sorting criterion and a means of calculating appropriate mechanical properties.

The philosophies contained in this chapter are used by a number of organizations to develop visual and mechanical stress grades. The exact procedures and the resulting allowable stresses are not detailed in the Wood Handbook, but references to them are given.

Developing Visual Grades

Visual Sorting Criteria

Visual grading is the oldest stress-grading method. It is based on the premise that mechanical properties of lumber differ from mechanical properties of clear wood because there are many growth characteristics that affect properties and can be seen and judged by eye. These growth characteristics are used to sort the lumber into stress grades. The following are typical visual sorting criteria:

Density

Strength is related to the weight per unit volume (density) of clear wood. Properties assigned to lumber are sometimes modified by using the rate of growth and the percentage of latewood as measures of density. Typically, selection for density requires that the number of annual rings per inch and the percent latewood be within a specified range. It is possible to eliminate some very low-strength pieces from a grade by excluding those that are exceptionally light in weight.

Decay

Decay in most forms should be prohibited or severely restricted in stress grades because the degree of decay is difficult to determine and its effect on strength is often greater than visual observation would indicate. Decay of the pocket type (e.g., *Fomes pini*) can be permitted to some degree in stress grades, as can decay that occurs in knots but does not extend into the surrounding wood.

Heartwood and Sapwood

Heartwood and sapwood of the same species have equal mechanical properties, and no heartwood requirement need be made in stress grading. Because heartwood of some species is more resistant to decay than sapwood, heartwood may be required if the untreated wood is to be exposed to a decay hazard. On the other hand, sapwood takes preservative treatment more readily and is preferred in lumber that is to be treated.

Slope of Grain

Slope of grain (cross grain) reduces the mechanical properties of lumber because the fibers are not parallel to the edges. Severely cross-grained pieces are also undesirable because they tend to warp with changes in moisture content. Stresses caused by shrinkage during drying are greater in structural lumber than in small, clear specimens and are increased in zones of sloping or distorted grain. To provide a margin of safety, the reduction in allowable properties due to cross grain in visually graded structural lumber should be considerably more than the reduction observed in tests of small, clear specimens that contain similar cross grain.

Knots

Knots cause localized cross grain with steep slopes. One of the most damaging aspects of knots in sawn lumber is that the continuity of the grain around the knot is interrupted by the sawing process.

In general, knots have a greater effect on strength in tension than compression; in bending it depends upon whether the knot is in the tension or compression side of the beam (those along the centerline have little or no effect). Intergrown (or live) knots resist (or transmit) some kinds of stress, but encased knots (unless very tight) or knotholes resist (or transmit) little or no stress. On the other hand, distortion of grain is greater around an intergrown knot than around an encased (or dead) knot of equivalent size. As a result, overall strength effects are roughly equalized, and often no distinction is made in stress grading between intergrown knots, dead knots, and knotholes.

The zone of distorted grain (cross grain) around a knot has

[*]Revision by B. Alan Bendtsen, Forest Products Technologist, and David W. Green, General Engineer.

less "parallel to piece" stiffness than straight-grained wood; thus, localized areas of low stiffness are often associated with knots. However, such zones generally comprise only a minor part of the total volume of a piece of lumber. Because overall piece stiffness reflects the character of all parts, stiffness is not greatly influenced by knots.

The presence of a knot in a piece has a greater effect on most strength properties than on stiffness. The effect of a knot on strength depends approximately on the proportion of the cross section of the piece of lumber occupied by the knot, upon knot location, and upon the distribution of stress in the piece. Limits on knot sizes are therefore made in relation to the width of and location on the face in which the knot appears. Compression members are stressed about equally throughout, and no limitation related to location of knots is imposed. In tension, knots along the edge of a member cause an eccentricity that induces bending stresses, and should therefore be more restricted than knots away from the edge. In structural members subjected to bending, stresses are greater in the middle part of the length and are greater at the top and bottom edges than at midheight. These facts are recognized by differing limitations on the sizes of knots in different locations.

Knots in glued-laminated structural members are not continuous as in sawed structural lumber, and different methods are used for evaluating their effect on strength (see ch. 10).

Shake

Shake is a separation or a weakness of fiber bond, between annual rings, that is presumed to extend lengthwise without limit. In members subjected to bending, shake reduces the resistance to shear and therefore grading rules restrict shake most closely in those parts of a bending member where shear stresses are highest. In members subjected only to tension or compression, shake does not greatly affect strength. Shake may be limited in a grade because of appearance and because it permits entrance of moisture that results in decay.

Checks and Splits

Checks and splits as opposed to shakes are rated only by the area of actual opening. An end split is considered equal to an end check that extends through the full thickness of the piece. The effects of checks and splits upon strength and the principles of their limitation are the same as for shake.

Wane

Requirements of appearance, fabrication, or the need for ample bearing or nailing surfaces generally impose stricter limitations on wane than does strength. Wane is therefore limited in structural lumber on those bases.

Pitch Pockets

Pitch pockets ordinarily have so little effect on structural lumber that they can be disregarded in stress grading if they are small and limited in number. The presence of a large number of pitch pockets, however, may indicate shake or weakness of bond between annual rings.

Deriving Properties for Visually Graded Lumber

The mechanical properties of visually graded lumber may be established by (1) appropriate modification of test results conducted on small, clear specimens (the small clear procedure) or (2) testing a representative sample of full-size members (in-grade testing procedure). Virtually all allowable properties given in current design standards in the United States have been derived from test results of small, clear specimens. However, a comprehensive re-evaluation of the mechanical properties of visually graded lumber is currently in progress in the United States and Canada that utilizes test results on full-size members (the so-called "in-grade" program). It is anticipated that within a few years allowable properties for most species in the United States and Canada will be based on tests of full-size specimens.

Small, Clear Procedure

The derivation of mechanical properties of visually graded lumber has historically been based on clear wood properties with appropriate modifications for the lumber characteristics allowed by the visual sorting criteria. Sorting criteria that influence mechanical properties are handled with "strength ratios" for the strength properties and with "quality factors" for the modulus of elasticity.

From piece to piece, there is variation both in the clear wood properties and in the occurrence of the growth characteristics. The influence of this variability on lumber properties is handled differently for strength than for modulus of elasticity.

Once the clear wood properties have been modified for the influence of sorting criteria and variability, additional modifications for size, moisture content, safety, and load duration are applied. The composite of these adjustments is an "allowable property," to be discussed in more detail later in the chapter.

Strength Properties—Each strength property of a piece of lumber is derived from the product of the clear wood strength for the species and the limiting strength ratio. The strength ratio is the hypothetical ratio of the strength of a piece of lumber with visible strength-reducing growth characteristics to its strength if those characteristics were absent.

The true strength ratio of a piece of lumber is never known and must be estimated. The strength ratio assigned to a growth characteristic, therefore, serves as a predictor of lumber strength. Strength ratio usually is expressed in percent, ranging from 0 to 100.

Estimated strength ratios for cross grain and density have been obtained empirically; strength ratios for other growth characteristics have been derived theoretically. For example, to account for the weakening effect of knots, the assumption is made that the knot is effectively a hole through the piece, reducing the cross section as shown in figure 6–1. For a beam containing an edge knot, the strength ratio is equal to the ratio of the bending moment that can be resisted by a beam with a reduced cross-section to that of a beam with a full cross-section:

$$SR = \left(1 - \frac{k}{h}\right)^2 \qquad (6-1)$$

where *SR* is strength ratio, *k* is the knot size, and *h* is the width of the face containing the knot. This is the basic expression for the effect of a knot at the edge of the vertical face of a beam that is deflected vertically. Figure 6–2 shows how strength ratio changes with knot size according to the formula.

Strength ratios for all knots, shakes, checks, and splits are derived using similar concepts. Strength ratio formulas are given in American Society for Testing and Materials (ASTM) D 245. The same reference contains rules for measuring the various growth characteristics.

An individual piece of lumber often will have several characteristics that can affect any particular strength property. Only the characteristic that gives the lowest strength ratio is used to derive the estimated piece strength. In theory, a visual stress grade contains lumber ranging from pieces having the minimum strength ratio permitted in the grade up to those pieces having the minimum strength ratio for the next higher grade. In practice, there are often pieces in a grade that have strength ratios of a higher grade. This is a result of grade reduction for factors such as wane that do not affect strength.

The range of strength ratios in a grade, and the natural variation in clear wood strength, give rise to variation in strength between pieces in the grade. To account for this variation, and to provide for safety in design, it is intended that the actual strength of at least 95 percent of the pieces in a grade exceed the allowable properties assigned to that grade. In visual grading according to ASTM D 245, this is handled by using a near-minimum clear wood strength as a base value, and multiplying it by the minimum strength ratio permitted in the grade to obtain the grade strength property. The near-minimum value is called the 5 percent exclusion limit. ASTM D 2555 provides clear wood strength data and gives a method for estimating the 5 percent exclusion limit.

Suppose a 5 percent exclusion limit for the clear wood bending strength for a species in the green condition is 7,000 psi. Suppose also that among the characteristics allowed in a grade of lumber, one characteristic (a knot, for example) provides the lowest strength ratio in bending—assumed in

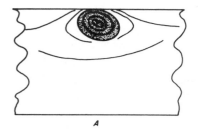
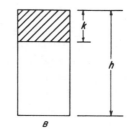

Figure 6–1—An edge knot in lumber, *A*, and the assumed loss of cross-section (indicated by cross-hatched area), *B*. (M139 125)

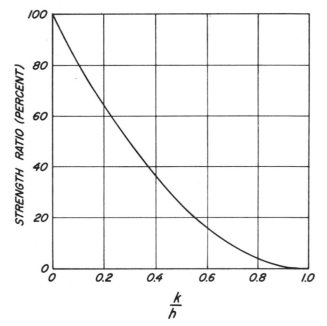

Figure 6–2—A relation between bending strength ratio and size of edge knot expressed as a fraction of face width. (M139 120)

this example as 40 percent. Using these numbers, the bending strength for the grade is estimated by multiplying the strength ratio (0.40) by 7,000 psi, equaling 2,800 psi (fig. 6–3). The bending strength in the green condition of 95 percent of the pieces in this species that have a strength ratio of 40 percent is expected to be 2,800 psi or more. Similar procedures are followed for other strength properties, using the appropriate clear wood property value and strength ratio. As noted, additional multiplying factors then are applied to produce allowable properties for design, as summarized later in this chapter.

Modulus of Elasticity—Modulus of elasticity (E) is a measure of the ability of a beam to resist deflection. The E assigned is an estimate of the average modulus of the lumber grade. The average modulus of elasticity for clear wood of the species, as recorded in ASTM D 2555, is used as a base. The clear wood average is multiplied by empirically derived "quality factors" to represent the reduction in modulus of elasticity that occurs by lumber grade. This procedure is outlined in ASTM D 245.

For example, assume a clear wood average modulus of elasticity of 1.8 million psi for the example shown earlier. The limiting bending strength ratio was 40 percent. ASTM D 245 assigns a quality multiplying factor of 0.80 for lumber with this bending strength ratio. The modulus of elasticity for that grade would be the product of the clear wood modulus and the quality factor; i.e., 1.8 x 0.8 = 1.44 million psi.

Actual modulus of elasticity of individual pieces of a grade varies from the average assumed for design (fig. 6—4). Small individual lots of lumber can be expected to deviate from the distribution shown by the histogram. The additional multiplying factors used to derive final design values of modulus of elasticity are discussed later in this chapter.

In-Grade Testing Procedure

To establish the mechanical properties of specified grades of lumber from tests of full-size specimens, a representative sample of the lumber population is obtained following guidelines given in ASTM D 2915. The specimens are then tested using appropriate procedures such as those given in ASTM D 198. An average value for the modulus of elasticity or a near-minimum estimate of strength properties are obtained using ASTM D 2915 guidelines. These properties are further modified for design use by consideration of moisture content, duration of load, and safety.

Developing Mechanical Grades

Mechanical stress rating (MSR) is based on an observed relation between modulus of elasticity (E) and bending strength of lumber. The modulus of elasticity is the sorting criterion used in this method of grading. Mechanical devices operating at high rates of speed measure the modulus of elasticity of individual pieces of lumber.

There are three basic components of MSR systems used in the United States and Canada: (1) The mechanical sorting and prediction of strength through a machine-measured nondestructive determination of E; (2) the assignment of allowable design stresses based upon strength predictions; and (3) quality control. The quality control procedures assure (a) proper operation of the machine used to measure E, (b) the appropriateness of the predictive parameter-bending strength relationship, and (c) the appropriateness of properties assigned for tension and compression (properties not directly related to E).

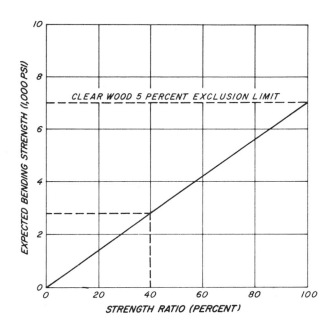

Figure 6—3—An example of the relation between strength and strength ratio. (M139 119)

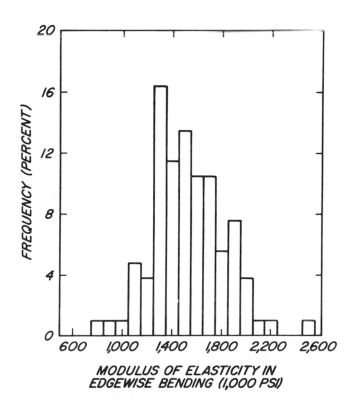

Figure 6—4—Histogram of modulus of elasticity observed in a single visual grade. From pieces selected over a broad geographical range. (M139 780)

6-5

Mechanical Sorting Criteria

The modulus of elasticity used as a sorting criterion for mechanical properties of lumber can be measured in a variety of ways. Usually the apparent E, or a stiffness-related deflection, is actually measured. Because lumber is heterogeneous, the apparent E depends upon span, orientation (edgewise or flatwise in bending), mode of test (static or dynamic), and method of loading (tension, bending, concentrated, uniform, etc.). Any of the apparent E's can be used, so long as the grading machine is properly calibrated to assign the appropriate design property. Most grading machines in the United States are designed to detect the lowest flatwise bending stiffness that occurs in any approximate 4-foot span.

In the United States and Canada, MSR lumber is also subjected to a visual override because the size of edge knots in combination with E is a better predictor of strength than E alone. Maximum edge knots are limited to a specified proportion of the cross section, depending upon grade level. Other visual restrictions, which are primarily appearance criteria rather than for control of strength, are on checks, shake, skips, splits, wane, and warp.

Deriving Properties of Mechanically Graded Lumber

Allowable Stress in Bending

A stress grade derived for mechanically graded lumber relates allowable strength in bending to the modulus of elasticity levels by which the grade is identified. Because E is an imperfect predictor of strength, lumber sorted solely by average E falls into one of four categories, two of which are sorted correctly and two incorrectly.

Consider, for example, a most simple case (sometimes referred to as "go," or "no go") where lumber is sorted into two groups; one has sufficient strength for a specific application, the other does not. In figure 6-5, a regression line relating E and strength is used as the prediction model. The "accept-reject" groups identified by the regression sort can be further classified into four categories. Material in category 1 has been accepted correctly, i.e., the pieces have adequate strength as defined. That in category 2 is accepted incorrectly, i.e., pieces do not have adequate strength. Pieces in category 3 are rejected correctly because they do not have acceptable strength, while those in category 4 are rejected incorrectly because they do have acceptable strength.

Thus the sort has worked correctly for categories 1 and 3 but incorrectly for 2 and 4. Pieces in category 4 present no problem other than that they are destined to be used inefficiently. Those in category 2, however, present a problem. They are accepted as strong enough, but are not, and they are mixed in with the accepted pieces of category 1. The number

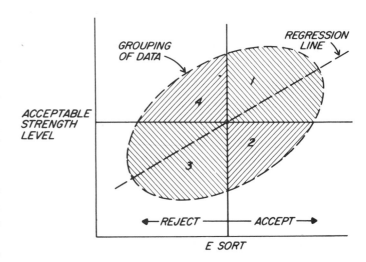

Figure 6-5—A schematic E sort, using a regression line as the predictor, showing four categories: 1—accepted correctly; 2—accepted incorrectly; 3—rejected correctly; and 4—rejected incorrectly. (M84 5724)

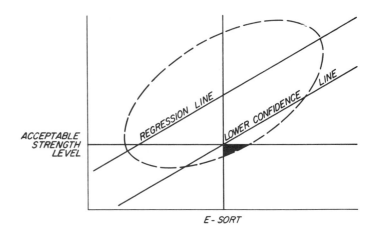

Figure 6-6—A schematic E sort using a lower confidence line as the predictor and showing the relatively low proportion of material in the accepted incorrectly category (lower right). (M149 577)

of problem pieces that fall in category 2 depends upon the variability in the prediction model.

Clearly, the E-versus-strength regression line is not an adequate sorting model for structural lumber because of the large proportion of material in the incorrectly accepted category. An appropriate model is one that minimizes the material in category 2 or at least reduces it to a lower risk level.

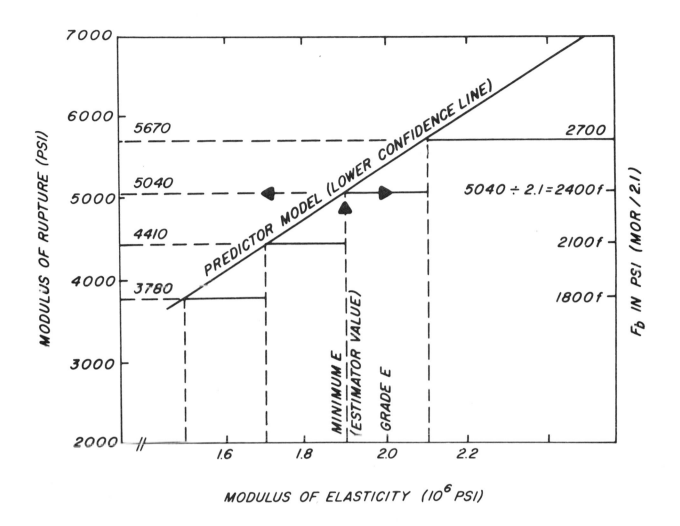

Figure 6–7—Typical assignment of F_b–E values for MSR lumber in the United States (dashed lines are the minimum E for each F_b–E classification and the bending strengths predicted by the minimum E's).

(ML84 5718)

Commonly, a lower confidence line is used as the prediction model (fig. 6–6). The number of pieces that fall into category 2 is now low compared to the regression line model. Further, the probability of a piece (and, thus, the number of pieces) falling into category 2 is controlled by the confidence line selected.

In actual MSR systems the lumber is sorted (graded) by E classes. In the United States and Canada, the number of grades has increased as specific market needs have developed for MSR lumber. Today, individual grading agencies list as many as 20 E classifications or grades. The grades are designated by the recommended allowable extreme fiber stress in bending[1], F_b , and the allowable modulus of elasticity (E).

For example, "2100 F–1.8 E" designates an MSR grade with an allowable F_b of 2,100 psi and an E of 1.8 million psi.

In theory, any F-E combination can be marketed that can be supported by test data. In practice, a mill will usually produce only a few of the possible F-E classifications depending upon the potential of the timber being harvested, mill production capabilities, and product or market demand. The allowable E for each classification or grade is the midpoint of each E class, and F_b is calculated from the bending strength or modulus of rupture (MOR) estimated by the minimum E in the grade.

Figure 6–7 is an example of a typical stress assignment for several F-E classifications. Here it is assumed that a mill is producing lumber with a range in E from 1.5 to 2.3 million psi and that there is a market for four grades with an assigned

---

[1]Design stresses are indicated by F_b = bending and F_t = tension.

E of 1.6, 1.8, 2.0, and 2.2 million psi. F_b values, as indicated in figure 6−7, are developed from the minimum E in each E class. For example, the minimum E in the 2.0 E class is 1.9 million psi which predicts a fifth percentile MOR value of 5,040 psi. The 5th percentile is a value that is expected to be exceeded by 95 percent of the pieces in a grade or class. This value is then adjusted by a factor (2.1) for an assumed 10-year duration of load and safety to obtain F_b. This factor applied to an estimated fifth percentile MOR value of 5,040 psi yields an F_b of 2400 for the 2.0 E grade.

Allowable Stresses for Other Properties

Properties in tension and compression are commonly developed from relationships with F_b rather than being estimated directly by the nondestructive parameter, E. In Canada and the United States, the relationships between bending and those in tension and compression are based upon limited lumber testing for the three properties but supported by years of successful experience in construction with visual stress grades of lumber. For tension, it is assumed that the relationships between F_b and allowable tensile stresses depend on the F_b level. A sliding scale factor ranging from 0.39 to 0.80, depending upon F_b level, is applied to F_b to obtain F_t (the factor is 0.8 for all F_b values of 2400 and higher). The relationship between F_b and F_c is assumed to be independent of quality level and F_c is $0.80\,F_b$ for all F_b−E grades or classifications.

Strengths in shear parallel to the grain and in compression perpendicular to the grain are poorly related to modulus of elasticity. Therefore, in mechanical stress grading these properties are assumed to be grade-independent and are assigned the same values as for visual lumber grades.

Quality Control

Quality control procedures are necessary to assure that stresses assigned by an MSR system reflect the actual properties of the lumber graded. These procedures must check for both correct machine operation and the appropriateness of the strength prediction model. Verification of the relationships between bending and other properties may also be required, particularly for F_t.

Daily or even more frequent certification of machine operation may be necessary. Depending upon machine principle, this may involve operating the machine on a calibration bar of known stiffness, comparing grading machine E's to those obtained on the same pieces of lumber by calibrated laboratory test equipment, or in some instances both. Machine operation should be certified for all sizes of lumber being produced and for the range of lumber quality anticipated in a mill because some machines show sensitivity to these factors. For example, a machine certified for operation on 2 by 4's in a quality range from 1200 F−1.2 E to 1800 F−1.6 E may not

necessarily operate correctly on 2 by 6's in a quality range from 1800 F−1.6 E to 2400 F−2.0 E.

Uniform quality control procedures of the MSR prediction model (the E-bending strength relationship) have been adopted in Canada and the United States. Daily or more frequently lumber production is representatively sampled and proof-loaded, primarily in bending with supplementary testing in tension. The pieces are proof-loaded to 2.1 times the allowable design stress (F_b or F_t) for the assigned F_b−E classification. In bending, the pieces are loaded on a random edge with the maximum edge defect within the maximum moment area (third-point loading) or as near so as possible.

If less than 95 percent of the pieces of the lumber sampled survive the proof test, a second sampling and proof test is conducted immediately. If the second sample also fails the 95 percent survival criterion, the MSR grading system is declared "out of control" and the operation is shut down to isolate and correct the problem.

Cumulative machine calibration records are useful for detecting trends or gradual change in machine operation that might accompany use and wear of machine parts. The proof-test results are also accumulated. Standard statistical quality control procedures (control charts, etc.) are used to monitor the appropriateness of the MSR model and to modify the model as needed in response to change in the timber resource.

Too many failures in one, or even consecutive samples, do not necessarily indicate that the system is "out of control." If the prediction line is based upon 95 percent confidence, it can be expected by chance alone that 1 sample in 20 will not meet the proof-test requirements. One or more "out of control" samples may also represent a temporary aberration in material properties (the E-strength relationship). In any event, it would call for inspection of the cumulative quality control records for trends to determine if machine adjustment might be needed. A "clean" record for a given period of time forgives a previously "out of control" system.

Adjusting Properties for Design Use

The mechanical properties associated with lumber quality are adjusted to give allowable unit stresses and an allowable modulus of elasticity suitable for engineering uses.

A composite adjustment factor is applied to each strength property to adjust for an assumed 10-year duration of full design load. In visual grades, the composite factor ranges from 1.9 to 4.5 and includes a safety factor of about 1.3. A factor of 2.1 is applied in deriving properties for mechanically graded lumber. Additional adjustments may be made for size and moisture content.

Discussion of these adjustment factors follows; specific adjustments are given in ASTM Designation D 245.

Size Factor

In bending, a size effect causes small members to have a greater unit strength than large members. In ASTM D 245, the bending strength for lumber is adjusted for a depth other than 2 inches using the formula

$$F = \left(\frac{2}{d}\right)^{1/9} \qquad (6-2)$$

where d = the net surfaced depth for which a bending strength is desired. This formula is based on an assumed center load and a span-to-depth ratio of 14. If bending strength is known for one size of lumber, it can be approximated for another size by using formulas given in chapter 8.

Moisture Adjustments

For lumber 4 inches thick or less that has been dried, properties are linearly related to moisture content. As an example, the relationship for modulus of elasticity at any moisture content, expressed as an increase above the modulus for green lumber is shown in figure 6−8. In typical practice, adjustments are made to correspond to average moisture contents of 15 and 12 percent with expected maximum moisture contents of 19 and 15 percent. The factors for adjusting properties to these two moisture levels are given in ASTM D 245. An equation applicable to adjusting properties to other moisture levels is given in ASTM D 2915.

For lumber thicker than 4 inches, often no adjustment for moisture content is made because allowable properties are assigned on the basis of wood in the green condition. Lumber in large sizes is usually put in place without drying and it is assumed that drying degrade offsets the increase in strength normally associated with loss in moisture.

Duration of Load

Allowable design stresses are based on an assumed 10-year loading period (called normal loading). If duration of loading, either continuously or cumulatively, is expected to exceed 10 years, allowable design stresses are reduced 10 percent. If for shorter periods, allowable stresses can be adjusted using figure 6−9. Intermittent loading causes cumulative effects on strength and should be treated as a continuous load of equivalent duration. The effects of cyclic loads of short duration must also be considered in design (see discussion of fatigue in ch. 4). These duration-of-load modifications are not applicable to modulus of elasticity.

In many design circumstances there are several loads on the structure, some acting simultaneously and each with a

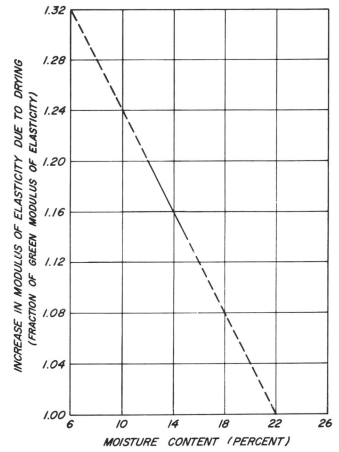

Figure 6−8—Modulus of elasticity as a function of moisture content for lumber. Solid line represents range of experimental data on which graph is based. (M139 118)

different duration. Each increment of time during which the total load is constant should be treated separately, and the most severe condition governs the design. Either the allowable stress or the total design load (but not both) can be adjusted using figure 6−9.

For example, suppose a structure is expected to support a load of 100 pounds per square foot (psf) off and on for a cumulative duration of 1 year. Also, it is expected to support its own dead load of 20 psf for the anticipated 50-year life of the structure. The adjustments to be made to arrive at an equivalent 10-year design load are listed in the tabulation.

Time	Total load	Load adjustment (from fig. 6−9)	Equivalent 10-year design load
(Yr)	(Psf)		(Psf)
1	100 + 20 = 120	0.93	112
50	20	1.04	21

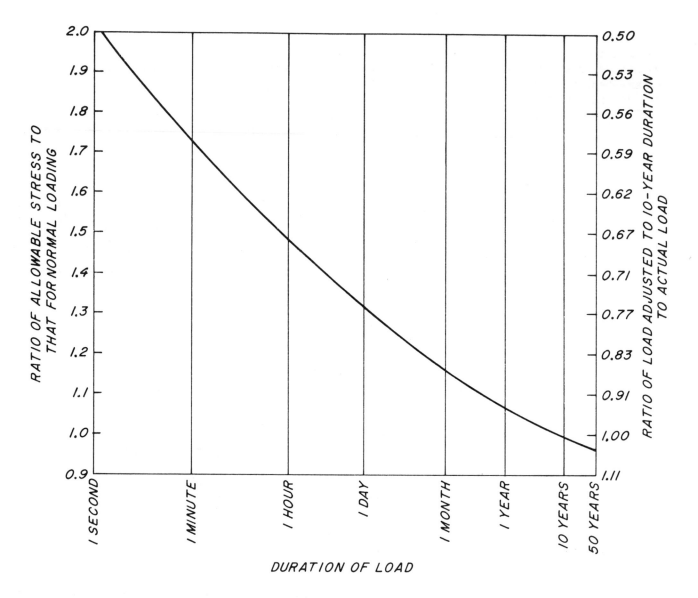

Figure 6–9—Relation of strength to duration of load.

(M139 123)

The more critical design load is 112 psf, and this load and the allowable stress for lumber would be used to pick members of suitable size. In this case, it was convenient to adjust the loads on the structure, although the same result can be obtained by adjusting the allowable stress.

Practice of Stress Grading

An orderly, voluntary, but circuitous, system of responsibilities has evolved in the United States for the development, manufacture. and merchandising of most stress-graded lumber.

The system is shown schematically in figure 6–10. Stress-grading principles are developed from research findings and engineering concepts, often within committees and subcommittees of the American Society for Testing and Materials.

The National Bureau of Standards cooperates with lumber producers, distributors, and users through an American Lumber Standards Committee (ALSC) to assemble a voluntary softwood standard of manufacture, called the American Softwood Lumber Standard (see ch. 5). The American Softwood Lumber Standard and the related National Grading Rule prescribe the ways in which stress-grading principles can be

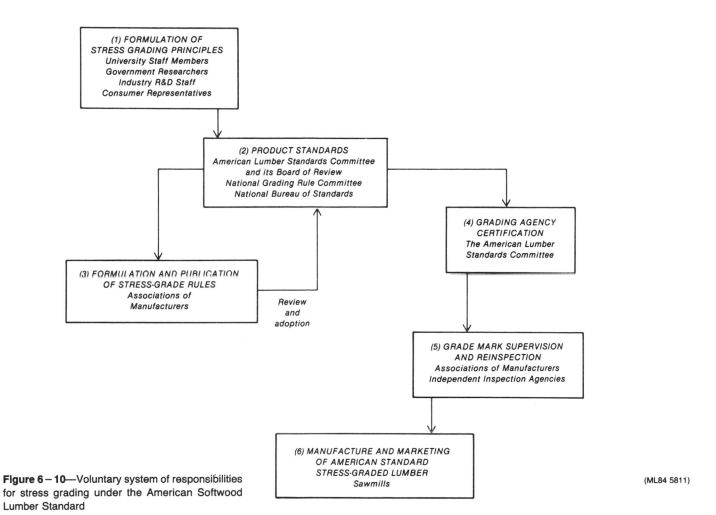

Figure 6–10—Voluntary system of responsibilities for stress grading under the American Softwood Lumber Standard

(ML84 5811)

used to formulate grading rules said to be American Standard.

Organizations of lumber manufacturers publish grading rule books containing stress-grade descriptions. If an organization wants its published rules to be American Standard, it submits them to the ALSC's Board of Review for review of conformance with the American Softwood Lumber Standard (ALS).

Organizations that write grading rules, as well as independent agencies, can be certified by ALSC to issue grade marks corresponding to published stress-grade rules and provide grade-marking supervision and reinspection services to individual lumber manufacturers. The performance of these organizations is then under the scrutiny of the Board of Review of the ALSC.

Most commercial softwood species sold in the United States are stress graded under ALS practice. The principles of stress grading are also applied to several hardwood species under provisions of the ALS.

Lumber found in the marketplace may be stress graded by grading rules developed in accordance with methods approved

by the ALSC, by some other stress grading rule, or it may not be stress graded. Stress grades that meet the requirements of the voluntary ALS are developed by the principles that have been described in this chapter, and only these stress grades are discussed here.

Stress grading under the auspices of the ALSC is applied to many of the sizes and several of the patterns of lumber meeting the provisions of the ALS. A majority of stress-graded lumber is dimension (2–4 inches thick), however, and a uniform procedure, the National Grading Rule, is used for writing grading rules for lumber of this size. Grade rules for other sizes may vary by grading agencies or species. Dimension should not be confused with hardwood dimension lumber, a name applied to hardwood stock intended for remanufacture into pieces of specific size called dimension parts.

National Grading Rule

The Product Standard for softwood lumber (American Softwood Lumber Standard), PS 20–70, provides for a National

Grading Rule for lumber from 2 inches up to, but not including, 5 inches in nominal thickness (dimension lumber). All American Standard dimension lumber in that thickness range is required to conform to the National Grading Rule, except for special products such as scaffold planks.

The National Grading Rule establishes the lumber classifications and grade names for visually stress-graded lumber shown in table 6−1. The minimum strength ratios permitted are also shown to provide a comparative index of quality. The corresponding visual descriptions of the grades can be found in the grading rule books of most of the softwood rule-writing agencies listed in chapter 5. Grades of lumber that meet these requirements should have about the same appearance regardless of species. They will not have the same allowable properties. The allowable properties for each species and grade are given in the appropriate rule books and in the National Design Specification.

The National Grading Rule also establishes limitations on sizes of permissible edge knots and other visual characteristics for American Standard lumber that is graded by a combination of mechanical and visual methods.

Table 6−1—Visual grades described in the National Grading Rule[1]

Lumber classification	Grade name	Strength ratio
		Pct
Light framing (2 to 4 in thick, 2 to 4 in wide)[2]	Construction	34
	Standard	19
	Utility	9
Structural light framing (2 to 4 in thick, 2 to 4 in wide)	Select Structural	67
	1	55
	2	45
	3	26
Studs (2 to 4 in thick, 2 to 6 in wide, 10 ft and shorter)	Stud	26
Structural joists and planks (2 to 4 in thick, 6 in and wider)	Select Structural	65
	1	55
	2	45
	3	26
Appearance framing (2 to 4 in thick, 2 in and wider)	Appearance	55

[1] Sizes shown are nominal.
[2] Widths narrower than 4 inches may have different strength ratio than shown. Contact rules-writing agencies for additional information.

Grouping of Species

Some species are grouped together and the lumber from them treated as equivalent. This is usually done for species that have about the same mechanical properties, where the wood of two or more species is very similar in appearance, or for marketing convenience. For visual stress grades, ASTM D 2555 contains rules for calculating clear wood properties for groups of species. The properties assigned to a group by such a procedure often will not be identical with those of any of the species that make up the group. The group will display a unique identity with nomenclature approved by the American Lumber Standards Committee. The grading association under whose auspices the lumber was graded should be contacted if the identities, properties, and characteristics of individual species of the group are desired.

In the case of mechanical stress grading, the inspection agency that supervises the grading certifies by test that the allowable properties in the grading rule are appropriate for the species or species group and the grading process.

Selected References

American Society for Testing and Materials. Standard methods for establishing structural grades for visually graded lumber. ASTM D 245. Philadelphia, PA: ASTM; (see current edition).

American Society for Testing and Materials. Standard methods for establishing clear wood strength values. ASTM D 2555. Philadelphia, PA: ASTM; (see current edition).

American Society for Testing and Materials. Standard method for evaluating properties for stress grades of structural lumber. ASTM D 2915. Philadelphia, PA: ASTM; (see current edition).

American Society for Testing and Materials. Standard methods of static tests of timbers in structural sizes. ASTM D 198. Philadelphia, PA: ASTM; (see current edition).

Galligan, W. L.; Green, D. W.; Gromala, D. S.; Haskell, J. H. Evaluation of lumber properties in the United States and their application to structural research. Madison, WI: Forest Products Journal. 30(10): 45−51; 1980.

Galligan, W. L.; Snodgrass, D. V. Machine stress rated lumber: Challenge to design. Madison, WI: Forest Products Journal. 20(9): 63−69; 1970.

Gerhards, C. C. Effect of duration and rate of loading on strength of wood and wood based materials. Res. Pap. FPL 283. Madison, WI: U.S. Department of Agriculture, Forest Service, Forest Products Laboratory; 1977.

Green, D. W. Adjusting the static strength of lumber for changes in moisture content. How the environment affects lumber design: assessments and recommendations. Proceedings of workshop sponsored by Society of Wood Science and Technology, U.S. Forest Products Laboratory, and Mississippi State University; 1980 May 28−30; Madison, WI.

National Forest Products Association. National design specification for wood construction. Washington, DC: NFPA; (see current edition).

National Forest Products Association. Design values for wood construction—a supplement to the national design specification for wood construction. Washington, DC: NFPA; (see current edition).

U.S. Department of Commerce. American softwood lumber standard. Prod. Stand. PS 20. Washington, DC: USDC; (see current edition).

Chapter 7

Fastenings

Fastenings*

The strength and stability of any structure depend heavily on the fastenings that hold its parts together. One prime advantage of wood as a structural material is the ease with which wood structural parts can be joined together with a wide variety of fastenings—nails, spikes, screws, bolts, lag screws, drift pins, staples, and metal connectors of various types. For utmost rigidity, strength, and service, each type of fastening requires joint designs adapted to the strength properties of wood along and across the grain, and to dimensional changes that may occur with changes in moisture content.

The information in this chapter represents research results. These results are often modified, based on judgment or experience, and thus design documents may differ somewhat from what is presented in this chapter.

Nails

Nails are the most common mechanical fastenings used in construction. There are many types, sizes, and forms of nails (fig. 7−1). The formulas presented in this chapter for loads apply for bright, smooth, common steel wire nails driven into wood when there is no visible splitting. For nails other than common wire nails, the loads can be adjusted by factors given later in the chapter.

Nails in use resist either withdrawal loads or lateral loads, or a combination of the two. Both withdrawal and lateral resistance are affected by the wood, the nail, and the condition of use. In general, however, any variation in these factors has a more pronounced effect on withdrawal resistance than on lateral resistance. The serviceability of joints with nails laterally loaded does not depend greatly on withdrawal resistance unless large joint distortion is tolerable.

Withdrawal Resistance

The resistance of a nail shank to direct withdrawal from a piece of wood depends on the density of the wood, the diameter of the nail, and the depth of penetration. The surface condition of the nail at the time of driving also influences the initial withdrawal resistance.

For bright, common wire nails driven into the side grain of seasoned wood or unseasoned wood that remained wet, the results of many tests have shown that the maximum withdrawal load is given by the empirical formula:

$$p = 7{,}850\ G^{5/2}DL \qquad (7-1)$$

where p is the maximum load in pounds; L, the depth, in inches, of penetration of the nail in the member holding the nail point; G, the specific gravity of the wood based on

* Revision by Thomas L. Wilkinson, General Engineer.

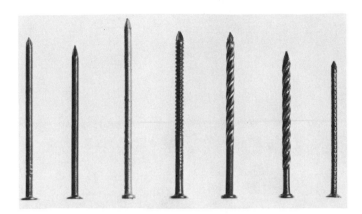

Figure 7−1—Various types of nails: (left to right), bright smooth wire nail; cement coated; zinc-coated; annularly threaded; helically threaded; helically threaded and barbed; and barbed. (M84 0274)

ovendry weight and volume at 12 percent moisture content (see table 4−2, ch. 4); and D, the diameter of the nail in inches.

The diameters of various penny or gauge sizes of bright common nails are given in table 7−1. Bright box nails are generally of the same length but slightly smaller diameter (table 7−2), while cement-coated nails such as coolers, sinkers, and coated box nails are slightly shorter (1/8 in) and of smaller diameter for the same penny size than common nails. Annularly and helically threaded nails generally have smaller diameters than common nails for the same penny size (table 7−3).

The loads expressed by formula (7−1) represent average data. Certain wood species give test values that are somewhat higher or lower than the formula values. A typical load-displacement curve for nail withdrawal (fig. 7−2) shows that maximum load occurs at relatively small values of displacement.

Although the formula for nail-withdrawal resistance indicates that the dense, heavy woods offer greater resistance to nail withdrawal than the lighter weight ones, lighter species should not be disqualified for uses requiring high resistance to withdrawal. As a rule, the less dense species do not split so readily as the denser ones and thus offer an opportunity for increasing the diameter, length, and number of the nails to compensate for the wood's lower resistance to nail withdrawal.

The withdrawal resistance of nail shanks is greatly affected by such factors as type of nail point, type of shank, time the nail remains in the wood, surface coatings, and moisture content changes in the wood.

Effect of Seasoning

With practically all species, nails driven into green wood and pulled before any seasoning takes place offer about the

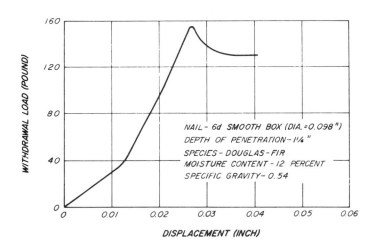

Figure 7–2—Typical load-displacement curve for direct withdrawal of a nail. (M139 045)

NAIL - 6d SMOOTH BOX (DIA.=0.098")
DEPTH OF PENETRATION- 1¼"
SPECIES - DOUGLAS-FIR
MOISTURE CONTENT - 12 PERCENT
SPECIFIC GRAVITY- 0.54

same withdrawal resistance as nails driven into seasoned wood and pulled soon after driving. If, however, common smooth-shank nails are driven into green wood that is allowed to season, or into seasoned wood that is subjected to cycles of wetting and drying before the nails are pulled, they lose a major part of their initial withdrawal resistance. The withdrawal resistance for nails driven into wood that is subjected to changes in moisture content may be as low as 25 percent of the values for nails tested soon after driving. On the other hand if the wood fibers deteriorate or the nail corrodes under some conditions of moisture variation and time, withdrawal resistance is erratic; resistance may be regained or even increased over the immediate withdrawal resistance. However, such sustained performance should not be relied on in the design of a nailed joint.

In seasoned wood that is not subjected to appreciable moisture content changes, the withdrawal resistance of nails may also diminish due to relaxation of the wood fibers with time. Under all these conditions of use, the withdrawal resistance of nails differs among species as well as showing variation within individual species.

Effect of Nail Form

The surface condition of nails is frequently modified during the manufacturing process to improve withdrawal resistance. Such modification is usually done by surface coating, surface roughening, or mechanical deformation of the shank. Other factors that affect the surface condition of the nail are the oil film remaining on the shank after manufacture or corrosion resulting from storage under adverse conditions; but these factors are so variable that their influence on withdrawal resistance cannot be adequately evaluated.

Table 7–1—*Sizes of bright, common wire nails*

Size	Gauge	Length	Diameter
		- - - - - - - - Inch - - - - - - - -	
6d	11-1/2	2	0.113
8d	10-1/4	2-1/2	.131
10d	9	3	.148
12d	9	3-1/4	.148
16d	8	3-1/2	.162
20d	6	4	.192
30d	5	4-1/2	.207
40d	4	5	.225
50d	3	5-1/2	.244
60d	2	6	.262

Table 7–2—*Sizes of smooth box nails*

Size	Gauge	Length	Diameter
		- - - - - - - Inch - - - - - - - -	
3d	14-1/2	1-1/4	0.076
4d	14	1-1/2	.080
5d	14	1-3/4	.080
6d	12-1/2	2	.098
7d	12-1/2	2-1/4	.098
8d	11-1/2	2-1/2	.113
10d	10-1/2	3	.128
16d	10	3-1/2	.135
20d	9	4	.148

Table 7–3—*Sizes of helically and annularly threaded nails*

Size	Length	Diameter
	- - - - - - - - - - Inch - - - - - - - - - -	
6d	2	0.120
8d	2-1/2	.120
10d	3	.135
12d	3-1/4	.135
16d	3-1/2	.148
20d	4	.177
30d	4-1/2	.177
40d	5	.177
50d	5-1/2	.177
60d	6	.177
70d	7	.207
80d	8	.207
90d	9	.207

Surface Modifications—A common surface treatment for nails is the so-called cement coating. Cement coatings, contrary to what the name implies, do not include cement as an ingredient; they generally are a composition of resin applied to the nail to increase the resistance to withdrawal by increasing the friction between the nail and the wood. If properly applied, they increase the resistance of nails to withdrawal immediately after the nails are driven into the softer woods. In the denser woods (such as hard maple, birch, or oak), however, cement-coated nails have practically no advantage over plain nails, because most of the coating is removed in driving. Some of the coating may also be removed in the cleat or facing member before the nail penetrates the foundation member.

Good-quality cement coatings are uniform, not sticky to the touch, and cannot be rubbed off easily. Different techniques of applying the cement coating and variations in its ingredients may cause large differences in the relative resistance to withdrawal of different lots of cement-coated nails. Some nails may show only a slight initial advantage over plain nails. The increase in withdrawal resistance of cement-coated nails is not permanent but drops off about one-half after a month or so in the softer woods. Cement-coated nails are used primarily in construction of boxes, crates, and other containers usually built for rough handling and relatively short service.

Nails that have special coatings, such as zinc, are intended primarily for uses where corrosion and staining are important factors in permanence and appearance. If the zinc coating is evenly applied, withdrawal resistance may be increased, but extreme irregularities of the coating may actually reduce it. The advantage that zinc-coated nails with a uniform coating may have over plain nails in resistance to initial withdrawal is usually reduced by repeated cycles of wetting and drying.

Nails have also been made with plastic coatings. The usefulness and characteristics of these coatings are influenced by the quality and type of coating, the effectiveness of the bond between the coating and base fastener, and the effectiveness of the bond between the coating and wood fibers. Some plastic coatings appear to resist corrosion or improve resistance to withdrawal, while others offer little improvement.

Fasteners with properly applied nylon coating tend to retain their initial resistance to withdrawal, as compared to other coatings which exhibit a marked decrease in withdrawal resistance within the first month after driving.

A chemically etched nail has somewhat higher withdrawal resistance than some coated nails, as the minutely pitted surface is an integral part of the nail shank. Under impact loading, however, the withdrawal resistance of etched nails is little different from that of plain or cement-coated nails under various moisture conditions.

Sand-blasted nails perform in much the same manner as do chemically etched nails.

Shape Modifications—Nail shanks may be varied from a smooth, circular form to give an increase in surface area without an increase in nail weight. Special nails with barbed, helically or annularly threaded, and other irregular shanks (fig. 7−1) are offered commercially.

The form and magnitude of the deformations along the shank influence the performance of the nails in the various wood species. The withdrawal resistance of these nails, except some types of barbed nails, is generally somewhat greater than that of common wire nails of the same diameter, in wood remaining at a uniform moisture content. For instance, annular-shank nails have about 40 percent greater resistance to withdrawal. Under conditions involving changes in the moisture content of the wood, however, some special nail forms provide considerably greater withdrawal resistance than the common wire nail—about four times greater for annular and helical shank nails of the same diameter. This is especially true of nails driven into green wood that subsequently dries. In general, annularly threaded nails sustain larger withdrawal loads, and helically threaded nails sustain greater impact withdrawal work values than the other nail forms.

Nails with deformed shanks are sometimes hardened by heat treatments for use where driving conditions are difficult, or to obtain improved performance, such as in pallet assembly. Hardened nails are brittle and care should be exercised to avoid injuries from fragments of nails broken during driving.

Nail Point—A smooth, round-shank nail with a long, sharp point will usually have a higher withdrawal resistance, particularly in the softer woods, than the common wire nail (which usually has a diamond point). Sharp points, however, accentuate splitting of certain species, which may reduce withdrawal resistance. A blunt or flat point without taper reduces splitting, but its destruction of the wood fibers when driven reduces withdrawal resistance to less than that of the common wire nail. A nail tapered at the end and terminating in a blunt point will cause less splitting. In the heavier woods, such a tapered, blunt-pointed nail will provide about the same withdrawal resistance, but in the less dense woods, its resistance to withdrawal is lower than the common nail.

Nailhead—Nailhead classifications include flat, oval, countersunk, deep-countersunk, and brad. Nails with all types of heads, except the deep-countersunk, brad, and some of the thin flathead nails, are sufficiently strong to withstand the force required to pull them from most woods in direct withdrawal. The deep-countersunk and brad nails are usually driven below the wood surface and are not intended to carry large withdrawal loads. In general, the thickness and diameter of the heads of the common wire nails increase as the size of the nail increases.

The development of some pneumatically operated portable nailers has introduced special headed nails such as T-nails and nails with a segment of the head cut off. Although the resistance of these heads to pulling through the wood might be less than for conventional nailheads, the performance of the modified heads appears adequate. It is preferable that the T-head be oriented so that its head is perpendicular to the grain of the adjoining wood.

Corrosion and Staining

In the presence of moisture, metals used for nails may corrode when in contact with material treated with certain preservative or fire-retardant salts (see chs. 15 and 18). Use of certain metals or metal alloys will reduce the amount of corrosion. Nails of copper, silicon bronze, and 304 and 316 stainless steel have performed well in wood treated with ammoniacal copper arsenate and chromated copper arsenate. The choice of metals for use with fire-retardant-treated woods depends upon the particular fire-retardant chemical.

Staining caused by the reaction of certain wood extractives (see ch. 3) and steel in the presence of moisture is a problem where appearance is important, such as naturally finished siding. Use of aluminum or hot dipped galvanized nails can alleviate staining. Care should be taken in driving galvanized nails to avoid breaking the zinc coating on the heads.

In general, the withdrawal resistance of copper and other alloy nails is somewhat comparable to that of common steel wire nails when pulled soon after driving.

Driving

The resistance of nails to withdrawal is generally greatest when they are driven perpendicular to the grain of the wood. When the nail is driven parallel to the wood fibers—that is, into the end of the piece—withdrawal resistance in the softer woods drops to 75 or even 50 percent of the resistance obtained when the nail is driven perpendicular to the grain. The difference between side- and end-grain withdrawal loads is less for dense woods than for softer woods. With most species, the ratio between the end- and side-grain withdrawal loads of nails pulled after a time interval, or after moisture content changes have occurred, is usually somewhat higher than that of nails pulled immediately after driving.

Toenailing, a common method of joining wood framework, involves slant driving a nail or group of nails through the end or edge of an attached member and into a main member. Toenailing requires greater skill in assembly than does ordinary end nailing but provides joints of greater strength and stability. Tests show that the maximum strength of toenailed joints under lateral and uplift loads is obtained by (1) using the largest nail that will not cause excessive splitting, (2) allowing an end distance (distance from the end of the attached member to the point of initial nail entry) of approximately one-third the length of the nail, (3) driving the nail at a slope of 30° with the attached member, and (4) burying the full shank of the nail but avoiding excessive mutilation of the wood from hammer blows.

In tests of stud-to-sill assemblies with the number and size of nails frequently used in toenailed and end-nailed joints, a joint toe-nailed with four eightpenny common nails was superior to a joint end nailed with two sixteenpenny common nails. With such woods as Douglas-fir, toenailing with tenpenny common nails gave greater joint strength than the commonly used eightpenny nails.

The results of withdrawal tests with multiple nail joints in which the piece attached is pulled directly away from the main member show that slant driving is usually superior to straight driving when nails are driven into dry wood and pulled immediately, and decidedly superior when nails are driven into green or partially dry wood that is allowed to season for a month or more. However, the loss in depth of penetration due to slant driving may, in some types of joints, offset the advantages of slant nailing. Cross slant driving of groups of nails is usually somewhat more effective than parallel slant driving.

Withdrawal resistance of nails driven with power nailers or nailing machines is no different than nails driven by hand.

Nails driven into lead holes with a diameter slightly smaller (approximately 90 percent) than the nail shank have somewhat higher withdrawal resistance than nails driven without lead holes. Lead holes also prevent or reduce splitting of the wood, particularly for dense species.

Clinching

The withdrawal resistance of smooth-shank, clinched nails is considerably higher than that of unclinched nails. The ratio between the loads for clinched and unclinched nails varies enormously, depending upon the moisture content of the wood when the nail is driven and withdrawn, the species of wood, the size of nail, and the direction of clinch with respect to the grain of the wood.

In dry or green wood, a clinched nail provides from 45 to 170 percent more withdrawal resistance than an unclinched nail when withdrawn soon after driving. In green wood that seasons after a nail is driven, a clinched nail gives from 250 to 460 percent greater withdrawal resistance than an unclinched nail. However, this improved strength of the clinched over the unclinched-nail joint does not justify the use of green lumber, because the joints may loosen as the lumber seasons. Furthermore, laboratory tests were made with single nails, and the effects of drying, such as warping, twisting, and splitting, may reduce the efficiency of a joint that has more than one nail. Clinching of nails is generally confined to such construction as boxes and crates and other container applications.

Nails clinched across the grain have approximately 20 percent more resistance to withdrawal than nails clinched along the grain.

Fastening Plywood

The nailing characteristics of plywood are not greatly different from those of solid wood except for plywood's greater resistance to splitting when nails are driven near an edge. The nail withdrawal resistance of plywood is from 15 to 30 percent less than that of solid wood of the same thickness. The reason is that fiber distortion is less uniform in plywood than in solid wood. For plywood less than 1/2 inch thick, the high splitting resistance tends to offset the lower withdrawal resistance as compared to solid wood. The withdrawal resistance per inch of penetration decreases as the number of plies per inch is increased. The direction of the grain of the face ply has little influence on the withdrawal resistance from the face near the end or edge of a piece of plywood. The direction of the grain of the face ply may influence the pullthrough resistance of staples or nails with severely modified heads, such as T-heads. Fastener design information for plywood is available from the American Plywood Association.

Allowable Loads

The preceding discussion has dealt with maximum withdrawal loads obtained in short-time test conditions. For design, these loads must be reduced to account for variability and duration of load effects. A value of one-sixth the average maximum load has usually been accepted as the allowable load for longtime loading conditions. For normal duration of load, this value may be increased by 10 percent.

Lateral Resistance

Test loads at joint slips of 0.015 inch for bright, common wire nails in lateral resistance driven into the side grain (perpendicular to the wood fibers) of seasoned wood are expressed by the following empirical formula:

$$p = KD^{3/2} \qquad (7-2)$$

where p is the lateral load in pounds per nail at a joint slip of 0.015 inch (approximate proportional limit load); K is a coefficient; and D is the diameter of the nail in inches. Values of the coefficient K are listed in table 7-4 for ranges of specific gravity of hardwoods and softwoods. The loads given by the formula apply only where the side member and the member holding the nail point are of approximately the same density. The thickness of the side member should be about one-half the depth of penetration of the nail in the member holding the point.

The ultimate lateral nail loads for softwoods may approach 3.5 times the loads expressed by the formula, and for hardwoods they may be seven times as great. The joint slip at

maximum load, however, is over 20 times 0.015 inch. This is demonstrated by the typical load-slip curve shown in figure 7-3. To maintain a sufficient ratio between ultimate load and the load at 0.015 inch, the nail should penetrate into the member holding the point by not less than 10 times the nail diameter for dense woods (specific gravity greater than 0.61) and 14 times the diameter for lightweight woods (specific gravity less than 0.42). For species having densities between these two ranges, the penetration may be found by straight-line interpolation.

End distance, edge distance, and spacing of nails should be such as to prevent unusual splitting. As a general rule, nails should be driven no closer to the edge of the side member than one-half its thickness and no closer to the end than the thickness of the piece. Smaller nails can be driven closer to the edges or ends than larger ones because they are less likely to split the wood.

When the side member is steel, an increase of about 25 percent can be applied to the lateral load because initiation of failure is forced to occur in the wood member holding the nail point.

Grain Direction Effects

The lateral load for side-grain nailing given by the empirical formula $p = KD^{3/2}$ applies whether the load is in a direction parallel to the grain of the pieces joined or at right angles to it. When nails are driven into the end grain (parallel with the wood fibers), limited data on softwood species indicate that their maximum resistance to lateral displacement is about two-thirds that for nails driven into the side grain. Although the average proportional limit loads appear to be about the same for end and side-grain nailing, the individual results are more erratic for end-grain nailing, and the minimum loads approach only 75 percent of corresponding values for side-grain nailing.

Moisture Content Effects

Nails driven into the side grain of unseasoned wood give maximum lateral resistance loads approximately equal to those obtained in seasoned wood, but the lateral resistance loads at 0.015-inch joint slip are somewhat less. To prevent excessive deformation, lateral loads obtained by the formula for seasoned wood should be reduced by 25 percent for unseasoned wood that will remain wet or be loaded before seasoning takes place.

When nails are driven into green wood, their lateral proportional limit loads after the wood has seasoned are also less than when they are driven into seasoned wood and loaded. The erratic behavior of a nailed joint that has undergone one or more moisture content changes makes it difficult to establish a lateral load for a nailed joint under these conditions. Structural joints should be inspected at intervals, and if it is apparent that the joint has loosened during drying, the joint should be reinforced with additional nails.

Deformed-Shank Nails

Deformed-shank nails carry somewhat higher maximum lateral loads than do common wire nails, but both perform similarly at small distortions in the joint.

Allowable Loads

The value of the lateral load at proportional limit obtained from tests (see formula (7−2)) must be reduced (to account for variability and duration of load effects) to arrive at allowable values. A reduction factor of 1.6 has been used to arrive at a value for longtime loading. For normal loading, the resulting value may be increased by 10 percent. In practice, an additional increase of 10 percent is used as an engineering judgment factor.

Lateral Load-Slip Models

A considerable amount of work has been done to describe, by mathematical models, the lateral load-slip curve of nails. These models have become important because of advanced methods of structural analysis.

One theoretical model, which considers the nail to be a beam supported on an elastic foundation (the wood), describes the initial slope of the curve. This expression is

$$\delta = P\left[2(L_1 + L_2) - \frac{(J_1 - J_2)^2}{(K_1 + K_2)}\right] \qquad (7-3)$$

where P is the lateral load and δ is the joint slip. The factors $L_1, L_2, J_1, J_2, K_1,$ and K_2 (table 7−5), are combinations of hyperbolic and trigonometric functions of the quantities $\lambda_1 a$ and $\lambda_2 b$ in which a and b equal the depth of penetration of the

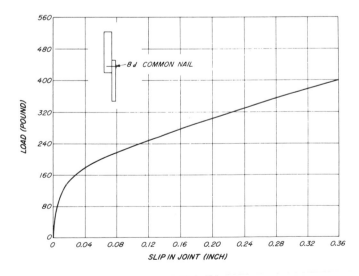

Figure 7−3—Typical relation between lateral load and slip in the joint.

(M125 339)

Table 7−4—Coefficients for computing test loads for fasteners in seasoned wood[1]

Specific gravity range[2]	Lateral load coefficient (K)		
	Nails[3]	Screws	Lag screws
HARDWOODS			
0.33−0.47	1,440	3,360	3,820
.48− .56	2,000	4,640	4,280
.57− .74	2,720	6,400	4,950
SOFTWOODS			
.29− .42	1,440	3,360	3,380
.43− .47	1,800	4,320	3,820
.48− .52	2,200	5,280	4,280

[1] Wood with a moisture content of 15 percent.
[2] Specific gravity based on ovendry weight and volume at 12 percent moisture content.
[3] Coefficients based on load at joint slip of 0.015 in.

Table 7−5—Expressions for factors in formula 7−3

Factor	Expression[1]
L_1	$\dfrac{\lambda_1}{k_1} \dfrac{\sinh \lambda_1 a \cosh \lambda_1 a - \sin \lambda_1 a \cos \lambda_1 a}{\sinh^2 \lambda_1 a - \sin^2 \lambda_1 a}$
L_2	$\dfrac{\lambda_2}{k_2} \dfrac{\sinh \lambda_2 b \cosh \lambda_2 b - \sin \lambda_2 b \cos \lambda_2 b}{\sinh^2 \lambda_2 b - \sin^2 \lambda_2 b}$
J_1	$\dfrac{\lambda_1^2}{k_1} \dfrac{\sinh^2 \lambda_1 a + \sin^2 \lambda_1 a}{\sinh^2 \lambda_1 a - \sin^2 \lambda_1 a}$
J_2	$\dfrac{\lambda_2^2}{k_2} \dfrac{\sinh^2 \lambda_2 b + \sin^2 \lambda_2 b}{\sinh^2 \lambda_2 b - \sin^2 \lambda_2 b}$
K_1	$\dfrac{\lambda_1^3}{k_1} \dfrac{\sinh \lambda_1 a \cosh \lambda_1 a + \sin \lambda_1 a \cos \lambda_1 a}{\sinh^2 \lambda_1 a - \sin^2 \lambda_1 a}$
K_2	$\dfrac{\lambda_2^3}{k_2} \dfrac{\sinh \lambda_2 b \cosh \lambda_2 b + \sin \lambda_2 b \cos \lambda_2 b}{\sinh^2 \lambda_2 b - \sin^2 \lambda_2 b}$

[1] $k_1 = k_{01}$ d and $k_2 = k_{02}$ d, where k_1 and k_2 are the foundation moduli of members 1 and 2 respectively.

nail in members 1 and 2, respectively. For smooth round nails,

$$\lambda = 2\sqrt[4]{\frac{k_o}{\pi E d^3}} \qquad (7-4)$$

where k_o is elastic bearing constant, d is nail diameter, and E is modulus of elasticity of the nail. For seasoned wood, the elastic bearing constant has been shown to be related to average species specific gravity, G, by

$$k_o = 2,144,000\ G \qquad (7-5)$$

if no lead hole is used or by

$$k_o = 3,200,000\ G \qquad (7-6)$$

if a prebored lead hole equal to 90 percent of the nail diameter is used.

Other models have been empirically derived which attempt to describe the entire load-slip curve. One such expression is

$$P = A\ \log_{10}(1 + B\delta) \qquad (7-7)$$

where the parameters A and B are empirically fitted.

Spikes

Common wire spikes are manufactured in the same manner as common wire nails. They have either a chisel point or a diamond point and are made in lengths of 3 to 12 inches. For corresponding lengths (3 to 6 in), they have larger diameters (table 7−6) than the common wire nails, and beyond the sixtypenny size they are usually designated by inches of length.

The withdrawal and lateral resistance formulas and limitations given for common wire nails are also applicable to spikes, except that in calculating the withdrawal load for spikes, the depth of penetration is taken as the length of the spike in the member receiving the point, minus two-thirds the length of the point.

Staples

Different types of staples have been developed with various modifications in points, shank treatment and coatings, gauge, crown width, and length. These fasteners are available in clips or magazines to permit their use in pneumatically operated portable staplers. Most of the factors that affect the withdrawal and lateral loads of nails similarly affect the loads on staples. The withdrawal resistance, for example, varies almost directly with the circumference and depth of penetration when the type of point and shank are similar to nails. Thus, formula (7−1) may be used to predict the withdrawal load for one leg of a staple, and the same factors used for nails may be used to arrive at an allowable load.

The load in lateral resistance varies about as the 3/2 power of the diameter when other factors, such as quality of metal, type of shank, and depth of penetration, are similar to nails. The diameter of each leg of a two-legged staple must therefore be about two-thirds the diameter of a nail to provide a comparable load. Formula (7−2) may be used to predict the lateral resistance of staples and the same factors as for nails may be used to arrive at allowable loads.

In addition to the immediate performance capability of staples and nails as determined by test, such factors as corrosion, sustained performance under service conditions, and durability in various uses should be considered in evaluating the relative usefulness of a stapled connection.

Drift Bolts

A drift bolt (or drift pin) is a long pin of iron or steel, with or without head or point. It is driven into a bored hole through one timber and into an adjacent one, to prevent the separation of the timbers connected and to transmit lateral load. The hole in the second member is drilled sufficiently deep to prevent the pin from hitting the bottom.

The ultimate withdrawal load of a round drift bolt or pin from the side grain of seasoned wood is given by the formula:

$$p = 6,600\ G^2 DL \qquad (7-8)$$

where p is the ultimate withdrawal load in pounds, G is the specific gravity based on the ovendry weight and volume at 12 percent moisture content of the wood, D is the diameter of the drift bolt in inches, and L is the length of penetration of the bolt in inches.

Table 7−6—*Sizes of common wire spikes*

Size	Length	Diameter	Size	Length	Diameter
		- - - - - *Inch* - - - - -			- - - - - *Inch* - - - - -
10d	3	0.192	40d	5	0.263
12d	3-1/4	.192	50d	5-1/2	.283
16d	3-1/2	.207	60d	6	.283
20d	4	.225	5/16 inch	7	.312
30d	4-1/2	.244	3/8 inch	8-1/2	.375

This formula provides an average relationship for all species, and the withdrawal load for some species may be above or below the formula values. It also presumes that the bolts are driven into prebored holes having a diameter one-eighth inch less than the bolt diameter.

There are no data available on lateral resistance of drift bolts. Designers have used bolt data and design methods based on experience. This suggests that the load for a drift bolt driven into the side grain of wood should not exceed, and ordinarily should be taken as less than, that for a bolt of the same diameter. Bolt design values are based on the thickness of the main member in a joint. Thus the depth of penetration of the drift bolt must be greater than or equal to the main-member thickness on which the bolt design value is based. However, the drift bolt should not fully penetrate its joint.

Wood Screws

The common types of wood screws have flat, oval, or round heads. The flathead screw is most commonly used if a flush surface is desired. Ovalhead and roundhead screws are used for appearance, and roundhead screws are used when countersinking is objectionable. Besides the head, the principal parts of a screw are the shank, thread, and core (fig. 7−4). Wood screws are usually made of steel or brass or other metals, alloys, or with specific finishes such as nickel, blued, chromium, or cadmium. They are classified according to material, type, finish, shape of head, and diameter or gauge of the shank.

Current trends in fastenings for wood also include tapping screws. Tapping screws have threads the full length of the shank and thus may have some advantage for certain specific uses.

Withdrawal Resistance

Experimental Loads
The resistance of wood screw shanks to withdrawal from the side grain of seasoned wood varies directly with the square of the specific gravity of the wood. Within limits, the withdrawal load varies directly with the depth of penetration of the threaded portion and the diameter of the screw, provided the screw does not fail in tension. The screw will fail in tension when its strength is exceeded by the withdrawal strength from the wood. The limiting length to cause a tension failure decreases as the density of the wood increases since the withdrawal strength of the wood increases with density. The longer lengths of standard screws are therefore superfluous in dense hardwoods.

The withdrawal resistance of type A tapping screws, commonly called sheet metal screws, is in general about 10 per-

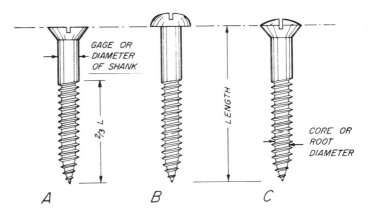

Figure 7−4—Common types of wood screws: A, flathead; B, roundhead; and C, ovalhead. (M139 044)

cent higher than for wood screws of comparable diameter and length of threaded portion. The ratio between the withdrawal resistance of tapping screws and wood screws varies from 1.16 in the denser woods, such as oak, to 1.05 in the lighter woods, such as redwood.

Ultimate test values for withdrawal loads of wood screws inserted into the side grain of seasoned wood may be expressed as:

$$p = 15{,}700G^2DL \qquad (7-9)$$

where p is the maximum withdrawal load in pounds, G is the specific gravity based on ovendry weight and volume at 12 percent moisture content, D is the shank diameter of the screw in inches, and L is the length of penetration of the threaded part of the screw in inches.

This formula is applicable when screw lead holes in soft-woods have a diameter of about 70 percent of the root diameter of the threads, and in hardwoods about 90 percent. The root diameter for most sizes of screws averages about two-thirds of the shank diameter.

The formula values are applicable to the following sizes of screws:

Screw length	Gauge limits
(In)	
1/2	1 to 6
3/4	2 to 11
1	3 to 12
1-1/2	5 to 14
2	7 to 16
2-1/2	9 to 18
3	12 to 20

For lengths and gauges outside these limits, the actual values are likely to be less than the formula values.

The withdrawal loads of screws inserted into the end grain of wood are somewhat erratic; but, when splitting is avoided, they should average 75 percent of the load sustained by screws inserted into the side grain.

Lubricating the surface of a screw with soap or similar lubricants is recommended to facilitate insertion, especially in the dense woods. It will have little effect on ultimate withdrawal resistance.

Allowable Loads

For allowable values, the practice has been to use one-sixth the ultimate load for longtime loading conditions. The one-sixth factor also includes a reduction for variability in test data. For normal duration of load, the allowable load may be increased 10 percent.

Fastening Particleboard

Tapping screws are commonly used in particleboard where withdrawal strength is important. Care must be taken when tightening screws in particleboard to avoid stripping the threads. The maximum amount of torque which can be applied to a screw before the threads in the particleboard are stripped is given by the formula:

$$T = 27.98 + 1.36X \qquad (7-10)$$

where T is torque in inch-pounds and X is density of the particleboard in pounds per cubic foot. Formula $(7-10)$ is for eight-gauge screws with a depth of penetration of 5/8 inch. The maximum torque is fairly constant for lead holes of zero to 90 percent of the root diameter of the screw.

Ultimate withdrawal loads of screws from particleboard can be predicted by the formula:

$$P = K\, D^{1/2}(L - D/3)^{5/4}G^2 \qquad (7-11)$$

where $K = 2,655$ for withdrawal from the face and $K = 2,055$ for withdrawal from the edge of the board. D is the shank diameter of the screw in inches, L is the depth of embedment of the threaded portion of the screw, and G is the specific gravity of the board based on ovendry weight and volume at current moisture content. Formula $(7-11)$ applies when the setting torque is between 60 to 90 percent of T (formula $7-10$).

Withdrawal resistance of screws from particleboard is not significantly different for lead holes of 50 to 90 percent of the root diameter. A higher setting torque will produce a somewhat higher withdrawal load, but there is only a slight difference (3 percent) in values between 60 to 90 percent setting torques (formula $7-10$). A modest tightening of screws in many cases provides an effective compromise between optimizing withdrawal resistance and stripping threads.

Formula $(7-11)$ can also predict the withdrawal of screws from fiberboard with $K = 3,700$ for the face and $K = 2,860$ for the edge of the board.

Lateral Resistance

Experimental Loads

The proportional limit loads obtained in tests of lateral resistance for wood screws in the side grain of seasoned wood are given by the empirical formula:

$$p = KD^2 \qquad (7-12)$$

where p is the lateral load in pounds, D is the diameter of the screw shank in inches, and K is a coefficient depending on the inherent characteristics of the wood species. Values of screw shank diameters for various screw gauges are:

Screw number or gauge	Diameter
	(In)
4	0.112
5	.125
6	.138
7	.151
8	.164
9	.177
10	.190
11	.203
12	.216
14	.242
16	.268
18	.294
20	.320
24	.372

Values of K are based on ranges of specific gravity of hardwoods and softwoods and are given in table $7-4$. They apply to wood at about 15 percent moisture content. Loads computed by substituting these constants in the formula are expected to have a slip of 0.007 to 0.010 inch, depending somewhat on the species and density of the wood.

Formula $(7-12)$ applies when the depth of penetration of the screw into the block receiving the point is not less than seven times the shank diameter and when the cleat (member not receiving the point) and the block holding the point are approximately of the same density. The thickness of the side member should be about one-half the depth of penetration of the screw in the member holding the point. The end distance should be no less than the side member thickness, and the edge distances no less than one-half the side member thickness.

This depth of penetration (seven times shank diameter) gives an ultimate load of about four times the load obtained by the formula. For a depth of penetration of less than seven times the shank diameter, the ultimate load is reduced about in proportion to the reduction in penetration and the load at the proportional limit is reduced somewhat less rapidly. When the depth of penetration of the screw in the holding block is four times the shank diameter, the maximum load will be less than three times the load expressed by the formula, and the proportional limit load will be approximately equal to that given by the formula. When the screw holds metal to wood, the load can be increased by about 25 percent.

For these lateral loads, the part of the lead hole receiving the shank should be the same diameter as the shank or slightly smaller; that part receiving the threaded portion should be the same diameter as the root of the thread in dense species or slightly smaller than the root in low-density species.

Screws should always be turned in. They should never be started or driven with a hammer because this practice tears the wood fibers and injures the screw threads, seriously reducing the loadcarrying capacity of the screw.

Allowable Loads

For allowable values, the practice has been to reduce the proportional limit load by a factor of 1.6 to account for variability in test data and reduce the load to a longtime loading condition. For normal duration of load, the allowable load may be increased by 10 percent.

Lag Screws

Lag screws are commonly used because of their convenience, particularly where it would be difficult to fasten a bolt or where a nut on the surface would be objectionable. Commonly available lag screws range from about 0.2 to 1 inch in diameter and from 1 to 16 inches in length. The length of the threaded part varies with the length of the screw and ranges from three-fourths inch with the 1- and 1-1/4-inch screws to half the length for all lengths greater than 10 inches. The formulas given here for withdrawal and lateral loads are based on lag screws having a base metal average tensile yield strength of about 45,000 psi and an average ultimate tensile strength of 77,000 psi. For lag screws having greater or lower yield and tensile strengths, the withdrawal loads should be adjusted in proportion to the tensile strength and the lateral loads in proportion to the square root of the yield-point stress.

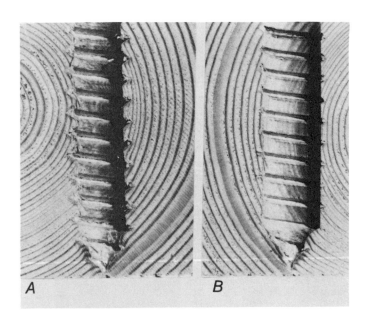

Figure 7–5—A, Clean-cut, deep penetration of thread made by lag screw turned into a lead hole of proper size, and B, rough, shallow penetration of thread made by a lag screw turned into oversized lead hole.

(M84 0276)

Withdrawal Resistance

Experimental Loads

The results of withdrawal tests have shown that the maximum load in direct withdrawal of lag screws from seasoned wood may be computed from the formula:

$$p = 8,100 \, G^{3/2} D^{3/4} L \qquad (7-13)$$

where p is the maximum withdrawal load in pounds, D is the shank diameter in inches, G is the specific gravity of the wood based on ovendry weight and volume at 12 percent moisture content, and L is the length, in inches, of penetration of the threaded part. Formula $(7-13)$ was developed independently of formula $(7-9)$ but gives approximately the same results.

Lag screws, like wood screws, require prebored holes of the proper size (fig. 7–5). The lead hole for the shank should be the same diameter as the shank. The diameter of the lead hole for the threaded part varies with the density of the wood: For the lightweight softwoods, such as the cedars and white pines, 40 to 70 percent of the shank diameter; for Douglas-fir and southern pine, 60 to 75 percent; and for dense hardwoods, such as the oaks, 65 to 85 percent. The smaller percentage in each range applies to lag screws of the smaller diameters, and the larger percentage to lag screws of larger diameters. Soap or similar lubricants should be used on the screws to facilitate turning, and lead holes slightly larger than those recommended for maximum efficiency should be used with long screws.

In determining the withdrawal resistance, the allowable tensile strength of the lag screw at the net (root) section should not be exceeded. Penetration of the threaded part to a distance about seven times the shank diameter in the denser species (specific gravity greater than 0.61) and 10 to 12 times the shank diameter in the less dense species (specific gravity less than 0.42) will develop approximately the ultimate tensile strength of the lag screw. Penetrations at intermediate densities may be found by straight-line interpolation.

The resistance to withdrawal of a lag screw from the end-grain surface of a piece of wood is about three-fourths as great as its resistance to withdrawal from the side-grain surface of the same piece.

Allowable Loads

For allowable values, the practice has been to use one-fifth the ultimate load to account for variability in test data and reduce the load to a longtime loading conditon. For normal duration of load, the allowable load may be increased by 10 percent.

Lateral Resistance

Experimental Loads

The experimentally determined lateral loads for lag screws inserted in the side grain and loaded parallel to the grain of a piece of seasoned wood can be computed from the formula:

$$p = KD^2 \qquad (7-14)$$

where p is the proportional limit lateral load (in pounds) parallel to the grain, K is a coefficient depending on the species specific gravity, and D is the shank diameter of the lag screw in inches. Values of K for a number of specific gravity ranges can be found in table 7−4. These coefficients are based on average results for several ranges of specific gravity for hardwoods and softwoods. The loads given by this formula apply when the thickness of the attached member is 3.5 times the shank diameter of the lag screw, and the depth of penetration in the main member is seven times the diameter in the harder woods and 11 times the diameter in the softer woods. For other thicknesses, the computed loads should be multiplied by the following factors:

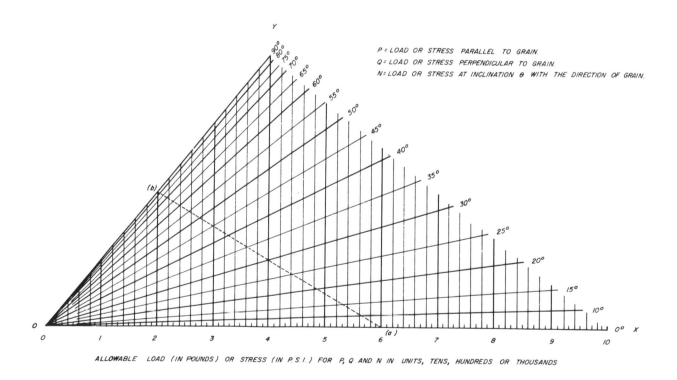

P = LOAD OR STRESS PARALLEL TO GRAIN.
Q = LOAD OR STRESS PERPENDICULAR TO GRAIN.
N = LOAD OR STRESS AT INCLINATION θ WITH THE DIRECTION OF GRAIN.

ALLOWABLE LOAD (IN POUNDS) OR STRESS (IN P.S.I.) FOR P, Q AND N IN UNITS, TENS, HUNDREDS OR THOUSANDS

Figure 7−6—Scholten nomograph for determining the bearing stress of wood at various angles to the grain. The dotted line *ab* refers to the example given in the text.

(M139 043)

Ratio of thickness of attached member to shank diameter of lag screw	Factor
2	0.62
2-1/2	.77
3	.93
3-1/2	1.00
4	1.07
4-1/2	1.13
5	1.18
5-1/2	1.21
6	1.22
6-1/2	1.22

The thickness of a solid wood side member should be about one-half the depth of penetration in the member holding the point.

When the lag screw is inserted into the side grain of wood and the load is applied perpendicular to the grain, the load given by the lateral resistance formula should be multiplied by the following factors:

Shank diameter of lag screw	Factor
(in)	
3/16	1.00
1/4	.97
5/16	.85
3/8	.76
7/16	.70
1/2	.65
5/8	.60
3/4	.55
7/8	.52
1	.50

For other angles of loading, the loads may be computed from the parallel and perpendicular values by the use of the Scholten nomograph for determining the bearing strength of wood at various angles to the grain (fig. 7−6). The nomograph provides values as given by the Hankinson formula:

$$N = \frac{PQ}{P \sin^2 \theta + Q \cos^2 \theta} \qquad (7-15)$$

where P represents the load or stress parallel to the grain, Q, the load or stress perpendicular to the grain, and N, the load or stress at an inclination θ with the direction of the grain.

Example: P, the load parallel to grain is 6,000 pounds, and Q, the load perpendicular to the grain is 2,000 pounds. N, the load at an angle of 40° to grain is found as follows: Connect with a straight line 6,000 pounds (a) on line OX of the nomograph with the intersection (b) on line OY of a vertical line through 2,000 pounds. The point where line (ab) intersects the line representing the given angle 40° is directly above the load, 3,300 pounds.

Values for lateral resistance as computed by the preceding methods are based on complete penetration of the shank into the attached member but not into the foundation member. When the shank penetrates the foundation member, the following increases in loads are permitted:

Ratio of penetration of shank into foundation member to shank diameter	Factor for increasing load
1	1.08
2	1.17
3	1.26
4	1.33
5	1.36
6	1.38
7	1.39

When lag screws are used with metal plates, the lateral loads parallel to the grain may be increased 25 percent, provided the plate thickness is sufficient so that the bearing capacity of the steel is not exceeded. No increase should be made when the applied load is perpendicular to the grain.

Lag screws should not be used in end grain, because splitting may develop under lateral load. If lag screws are so used, however, the loads should be taken as two-thirds those for lateral resistance when lag screws are inserted into side grain and the loads act perpendicular to the grain.

The spacings, end and edge distances, and net section for lag screw joints should be the same as those for joints with bolts of a diameter equal to the shank diameter of the lag screw.

Lag screws should always be inserted by turning with a wrench, not by driving with a hammer. Soap, beeswax, or other lubricants applied to the screw, particularly with the denser wood species, will facilitate insertion and prevent damage to the threads but will not affect the lag screw's performance.

Allowable Loads

For allowable loads, the accepted practice has been to reduce the proportional limit load by dividing by a factor of 2.25 to account for variability in test data and reduce the load to a longtime loading condition. For normal duration of load, the allowable load may be increased by 10 percent.

Bolts

Bearing Stress of Wood Under Bolts

The bearing stress under a bolt is computed by dividing the load on a bolt by the product LD where L is the length of a bolt in the main member and D is the bolt diameter. Basic parallel-to-grain and perpendicular-to-grain bearing stresses have been obtained from tests of three-member wood joints where each side member is half the thickness of the main member. The side members were loaded parallel to grain for both parallel- and perpendicular-to-grain tests.

The bearing stress at proportional limit load is largest when the bolt does not bend, i.e., for joints with small L/D values. The curves of figures 7−7 and 7−8 show the reduction in proportional limit bolt-bearing stress as L/D ratio increases. The bearing stress at maximum load does not decrease as L/D increases, but remains fairly constant, which means that the ratio of maximum load to proportional limit load increases as L/D increases. To maintain a fairly constant ratio between maximum load and design load for bolts, the relations between bearing stress and L/D ratio have been adjusted as indicated in figures 7−7 and 7−8.

The proportional limit bolt-bearing stress parallel to grain for small L/D ratios is approximately 50 percent of the small clear crushing strength for softwoods and approximately 60 percent for hardwoods. For bearing stress perpendicular to the grain, the ratio between bearing stress at proportional limit load and the small clear proportional limit stress in compression perpendicular to grain depends upon bolt diameter, figure 7−9, for small L/D ratios.

Species compressive strength also affects the L/D ratio relationship as indicated in figure 7−8. Relatively higher bolt proportional-limit stress perpendicular to grain is obtained with wood low in strength (proportional limit stress of 570 psi) than with material of high strength (proportional limit stress of 1,140 psi). This effect also occurs for bolt-bearing stress parallel to grain, but not to the same extent as for perpendicular-to-grain loading.

The proportional limit bolt load for a three-member joint with side members half the thickness of the main member may be estimated by the following procedures.

For parallel-to-grain loading: (1) multiply the species small clear crushing strength (table 4−2) by 0.50 for softwoods or 0.60 for hardwoods, (2) multiply this product by the appropriate factor from figure 7−7 for the L/D ratio of the bolt, and (3) multiply this product by LD.

For perpendicular-to-grain loading: (1) multiply the species compression perpendicular-to-grain proportional limit stress (table 4−2) by the appropriate factor from figure 7−9, (2) multiply this product by the appropriate factor from figure 7−8, and (3) multiply this product by LD.

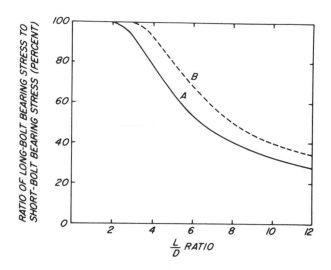

Figure 7−7—Variation in bolt-bearing stress at the proportional limit parallel to grain with L/D ratio. Curve A, relation obtained from experimental evaluation; curve B, modified relation used for establishing design loads. (ML84 5725)

Loads at an Angle to the Grain

For loads applied at an angle intermediate between those parallel to the grain and perpendicular to the grain, the bolt-bearing stress may be obtained from the nomograph in figure 7−6.

Steel Side Plates

When steel side plates are used, the bolt-bearing stress parallel to grain at joint proportional limit is approximately 25 percent higher than for wood side plates. The joint deformation at proportional limit is much smaller with steel side plates. If loads at equivalent joint deformation are compared, the load for joints with steel side plates is approximately 75 percent higher than for wood side plates.

For perpendicular-to-grain loading, the same loads are obtained for wood and steel side plates.

Bolt Quality

Both the properties of the wood and the quality of the bolt are factors in determining the proportional limit strength of a bolted joint. The percentages given in figures 7−7 and 7−8 for calculating bearing stress apply to steel machine bolts with a yield stress of 45,000 psi. Figure 7−10 indicates the

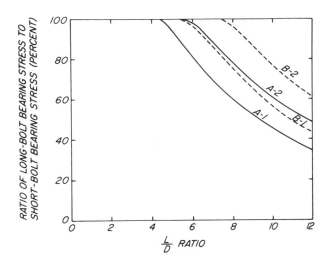

Figure 7—8—Variation in bolt-bearing stress at the proportional limit perpendicular to grain with L/D ratio. Relations obtained from experimental evaluation for materials with average compression perpendicular stress of 1,140 psi, A—1 and 570 psi, A—2. Curves B—1 and B—2, modified relations used for establishing design loads.

(ML84 5726)

increase in bearing stress parallel to grain for bolts with a yield stress of 125,000 psi.

Effect of Member Thickness

The proportional limit load is affected by the ratio of the side member thickness to the main member thickness, figure 7—11.

Design values for bolts are based on joints with side members half the thickness of the main member. The usual practice in design of bolted joints is to take no increase in design load when the side members are thicker than half the main member. When the side members are less than half the thickness of the main member, a design load for a main member which is twice the thickness of the side member is used.

Two-Member, Multiple-Member Joints

For two-member joints, the proportional limit load can be taken as half the load for a three-member joint with a main member the same thickness as the thinnest member.

For four or more members in a joint, the proportional limit load can be taken as the sum of the loads for the individual shear planes by treating each shear plane as an equivalent two-member joint.

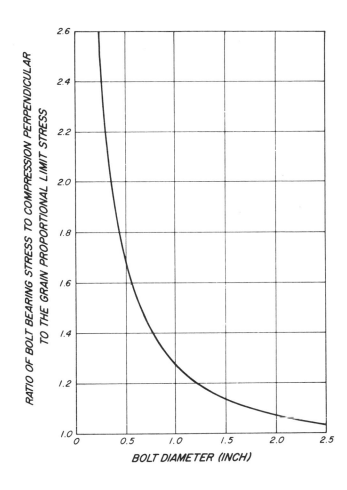

Figure 7—9—Bearing stress perpendicular to the grain as affected by bolt diameter.

(M139 050)

Spacing, Edge, and End Distance

The center-to-center distance along the grain should be at least four times the bolt diameter for parallel-to-grain loading. The minimum center-to-center spacing of bolts in the across-the-grain direction for loads acting through metal side plates and parallel to the grain need only be sufficient to permit the tightening of the nuts. For wood side plates, the spacing is controlled by the rules applying to loads acting parallel to grain if the design load approaches the bolt-bearing capacity of the side plates. When the design load is less than the bolt-bearing capacity of the side plates, the spacing may be reduced below that required to develop their maximum capacity.

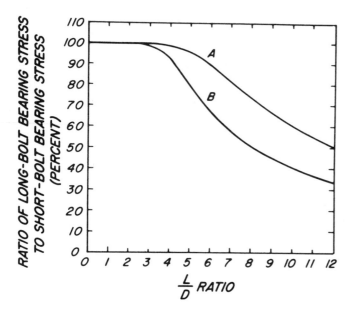

Figure 7–10—Variation in the proportional limit bolt-bearing stress parallel to grain with L/D ratio. Curve A for bolts with yield stress of 125,000 psi; curve B for bolts with a yield stress of 45,000 psi.

(ML84 5814)

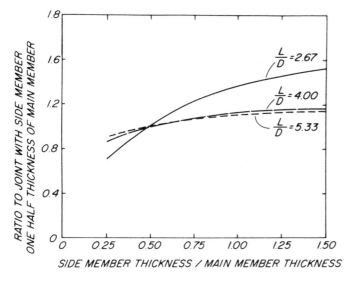

Figure 7–11—Proportional limit load versus side member thickness for three-member joints. Center-member thickness was 2 inches.

(ML84 5727)

When a joint is in tension, the bolt nearest the end of a timber should be at a distance from the end of at least seven times the bolt diameter for softwoods and five times for hardwoods. When the joint is in compression, the end margin may be four times the bolt diameter for both softwoods and hardwoods. Any decrease in these spacings and margins will decrease the load in about the same ratio.

For bolts bearing parallel to the grain, the distance from the edge of a timber to the center of a bolt should be at least 1.5 times the bolt diameter. This margin, however, will usually be controlled by (a) the common practice of having an edge margin equal to one-half the distance between bolt rows and (b) the area requirements at the critical section. (The critical section is that section of the member taken at right angles to the direction of load, which gives the maximum stress in the member based on the net area remaining after reductions are made for bolt holes at that section.) For parallel-to-grain loading in softwoods, the net area remaining at the critical section should be at least 80 percent of the total area in bearing under all the bolts in the particular joint under consideration; in hardwoods it should be 100 percent.

For bolts bearing perpendicular to the grain, the margin between the edge toward which the bolt pressure is acting and the center of the bolt or bolts nearest this edge should be at least four times the bolt diameter. The margin at the opposite edge is relatively unimportant.

Effect of Bolt Holes

The bearing strength of wood under bolts is affected considerably by the size and type of bolt hole into which the bolts are inserted. A bolt hole that is too large causes nonuniform bearing of the bolt; if the bolt hole is too small, the wood will split when the bolt is driven. Normally, bolts should fit so that they can be inserted by tapping lightly with a wood mallet. In general, the smoother the hole, the higher the bearing values will be (fig. 7–12). Deformations accompanying the load are also less with a smoother bolt-hole surface (fig. 7–13).

Rough holes are caused by using dull bits and improper rates of feed and drill speed. A twist drill operated at a peripheral speed of approximately 1,500 inches per minute produces uniformly smooth holes at moderate feed rates. The rate of feed depends upon the diameter of the drill and the speed of rotation but should enable the drill to cut rather than tear the wood. The drill should produce shavings, not chips.

Proportional limit loads for joints with bolt holes the same diameter as the bolt will be slightly higher than for joints with a 1/16-inch oversized hole. However, if drying takes place after assembly of the joint, the proportional limit load for snug-fitting bolts will be considerably less due to the effects of shrinkage.

Allowable Loads

The following procedures are used to calculate allowable bolt loads for joints with wood side members, each half the thickness of the main member.

Parallel to Grain

The starting point for parallel-to-grain bolt values is the maximum green crushing strength for the species or group of species. Procedures outlined in ASTM D 2555 are used to establish a 5 percent exclusion value. The exclusion value is divided by a factor of 1.9 to adjust to a normal duration of load and provide a factor of safety. This value is multiplied by 1.20 to adjust to a seasoned strength. The resulting value is called the basic bolt-bearing stress parallel to grain.

The basic bolt-bearing stress is then adjusted for the effects of L/D ratio. Table 7–8 gives the percentage of basic stress for three classes of species. The particular class for the species is determined from the basic bolt-bearing stress as indicated in table 7–7. The adjusted bearing stress is further multiplied by a factor of 0.80 to adjust to wood side plates. The allowable bolt load in pounds is then determined by multiplying by the projected bolt area, LD.

Perpendicular to Grain

The starting point for perpendicular-to-grain bolt values is the average green proportional limit stress in compression perpendicular to grain. Procedures in ASTM D 2555 are used to establish compression perpendicular values for groups of species. The average proportional limit stress is divided by 1.5 for ring position (growth rings neither parallel nor perpendicular to load during test) and a factor of safety. This value is then multiplied by 1.20 to adjust to a seasoned strength and by 1.10 to adjust to a normal duration of load. The resulting value is called the basic bolt-bearing stress perpendicular to grain.

The basic bolt-bearing stress is then adjusted for the effects of bolt diameter, table 7–9, and L/D ratio, table 7–8. The allowable bolt load in pounds is then determined by multiply-

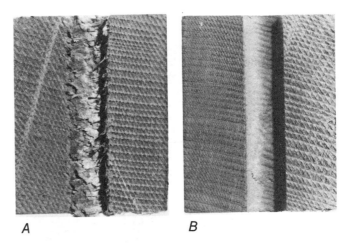

Figure 7–12—Effect of rate of feed and drill speed on the surface condition of bolt holes drilled in Sitka spruce. The hole on the left was bored with a twist drill rotating at a peripheral speed of 300 inches per minute; the feed rate was 60 inches per minute. The hole on the right was bored with the same drill at a peripheral speed of 1,250 inches per minute; the feed rate was 2 inches per minute.

(M84 0275)

ing the adjusted basic bolt-bearing stress by the projected bolt area, LD.

Connector Joints

Several types of connectors have been devised that increase joint bearing and shear areas by utilizing rings or plates around bolts holding joint members together. The primary load-carrying portions of these joints are the connectors; the bolts

Table 7–7—L/D adjustment class associated with basic bolt-bearing stress

Loading direction	Basic bolt-bearing stress for species group		L/D adjustment class (table 7–8)
	Softwoods	Hardwoods	
	----------- Pounds per Square Inch -----------		
Parallel	Less than 1,150	Less than 1,063	1
	1,150–1,504	1,063–1,389	2
	Greater than 1,504	Greater than 1,389	3
Perpendicular	Less than 190	Less than 209	1
	190–290	209–319	2
	291–375	320–412	3
	Greater than 375	Greater than 412	4

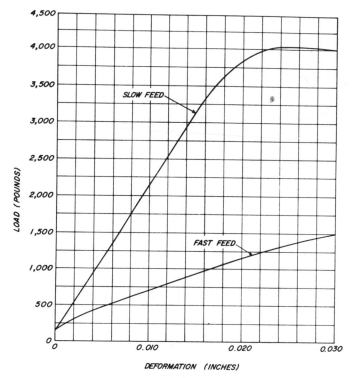

DEFORMATION (INCHES)

Figure 7–13—Typical load-deformation curves
showing the effect of surface condition of bolt holes,
resulting from a slow feed rate and a fast feed rate,
on the deformation in a joint when subjected to loading under bolts. The surface conditions of the bolt
holes were similar to those illustrated in figure 7–12.

(M139 042)

Table 7–8—*Percentage of basic bolt-bearing stress used
for calculating allowable bolt loads*

Ratio of bolt length to diameter (L/D)	L/D adjustment factor by class[1]						
	Parallel to grain			Perpendicular to grain			
	1	2	3	1	2	3	4
	-------------- Percent --------------						
1	100.0	100.0	100.0	100.0	100.0	100.0	100.0
2	100.0	100.0	100.0	100.0	100.0	100.0	100.0
3	100.0	100.0	99.0	100.0	100.0	100.0	100.0
4	99.5	97.4	92.5	100.0	100.0	100.0	100.0
5	95.4	88.3	80.0	100.0	100.0	100.0	100.0
6	85.6	75.8	67.2	100.0	100.0	100.0	96.3
7	73.4	65.0	57.6	100.0	100.0	97.3	86.9
8	64.2	56.9	50.4	100.0	96.1	88.1	75.0
9	57.1	50.6	44.8	94.6	86.3	76.7	64.6
10	51.4	45.5	40.3	85.0	76.2	67.2	55.4
11	46.7	41.4	36.6	76.1	67.6	59.3	48.4
12	42.8	37.9	33.6	68.6	61.0	52.0	42.5
13	39.5	35.0	31.0	62.2	55.3	45.9	37.5

[1] Class determined from basic bolt-bearing stress according to table 7–7.

Table 7–9—*Factors for adjusting basic bolt-bearing stress
perpendicular to grain for bolt diameter when
calculating allowable bolt loads*

Bolt diameter	Adjustment factor
In	
1/4	2.50
3/8	1.95
1/2	1.68
5/8	1.52
3/4	1.41
7/8	1.33
1	1.27
1-1/4	1.19
1-1/2	1.14
1-3/4	1.10
2	1.07
2-1/3	1.03
3 or over	1.00

usually serve to prevent sideways separation of the members,
but do contribute some load-carrying capacity.

The strength of the connector joint depends on the type and
size of the connector, the species of wood, the thickness and
width of the member, the distance of the connector from the
end of the member, the spacing of the connectors, the direction of application of the load with respect to the direction of
the grain of the wood, and other factors. Loads for wood
joints with steel connectors—split ring (fig. 7–14) and shear
plate (fig. 7–15)—are discussed in this section. These connectors require closely fitting machined portions in the wood
members.

Parallel-to-Grain Loading

Tests have demonstrated that the density of the wood is a
controlling factor in the strength of connector joints. For
split-ring connectors, both maximum load and proportional
limit load parallel to grain vary linearly with specific gravity,
figures 7–16 and 7–17. For shear plates, the maximum load
and proportional limit load vary linearly with specific gravity

for the less dense species, figures 7—18 and 7—19. In the higher density species, the shear strength of the bolts becomes the controlling factor. These relations were obtained for seasoned members, approximately 12 percent moisture content.

Perpendicular-to-Grain Loading

Loads for perpendicular-to-grain loading have been established using three-member joints with the side members loaded parallel to grain. Specific gravity has been found to be a good indicator of perpendicular-to-grain strength of timber connector joints. For split-ring connectors, the proportional limit loads perpendicular to grain have been found to be 58 percent of the parallel-to-grain proportional limit loads. The joint deformation at proportional limit is 30 to 50 percent more than for parallel-to-grain loading.

For shear-plate connectors, the proportional limit and maximum loads vary linearly with specific gravity, figures 7—20 and 7—21. The wood strength controls the joint strength for all species.

Design Loads

Design loads for parallel-to-grain loading have been established by dividing ultimate test loads by an average factor of 4. This gives values that do not exceed five-eighths of the proportional limit loads. The reduction accounts for variability in material, a reduction to long-time loading, and a factor of safety. Loads for normal duration of load are 10 percent higher.

For perpendicular-to-grain loading, ultimate load was given less consideration and greater dependence placed on load at proportional limit. For split rings, the proportional limit load was reduced by approximately half. For shear plates, the design loads are approximately five-eighths of the proportional limit test loads. These reductions again account for material variability, a reduction to long-time loading, and a factor of safety.

Design loads are presented in figures 7—16 through 7—21. In practice, four wood species groups have been established, based primarily on specific gravity, and design loads assigned for each group. Species groupings for connectors are presented in table 7—10. The corresponding design loads (for long-continued load) are given in table 7—11. The National Design Specification gives design values for normal-duration load for these and additional species.

Modifications

Some of the factors that affect the loads of connectors were taken into account in deriving the tabular values. Other var-

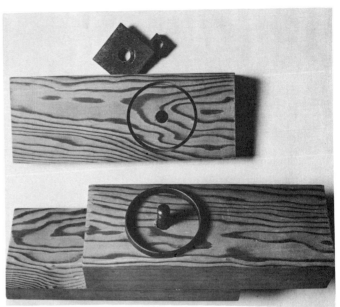

Figure 7—14—Joint with split-ring connector showing connector, precut groove, bolt, washer, and nut. (M93 396)

Figure 7—15—Joints with shear-plate connectors with A, wood side plates; and B, steel side plates. (M92 355)

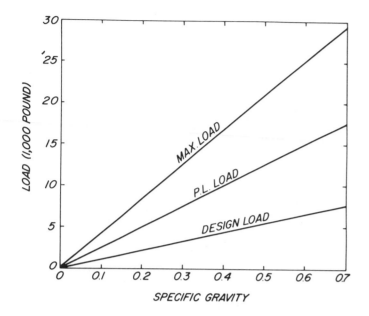

Figure 7–16—Relation between load bearing parallel to grain and specific gravity (ovendry weight, volume at test) for two 2-1/2-inch split rings with a single 1/2-inch bolt in air-dry material. Thickness of the center member was 4 inches and side member thickness was 2 inches.

(ML84 5728)

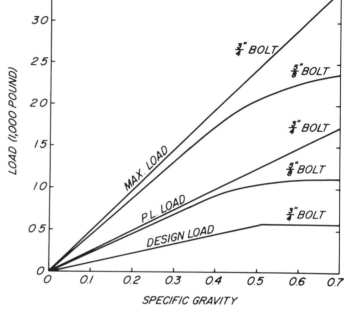

Figure 7–18—Relation between load bearing parallel to grain and specific gravity (ovendry weight, volume at test) for two 2-5/8-inch shear plates in air-dry material with steel side plates. Center member thickness was 3 inches.

(ML84 5730)

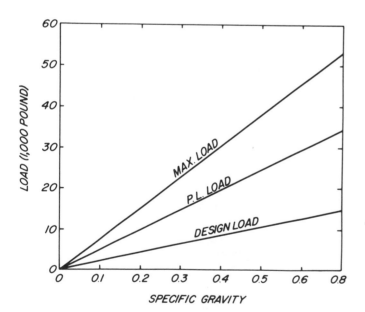

Figure 7–17—Relation between load bearing parallel to grain and specific gravity (ovendry weight, volume at test) for two 4-inch split rings and a single 3/4-inch-diameter bolt in air-dry material. Thickness of the center member was 5 inches and side member thickness was 2-1/2 inches.

(ML84 5729)

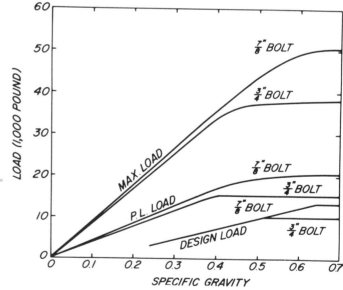

Figure 7–19—Relation between load bearing parallel to grain and specific gravity (ovendry weight, volume at test) for two 4-inch shear plates in air-dry material with steel side plates. Center member thickness was 3-1/2 inches.

(ML84 5805)

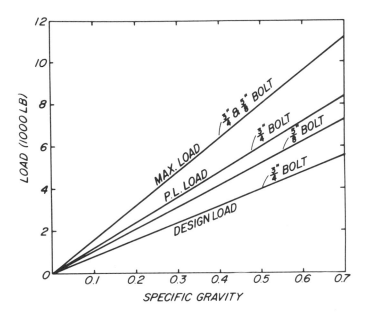

Figure 7–20—Relation between load bearing perpendicular to grain and specific gravity (ovendry weight, volume at test) for two 2-5/8-inch shear plates in air-dry material with steel side plates. Center member thickness was 3 inches.

(ML84 5806)

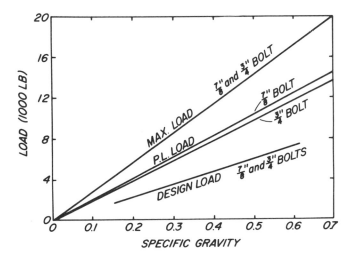

Figure 7–21—Relation between load bearing perpendicular to grain and specific gravity (ovendry weight, volume at test) for two 4-inch shear plates in air-dry material with steel side plates. Center member thickness was 3-1/2 inches.

(ML84 5731)

ied and extreme conditions require modification of the values.

Steel Side Plates

Steel side plates are often used with shear-plate connectors. The loads parallel to grain have been found to be approximately 10 percent higher than with wood side plates. The perpendicular-to-grain loads are unchanged.

Exposure and Moisture Condition of Wood

The loads listed in table 7–11 apply to seasoned members used where they will remain dry. If the wood will be more or less continuously damp or wet in use, two-thirds of the tabulated values should be used. The amount by which the loads should be reduced to adapt them to other conditions of use depends upon the extent to which the exposure favors decay, the required life of the structure or part, the frequency and thoroughness of inspection, the original cost and the cost of replacements, the proportion of sapwood and the durability of the heartwood of the species (if untreated), and the character and efficiency of any treatment. These factors should be evaluated for each individual design. Industry recommendations for the use of connectors when the condition of the lumber is other than continuously wet or continuously dry are given in the National Design Specification for Wood Construction.

Ordinarily, before fabrication of connector joints, members should be seasoned to a moisture content corresponding as nearly as practical to that which they will attain in service. This is particularly desirable for lumber for roof trusses and other structural units used in dry locations and in which shrinkage is an important factor. Urgent construction needs sometimes result in the erection of structures and structural units employing green or inadequately seasoned lumber with connectors. Since such lumber subsequently dries out in most buildings, causing shrinkage and opening the joints, it is essential that adequate maintenance measures be adopted. The maintenance for connector joints in green lumber should include inspection of the structural units and tightening of all bolts as needed during the time the units are coming to moisture equilibrium, which is normally during the first year.

Grade and Quality of Lumber

The lumber for which the loads for connectors are applicable should conform to the general requirements in regard to the quality of structural lumber given in the grading rule books of lumber manufacturers' associations for the various commercial species.

The loads for connectors were obtained from tests of joints whose members were clear and free from checks, shakes, and splits. Cross grain at the joint should not be steeper than 1 in 10, and knots in the connector area should be accounted for as explained under "Net Section."

Loads at Angle With Grain

The loads for the split-ring and shear-plate connectors for

Table 7–11—Design loads for one connector in a joint[1]

Connector	Minimum thickness of wood member — With one connector only	With two connectors in opposite faces, one bolt[3]	Minimum width all members	Group 1 woods[2]		Group 2 woods[2]		Group 3 woods[2]		Group 4 woods[2]	
				Load at 0° angle to grain	Load at 90° angle to grain	Load at 0° angle to grain	Load at 90° angle to grain	Load at 0° angle to grain	Load at 90° angle to grain	Load at 0° angle to grain	Load at 90° angle to grain
	- - - - - - - - In - - - - - - - -			- - - - - - - - - - - - - - - - - - - Lb - - - - - - - - - - - - - - - - - - -							
Split ring:											
2-1/2-inch diameter, 3/4 inch wide with 1/2-inch bolt	1	2	3-1/2	1,785	1,055	2,085	1,230	2,480	1,475	2,875	1,725
4-inch diameter, 1 inch wide, with 3/4-inch bolt	1-1/2	3	5-1/2	3,445	1,995	3,985	2,310	4,780	2,775	5,580	3,235
Shear plate:											
2-5/8-inch diameter, 0.42 inch wide with 3/4-inch bolt	1-1/2	2-5/8	3-1/2	1,890	1,095	2,190	1,270	2,630	1,525	2,665	1,780
4-inch diameter, 0.64 inch wide with 3/4- or 7/8-inch bolt	1-3/4	3-5/8	5-1/2	2,850	1,655	3,305	1,915	3,965	2,300	4,625	2,685

[1] The loads apply to seasoned timbers in dry, inside locations for a long-continued load. It is assumed also that the joints are properly designed with respect to such features as centering of connectors, adequate end distance, and suitable spacing.

[2] Group 1 woods provide the weakest connector joints, and group 4 woods the strongest. Groupings are given in table 7–10.

[3] A 3-member assembly with 2 connectors takes double the loads indicated in fifth to twelfth columns.

angles of 0 to 90° between direction of load and grain may be obtained by Hankinson's formula (7–15) or by the nomograph in figure 7–6.

Thickness of Member

The relationship between the loads for the different thicknesses of lumber is based on test results for connector joints. The least thickness of member given in table 7–11 for the various sizes of connectors is the minimum to obtain optimum load. The loads listed for each type and size of connector are the maximum loads to be used for all thicker lumber. The loads for wood members of thicknesses less than those listed can be obtained by the percentage reductions indicated in figure 7–22. Thicknesses below those indicated by the curves should not be used.

When one member contains a connector in only one face, loads for thicknesses less than those listed in table 7–11 can be obtained by the percentage reductions indicated in figure 7–22 using an assumed thickness equal to twice the actual member thickness.

Width of Member

The width of member listed for each type and size of connector is the minimum that should be used. When the connectors are bearing parallel to the grain, no increase in load occurs with an increase in width. When they are bearing perpendicular to the grain, the load increases about 10 percent for each 1-inch increase in width of member over the minimum widths required for each type and size of connector, up to twice the diameter of the connectors. When the connector is placed off center and the load is applied continuously in

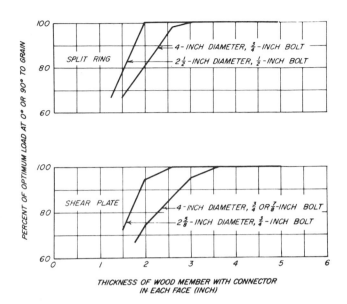

Figure 7–22—Effect of thickness of wood member on the optimum load capacity of a timber connector. (M139 049)

one direction only, the proper load can be determined by considering the width of member as equal to twice the edge distance (the distance between the center of the connector and the edge of the member toward which the load is acting). The distance between the center of the connector and the opposite edge should not, however be less than one-half the permissible minimum width of the member.

Table 7–10—*Species groupings for connector loads[1]*

Connector	Species or species group		
Group 1	Aspen	Basswood	Cottonwood
	Western redcedar	Balsam fir	White fir
	Eastern hemlock	Eastern white pine	Ponderosa pine
	Sugar pine	Western white pine	Engelmann spruce
Group 2	Chestnut	Yellow-poplar	Baldcypress
	Alaska-cedar	Port-Orford-cedar	Western hemlock
	Red pine	Redwood	Red spruce
	Sitka spruce	White spruce	
Group 3	Elm, American	Elm, slippery	Maple, soft
	Sweetgum	Sycamore	Tupelo
	Douglas-fir	Larch, western	Southern pine
Group 4	Ash, white	Beech	Birch
	Elm, rock	Hickory	Maple, hard
	Oak		

[1] Group 1 woods provide the weakest connector joints and group 4 woods the strongest.

Net Section

The net section is the area remaining at the critical section after subtracting the projected area of the connectors and bolt from the full cross-sectional area of the member. For sawed timbers, the stress in the net area (whether in tension or compression) should not exceed the stress for clear wood in compression parallel to the grain. In using this stress, it is assumed that knots do not occur within a length of one-half the diameter of the connector from the net section. If knots are present in the longitudinal projection of the net section within a length from the critical section of one-half the diameter of the connector, the area of the knots should be subtracted from the area of the critical section.

In laminated timbers, knots may occur in the inner laminations at the connector location without being apparent from the outside of the member. It is impractical to assure that there are no knots at or near the connector. In laminated construction, therefore, the stress at the net section is limited to the compressive stress for the member, accounting for the effect of knots.

End Distance and Spacing

The load values in table 7–11 apply when the distance of the connector from the end of the member (end distance e) and the spacing (s) between connectors in multiple joints are not factors affecting the strength of the joint (fig. 7–23, A). When the end distance or spacing for connectors bearing parallel to the grain is less than that required to develop the full load, the proper reduced load may be obtained by multiplying the loads in table 7–11 by the appropriate strength ratio given in table 7–12. For example, the load for a 4-inch split-ring connector bearing parallel to the grain, when placed 7 or more inches from the end of a Douglas-fir tension member that is 1-1/2 inches thick, is 4,780 pounds. When the end distance is only 5-1/4 inches, the strength ratio obtained by direct interpolation between 7 and 3-1/2 inches in table 7–12 is 0.81, and the load equals 0.81 times 4,780 or 3,870 pounds.

Placement of Multiple Connectors

Preliminary investigations of the placement of connectors in a multiple joint, together with the observed behavior of single-connector joints tested with variables that simulate those in a multiple joint, furnish a basis for some suggested design practices.

When two or more connectors in the same face of a member are in a line at right angles to the grain of the member and are bearing parallel to the grain (fig. 7–23, C), the clear

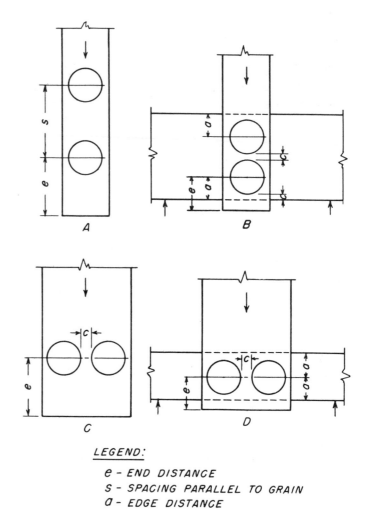

LEGEND:

e - END DISTANCE
S - SPACING PARALLEL TO GRAIN
a - EDGE DISTANCE
C - CLEAR DISTANCE

Figure 7–23—Types of multiple-connector joints: (M39 253) A, joint strength depends on end distance e and connector spacing s; B, joint strength depends on end e, clear c, and edge a distances; C, joint strength depends on end e and clear c distances; D, joint strength depends on end e, clear c, and edge a distances.

distance (c) between the connectors should not be less than 1/2 inch.

When two or more connectors are acting perpendicular to the grain and are spaced on a line at right angles to the length of the member (fig. 7–23, B), the rules for the width of member and edge distances used with one connector are applicable to the edge distances for multiple connectors. The clear distance between the connectors (c) should be equal to the clear distance from the edge of the member toward which the load is acting to the connector nearest this edge (c).

In a joint with two or more connectors spaced on a line

Table 7-12—Strength ratio for connectors for various longitudinal spacings and end distances[1]

Connector diameter	Spacing[2]	Spacing strength ratio	End distance[3]		End distance strength ratio
			Tension member	Compression member	
---------- In ----------		Pct	---------- In ----------		Pct
SPLIT-RING					
2-1/2	6-3/4 +	100	5-1/2 +	4 +	100
2-1/2	3-3/8	50	2-3/4	2-1/2	62
4	9 +	100	7 +	5-1/2 +	100
4	4-7/8	50	3-1/2	3-1/4	62
SHEAR-PLATE					
2-5/8	6-3/4 +	100	5-1/2 +	4 +	100
2-5/8	3-3/8	50	2-3/4	2-1/2	62
4	9 +	100	7 +	5-1/2 +	100
4	4-1/2	50	3-1/2	3-1/4	62

[1] Strength ratio for spacings and end distances intermediate to those listed may be obtained by interpolation, and multiplied by the loads in table 7-11 to obtain design load. The strength ratio applies only to those connector units affected by the respective spacings or end distances. The spacings and end distances should not be less than the minimum shown.

[2] Spacing is distance from center to center of connectors (fig. 7-23, A).

[3] End distance is distance from center of connector to end of member (fig. 7-23, A).

Table 7-13—Modification factors for timber connector, bolt, and laterally loaded lag-screw joints with wood side plates as given in the National Design Specification for Wood Construction

A_1/A_2[1]	A_1[2]	Number of fasteners in a row										
		2	3	4	5	6	7	8	9	10	11	12
	In^2											
0.5[3,4]	<12	1.00	0.92	0.84	0.76	0.68	0.61	0.55	0.49	0.43	0.38	0.34
	12-19	1.00	.95	.88	.82	.75	.68	.62	.57	.52	.48	.43
	>19-28	1.00	.97	.93	.88	.82	.77	.71	.67	.63	.59	.55
	>28-40	1.00	.98	.96	.92	.87	.83	.79	.75	.71	.69	.69
	>40-64	1.00	1.00	.97	.94	.90	.86	.83	.79	.76	.74	.72
	>64	1.00	1.00	.98	.95	.91	.88	.85	.82	.80	.78	.76
1.0[3,4]	<12	1.00	.97	.92	.85	.78	.71	.65	.59	.54	.49	.44
	12-19	1.00	.98	.94	.89	.84	.78	.72	.66	.61	.56	.51
	>19-28	1.00	1.00	.97	.93	.89	.85	.80	.76	.72	.68	.64
	>28-40	1.00	1.00	.99	.96	.92	.89	.86	.83	.80	.78	.75
	>40-64	1.00	1.00	1.00	.97	.94	.91	.88	.85	.84	.82	.80
	>64	1.00	1.00	1.00	.99	.96	.93	.91	.88	.87	.86	.85

[1] A_1 = Cross-sectional area of main members before boring or grooving; A_2 = sum of the cross-sectional area of side members before boring or grooving.

[2] When A_1/A_2 exceeds 1.0, use A_2 instead of A_1.

[3] When A_1/A_2 exceeds 1.0, use A_2/A_1.

[4] For A_1/A_2 between 0 and 1.0, interpolate or extrapolate from the tabulated values.

parallel to the grain and with the load acting perpendicular to the grain (fig. 7−23, D), the available data indicate that the load for multiple connectors is not equal to the sum of the loads for individual connectors. Somewhat more favorable results can be obtained if the connectors are staggered so that they do not act along the same line with respect to the grain of the transverse member. Industry recommendations for various angle-to-grain loadings and spacings are given in National Design Specifications.

Cross Bolts

Cross bolts or stitch bolts placed at or near the end of members joined with connectors or at points between connectors will provide additional safety. They may also be used to reinforce members that have, through change in moisture content in service, developed splits to an undesirable degree.

Multiple-Fastener Joints

When fasteners are used in rows parallel to the direction of loading, there is an unequal distribution of the total joint load among fasteners in the row. Simplified methods of analysis have been developed to predict the load distribution among the fasteners in a row. These analyses indicate that the load distribution is a function of (1) the extensional stiffness, EA, of the joint members, where E is the modulus of elasticity and A is the gross cross-sectional area, (2) the fastener spacing, (3) the number of fasteners, and (4) the single fastener load-

Table 7−14—Modification factors for timber connector, bolt, and laterally loaded lag-screw joints with metal side plates as given in the National Design Specification for Wood Construction

A_1/A_2[1]	A_1	Number of fasteners in a row										
		2	3	4	5	6	7	8	9	10	11	12
	In²											
2−12	25−39	1.00	0.94	0.87	0.80	0.73	0.67	0.61	0.56	0.51	0.46	0.42
	40−64	1.00	.96	.92	.87	.81	.75	.70	.66	.62	.58	.55
	65−119	1.00	.98	.95	.91	.87	.82	.78	.75	.72	.69	.66
	120−199	1.00	.99	.97	.95	.92	.89	.86	.84	.81	.79	.78
12−18	40−64	1.00	.98	.94	.90	.85	.80	.75	.70	.67	.62	.58
	65−119	1.00	.99	.96	.93	.90	.86	.82	.79	.75	.72	.69
	120−199	1.00	1.00	.98	.96	.94	.92	.89	.86	.83	.80	.78
	200	1.00	1.00	1.00	.98	.97	.95	.93	.91	.90	.88	.87
18−24	40−64	1.00	1.00	.96	.93	.89	.84	.79	.74	.69	.64	.59
	65−119	1.00	1.00	.97	.94	.93	.89	.86	.83	.80	.76	.73
	120−199	1.00	1.00	.99	.98	.96	.94	.92	.90	.88	.86	.85
	200	1.00	1.00	1.00	1.00	.98	.96	.95	.93	.92	.92	.91
24−30	40−64	1.00	.98	.94	.90	.85	.80	.74	.69	.65	.61	.58
	65−119	1.00	.99	.97	.93	.90	.86	.82	.79	.76	.73	.71
	120−199	1.00	1.00	.98	.96	.94	.92	.89	.87	.85	.83	.81
	200	1.00	1.00	.99	.98	.97	.95	.93	.92	.90	.89	.89
30−35	40−64	1.00	.96	.92	.86	.80	.74	.68	.64	.60	.57	.55
	65−119	1.00	.98	.95	.90	.86	.81	.76	.72	.68	.65	.62
	120−199	1.00	.99	.97	.95	.92	.88	.85	.82	.80	.78	.77
	200	1.00	1.00	.98	.97	.95	.93	.90	.89	.87	.86	.85
35−42	40−64	1.00	.95	.89	.82	.75	.69	.63	.58	.53	.49	.46
	65−119	1.00	.97	.93	.88	.82	.77	.71	.67	.63	.59	.56
	120−199	1.00	.98	.96	.93	.89	.85	.81	.78	.76	.73	.71
	200	1.00	.99	.98	.96	.93	.90	.87	.84	.82	.80	.78

[1] A_1 = Cross-sectional area of main member before boring or grooving; A_2 = sum of cross-sectional areas of metal side plates before drilling.

deformation characteristics.

Theoretically, the two end fasteners carry a majority of the load. For example, in a row of six bolts, the two end bolts will carry over 50 percent of the total joint load. Adding bolts to a row tends to reduce the load on the less heavily loaded interior bolts. The most even distribution of bolt loads occurs in a joint where the extensional stiffness of the main member is equal to that of both splice plates. Increasing the fastener spacing tends to put more of the joint load on the end fasteners. Load distribution tends to be worse for stiffer fasteners.

The actual load distribution in field-fabricated joints is difficult to predict. Small misalignment of fasteners, variations in spacing between side and main members, as well as variations in single fastener load-deformation characteristics can cause the load distribution to be different than predicted by the theoretical analyses.

For design purposes, modification factors for application to a row of bolts, lag screws, or timber connectors have been developed based on the theoretical analyses. Table 7−13 gives the modification factors for joints with wood side plates as given in the National Design Specification for Wood Construction and table 7−14 gives the factors for joints with steel side plates.

Metal Plate Connectors

Metal plate connectors, commonly called truss plates, have become a popular means of joining, especially in trussed rafters and joists. These connectors transmit loads by means of teeth, plugs, or nails which vary from manufacturer to manufacturer. Some examples of such plates are shown in figure 7−24. Plates are usually made of 20-gauge galvanized steel and have an area and shape necessary to transmit the forces on the joint. Installation of plates usually requires a hydraulic press or other heavy equipment although some plates can be installed by hand.

Basic strength values for plate connectors are determined from tension tests of two butted wood members joined with two plates. Load-slip curves are determined. Some typical curves are shown in figure 7−25.

Design values are expressed as load per tooth, nail, plug, or square inch of plate. The smallest value as determined by two different means is the design load for normal duration of load: (1) the average load of at least five specimens at 0.015-inch slip from plate to wood member or 0.030-inch slip from member to member is divided by 1.6; (2) the average ultimate load of at least five specimens is divided by 3.0.

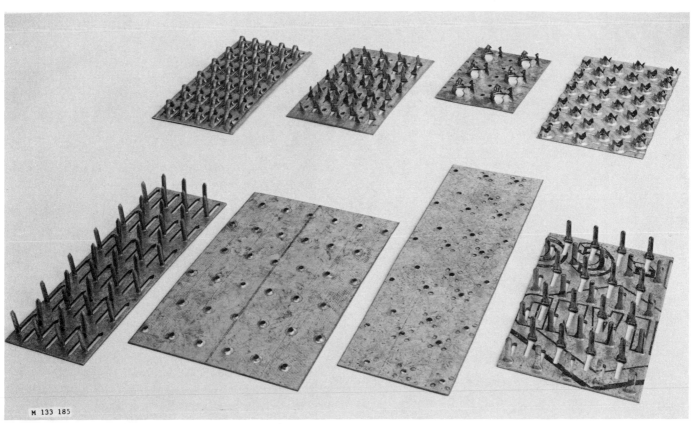

M 133 185

Figure 7−24—Some typical metal plate connectors.

(M133 185)

7-27

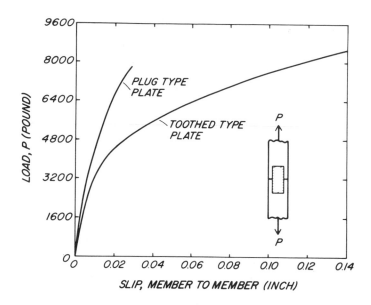

Figure 7−25—Typical load-slip curves for two types of metal plate connectors loaded in tension.

(ML84 5812)

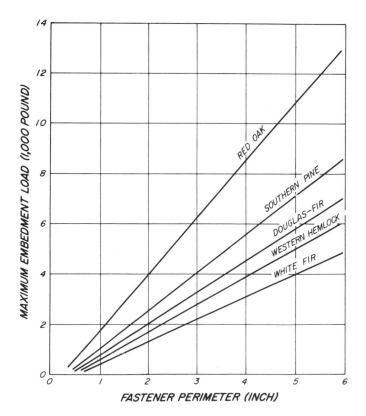

Figure 7−26—Relation between maximum embedment load and fastener perimeter for several species of wood.

(M139 048)

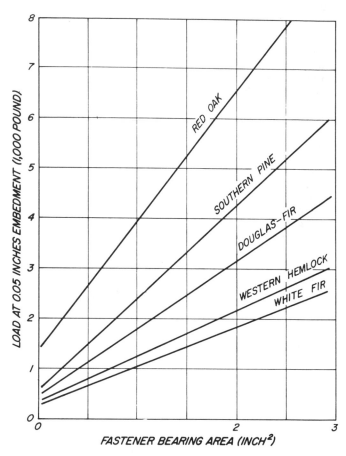

Figure 7−27—Relation between load at 0.05-inch embedment and fastener bearing area for several species.

(M139 047)

The strength of a metal plate joint may also be controlled by the tensile or shear strength of the plate.

Fastener Head Embedment

The bearing strength of wood under fastener heads is important in such applications as the anchorage of building framework to foundation structures. When pressure tends to pull the framing member away from the foundation, the fastening loads could cause tensile failure of the fastenings, withdrawal of the fastenings from the framing member, or embedment of the fastener heads into the member. Possibly the fastener head could even be pulled completely through.

The maximum load for fastener head embedment is related to the fastener head perimeter, while loads at low embedments (0.05 in) are related to the fastener head bearing area. These relations for several species at 10 percent moisture content are shown in figures 7−26 and 7−27.

Selected References

American Society for Testing and Materials. Tentative methods of testing metal fasteners in wood. ASTM D 1761. Philadelphia, PA: ASTM; (see current edition).

American Society for Testing and Materials. Standard methods for establishing clear wood strength values. ASTM D 2555. Philadelphia, PA: ASTM; (see current edition).

Anderson, L. O. Nailing better wood boxes and crates. Agric. Handb. 160. Washington, DC: U.S. Department of Agriculture; 1959.

Anderson, L. O. Wood-frame house construction. Agric. Handb. 73. Washington, DC: U.S. Department of Agriculture; rev. 1970.

Cramer, C. O. Load distribution in multiple-bolt tension joints. Journal of Structural Division, American Society of Civil Engineers. 94(ST5): 1101−1117. (Proc. Pap. 5939); 1968.

Doyle, D. V.; Scholten, J. A. Performance of bolted joints in Douglas-fir. Res. Pap. FPL 2. Madison, WI: U.S. Department of Agriculture, Forest Service, Forest Products Laboratory; 1963.

Eckelman, C. A. Screw holding performance in hardwoods and particleboard. Madison, WI: Forest Products Journal. 25(6); 1975.

Fairchild, I. J. Holding power of wood screws. Technol. Pap. 319. Washington, DC: U.S. National Bureau of Standards; 1926.

Forest Products Laboratory. General observations on the nailing of wood. FPL Tech. Note 243. Madison, WI: U.S. Department of Agriculture, Forest Service, Forest Products Laboratory; 1962.

Forest Products Laboratory. Nailing dense hardwoods. Res. Note FPL-037. Madison, WI: U.S. Department of Agriculture, Forest Service, Forest Products Laboratory; 1964.

Forest Products Laboratory. Nail withdrawal resistance of American woods. Res. Note FPL−033. Madison, WI: U.S. Department of Agriculture, Forest Service, Forest Products Laboratory; 1965.

Goodell, H. R.; Philipps, R. S. Bolt-bearing strength of wood and modified wood: Effect of different methods of drilling bolt holes in wood and plywood. FPL Rep. 1523. Madison, WI: U.S. Department of Agriculture, Forest Service, Forest Products Laboratory; 1944.

Jordan, C. A. Response of timber joints with metal fasteners to lateral impact loads. FPL Rep. 2263. Madison, WI: U.S. Department of Agriculture, Forest Service, Forest Products Laboratory; 1963.

Kuenzi, E. W. Theoretical design of a nailed or bolted joint under lateral load. FPL Rep. 1951. Madison, WI: U.S. Department of Agriculture, Forest Service, Forest Products Laboratory; 1951.

Kurtenacker, R. S. Performance of container fasteners subjected to static and dynamic withdrawal. Res. Pap. FPL 29. Madison, WI: U.S. Department of Agriculture, Forest Service, Forest Products Laboratory; 1965.

Lantos, G. Load distribution in a row of fasteners subjected to lateral load. Madison, WI: Wood Science. 1(3): 129−136; 1969.

Markwardt, L. J. How surface condition of nails affects their holding power in wood. FPL Rep. D1927. Madison, WI: U.S. Department of Agriculture, Forest Service, Forest Products Laboratory; 1952.

Markwardt, L. J.; Gahagan, J. M. Effect of nail points on resistance to withdrawal. FPL Rep. 1226. Madison, WI: U.S. Department of Agriculture, Forest Service, Forest Products Laboratory; 1930.

Markwardt, L. J.; Gahagan, J. M. Slant driving of nails. Does it pay? Packing and Shipping. 56(10): 7−9,23,25; 1952.

McLain, Thomas E. Curvilinear load-slip relations in laterally-loaded nailed joints. Thesis, Department of Forestry and Wood Science, Colorado State University, Fort Collins, CO; 1975.

National Forest Products Association. National design specification for wood construction. NFPA, Washington, DC; (see current edition).

National Particleboard Association. Screw holding of particleboard. Tech. Bull. No. 3; NPA: Washington, DC; 1968.

Newlin, J. A.; Gahagan, J. M. Lag screw joints: Their behavior and design. Tech. Bull. 597. Washington, DC: U.S. Department of Agriculture; 1938.

Perkins, N. S.; Landsem, P.; Trayer, G. W. Modern connectors for timber construction. U.S. Department of Commerce, National Committee on Wood Utilization, and U.S. Department of Agriculture, Forest Service; 1933.

Scholten, J. A. Timber-connector joints, their strength and design. Tech. Bull. 865. Washington, DC: U.S. Department of Agriculture; 1944.

Scholten, J. A. Strength of bolted timber joints. FPL Rep. R1202. Madison, WI: U.S. Department of Agriculture, Forest Service, Forest Products Laboratory; 1946.

Scholten, J. A. Nail-holding properties of southern hardwoods. Southern Lumberman. 181(2273): 208−210; 1950.

Scholten, J. A.; Molander, E. G. Strength of nailed joints in frame walls. Agricultural Engineering. 31(11): 551−555; 1950.

Stern, E. G. A study of lumber and plywood joints with metal split-ring connectors. Bull. 53. State College, PA: Pennsylvania Engineering Experiment Station; 1940.

Stern, E. G. Nails in end-grain lumber. Timber News and Machine Woodworker. 58(2138): 490−492; 1950.

Trayer, G. W. Bearing strength of wood under bolts. Tech. Bull. 332. Washington, DC: U.S. Department of Agriculture; 1932.

Truss Plate Institute. Design specification for metal plate connected wood trusses. TPI−78. TPI: Madison, WI; n.d.

Wilkinson, T. L. Bearing strength of wood under embedment loading of fasteners. Res. Pap. FPL 163. Madison, WI: U.S. Department of Agriculture, Forest Service, Forest Products Laboratory; 1971.

Wilkinson, T. L. Theoretical lateral resistance of nailed joints. Journal of Structural Division, Proceedings of American Society of Civil Engineers. ST5(97): (Pap. 8121): 1381−1398; 1971.

Wilkinson, T. L. Strength of bolted wood joints with various ratios of member thicknesses. Res. Pap. FPL 314. Madison, WI: U.S. Department of Agriculture, Forest Service, Forest Products Laboratory; 1978.

Wilkinson, T. L. Assessment of modification factors for a row of bolts or timber connectors. Res. Pap. FPL 376. Madison, WI: U.S. Department of Agriculture, Forest Service, Forest Products Laboratory; 1980.

Wilkinson, T. L.; Laatsch, T. R. Lateral and withdrawal resistance of tapping screws in three densities of wood. Madison, WI: Forest Products Journal. 20(7): 34−41; 1970.

Chapter 8

Structural Analysis Formulas

Structural Analysis Formulas [*]

Formulas for deformation and stress, which are the basis for beam and column design, are discussed in this chapter. The first two sections cover curved and tapered members, as well as straight members, and discuss various special considerations such as notches and size effect. A third section presents stability criteria for members subject to buckling and for members subject to special conditions.

Deformation Formulas

Formulas for deformation of wood members are presented as functions of applied loads, moduli of elasticity, and member dimensions. They may be solved for actual cross-sectional dimensions required to meet deformation limitations imposed in design. Moduli of elasticity are given in chapter 4. Due consideration must be given to variability in material properties and uncertainties in applied loads in order to control reliability of the design.

Axial Load

The deformation of an axially loaded member is not usually an important design consideration. More important considerations will be presented in later sections dealing with combined loads or stability. Axial load produces a change of length given by

$$\delta = \frac{PL}{AE} \qquad (8-1)$$

where δ is change of length, L is length, A is cross-sectional area, E is modulus of elasticity (E_L when grain runs parallel to member axis), and P is axial force.

Bending

Straight Beam Deflection
The deflection of straight beams (or long, slightly curved beams with the radius of curvature in the plane of bending), elastically stressed, and having a constant cross section throughout their length is given by the formula:

$$\delta = \frac{k_b W L^3}{EI} + \frac{k_s W L}{GA'} \qquad (8-2)$$

where δ is deflection, W is total beam load acting perpendicular to beam neutral axis, L is beam span, k_b and k_s are constants

*Revision by John Zahn, General Engineer.

dependent upon beam loading, support conditions, and location of point whose deflection is to be calculated, I is beam moment of inertia, A' is a modified beam area, E is beam modulus of elasticity (for beams having grain direction parallel to their axis $E = E_L$), and G is beam shear modulus (for beams with flat-grained vertical faces $G = G_{LT}$ and for beams with edge-grained vertical faces $G = G_{LR}$). Elastic property values are given in chapter 4.

The first term on the right side of formula (8−2) gives the bending deflection and the second term the shear deflection. Values of k_b and k_s for several cases of loading and support are given in table 8−1.

The moment of inertia, I, of the beams is given by the formulas:

$$I = \frac{bh^3}{12} \quad \text{for beam of rectangular cross section}$$
$$(8-3)$$

$$I = \frac{\pi d^4}{64} \quad \text{for beam of circular cross section}$$

where b is beam width, h is beam depth, and d is beam diameter. The modified area A' is given by the formulas:

$$A' = \frac{5}{6} bh \text{ for beam of rectangular cross section}$$

$$(8-4)$$

$$A' = \frac{9}{40} \pi d^2 \text{ for beam of circular cross section}$$

If the beam has initial deformations such as bow (lateral bend) or twist, these deformations will be increased by the bending loads considered above. It may be necessary to provide lateral or torsional restraints to hold such members in line. See Interaction of Buckling Modes, Amplification of Prebuckling Deflections and Stresses.

Tapered Beam Deflection
The graphs of figures 8−1 and 8−2 are useful in the design of tapered beams. The ordinates are based on design criteria such as the span, the difference in beam height (h_c −h_o) as required by roof slope or architectural effect, and the maximum allowable deflection, together with material properties. From this, the value of the abscissa can be determined as shown by the example line on the graph and the smallest beam depth, h_o, can be calculated for comparison with that given by the design criteria. On the other hand, the deflection of a beam can be calculated if the value of the

abscissa is known. Tapered beams deflect due to shear deflection in addition to bending deflections, and this shear deflection Δ_s can be closely approximated by the formulas:

$$\Delta_s = \frac{3WL}{20Gbh_o} \quad \text{for uniformly distributed load}$$

$$(8-5)$$

$$\Delta_s = \frac{3PL}{10Gbh_o} \quad \text{for midspan concentrated load}$$

The final beam design should consider shear as well as bending deflection, and it may be necessary to iterate to arrive at final beam dimensions. Formulas (8−5) are applicable to either single-tapered or double-tapered beams.

Effect of Notches and Holes

The deflection of beams is increased if reductions in cross-section dimensions occur such as are caused by holes or notches. The deflection of such beams can be determined by considering them of variable cross section along their length and appropriately solving the general differential equations of

the elastic curves, $EI\dfrac{d^2y}{dx^2} = M$, to obtain

deflection expressions or by the application of Castigliano's theorem. (These procedures are given in most texts on strength of materials.) In calculating deflection of beams containing notches, the length of the notch should be considered to be equal to the actual length plus twice the depth of the notch.

Effect of Time: Creep Deflections

In addition to the elastic deflections previously discussed, wood beams usually sag in time; that is, the deflection increases beyond what it was immediately after the load was

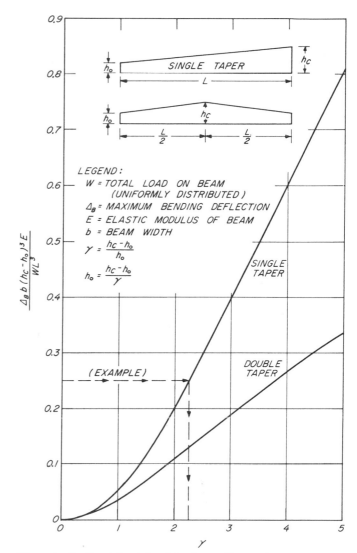

Figure 8−1—Graph for determining tapered beam size based on deflection under uniformly distributed load. (M128 982)

Table 8−1—*Values of k_b and k_s for several beam loadings*

Loading	Beam ends	Deflection at	k_b	k_s
Uniformly distributed	Both simply supported	Midspan	5/384	1/8
	Both clamped	do.	1/384	1/8
Concentrated at midspan	Both simply supported	do.	1/48	1/4
	Both clamped	do.	1/192	1/4
Concentrated at outer quarter span points	Both simply supported	do.	11/768	1/8
	do.	Load point	1/96	1/8
Uniformly distributed	Cantilever, one free, one clamped	Free end	1/8	1/2
Concentrated at free end	do.	do.	1/3	1

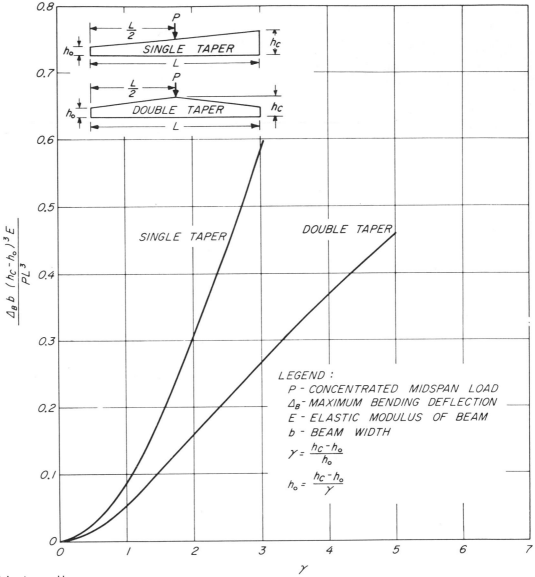

Figure 8-2—Graph for determining tapered beam size based on deflection under concentrated midspan load.

(M128 978)

first applied. See the discussion of creep in Time Under Load in chapter 4.

Green timbers, especially, will sag if allowed to dry under load, although partially dried material will also sag to some extent. In thoroughly dried beams, there are small changes in deflection with changes in moisture content but little permanent increase in deflection. If deflection under longtime load is to be limited, it has been customary to design for an initial deflection of about one-half the value permitted for longtime deflection. This can be done by doubling the longtime load value when calculating deflection, by using one-half of the usual value for modulus of elasticity or any equivalent method.

Water Ponding

Ponding of water on roofs already deflected by other loads can cause large increases in deflection. The total deflection Δ due to design load plus ponded water can be closely estimated by

$$\Delta = \frac{\Delta_o}{1 - \frac{S}{S_{cr}}} \qquad (8-6)$$

where Δ_o is the deflection due to design load alone, S is the beam spacing, and S_{cr} is the critical beam spacing given by formula (8-29) under Stability Formulas—Bending.

Combined Bending and Axial Load

Concentric Load

Addition of a concentric axial load to a beam under loads acting perpendicular to the beam neutral axis causes increase in deflection for added axial compression and decrease in deflection for added axial tension. The deflection under combined loading can be estimated closely by the formula:

$$\Delta = \frac{\Delta_o}{1 \pm \dfrac{P}{P_{cr}}} \qquad (8-7)$$

where the plus sign is chosen if the axial load is tension and the minus sign is chosen if the axial load is compression; Δ is deflection under combined loading; Δ_o is beam deflection without axial load; P is axial load; and P_{cr} is a constant equal to the buckling load of the beam under axial compressive load only (see Stability Formulas—Axial Compression) based on flexural rigidity about the neutral axis perpendicular to the direction of bending loads. This constant appears regardless of whether P is tension or compression. If P is compression it must be less than P_{cr} in order to avoid collapse. When the axial load is tension, it is conservative to ignore the P/P_{cr} term. If the beam is not supported against lateral deflection its buckling load should be checked using formula (8–33) under Stability Formulas—Interaction of Buckling Modes.

Eccentric Load

If an axial load is eccentrically applied, it will induce bending deflections as well as the change in length given by formula (8–1). Formula (8–7) can be applied to find the bending deflection by writing it in the form

$$\delta_b + \epsilon_o = \frac{\epsilon_o}{1 \pm \dfrac{P}{P_{cr}}} \qquad (8-8)$$

where δ_b is the induced bending deflection at midspan, ϵ_o is the eccentricity of P from the centroid of the cross section, and other terms are as defined above for use with formula (8–7).

Torsion

The angle of twist of wood members can be computed by the formula:

$$\theta = \frac{TL}{GK} \qquad (8-9)$$

where θ is angle of twist in radians, T is applied torque, L is member length, G is shear modulus (use $\sqrt{G_{RL}G_{TL}}$ or approximate G by $E_L/16$ if measured G not available), and K is a cross-section shape factor. For circular cross section K is the polar moment of inertia:

$$K = \frac{\pi D^4}{32} \text{ (circular section)} \qquad (8-10)$$

where D is diameter. For rectangular cross section

$$K = \frac{hb^3}{\gamma} \text{ (rectangular section)} \qquad (8-11)$$

where h is larger cross-section dimension, b is smaller cross-section dimension, and γ is presented in figure 8–3.

Stress Formulas

The formulas presented here are limited by the assumption that stress and strain are directly proportional (Hooke's law) and by the fact that local stresses in the vicinity of points of support or points of load application are correct only to the extent of being statically equivalent to the true stress distribution (St. Venant's principle). Local stress concentrations must be separately accounted for if they are to be limited in design.

Axial Load

Tensile Stress

Concentric axial load (along the line joining the centroids of the cross sections) produces a uniform stress:

$$f_t = \frac{P}{A} \qquad (8-12)$$

where f_t is tensile stress, P is axial load, and A is cross-sectional area.

Short-Block Compressive Stress

Formula (8–12) can be used in compression as well if the member is short enough to fail without deflecting laterally, that is, by simple crushing. Such fiber crushing produces a local "wrinkle" caused by microstructural instability. The member as a whole remains structurally stable and able to bear load.

Bending

The strength of beams is determined by flexural stresses caused by bending moment, shear stresses caused by shear

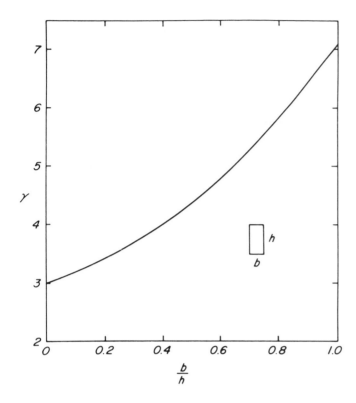

Figure 8–3—Coefficient, γ, for determining torsional rigidity of rectangular member.　(ML84 5808)

load, and compression across the grain at the end bearings and load points.

Straight Beam Stresses

The stress due to bending moment is a maximum at the top and bottom edges. The concave edge is compressed, and the convex edge is under tension. The maximum stress is given by

$$f_b = \frac{M}{Z} \qquad (8-13)$$

where f_b is bending stress, M is bending moment, and Z is beam section modulus (for a rectangular cross section $Z = bh^2/6$ and for a circular cross section $Z = \pi D^3/32$).

This formula is also used beyond the limits of Hooke's law with M as the ultimate moment at failure. The resulting pseudo-stress is called the "modulus of rupture," values of which are tabulated in chapter 4. The modulus of rupture has been found to decrease with increasing size of member. See Size Effect under Stress Formulas—Bending.

The shear stress due to bending is a maximum at the centroidal axis of the beam, where the bending stress happens to

be zero. (This statement is no longer true if the beam is tapered. See next section.) In wood beams this shear stress may produce a failure crack near mid-depth running along the axis of the member. Unless the beam is sufficiently short and deep it will fail in bending before shear failure can develop, but wood beams are relatively weak in shear, and shear strength can sometimes govern a design. The maximum shear stress is

$$f_s = k \frac{V}{A} \qquad (8-14)$$

where f_s is shear stress, V is vertical shear force on cross section, A is cross-sectional area, and k is 3/2 for a rectangular cross section or 4/3 for circular cross section.

Tapered Beam Stresses

For beams of constant width that taper in depth at a slope less than 25° the bending stress can be obtained from formula (8–13) with an error of less than 5 percent. The shear stress, however, differs markedly from that found in uniform beams. It can be determined from the basic theory presented by Maki and Kuenzi. The shear stress at the tapered edge can reach a maximum value as great as that at the neutral axis at a reaction.

Consider the example shown in figure 8–4, where concentrated loads farther to the right have produced a support reaction V at the left end. In this case the maximum stresses occur at the cross section which is double the depth of the beam at the reaction. For other loadings, the location of the cross section with maximum shear stress at the tapered edge will be somewhat different.

For the beam shown in figure 8–4 the bending stress is also a maximum at the same cross section where the shear stress is maximum at the tapered edge. This stress situation also causes a stress in the direction perpendicular to the neutral axis that is maximum at the tapered edge. The effect of combined stresses at a point can be approximately accounted for by an interaction formula based on the Henky-von Mises theory of energy due to the change of shape. This theory applied to wood by Norris results in the formula:

$$\frac{f_x^2}{F_x^2} + \frac{f_{xy}^2}{F_{xy}^2} + \frac{f_y^2}{F_y^2} = 1 \qquad (8-15)$$

where f_x is the bending stress, f_y is the stress perpendicular to the neutral axis, and f_{xy} is the shear stress. Values of F_x, F_y, and F_{xy} are corresponding stresses chosen at design allowable values or maximum values in accordance with allowable or maximum values being determined for the tapered beam. Maximum stresses in the beam shown in figure 8–4 are given by the formulas:

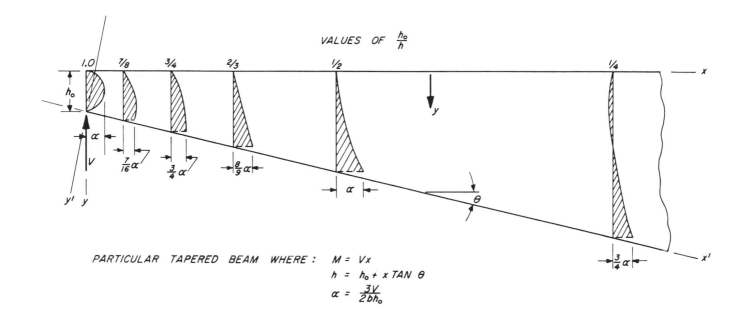

$$M = Vx$$

PARTICULAR TAPERED BEAM WHERE : $h = h_o + x \, TAN \, \theta$

$$\alpha = \frac{3V}{2bh_o}$$

Figure 8–4—Shear stress distribution for a tapered beam.

(M128 964)

$$f_x = \frac{3M}{2bh_o^2}$$

$$f_{xy} = f_x \tan \theta \qquad (8-16)$$

$$f_y = f_x \tan^2 \theta$$

Substitution of these formulas into the interaction formula (8–15) will result in an expression for the moment capacity M of the beam. If the taper is on the beam tension edge, the values of f_x and f_y are tensile stresses.

Example: Determine the moment capacity of a tapered beam of width $b = 5$ inches depth $h_o = 10$ inches, and taper $tan \, \theta = 1/10$. Substitution of these dimensions into formulas (8–16) results in

$$f_x = \frac{3M}{1,000}$$

$$f_{xy} = \frac{3M}{10,000}$$

$$f_y = \frac{3M}{100,000}$$

and substitution of these expressions into formula (8–15) and solving for M results finally in:

$$M = \frac{10^5}{3 \left[\dfrac{10^4}{F_x^2} + \dfrac{10^2}{F_{xy}^2} + \dfrac{1}{F_y^2} \right]^{1/2}}$$

where appropriate allowable or maximum values of the F stresses are chosen.

Size Effect

It has been found that the modulus of rupture (maximum bending stress) of wood beams depends on beam size and method of loading and that the strength of clear, straight-grained beams decreases as size increases. These effects were found to be describable by statistical strength theory involving "weakest link" hypotheses and can be summarized as: For two beams under two equal concentrated loads applied symmetrical to the midspan points, the ratio of the modulus of rupture of beam 1 to the modulus of rupture of beam 2 is given by the formula:

$$\frac{R_1}{R_2} = \left[\frac{h_2 L_2 \left(1 + \dfrac{ma_2}{L_2} \right)}{h_1 L_1 \left(1 + \dfrac{ma_1}{L_1} \right)} \right]^{1/m} \qquad (8-17)$$

where subscripts 1 and 2 refer to beam 1 and beam 2; R is modulus of rupture; h is beam depth; L is beam span; a is distance between loads placed $a/2$ each side of midspan; and m is a constant. For clear, straight-grained Douglas-fir beams $m = 18$. If formula (8–17) is used for beam 2 of standard size (see ch. 4) loaded at midspan then $h_2 = 2$ inches, $L_2 = 28$ inches, and $a_2 = 0$ and formula (8–17) becomes

$$\frac{R_1}{R_2} = \left[\frac{56}{h_1 L_1 \left(1 + \dfrac{ma_1}{L_1}\right)}\right]^{1/m} \qquad (8-18)$$

Example: Determine modulus of rupture for a beam 10 inches deep, spanning 18 feet, and loaded at one-third span points compared with a beam 2 inches deep, spanning 28 inches, and loaded at midspan that had a modulus of rupture of 10,000 psi. Assume $m = 18$. Substitution of the dimensions into formula (8–18) produces

$$R_1 = 10,000 \left[\frac{56}{2,160 \, (1 + 6)}\right]^{1/18}$$

$$= 7,340 \text{ psi}$$

Application of the statistical strength theory to beams under uniformly distributed load resulted in the following relationship between modulus of rupture of beams under uniformly distributed load and modulus of rupture of beams under concentrated loads:

$$\frac{R_u}{R_c} = \left[\frac{\left(1 + 18 \, \dfrac{a_c}{L_c}\right) h_c L_c}{3.876 h_u L_u}\right]^{1/18} \qquad (8-19)$$

where subscripts u and c refer to beams under uniformly distributed and concentrated loads, respectively, and other symbols are as previously defined.

Effect of Notches and Holes

In beams having notches or holes with sharp interior corners, large stress concentrations exist at the corners. The local stresses include shear parallel to grain and tension perpendicular to grain. As a result, even moderately low loads can cause a crack to initiate at the sharp corner and propagate along the grain. An estimate of the crack-initiation load can be obtained by the fracture mechanics analysis of Murphy, but it is generally more economical to avoid sharp notches entirely in wood beams, especially large wood beams, since there is a size effect: sharp notches cause greater reductions in strength for larger beams.

A conservative criterion for crack initiation is

$$\sqrt{h} \left[A \left(\frac{6M}{bh^2}\right) + B \left(\frac{3V}{2bh}\right)\right] = 1 \qquad (8-20)$$

where h is beam depth, b is beam width, M is bending moment, V is vertical shear force, and A and B are presented in figure 8–5 as functions of a/h where a is notch depth. Note that the value of A depends on whether the notch is on the tension edge or the compression edge. Therefore, use either A_t or A_c as appropriate. Strictly speaking, the values of A and B are dependent upon species; the values given in figure 8–5 are conservative for most species.

Effects of Time: Creep Rupture, Fatigue, and Aging

See chapter 4 for a discussion of these effects. Creep rupture is accounted for by a duration-of-load adjustment in the setting of allowable stresses, as discussed in chapter 6.

Water Ponding

Ponding of water on roofs can cause increases in bending stresses that can be computed by the same amplification factor (formula 8–6) used with deflection. See Water Ponding under Deformation Formulas—Bending.

Combined Bending and Axial Load

Concentric Load

Formula (8–7) gives the effect, on deflection, of adding an end load to a beam already bent by transverse loads. The bending stress in the member is modified by the same factor as the deflection:

$$f_b = \frac{f_{bo}}{1 \pm \dfrac{P}{P_{cr}}} \qquad (8-21)$$

where the plus sign is chosen if the axial load is tension and the minus sign is chosen if the axial load is compression, f_b is net bending stress from combined bending and axial load, f_{bo} is bending stress without axial load, P is axial load, and P_{cr} is the buckling load of the beam under axial compressive load only (see Stability Formulas—Axial Compression), based on flexural rigidity about the neutral axis perpendicular to the direction of the bending loads. Note that this P_{cr} is not necessarily the minimum buckling load of the member. If P is compressive, the possibility of buckling under combined loading must be checked. See Interaction of Buckling Modes, General Buckling Criterion.

The total stress under combined bending and axial load is obtained by superposition of the stresses given by formulas (8–12) and (8–21).

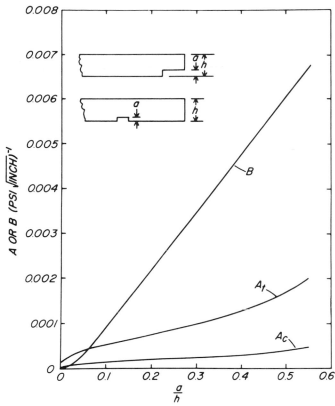

Figure 8−5—Coefficients, *A* and *B*, for crack initiation criterion (8−20).

(ML84 5802)

Example: Suppose transverse loads produce a bending stress, f_{bo}, tensile on the convex edge and compressive on the concave edge of the beam. Then the addition of a tensile axial force P at the centroids of the end sections will produce a maximum tensile stress on the convex edge of

$$f_{t\ max} = \frac{f_{bo}}{1 + \dfrac{P}{P_{cr}}} + \frac{P}{A}$$

and a maximum compressive stress on the concave edge of

$$f_{c\ max} = \frac{f_{bo}}{1 + \dfrac{P}{P_{cr}}} - \frac{P}{A}$$

where a negative result would indicate that the stress was in fact tensile.

Eccentric Load

If the axial load is eccentrically applied, then the bending stress f_{bo} should be augmented by $\pm P\epsilon_o/Z$ where ϵ_o is eccentricity of the axial load.

Example: In the preceding example let the axial load be eccentric toward the concave edge of the beam. Then the maximum stresses become

$$f_{t\ max} = \frac{f_{bo} - \dfrac{P\epsilon_o}{Z}}{1 + \dfrac{P}{P_{cr}}} + \frac{P}{A}$$

$$f_{c\ max} = \frac{f_{bo} - \dfrac{P\epsilon_o}{Z}}{1 + \dfrac{P}{P_{cr}}} - \frac{P}{A}$$

Torsion

The shear stress induced by torsion is, for a circular cross section:

$$f_s = \frac{16\ T}{\pi D^3} \tag{8−22}$$

where T is applied torque and D is diameter. For a rectangular cross section

$$f_s = \frac{T}{\beta h b^2} \tag{8−23}$$

where T is applied torque, h is larger cross-section dimension, b is smaller cross-section dimension, and β is presented in figure 8−6.

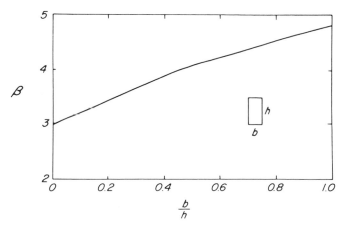

Figure 8−6—Coefficient, β, for computing maximum shear stress in torsion of rectangular member.

(ML84 5815)

Stability Formulas

Axial Compression

For slender members under axial compression, stability is the principal design criterion. The following formulas are for concentrically loaded members. For eccentrically loaded columns see General Buckling Criterion.

Long Columns

A column long enough to buckle before the compressive stress P/A exceeds the proportional limit stress is called a "long column." The critical stress at buckling is

$$f_{cr} = \frac{\pi^2 E_L}{\left(\frac{L}{r}\right)^2} \tag{8-24}$$

where E_L is the elastic modulus parallel to the axis of the member, L is the length, and r is the least radius of gyration (for a rectangular cross section with b as its least dimension $r = \dfrac{b}{\sqrt{12}}$ and for a circular cross section $r = \dfrac{d}{4}$). Formula (8-24) is based on a pinned end condition but may be used also for square ends.

Short Columns

Columns that buckle at a compressive stress P/A beyond the proportional limit stress are called short columns. Usually the short column range is explored empirically and appropriate design formulas are proposed. Material of this nature is presented in USDA Technical Bulletin 167. The final formula is a fourth-power parabolic function which can be written as:

$$f_{cr} = F_c \left\{ 1 - \frac{4}{27\pi^4} \left[\frac{L}{r} \sqrt{\frac{F_c}{E_L}} \right]^4 \right\} \tag{8-25}$$

where F_c is compressive strength, and remaining terms are defined as in formula (8-24). A graphical presentation of formulas (8-24) and (8-25) is given in figure 8-7.

Short columns can be analyzed by fitting a nonlinear function to compressive stress-strain data and using it in place of Hooke's law. One such nonlinear function proposed by Ylinen is

$$\epsilon = \frac{F_c}{E_L} \left[c\frac{f}{F_c} - (1-c) \, \text{Log}_e \left(1 - \frac{f}{F_c} \right) \right] \tag{8-26}$$

where ϵ is compressive strain, f is compressive stress, c is a constant between zero and one, and E_L and F_c are defined above. Using the slope of formula (8-26) in place of E_L in Euler's formula (8-24) leads to Ylinen's buckling formula

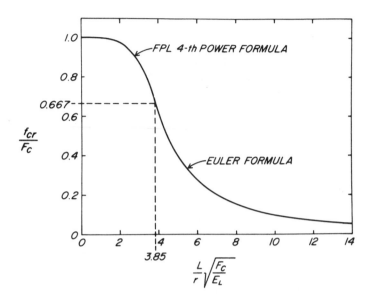

Figure 8-7—Graph for determining critical buckling stress of wood columns. (ML84 5798)

$$f_{cr} = \frac{F_c + f_e}{2c} - \sqrt{\left(\frac{F_c + f_e}{2c} \right)^2 - \frac{F_c f_e}{c}} \tag{8-27}$$

where F_c is compressive strength and f_e is the buckling stress given by Euler's formula (8-24). Formula (8-27) can be made to agree closely with figure 8-7 by choosing c = 0.957.

Built-Up Columns

Built-up columns of nearly square cross section will not support loads as high if the lumber is nailed together as if it were glued together. The reason is that shear distortions can occur in the nailed joints. The shearing resistance of the column can be improved, so that previously presented formulas may be used, by nailing cover plates of lumber to the edges of the built-up layers. If the built-up column is of several spaced pieces, the spacer blocks should be placed close enough together, lengthwise in the column, so that the unsupported portion of the spaced member will not buckle at the same or lower stress than that of the complete member. "Spaced columns" are designed with previously presented column formulas, considering each compression member as an unsupported simple column; the sum of column loads for all the members is taken as the column load for the spaced column. Sufficient net area should be provided in short columns so that compression failure does not occur. The net area is, of course, that area remaining after subtracting portions for connectors or bolts used to fasten the members together at the spacer blocks.

Columns With Flanges

Columns with thin, outstanding flanges can fail by elastic instability of the outstanding flange, causing wrinkling of the flange and twisting of the column at stresses less than those for general column instability as given by formulas (8−24) and (8−25). For outstanding flanges of sections such as I, H, +, and L the flange instability stress can be estimated by the formula:

$$f_{cr} = 0.044E \frac{t^2}{b^2} \qquad (8-28)$$

where E is the column modulus of elasticity; t is the thickness of the outstanding flange; and b is the width of the outstanding flange. If the joints between the column members are glued and reinforced with glued fillets, the instability stress increases to as much as 1.6 times that given by formula (8−28).

Bending

Beams are subject to two kinds of instability: lateral-torsional buckling and progressive deflection under water ponding, both of which are determined by member stiffness.

Water Ponding

Roof beams that are insufficiently stiff or spaced too far apart for their given stiffness can fail by progressive deflection under the weight of water from steady rain or other continuous source. The critical beam spacing S_{cr} is given by

$$S_{cr} = \frac{m\pi^4 EI}{\rho L^4} \qquad (8-29)$$

where E is beam modulus of elasticity, I is beam moment of inertia, ρ is density of water (0.0361 lb/in.^{3}), L is beam length, and $m = 1$ for simple support or $m = 16/3$ for fixed end condition. To prevent ponding, the beam spacing must be less than S_{cr}.

Lateral-Torsional Buckling

Since beams are compressed on the concave edge when bent under load, they can buckle by a combination of lateral deflection and twist. Because most wood beams are rectangular in cross section, the formulas presented below are for rectangular members only. Beams of I, H, or other built-up cross section exhibit a more complex resistance to twisting and are more stable than the following formulas would predict.

Long Beams—Long slender beams which are restrained against axial rotation at their points of support but are otherwise free to twist and to deflect laterally will buckle when the maximum bending stress f_b equals or exceeds the following critical value:

$$f_{b\ cr} = \frac{\pi^2 E_L}{\rho^2} \qquad (8-30)$$

where ρ is the slenderness factor given by

$$\rho = \sqrt{2\pi} \ \sqrt[4]{\frac{EI_2}{GK}} \sqrt{\frac{L_e h}{b}} \qquad (8-31)$$

where EI_2 is the lateral flexural rigidity equal to $\dfrac{E_L h b^3}{12}$, h is beam depth, b is beam width, GK is torsional rigidity defined in formula (8−9) under Deformation Formulas—Torsion, and L_e is the effective length determined by type of loading and support as given in table 8−2. Formula (8−30) is valid for bending stresses below the proportional limit.

Short Beams—Short beams can buckle at stresses beyond the proportional limit. In view of the similarity of formula (8−30) to Euler's formula (8−24) for column buckling, it is recommended that short-beam buckling be analyzed by using the column buckling criterion in figure 8−7 applied with ρ in place of $\dfrac{L}{r}$ on the abscissa and $\dfrac{f_{b\ cr}}{F_b}$ in place of $\dfrac{f_{cr}}{F_c}$ on the ordinate. Here F_b is beam bending strength (modulus of rupture).

Table 8−2—Effective length for checking lateral-torsional stability of beams[1]

Support	Load	Effective length, L_e
Simple support	Equal end moments	L
Do.	Concentrated force at center	$\dfrac{0.742\,L}{1 - 2\dfrac{h}{L}}$
Do.	Uniformly distributed force	$\dfrac{0.887\,L}{1 - 2\dfrac{h}{L}}$
Cantilever	Concentrated force at end	$\dfrac{0.783\,L}{1 - 1.15\dfrac{h}{L}}$
Do.	Uniformly distributed force	$\dfrac{0.489\,L}{1 - 2\dfrac{h}{L}}$

[1] These values are conservative for beams whose width-to-depth ratio is less than 0.4. The load is assumed to act at the top edge of the beam.

Effect of Deck Support—The most common form of support against lateral deflection is a deck continuously attached to the top edge of the beam. If this deck is rigid against shear in the plane of the deck and is attached to the compression edge of the beam, the beam cannot buckle. In regions where the deck is attached to the tension edge of the beam, as where a beam is continuous over a support, the deck cannot be counted on to prevent buckling and restraint against axial rotation should be provided at the support point.

If the deck is not very rigid against in-plane shear, as for example nominal 2-inch wood decking, the formula (8−30) and figure 8−7 can still be used to check stability except that now the effective length is modified by dividing by θ as given in figure 8−8. The abscissa of this figure is a deck shear stiffness parameter τ given by

$$\tau = \frac{SG_D L^2}{EI_2} \qquad (8-32)$$

where EI_2 is lateral flexural rigidity as in formula (8−31) above, S is beam spacing, G_D is in-plane shear rigidity of deck (ratio of shear force per unit length of edge to shear strain), and L is actual beam length. This figure applies only to simply supported beams. Cantilevers with deck on top have their tension edge supported and do not derive much support from the deck.

Interaction of Buckling Modes

When two or more loads are acting and each of them has a critical value associated with a mode of buckling, the combination can produce buckling even though each load is less than its own critical value.

General Buckling Criterion

The critical combination of loads that will produce buckling is given by the following formula:

$$\frac{P}{P_{cr}} + \left(\sum_i \frac{f_{bi}}{f_{bicr}}\right)^2 = 1 \qquad (8-33)$$

where P is concentric end load; P_{cr} is the buckling load when P acts alone, based on the minimum flexural rigidity of the member; f_{bi} is a bending stress due to the ith bending load; and f_{bicr} is its critical value as determined by the formulas under Lateral-Torsional Buckling.

If the end load P is eccentric, one of the f_{bi} should be for the case of constant bending moment equal to $P\epsilon$ where ϵ is the eccentricity of P.

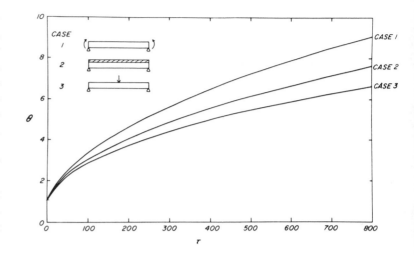

Figure 8−8—Increase in buckling stress due to attached deck; simply supported beams. To apply this figure, divide the effective length by θ. (ML84 5823)

Amplification of Prebuckling Deflections and Stresses

If the member has lateral loads that produce lateral deflection δ and angle of twist φ (both measured at midspan), then these deflections and their associated stresses are amplified by the addition of axial load and primary bending loads (loads which produce bending about the axis with greatest moment of inertia). The amplification of lateral deflection δ and lateral bending stresses is roughly approximated by the factor

$$\frac{1}{1 - \dfrac{P}{P_{cr}} - \left(\sum_i \dfrac{f_{bi}}{f_{bicr}}\right)^2} \qquad (8-34)$$

and the amplification of twist φ, and torsional shear stress is roughly approximated by the factor

$$\frac{1 - \dfrac{P}{P_{cr}}}{1 - \dfrac{P}{P_{cr}} - \left(\sum_i \dfrac{f_{bi}}{f_{bicr}}\right)^2} \qquad (8-35)$$

where all terms are defined as in formula (8−33).

If the deflections δ and φ are initial imperfections with no associated stresses, they still are amplified by the factors given above, but lateral bending stress and torsional shear stress are only proportional to the *increase* in their respective associated deflections.

Selected References

Bohannan, Billy. Effect of size on bending strength of wood members. Res. Pap. FPL 56. Madison, WI: U.S. Department of Agriculture, Forest Service, Forest Products Laboratory; 1966.

Kuenzi, E. W.; Bohannan, Billy. Increases in deflection and stress caused by ponding of water on roofs. Madison, WI: Forest Products Journal. 14(9): 421−424; 1964.

Liu, J. Y. A Weibull analysis of wood member bending strength. Journal of Mechanical Design, Transactions ASME. 104: 572−579; 1982.

Maki, A. C.; Kuenzi, E. W. Deflection and stresses of tapered wood beams. Res. Pap. FPL 34. Madison, WI: U.S. Department of Agriculture, Forest Service, Forest Products Laboratory; 1965.

Murphy, J. F. Using fracture mechanics to predict failure of notched wood beams. Proceedings of First International Conference on Wood Fracture; Aug. 14−16, 1978; Banff, AB. Vancouver, BC: Forintek Canada Corp.; 1979: 159, 161−173.

Banff, AB; Aug. 14-16, 1978. p. 159−173.

Newlin, J. A.; Gahagan, J. M. Tests of large timber columns and presentation of the Forest Products Laboratory column formula. Tech. Bull. 167. U.S. Department of Agriculture; 1930.

Newlin, J. A.; Trayer, G. W. Deflection of beams with special reference to shear deformations. Rep. 180. Washington, DC: U.S. National Advisory Committee on Aeronautics; 1924.

Norris, C. B. Strength of orthotropic materials subjected to combined stresses. Rep. 1816. Madison, WI: U.S. Department of Agriculture, Forest Service, Forest Products Laboratory; 1950.

Trayer, G. W. The torsion of members having sections common in aircraft construction. Rep. 334. Washington, DC: U.S. National Advisory Committee on Aeronautics; 1930.

Trayer, G. W.; March, H. W. Elastic instability of members having sections common in aircraft construction. Rep. 382. Washington, DC: U.S. National Advisory Committee on Aeronautics; 1931.

Ylinen, A. A method of determining the buckling stress and the required cross-sectional area for centrally loaded straight columns in elastic and inelastic range. Publications of the International Association for Bridge and Structural Engineering, Zurich. Vol. 16; 1956.

Zahn, J. J. Lateral stability of wood beam-and-deck systems. Journal of the Structural Division, ASCE. 99(ST7): 1391−1408; 1973.

Chapter 9

Bonding Wood and Wood Products

Bonding Wood and Wood Products *

Adhesive bonding is a key factor for efficient utilization of wood and is essential to the modern forest products industry. The ability of an adhesive to transfer stress from one member or particle to another through a thin layer or droplet has revolutionized wood construction.

Adhesives transfer load from one member (adherend) to another by surface attachment (adhesive bonding). The strength of adhesive bonded joints depends on the strength of each link (fig. 9−1) in the joint; the adherends, the interphases, and the adhesive. The performance of adhesive bonded joints then depends upon how well we understand and control the factors determining the strength of each link. The factors discussed in this chapter are:

- The type of wood or wood product.
- The surface quality of the wood or wood product.
- The adhesive.
- The bonding process.
- The type of joint.
- The service environment.

Bondability of Wood and Wood-Based Products

Adhesives are used to bond wood in solid form (lumber), and as veneer, flakes or particles, and fibers. They are also used in secondary manufacture to fasten bonded panel products to lumber and to other materials. Wood density is a crude but useful indicator of the ease of bonding. In general the strength of wood increases with its density. The strength of rigid-adhesive joints also increases with wood density up to about 0.70 to 0.80 gram per cubic centimeter. Above 0.70 to 0.80, joint strength decreases again with our present adhesive systems but for different reasons. The wood failure on the surfaces of tested joints decreases gradually up to density of 0.70 to 0.80, then decreases more rapidly with further density increase. High-quality joints are more difficult to achieve consistently as wood density increases because: (1) Extractives that interfere with the development of adhesion and cohesion are more likely to be present, (2) mechanical interlocking of the wood and adhesive is reduced, (3) adequate surface mating is harder to attain, even with extra pressure, and (4) shrinkage and springback stresses in the joint are higher.

Table 9−1 roughly classifies the bonding properties of domestic and imported woods often used for bonded products. The classifications for domestic woods are based on the average quality of side-to-side-grain joints of lumber (of average density for the species) determined in laboratory tests or based on industrial experience. The laboratory tests included animal, casein, starch, urea formaldehyde, and resorcinol formalde-

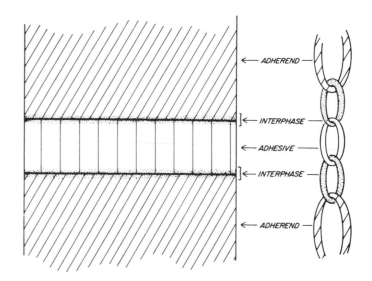

Figure 9−1—Five links in an adhesive bonded joint; the two members being bonded (adherends), the two adhesive bonding regions (interphases), and the adhesive itself.　(ML84 5347)

hyde adhesives. The classifications for imported woods are based on descriptions found in the literature related to bondability, species properties, and extractives content and on industrial experience. In most cases, information is not available to make the same objective classification as with the domestic woods. It should be pointed out, however, that woods which are difficult to bond satisfactorily with one adhesive type may bond more easily with another, depending on their chemical and rheological harmony. Adhesives manufacturers are often able to formulate special adhesives to solve specific wood bonding problems.

The bondability of wood products like plywood, particleboard, fiberboard, flakeboard, and hardboard is hard to classify categorically. These materials often have polished surfaces from hot-pressing which are difficult to bond. Extractives may concentrate on the surface during veneer drying and hot-pressing, or caul release agents may be left on the surface. Removal of the pressed surface by light sanding almost invariably improves the bondability. If a good surface is already present, the strength of joints with panel products is then usually limited by the tensile strength of the surface layers of the panels themselves. Lathe checks, particle or flake surface damage, and resin content strongly affect this strength and thus joint quality.

Plastic films, metal foils, and cellulosic overlays are generally laminated to wood as they are received from the manufacturer. High cohesive strength is not required of the adhesive, but it must be carefully chosen for good adhesion to both materials.

*Revision by Bryan H. River, Forest Products Technologist.

Structural joints between wood and rigid plastic or metal require special preparation of the plastic or metal to remove contaminants and present a chemically active surface for bonding. Some treatments are solvent wipe, abrasion, alkaline clean, plasma etch, and chemical etch.

Surface Quality

Because adhesives work by surface attachment, the adherend's surface qualities are extremely important to satisfactory joint performance. One aspect of surface quality is its physical condition. Wood surfaces to be bonded should be smooth, true, and free of machine marks or other surface irregularities such as torn or chipped grain or planer skip. There should be no crushing or glazing and the surface should be free of extractives, dirt, and other debris. The other aspect of surface quality is its chemical condition. The surface may contain airborne contaminants, or extractives or preservatives, or have undergone atmospheric reaction which interferes with bonding. Unfortunately some of these surface conditions are often difficult to detect. Contaminants not only interfere with wetting and adhesion but can also interfere with development of cohesive strength (cure) within the adhesive film.

Wood should be surfaced or resurfaced just before bonding to remove all materials that interfere with bonding and also to reduce the chance for a change in moisture content that could distort the surfaces from a flat condition. Parallel flat surfaces are required to obtain uniformly thin adhesive layers necessary for the best performance of most liquid-dispersed wood adhesives.

Two simple water tests can reveal a great deal about the surface quality and how difficult it may be to wet and bond. The first test is to place a small drop of water on the surface and observe how it spreads on or is absorbed by the surface. If the drop of water beads up and is still beaded after 30 seconds, this surface will likely be resistant to adhesive wetting (fig. 9-2). An alternate water drop test is used on the tight side of dry flat grain earlywood veneer for hot press bonded plywood. In this case good wettability is indicated if the drop is totally absorbed or dispersed in less than 20 minutes, while it is almost certain that bonding problems can be expected if more than 40 minutes is required for complete adsorption or dispersion. The second test is to wipe a very wet rag over a portion of the surface. After waiting for a minute or so, if all the water has not been absorbed, remove the remaining water by blotting with a paper towel. Then compare the roughness of the wet and dry surfaces. If the wetted area is much rougher than the dry area, the surface may have suffered damage in machining which will significantly damage its strength or interfere with adhesion.

Experience has shown that a smooth knife-cut wood surface is the best for bonding. Surfaces made by saws are usually rougher than those made by planers, jointers, and other machines equipped with cutterheads. However, surfaces cut with special blades on properly set straight-line ripsaws are satisfactory for both structural and nonstructural joints. Such surfaces are often bonded in furniture manufacture to save labor and materials. Unless the saws and feed works are very well maintained, however, joints between sawed surfaces are generally weaker and less uniform in quality than those between well-planed or jointed surfaces. Dull cutting edges of any type crush and glaze the wood surface. If the adhesive does not completely penetrate crushed cells and restore their original strength, a weak joint results. A glazed surface simply resists wetting and adhesive penetration. Abrasive planing with grit sizes from 24 to 80 causes surface and subsurface crushing damage. Figure 9-3 shows cross sections of bondlines between undamaged Douglas-fir lumber surfaces and surfaces damaged by abrasive planing. Such damaged surfaces are inherently weak and result in poor quality joints. Similar damage can be caused by dull planer knives or saws. There is some evidence that finish sanding with grits finer than 100 mesh may improve the abrasive-planed surface. However, abrasive planing is not recommended for structural joints or joints subjected to high swelling and shrinkage stress. When used for surfacing before bonding, abrasive belts must be kept clean and sharp and sander dust completely removed from the sanded surface.

Extractives have three main effects on adhesive joint performance. First, in some species and under certain drying conditions, extractives migrate to the surface. There they may concentrate and block the adhesive from contact with the wood. Surfacing before bonding usually alleviates this problem. Second, certain resinous or oily extractives are naturally resistant to adhesion, or may become so during drying or storage. Third, pH or chemical reactivity of an extractive may inhibit the normal hardening process of the adhesive so that its full cohesive strength does not develop. The reverse is also encountered occasionally in the case where the pH or chemical reactivity of an extractive substance may catalyze the cure rate of the adhesive so that it hardens prematurely and loses its ability to wet and flow.

Veneer is cut by rotary, slicing, or sawing processes. Well-cut veneer from any of the three processes will yield products with no appreciable difference in any property except appearance. Because veneer usually is not resurfaced before it is bonded, it must be cut with precision and carefully dried.

Sawed veneer is produced in long narrow strips, usually from flitches selected for figure and grain. The two sides of

Table 9 – 1—*Classification of selected wood species according to ease of bonding*

U. S. Hardwoods	U. S. Softwoods	Imported Woods	

BOND EASILY[1]

U. S. Hardwoods	U. S. Softwoods	Imported Woods	
Alder	Fir:	Balsa	Hura
Aspen	White	Cativo	Purpleheart
Basswood	Grand	Courbaril	Roble
Cottonwood	Noble	Determa[2]	
Chestnut, American	Pacific		
Magnolia	Pine:		
Willow, black	Eastern white		
	Western white		
	Redcedar, western		
	Redwood		
	Spruce, Sitka		

BOND WELL[3]

U. S. Hardwoods	U. S. Softwoods	Imported Woods	
Butternut	Douglas-fir	Afrormosia	Meranti (lauan):
Elm:	Larch, western[4]	Andiroba	Light red
American	Pine:	Angelique	White
Rock	Sugar	Avodire	Yellow
Hackberry	Ponderosa	Banak	Obeche
Maple, soft	Redcedar, eastern	Iroko	Okoume
Sweetgum		Jarrah	Opepe
Sycamore		Limba	Peroba rosa
Tupelo		Mahogany:	Sapele
Walnut, black		African	Spanish-cedar
Yellow-poplar		American	Sucupira
			Wallaba

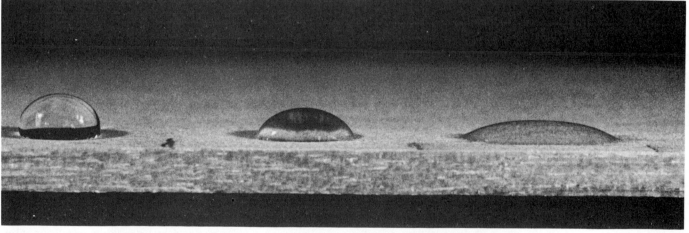

Figure 9 – 2—Differences in the wettability of wood surfaces shown by a simple water drop test. The shapes of three drops applied at the same time and photographed after 30 seconds illustrate differences in the wettability of yellow birch veneer with an old surface (left), a surface renewed by two light passes with 320 grit sandpaper (center), and four light passes with 320 grit sandpaper (right).

(M84 0270-15)

U. S. Hardwoods	U. S. Softwoods	Imported Woods	

BOND SATISFACTORILY[5]

U. S. Hardwoods	U. S. Softwoods	Imported Woods	
Ash, white	Alaska-cedar	Angelin	Meranti (lauan), dark red
Beech, American	Port-Orford-cedar	Azobe	Pau marfim
Birch:	Pine, southern	Benge	Parana-pine
Sweet		Bubinga	Pine:
Yellow		Karri	Caribbean
Cherry			Radiata
Hickory:			Ramin
Pecan			
True			
Madrone			
Maple, hard			
Oak:			
Red[2]			
White[2]			

BOND WITH DIFFICULTY[6]

U. S. Hardwoods	U. S. Softwoods	Imported Woods	
Osage-orange		Balata	Keruing
Persimmon		Balau	Lapacho
		Greenheart	Lignumvitae
		Kaneelhart	Rosewood
		Kapur	Teak

[1] Bond very easily with adhesives of a wide range of properties and under a wide range of bonding conditions.

[2] Difficult to bond with phenol-formaldehyde adhesive.

[3] Bond well with a fairly wide range of adhesives under a moderately wide range of bonding conditions.

[4] Wood from butt logs with high extractive content are difficult to bond.

[5] Bond satisfactorily with good-quality adhesives under well-controlled bonding conditions.

[6] Satisfactory results require careful selection of adhesives and very close control of bonding conditions; may require special surface treatment.

the sawn sheet are equally firm and strong, and either surface may be bonded or exposed to view with the same results.

Sliced veneer is also cut in the form of long strips by moving a flitch or block against a heavy knife. Because the veneer is forced abruptly away from the flitch at a fairly sharp angle by the knife, it tends to have fine checks or breaks on the knife side. This checked surface is likely to show imperfections in finishing and therefore should be the bonded side whenever possible. For book-matched face stock, where the checked side of half the sheets must be the finish side, the veneer simply must be well cut. Fancy hardwood face veneers are generally sliced.

The rotary process produces continuous sheets of flat-grained veneer by revolving a log against a knife. Again the veneer is forced away from the log surface at a fairly sharp angle. When rotary-cut veneer is used for faces, the knife or checked side should be the bonded side.

Veneer selected to be bonded in sanded grades of plywood should be (1) uniform in thickness, (2) smooth and flat, (3) free from deep checks (tightly cut), decay, knots, or other quality-reducing features, and (4) have grain suitable for the intended product. For plywood of the lower grades, most of these requirements are relaxed. For example, rough-cut (loose) veneers with frequent deep checks and larger defects are suitable for structural plywood, but a heavier adhesive spread is generally required than with tight veneers to prevent starved or dried out joints. When rotary-cut veneers are being bonded together, they are usually bonded tight side to loose side

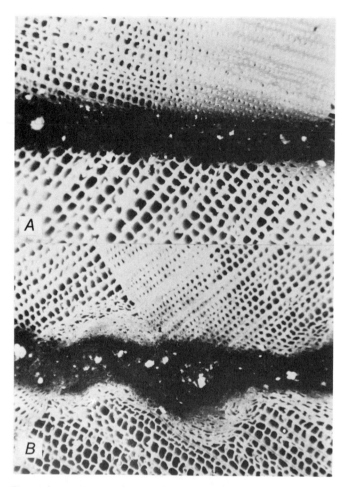

Figure 9—3—(A) A cross section of a portion of a bonded joint between two undamaged Douglas-fir surfaces prepared by planing with a sharp knife (magnification 120X). The dark area at the center of the photomicrograph is the adhesive bondline. The wood cells are open, and their walls are distinct without evidence of damage. (B) A cross section of a portion of a bonded joint between two damaged Douglas-fir surfaces prepared by abrasive planing with 36 grit (magnification 120X). The cells in and adjacent to the bondline are crushed and their walls indistinct. At higher magnifications (not shown) cracks in the cell walls can be seen.

(MC84 9071)

except for one loose-to-loose bondline which permits the second outside veneer to present its tight side out for wear and sanding.

Veneers are normally dried promptly after cutting, using continuous high-temperature dryers, heated either with steam or with hot gases from wood-residue or gas-fired burners. Drying temperatures are usually from 340 to 440 °F for limited periods of time. Overdrying veneer to very low moisture content, use of too-high temperatures, or prolonged drying at moderate temperatures are known to change the characteristics of the wood surfaces and result in unsatisfactory joints.

Adhesives

Adhesives have been used to bond wood for centuries, but until about 1930 the choice was limited to a relatively few resinous substances with adhesive properties derived from plants and animals. None of these natural "resins" is very resistant to water. Early in this century the first synthetic resin adhesive was developed that was not only as strong and rigid as wood but less affected by water and more durable than wood. Since then resin synthesis has produced many new adhesive families such as: urea and melamine formaldehyde; phenol and resorcinol formaldehyde; polyvinyl acetate and copolymer emulsions; elastomer-based solution and latex; hot melt; epoxy; isocyanate; and acrylic.

Table 9—2 describes the form, properties, preparation and typical uses of these families and others. There are many variations of each family, each with unique and useful properties for the user. With so many types of adhesives available, the choice of an adhesive for a given job can be quite confusing.

Most adhesives will adhere to wood, but satisfactory performance depends on careful consideration of these factors: physical and chemical compatibility of the adhesive and the adherends, processing requirements, mechanical properties, durability, ease of use, color, and cost.

All adhesives must go through a liquid phase at some point during bonding to establish adhesion. Consequently many adhesives are sold as solutions or emulsions. Water is the most common liquid carrier because it is cheap and has a natural affinity for wood. Organic solvents are necessary to dissolve some resins. Such solvents may have the added ability to disperse extractive layers, thus promoting better adhesion to certain difficult-to-bond woods. These solvents do not harm the wood, but care must be taken to check their compatibility if one adherend is a plastic.

The adhesive pH is another important factor in wood bonding. The adhesive may be acidic, neutral, or alkaline. Alkaline adhesives (such as hot-pressed phenol formaldehyde resin) or an alkaline pretreatment of the wood surface are sometimes favored for difficult-to-bond woods containing adhesive-resistant extractives and also for certain preservative treatments. Some woods (like oak) are quite acidic and may interfere with the cure of alkaline- or neutral-curing adhesives.

Plastics and metals are generally more difficult to bond successfully than wood. When they are bonded to wood, it is necessary to choose an adhesive capable of bonding the more difficult material, but which has also shown success in bond-

Table 9 – 2—Characteristics, preparation, and uses of the adhesives most commonly used for bonding wood

Class	Form	Properties	Preparation and application	Typical uses for wood bonding
Natural sources:				
Animal	Many grades sold in dry form; liquid adhesives available.	High dry strength; low resistance to moisture and damp conditions; brown to white.	Dry form mixed with water, soaked, and melted; solution kept warm during application; liquid forms applied as received; both pressed at room temperatures; adjustments in bonding procedures must be made for even minor changes in temperature.	Furniture assembly, use is declining. String instrument assembly.
Blood protein	Primarily, dry soluble or partially insolubilized whole blood. Commonly now handled and used like soybean adhesives.	Moderate resistance to water and damp atmospheres; moderate resistance to intermediate temperature and to micro-organisms; dark red, black.	Mixed with cold water, lime, caustic soda, and other chemicals; applied at room temperature; and pressed either at room temperature or in hot presses at 240 °F or higher.	Primarily for interior-type softwood plywood. Sometimes in combination with soybean protein. Now largely displaced by phenolic resin.
Casein	Several brands sold in dry powder form; may also be prepared from raw materials by user.	Moderately high dry strength; moderate resistance to water, damp atmospheres, and intermediate temperatures; not suitable for exterior uses; white to tan.	Mixed with water; applied and generally pressed at room temperature.	Laminated timbers and doors for interior use.
Vegetable protein (mainly soybean)	Protein sold in dry powder form (generally with small amounts of dry chemicals added) to be prepared for use by user.	Moderate to low dry strength; moderate to low resistance to water and damp atmospheres; moderate resistance to intermediate temperatures; white to tan.	Mixed with cold water, lime, caustic soda, and other chemicals; applied and pressed at room temperatures, but more frequently hot-pressed especially when blended with blood protein.	Bonding softwood plywood for interior use. Now largely displaced by phenolic resin in the softwood plywood industry.

Table 9–2—*Characteristics, preparation, and uses of the adhesives most commonly used for bonding wood*—Continued

Class	Form	Properties	Preparation and application	Typical uses for wood bonding
Natural sources—con.				
Lignocellulosic residues and extracts	Powder or liquid.	Good dry strength, moderate to good wet strength; used with phenolic resin for best performance; dark brown.	Generally used with phenolic resin for best durability. Use alone generally requires acid curing environment which is detrimental to the wood.	In developmental stage. Some use in panel products such as particleboard and partial replacement for phenol resin; continued development and use linked to cost and availability of phenolic resins; future potential for panel products.
Synthetic sources:				
Urea resin	Many brands sold as dry powders, others as liquids; may be blended with melamine or other resins.	High in both wet and dry strength; moderately durable under damp conditions; moderate to low resistance to temperatures in excess of 120 °F; white or tan.	Dry form mixed with water; hardeners, fillers, and extenders may be added by user to either dry or liquid form; applied at room temperatures, some formulas cure at room temperatures, others require hot-pressing at about 250 °F.	Hardwood plywood for interior use and furniture; interior particleboard; underlayment, flush doors, furniture core stock.
Melamine resin	Comparatively few brands available; usually marketed as a powder with or without catalyst.	High in both wet and dry strength; very resistant to moisture and damp conditions depending on type and amount of catalyst; white to tan.	Mixed with water and applied at room temperatures; heat required to cure (250 to 300 °F).	Primarily a fortifier for urea resins for hardwood plywood, end-jointing and edge-gluing of lumber, and scarf joining softwood plywood. High-frequency cure compatible.

Table 9 – 2.—*Characteristics, preparation, and uses of the adhesives most commonly used for bonding wood*—Continued

Class	Form	Properties	Preparation and application	Typical uses for wood bonding
Synthetic sources—con.				
Phenol resin[1]	Many brands available, some dry powders, others as liquids, and at least one as dry film. Most commonly sold as aqueous, alkaline dispersions for plywood, and as powder for waferboard manufacture.	High in both wet and dry strength; very resistant to moisture and damp conditions, more resistant than wood to high temperatures; often combined with neoprene, polyvinyl butyral, nitrile rubber, or epoxy resins for bonding metals. More resistant than wood to chemical aging; dark red.	Film form used as received; powder form mixed with solvent, often alcohol and water, at room temperature; with liquid forms, modifiers and fillers are added by users; most common types require hot-pressing at about 260 to 300 °F. Temperatures as high as 400 °F may be used for waferboard.	Primary adhesive for exterior softwood plywood and flakeboard.
Resorcinol resin and phenol-resorcinol resins	Several brands available in liquid form; hardener supplied separately; some brands are combinations of phenol and resorcinol resins.	High in both wet and dry strength; very resistant to moisture and damp conditions; more resistant than wood to high temperature and chemical aging; dark red.	Mixed with hardener and applied at room temperatures; resorcinol adhesives cure at room temperatures on most species; phenol-resorcinols cure at temperatures from 70 to 150 °F, depending on curing period and species.	Primary adhesives for laminated timbers and assembly joints that must withstand severe service conditions.
Polyvinyl acetate resin emulsions	Several brands are available, varying to some extent in properties; often copolymerized with other polymers; marketed in liquid form ready to use.	Generally high in dry strength; low resistance to moisture and elevated temperatures; joints tend to yield under continued stress; white or yellow.	A liquid ready to use; applied and pressed at room temperatures.	Furniture assembly, flush doors, bonding plastic laminates. Assembly of manufactured homes for transportation shock resistance.
Crosslinkable polyvinyl acetate resin emulsion	Similar to polyvinyl acetate resin emulsions but includes a resin capable of forming linkages when a catalyst is added.	Improved resistance to moisture and elevated temperatures; improved long-term performance in moist or wet environment; color varies.	Mix resin emulsion and catalyst; cure at room temperature or at elevated temperature.	Interior and exterior doors, molding and architectural woodwork, cellulosic overlays.

Table 9–2—*Characteristics, preparation, and uses of the adhesives most commonly used for bonding wood*—Continued

Class	Form	Properties	Preparation and application	Typical uses for wood bonding
Synthetic sources—con.				
Elastomer-base adhesives A. Contact adhesives	Typically a neoprene or styrene-butadine-rubber elastomer base in organic solvents or water emulsion. Other elastomer systems are also available.	Initial joint strength develops immediately upon pressing, increases slowly over a period of weeks; dry strengths generally lower than those of conventional woodworking glues; water resistance and resistance to severe conditions variable; color varies.	Used as received; both surfaces spread and partially dried before pressing; commonly used in roller presses for instantaneous bonding.	For some nonstructural bonds, as on-the-job bonding of decorative tops to kitchen counters. Useful for low-strength metal and some plastic bonding.
B. Mastics (elastomeric construction adhesives)	Puttylike consistency Synthetic or natural elastomer base usually in organic solvents; others solvent-free or latex emulsions.	Gap filling; develop strength slowly over several weeks. Water resistance and resistance to severe conditions variable; color varies.	Used as received; extruded by caulking guns in beads and ribbons, with and without supplemental nailing.	Lumber and plywood to floor joist and wall studs; laminating gypsum board, styrene and urethane foams, and other materials; assembly of panel systems; manufactured homes.
Thermoplastic synthetic resins (hot melts)	Solid chunks, pellets, ribbons, rods, or films; solvent-free.	Rapid bonding; gap filling; lower strength than conventional wood adhesives; minimal penetration; moisture resistant; white to tan.	Melted for spreading; bond formation by cooling and solidification; requires special equipment for controlling bonding conditions.	Edge banding of panels; plastic lamination; patching; films and paper overlays; furniture assembly.
Isocyanate	Liquid resins or water emulsions or with other resins to form polyurethane adhesives. Great versatility in formulating wide variety of adhesives; some forms may give toxic vapors.	Excellent adhesion to wood and many other materials. Resistance to moisture and elevated temperature varies widely; excellent chemical aging resistance.	One-part adhesives cure on application of heat or in presence of moisture; two-part resins cure upon mixing at either room temperature or elevated temperatures; very rapid cures are possible.	Limited use in structural flakeboard, film laminating, and as an assembly adhesive; use has been limited by high relative cost and in some instances by sensitivity to heat and moisture.

Table 9–2—Characteristics, preparation, and uses of the adhesives most commonly used for bonding wood—Continued

Class	Form	Properties	Preparation and application	Typical uses for wood bonding
Synthetic sources—con.				
Crosslinkable acrylic resin	A two-part adhesive of completely reactive components. Toxic vapors.	Excellent adhesion to wide variety of metals and plastics; long assembly time; fast cure; good heat and moisture resistance, but quality and durability of wood bonds not well established.	Resin applied to one adherend, hardener to the other; adhesive cure begins on contact of two adherends; cures at room temperature; two components may also be mixed before application.	Limited use with wood. Future potential for composite products with metal and plastics.
Epoxy resins	Several different chemical polymers of the general type available or possible; usually in two parts, both liquid. Completely reactive; no solvent or other volatiles in the liquid adhesives or evolved in curing.	Good adhesion to metals, glass, certain plastics, and wood products; permanence in wood joints not adequately established; gap-filling capability.	Marketed in two parts, resin and curing agent, both liquid; mixed at the point of use; applied at room temperatures; cured at room or elevated temperatures, depending on formulation. Potlife and cure conditions vary widely with composition; can be formulated for curing at either room or elevated temperatures.	Most common use in combination with other resins for bonding metals, plastics, and materials other than wood; bonding wood-to-wood specialty items; repair of laminated beams; fabrication of cold molded wood boats and wind generator blades.

[1] Most types used in the United States are alkaline-catalyzed. The general statements refer to this type. Acid-catalyzed systems are also available. primarily for use at curing temperatures of 70 to 140 °F. but are little used in the United States. Their principal limitation is the possible damage to wood by the acid catalyst.

ing wood. Generally 100 percent-reactive adhesives like epoxy, isocyanate, or elastomer-based contact adhesives (after solvent evaporation) are used.

The user must choose an adhesive that can be mixed, applied, and cured with available equipment accepting the performance properties it offers or consider the cost of purchasing equipment to meet specific adhesive requirements. Important bonding performance considerations are:

Working life—If the adhesive hardens chemically and has a catalyst, how much time is available after mixing before it can no longer be spread?

Consistency—Is the consistency suitable for the type of application equipment available (brush, roller, roll spreader, extruder, curtain coater, spray, powder metering device)?

Minimum cure temperature—What is the minimum cure temperature that will adequately harden the adhesive?

Pressure—Will high pressure be required to produce a thin adhesive layer or is the adhesive gap-filling?

Fixture time—At any given temperature what is the minimum curing time during which the adhesive must be undisturbed before the joint can be safely worked (sawn, sanded, etc.) or the assembly moved?

There are frequently tradeoffs between the bonding requirements of adhesives and their resistance to stress, time, and environmental conditions after cure, as illustrated in figure 9–4. Adhesives that are the strongest, most rigid, and most resistant to environmental conditions are often those least flexible in terms of bonding conditions such as wood moisture content, surface cleanliness, bondline thickness, plus the need for high pressure and a minimum temperature during cure. Adhesives that are the weakest, least rigid and least resistant to service conditions are usually the most flexible in terms of bonding conditions, especially temperature and bondline thickness. Many other adhesives have properties between these extremes as suggested by figure 9–4.

The ability of an adhesive to transfer load from one adherend to another under various conditions tends to govern its end use. Historically adhesives have been classified according to how well they compare to wood in this regard. Adhesives stronger and more rigid than wood are usually classified as structural adhesives. Adhesives less strong and less rigid than wood are usually classified as semistructural or nonstructural.

Adhesives are further classified as suitable for exterior or interior service depending on how rapidly they lose their ability to transfer load in comparison with wood as the service conditions worsen. Structural adhesives that maintain superior properties under all service conditions are classed as fully exterior. Those which lose their ability to transfer stress faster than wood as conditions worsen, particularly with regard to water, are classed as interior. A third class is made up of adhesives that maintain superior strength and rigidity in short-

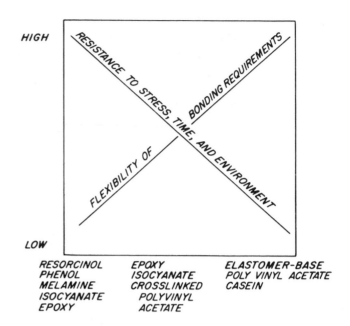

Figure 9–4—Schematic relationship between the degree of control required during bonding and the resistance of adhesive bonds to stress, load-time, and environmental factors affecting joint performance.

(ML84 5820)

term wetting, but degrade faster than wood in long term exposures to water or to high heat and or high moisture environments.

Semistructural adhesives may be suitable for exterior use depending on the structural demands and degree of assurance against failure required of the bonded joint. Many semistructural adhesives have the added capability of forming good joints between surfaces that are not in close continuous contact (gap-filling property) such as frequently occurs in building construction. Nonstructural adhesives may or may not be as strong and rigid as wood in dry service conditions but they very quickly lose their strength and rigidity when they become wet.

Based on these distinctions there are five end-use categories:
- structural—fully exterior.
- structural—limited exterior.
- structural—interior.
- semistructural—limited exterior.
- nonstructural—interior.

Examples of adhesives in each category are given in table 9–3. Some adhesives appear in more than one category to acknowledge that the formulation and consequent properties of those particular types vary greatly.

The classification of adhesives for end use is in a state of development. New methods are available to quantify the effects of stress, temperature, moisture, oxygen, and certain chemicals on adhesive strength and rigidity. There is a positive trend toward the use of performance specifications in place of prescribed adhesives for bonded products. These developments are certain to encourage the development of new adhesives and the growth of adhesive applications in building construction.

In some applications, particularly furniture and interior molding, the adhesive color, bleeding characteristics, and stain acceptance are important factors. Generally, adhesives sold for furniture manufacture, such as urea formaldehyde and polyvinyl acetate, are formulated for a tan or colorless joint and good acceptance of stain. Ease and simplicity of use can be important factors. One-part adhesives, like liquid animal glue, polyvinyl acetate, hot melt, and phenol formaldehyde film, are the simplest, most foolproof adhesives to use because there is no chance for error in weighing and mixing components. Water-dispersed and film adhesives are easy to clean up and films are the least messy to use. Two-, three-, or more-part adhesives require careful measuring and mixing of the various components. They often require solvents for cleanup after bonding. Frequently they are toxic to the skin, give off toxic fumes, or are skin sensitizers. The hardeners for resorcinol, phenol, melamine, and urea resin adhesives contain formaldehyde which is an irritant to some people. Amine hardeners used with some epoxy-resins are strong skin sensitizers.

The cost of an adhesive and the equipment required for its use must be balanced against similar factors for other adhesives or methods of fastening. In recent years the rising cost of organic solvents and the cost of their recovery to prevent air pollution have increased the cost of using adhesives containing them. Thus there is a positive trend toward substitution of water-based systems which may also be cheaper due to the low cost of the solvent. However, other effects such as wood

Table 9–3—*Classification of adhesives for various uses according to their strength; rigidity; and resistance to stress, time, temperature, moisture, and oxygen and other chemicals*

Classification		Adhesive
Structural	Fully exterior	Crosslinked polyvinyl acetate
		Epoxy
		Isocyanate
		Melamine formaldehyde
		Phenol formaldehyde
		Phenol-resorcinol formaldehyde
		Resorcinol formaldehyde
	Limited exterior (withstands short-term soaking)	Melamine-urea formaldehyde
		Urea formaldehyde
	Interior (withstands high humidity)	Casein
		Blood
Semistructural	Limited exterior	Crosslinked polyvinyl acetate
		Elastomer-based mastic
		Epoxy
		Isocyanate
Nonstructural	Interior	Animal (bone, hide)
		Elastomer-base mastic
		Elastomer-base contact
		Thermoplastic synthetic resin (hot melt)
		Polyvinyl acetate
		Soybean
		Starch
Unclassified		Lignocellulosic
		Modified acrylic

grain raising and slower drying must be considered for their effects on performance and overall cost.

Bonding Process

Drying and Conditioning Wood for Bonding

The moisture content of wood at the time of bonding has much to do with the final strength and durability of joints, the development of checks in the wood, and the dimensional stability of the bonded assembly. Large changes in the moisture content of the wood after bonding cause shrinking or swelling stresses that may seriously weaken both the wood and the joints, while also causing warping, twisting, and other undesirable effects. It is generally unwise to bond green wood or wood at high moisture content, particularly the higher density hardwoods that have high coefficients of shrinkage, unless the end use is also at high moisture content.

Expanding this thought to a general rule, the wood should be dry enough so that, even if moisture is added in bonding, the moisture content is at about the level desired for service.

The choice of moisture content conditions for bonding according to this principle depends heavily on whether the bonding process involves heating, as in hot-pressing, or merely pressing at room temperature. For example, in bonding 1-inch boards or thicker pieces at room temperature, the desired final moisture content is mainly controlled by proper drying of the wood. A waterborne adhesive adds only 1 to 2 percent. However, in bonding veneer or other thin pieces pressed at room temperature, the moisture added by the adhesive itself may increase the moisture content of the dry wood as much as 45 percent, or well above the desired inservice level. Thus thickness of the laminae, number of laminae, density of the wood, adhesive water content, quantity of adhesive spread, and bonding procedure (hot-pressing or cold-pressing) all affect the cumulative moisture content of the wood. In hot-pressing, a moderate amount of water is evaporated reducing the moisture content of the product when removed from the press. However, to minimize plastic flow of the wood when hot and wet, and the development of steam-caused blisters, the total moisture content of hot-pressed wood assemblies should not exceed 10 percent.

In practical terms lumber with a moisture content of 6 to 7 percent is generally satisfactory for cold-press bonding into furniture, interior millwork, and similar items. Lumber for outside use should generally contain 10 to 12 percent moisture before bonding. A moisture content of 3 to 5 percent in veneer at the time of hot-pressing is satisfactory for thin plywood to be used in furniture, interior millwork, softwood plywood for construction and industrial uses, and similar products. For such uses as plywood for boxes, veneer at a moisture content of about 8 to 10 percent is acceptable for cold-pressing.

Lumber that has been dried to the approximate average moisture content desired for bonding may still show moisture content differences between various boards and between the interior and the surfaces of individual pieces. Large differences in the moisture content of pieces bonded together result in considerable stress on joints as the pieces tend toward a common moisture content and may result in warping or delamination of the product. Differences should be no greater than about 5 percent for lower density species or about 2 percent for higher density species. Lumber that is to be bonded should also be free from casehardening, warp, checks, and splits in order to produce the highest quality bonded product.

Adhesive Application

Adhesives are applied in many ways, from simply brushing to hand roller, roll spreader, extruder, spray, and curtain coater. Satisfactory performance with any system depends on applying a continuous film of adequate and uniform thickness or continuous beads (fig. 9−5) that will squeeze to a uniform, thin layer under pressure. The amount of spread required varies with the type of adhesive, the wood, the temperature and humidity of the air, and the assembly time (the time between adhesive application and pressing). For good adhesion, the adhesive must wet the wood surface completely. The positive scrubbing action of certain methods, like brush or roller, can be important in wetting difficult-to-bond wood. Applying adhesive to both surfaces also improves the bonding of difficult wood surfaces.

Penetration

Because different wood species vary in their absorptivity, a given adhesive mixture may penetrate more into one wood than into another under the same bonding conditions. A moderate amount of such penetration is desirable, especially if the wood surfaces tend to be somewhat torn and damaged. Excessive penetration, however, wastes adhesive and may result in starved (totally absorbed) bondlines.

Adhesive Consistency and Pressure

Making strong joints with an adhesive applied as a liquid depends primarily upon proper correlation between bonding pressure and adhesive consistency during pressing as suggested by figure 9−6.

Pressure serves several purposes. It forces air from the joint, brings the wood surfaces into intimate contact with the adhesive, squeezes the adhesive into a thin continuous film, and holds the assembly in position during setting or curing of the adhesive. But pressure, particularly high pressure on dense or porous woods, also tends to squeeze the adhesive into the wood and out of the joint. Excessive squeeze-out and penetration as mentioned previously will result in an inferior starved joint. The strongest joints usually result when the consistency of the adhesive permits the use of moderately high pressures (100 to 250 psi).

Consistency once the adhesive mixture is spread on the wood surface may vary appreciably. It depends upon such factors as: the kind of adhesives; the adhesive-liquid proportion of the mixture; the age of the mixed adhesive; the quantity of adhesive spread; the moisture content and species of wood; the time elapsed after spreading; and the extent to which the adhesive coated surfaces are exposed to air. Room-temperature-setting adhesives usually thicken and harden steadily after spreading until they are fully cured, while hot press adhesives often thin out during the first stage of heating and then thicken and harden as curing progresses.

Low pressures (100 psi) are suitable for low density wood because the adherend surfaces easily conform to each other thus assuring intimate contact between the adhesive and the wood. The highest pressures (up to 250 psi) are required with the highest density woods to achieve the desired surface conformation and adhesive-wood contact. (Small flat well-planed surfaces can be bonded satisfactorily at lower pressures, however.) Because high pressure tends to squeeze the adhesive into the wood or out of the joint, adhesives of thicker consistency are required with denser woods as suggested by figure 9-6. Usually the thicker consistency is achieved by using longer assembly times with dense woods than with light woods. The longer assembly time increases adsorption of the liquid solvent (or dispersant) by the wood and its evaporation into the air. The loss of liquid increases the adhesive consistency before pressure is applied.

Care is required, regardless of wood density, so that the assembly period is not excessive, lest the adhesive become dried out or even precured before bonding pressure is applied. Predried or precured adhesive will result in poor adhesive transfer and will produce a thick weak joint as suggested by figure 9-6.

Hardening (Curing)

A basic understanding of how adhesives harden will help assure success in bonding. Solid resins, both natural and synthetic, are formed by chemical action which creates very large molecules out of smaller ones. The resin is solid or semisolid because the large molecules are entangled (like spaghetti) and, because of physical attractive forces or chemical bonds (crosslinks) between them. Some adhesives

Figure 9-5—An extruding device coupled with a feed mechanism is being used to apply phenol-resorcinol adhesive to thick veneer in the manufacture of laminated veneer lumber (LVL). (M147 291)

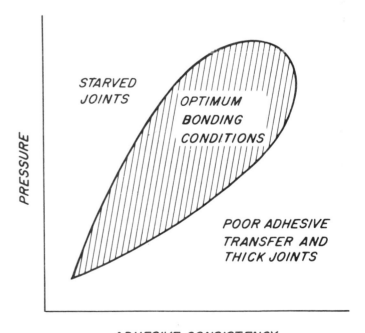

ADHESIVE CONSISTENCY

Figure 9-6—Schematic diagram of the relationship between the quality of a joint bonded with thermo-setting resin adhesives, and the adhesive consistency and bonding pressure, showing the tendency toward starved joints when pressure is too high in relation to adhesive consistency and the tendency toward poor adhesive transfer/penetration and thick joints when consistency is too high in relation to pressure. (ML84 5803)

9-15

are sold with the large resin molecules already fully developed. These are liquefied (for application) by heating or by dispersing the molecules in a fluid solvent, or by a combination of these. To harden the adhesive, it is cooled, or the liquid is removed, or both. For example: Hot-melt adhesives harden as they cool, polyvinyl acetate emulsion adhesives harden as they lose water and coalesce into a film, and animal glue hardens by both cooling and water loss. Other adhesives are sold with the large molecules partially formed. The chemical reaction which created the large molecules has been interrupted by cooling or changing the chemical environment. These adhesives may be 100 percent-reactive liquids, like epoxy, or solutions like resorcinol formaldehyde. Others like spray-dried urea formaldehyde resins are sold as powders and may be dispersed in a liquid solvent for application or dusted on as a dry powder. The chemical hardening reaction is restarted by heat in almost any form, restoring the proper chemical environment for curing (i.e. pH adjustment) or, in the case of many powder adhesives, by adding water which releases and activates a hardening agent.

Lumber joints should be kept under pressure at least until they have enough strength to withstand the internal and handling stresses that tend to separate the wood pieces. In cold-pressing lumber under favorable bonding conditions this stage will be reached in as little as 15 minutes or as long as 24 hours depending upon the temperature of the room and the wood, the curing characteristics of the adhesive, and the thickness, density, and absorptive characteristics of the wood. A longer than minimum pressing period is generally advisable, when operating conditions permit.

In hot-pressing operations the time required varies with temperature of platens or air, thickness and kind of material being pressed, and adhesive formulation. In actual practice the variation is from about 2 minutes to as much as 30 minutes. The time under pressure may be reduced to a few seconds by heating the joint with high-frequency electrical energy because it is possible to raise the temperature of only the adhesive film itself very rapidly. High-frequency heating is often used when bonding lumber end joints; in patching, scarfing, or fingerjointing plywood; and in limited cases in the manufacture of other panel products.

Conditioning Bonded Joints for Furniture

When boards are bonded edge to edge, the wood at the joint absorbs moisture from the adhesive and swells. If the bonded assembly is surfaced before this excess moisture is removed or distributed uniformly, more wood is removed along the swollen joints than elsewhere. Later, when the added moisture is removed or redistributed, the joint areas shrink and permanent depressions (sunken bondlines) are formed that may be very conspicuous under a high gloss finish. This is particularly important when using adhesives that contain large amounts of water (such as casein), is less important for urea formaldehyde, and is not at all important for epoxy or hot melt adhesives because they do not contain water.

Approximately uniform redistribution of the moisture added by the adhesive can usually be obtained by conditioning the stock after bonding for 24 hours at 160 °F, 4 days at 120 °F, or at least 7 days at room temperature. In each case the relative humidity must be adjusted to prevent drying the wood itself to below the level desired.

In plywood, veneered panels, and other constructions made by bonding together thin layers of wood, it is advisable to condition the finished panels to the average moisture content they are likely to encounter in service. In room-temperature bonding operations, it is frequently necessary to remove at least a part of the moisture added by the adhesive. The drying is most advantageously done under controlled conditions and time schedules. Drying room-temperature-bonded products to excessively low moisture content materially increases warping, opening of joints, and checking. Panels will often be very dry after hot-press operations. A growing practice among softwood plywood producers is to replace moisture by spraying the panels heavily with water and tightly stacking them to allow moisture to distribute uniformly. This practice restores some of the panel thickness lost by compression in hot-pressing and apparently minimizes warping in service.

Bonding Treated Wood

The advent of durable adhesives that will outlast wood itself under severe conditions has made it possible to bond wood treated with wood preservatives and fire retardants. Perhaps the most important applications have been in laminated beams in which the lumber was treated with preservatives before bonding, and in bonding fire-retardant-treated veneers and lumber for fire doors. Experience has shown that many types of preservative-treated lumber can be bonded successfully with phenol-resorcinol formaldehyde, resorcinol formaldehyde, and melamine formaldehyde adhesives under properly controlled bonding conditions. Generally, the preservative-treated wood surfaces should be resurfaced just before bonding to reduce interferences by oily solvents or other surface concentrations of preservative material. Fire-retardant-treated material should also be resurfaced just before bonding for the same reason. It is usually necessary to control the bonding conditions more carefully in bonding treated wood than untreated wood of the same species; it may also be desirable to use a somewhat higher curing temperature, or a longer curing period, with the treated wood.

The acidic nature of some waterborne salt preservatives hinders the cure of neutral- or alkaline-curing adhesives like resorcinol formaldehyde or phenol-resorcinol formaldehyde. Research has shown improved success in bonding wood and plywood treated with these salts by washing the surfaces to be bonded with a 10 percent caustic solution and redrying before bonding.

Types of Bonded Joints

Side-Grain Surfaces

With most species of wood, straight, plain joints between side-grain surfaces (fig. 9–7, A) can be made substantially as strong as the wood itself in shear parallel to the grain, tension across the grain, and cleavage. The tongued-and-grooved joint (fig. 9–7, B) and other shaped joints have the theoretical advantage of larger bonding surfaces than the straight joints, but in practice they do not give higher strength with most woods. The theoretical advantage is often lost, wholly or partly, because the shaped joints are difficult to machine to obtain a perfect fit of the parts and the joint wastes wood. Because of poor contact, the effective bonding area and strength may actually be less on a shaped joint than on a flat surface. The principal advantage of the tongued-and-grooved and other shaped joints is that the parts can be more quickly aligned in the clamps or press. A shallow tongue-and-groove is usually as useful in this respect as a deeper cut and is less wasteful of wood.

End-Grain Surfaces

It is practically impossible with present liquid-based adhesives and techniques to make end-butt joints (fig. 9–8, A) sufficiently strong or permanent to meet the requirements of ordinary service. With the most careful bonding technique possible, not more than about 25 percent of the tensile strength of the wood parallel with the grain can be obtained in butt joints using conventional water-based adhesives. To approximate the tensile strength of solid wood, a scarf, finger, or other type of joint that approaches as closely as possible parallel to the grain direction (a side-grain surface) must be used. This side-grain area should be at least 10 times as large as the cross-sectional area of the piece, because wood is approximately 10 times stronger in tension than in shear. In plywood scarfs and finger joints, a slope of 1 in 8 has been found adequate. For nonstructural, low-strength joints these requirements need not necessarily be met.

Finger joints may be cut with the profile showing either on the edge (horizontal joint) (fig. 9–8, D) or on the wide face (vertical joint) (fig. 9–8, C) of boards. There is greater

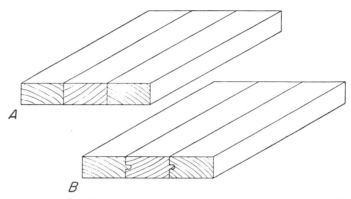

Figure 9–7—Side-to-side-grain joints: A, plain; B, tongued-and-grooved.   (M93 137)

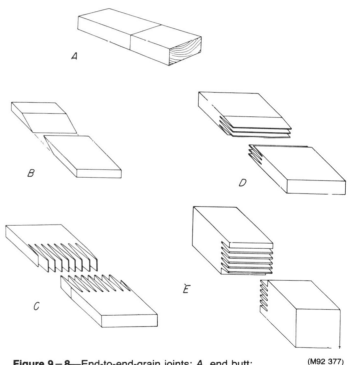

Figure 9–8—End-to-end-grain joints: A, end butt; B, plain scarf; C, vertical structural finger joint; D, horizontal structural finger joint; E, nonstructural finger joint.   (M92 377)

leeway in design of a finger joint when a vertical joint is used, but a longer cutting head with more knives is needed. When the adhesive curing is done by high-frequency electrical energy, it can generally be done more rapidly with the vertical joint than with the horizontal joint.

The efficiencies of scarf joints of different slopes are discussed in chapter 10. Slopes 1:12 or flatter generally give the highest strength. This also holds true for finger joints, but in these the tip thickness also has to be as small as practical (not greater than 1/32 inch). A thickness of 0.015 to 0.031 inch is about the minimum practical for machined tips depending on wood machinability and other production variables. Sharper tips can be created using various types of dies which are forced into the end of the board.

A well-manufactured adhesive-bonded end joint (scarf, finger, or lap joint), can achieve up to 90 percent of the strength of clear wood and exhibits fatigue behavior much like the wood that is joined. However, curves showing repeated stress versus cycles to failure are somewhat lower (inferior) to similar curves for unjointed wood when the repeated stress is expressed as a percentage of static strength.

End-to-Side-Grain Surfaces

Plain end-to-side-grain bonded joints (fig. 9−9, A) are difficult to design so that they can carry an appreciable load. Also, joints in service face severe internal stresses in the members from unequal dimensional changes as moisture content changes; such stresses may be high enough to cause failure. It is therefore necessary to use joints with interlocking surfaces (such as dowels, tenons, rabbets, or other devices) to reinforce a plain joint and bring side grain into contact with side grain or to secure larger bonding surfaces (fig. 9−9). All end-to-side-grain joints should be carefully protected from appreciable changes in moisture content in service.

Performance of Bonded Joints and Products

Performance of an adhesive-bonded joint or product can be optimized by careful attention to adhesive selection, adherend surface preparation, and control of the bonding process. Whatever the optimum conditions they must be quantified and used without variation in production to ensure high joint strength. Uniform optimal performance is important to: The adhesive manufacturer and user, who must stand behind their products; the product designer; the consumer; and possibly the lender and insurer.

The performance criteria involve levels of some mechanical property like joint strength, joint rigidity, panel bending strength, or bending stiffness. These properties are important because they are used to design safe, efficient composite structures. Performance according to one or more of these criteria is usually evaluated at three levels—short term, long term, and product quality assurance.

Short Term

In the short term, the properties of wood, adhesives, and bonded products vary with the environment. In most cases

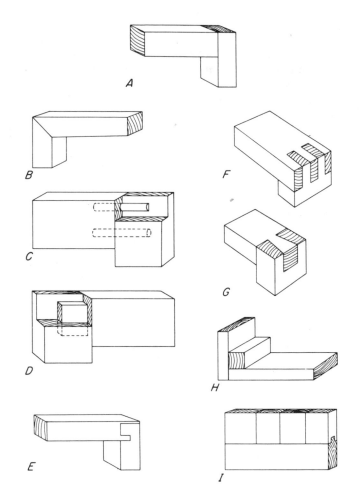

Figure 9−9—End-to-side-grain joints: *A*, plain; *B*, miter; *C*, dowels; *D*, mortise and tenon; *E*, dado tongue and rabbet; *F*, slip or lock corner; *G*, dovetail; *H*, blocked; *I*, tongued-and-grooved.   (M92 376)

the properties decrease as the temperature or moisture level increase. Strength and rigidity or stiffness may return to their original levels if the yield point of the material has not been exceeded while under load.

As noted earlier, the relationship between an adhesive's properties and the temperature, moisture, and stress levels is important for selecting an adhesive and in designing the joint or structure. The designer must know what the strength or stiffness of the joint or product will be if wet or at an elevated temperature. The properties of rigid adhesives like resorcinol formaldehyde, phenol formaldehyde, melamine formaldehyde, and urea formaldehyde all change less than do wood properties under equivalent temperature and moisture change. The short-term performance of joints and products made with these adhesives is simply evaluated by tests at room temperature, dry and wet. Other adhesives like casein, polyvinyl acetate, and elastomer-base, whose properties change more rapidly than wood's with changes in the environment, are tested dry (or dry after soaking) and after exposure to high humidity environments. In addition, some specifications require testing of bonded structural products at elevated temperatures such as may occur in roofs. In at least one case, a short-term dead load (creep) test at elevated temperatures is also required.

Performance specifications for adhesives for structural products like laminated beams and plywood require conformance to fairly high minimum strength and wood failure values after several different water exposure tests.

Specifications for particleboards and flakeboards require minimum property values when tested as prescribed.

Long Term

In the long term, wood, adhesives, and bonded products degrade at a rate determined by the temperature and moisture levels, the concentration of certain chemicals which can have serious effects, the presence of micro-organisms, and the stress level. Long-term performance is equated with the ability of a joint or product to withstand these factors. Resistance is measured by the loss or rate of loss of a mechanical property against the time of exposure. A durable joint or product is one that shows no greater loss of properties during its life in the service environment than solid wood of the same species and grade.

Many adhesives and bonded products have decades of documented performance in many environments. Thus it is possible to say with a high degree of certainty what the long-term performance of similar products will be. Joints well designed and well made with any of the commonly used woodworking adhesives will retain their strength indefinitely if the moisture content of the wood does not exceed approximately 15 percent and the temperature remains within the range of human comfort. However, some adhesives deteriorate when exposed either intermittently or continuously to temperatures much above 100 °F for long periods. Low temperatures seem to have no significant effect on strength of bonded joints.

Joints and products that were well made with phenol formaldehyde, resorcinol formaldehyde, or phenol-resorcinol formaldehyde adhesives have proved more durable than the wood when exposed to warmth and dampness, to water, to alternate wetting and drying, and to temperatures sufficiently high to char the wood. These glues are entirely adequate for use in products that are exposed indefinitely to the weather (fig. 9−10).

Joints and products well made with melamine formaldehyde, melamine-urea formaldehyde, and urea formaldehyde resin adhesives have proven less durable than wood. Melamine formaldehyde is only slightly less durable than phenol formaldehyde or resorcinol formaldehyde and is still considered acceptable for structural products. Melamine-urea formaldehyde is significantly less durable, and urea formaldehyde is quite susceptible to degradation by heat and moisture (fig. 9−10).

Joints or products bonded with polyvinyl acetate, hot melts, and natural resins will not withstand prolonged exposure to water or high moisture content, or repeated high-low moisture content cycling in bonds of high-density woods. However, if they are properly formulated these adhesives are durable in a normal interior environment.

At present it appears that some isocyanate, epoxy, and crosslinked polyvinyl acetate adhesives are durable enough to use on lower density species even under exterior conditions. Some elastomer-based adhesives may be durable enough for exterior use with lower density species in nonstructural applications, and in structural applications when used in conjunction with approved nailing schedules. Those that chemically cure and still remain flexible seem the most durable.

New adhesives or products do not have a history of long-term performance in service environments. For the sake of economics, an estimate of their long-term performance is usually required in the shortest possible time. Accelerated-aging tests have been developed to help make these estimates. Most accelerated laboratory exposures include the major elements of the service environment—heat, moisture, and stress. These tests accelerate degradation by using elevated temperatures and rapid forced wetting and drying to produce cyclic stress. However, laboratory exposures can never duplicate the actual conditions of a service environment. Estimates of long-term performance are also obtained by placing specimens in outdoor exposure for 1 to 20 years. Outdoor exposures are intensified by facing the specimens south at an angle perpendicular to the noonday sun and by exposing the specimens without protective covering or coating.

Only four long-term laboratory aging methods have been standardized. Minimum performance levels in these methods have not been established for interior or exterior exposures. Therefore performance of the new or unknown adhesive or bonded product is usually compared to the performance of established adhesives or products tested in the same laboratory exposure.

Treatments that can be used to increase the durability of glued products include: (1) Coatings that reduce the moisture content changes in the wood and (2) impregnation of the wood with preservatives. Moisture-excluding coatings reduce the shrinking and swelling stresses that occur during varying exposure conditions; however the coatings do not protect

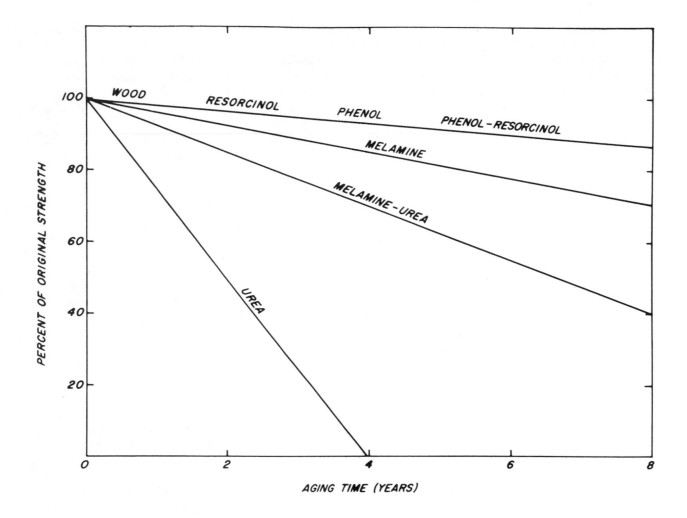

Figure 9–10—Schematic illustration of the relative rates of degradation of small specimens bonded with several rigid thermosetting adhesives fully exposed outdoors.

(ML84 5824)

wood effectively during prolonged exposure to damp conditions. By impregnating glued members with preservatives, particularly those that will also prevent rapid moisture changes, the deteriorating effects of prolonged exposure to outdoor or damp conditions can be greatly reduced. While reducing the rate of moisture exchange with the atmosphere these preservatives also protect against attack by microorganisms.

Product Quality Assurance

After the short- and long-term performance of a well-made joint or product have been established, then maintenance of the manufacturing process to assure that the product will perform to the desired level is the major concern. To evaluate the manufacturing process a quality-assurance program consisting of three principal parts is required:

(1) Determining the limits of the bonding process factors that will result in satisfactory joints or product.
(2) Monitoring the process and the resulting joint or product performance as it is produced.
(3) Detecting unacceptable joints or product, determining the cause, and correcting it.

The plywood, laminated beam, particleboard, millwork, and other industrial trade associations have established quality-assurance programs which effectively monitor the joint or product performance at the time of manufacture for compliance with voluntary product standards. Usually the immediate performance is evaluated by subjecting specimens cut from the newly manufactured joint or product to a series of swell-shrink cycles. The treatments are more rigorous for joints and products intended for exterior exposure. For example, exterior softwood plywood is subjected to two boil-dry cycles while interior plywood is subjected to a single soak-dry cycle at lower temperature. After exposure the specimens are tested for some mechanical property, examined for delamination, or evaluated for percent wood failure remaining on the tested joint area. The test result after exposure is compared to a minimum requirement in the trade association's standards. Long experience has shown that joints or products with at least the minimum values will probably perform satisfactorily in service. If the joint or product meets the requirement, it is certified by the association as meeting the standard for satisfactory performance.

Selected References

American Institute of Timber Construction. Laminators quality control system AITC 201. In: Inspection manual AITC 200. Englewood, CO: AITC; (see current edition).

American Institute of Timber Construction. American national standard for wood products: Structural glued laminated timber ANSI/AITC A 190.1. New York, NY: American National Standards Institute, Inc.; (see current edition).

American Plywood Association. Construction and industrial plywood. U.S. Product Standard PS 1. Washington, DC: U.S. Department of Commerce, National Bureau of Standards; (see current edition).

American Society for Testing and Materials. Standard test method for strength properties of adhesive bonds in shear by compression loading. ASTM D 905. Philadelphia, PA: ASTM; (see current edition).

American Society for Testing and Materials. Standard test method for strength properties of adhesives in plywood type construction in shear by tension loading. ASTM D 906. Philadelphia, PA: ASTM; (see current edition).

American Society for Testing and Materials. Standard specification for adhesives for field-gluing plywood to lumber framing for floor systems. ASTM D 3498. Philadelphia, PA: ASTM; (see current edition).

American Society for Testing and Materials. Standard practice for accelerated aging of adhesive joints by the oxygen-pressure method. ASTM D 3632. Philadelphia, PA: ASTM; (see current edition).

American Society for Testing and Materials. Standard specification for adhesives for wood-based materials for construction of mobile homes. ASTM D 3930. Philadelphia, PA: ASTM; (see current edition).

American Society for Testing and Materials. Standard practice for measuring strength and shear modulus of nonrigid adhesives by the thick adherend tensile lap specimen. ASTM D 3983. Philadelphia, PA: ASTM; (see current edition).

American Society for Testing and Materials. Standard practice for measuring shear properties of structural adhesives by the modified rail test. ASTM D 4027. Philadelphia, PA: ASTM; (see current edition).

American Society for Testing and Materials. Standard specification for adhesives for structural laminated wood products for use under exterior (wet use) exposure conditions. ASTM D 2559. Philadelphia, PA: ASTM; (see current edition).

American Society for Testing and Materials. Standard specification for adhesives used in nonstructural glued lumber products. ASTM D 3110. Philadelphia, PA: ASTM; (see current edition).

Blomquist, R. F. Adhesives—an overview. In: Blomquist, R. F., ed. Adhesive bonding of wood and other structural materials. Clark C. Heritage Memorial Series on Wood, Vol. 3. Educational Modules for Materials Science and Engineering (EMMSE). University Park, PA: Pennsylvania State University; 1984.

Caster, R. Testing and evaluation of adhesives in bonded products. In: Blomquist, R. F., ed. Adhesive bonding of wood and other structural materials. Clark C. Heritage Memorial Series on Wood, Vol 3. Educational Modules for Materials Science and Engineering (EMMSE). University Park, PA: Pennsylvania State University; 1984.

Caster, R. W. Correlation between exterior exposure and automatic boil test results. In: Proceedings, Wood adhesives—research, application, needs. Madison, WI: Symposium sponsored by Forest Products Laboratory and Washington State University; 1980.

Collett, B. M. A review of surface and interfacial adhesion in wood science and related fields. Wood Science and Technology. 6(1): 1−42; 1972.

Freas, A. D.; Selbo, M. L. Fabrication and design of glued laminated wood structural members. U.S. Dep. Agric. Tech. Bull. 1069. 1954. Available from American Institute of Timber Construction, Englewood, CO.

Gent, A. N.; Hamed, G. R. Fundamentals of adhesion. In: Blomquist, R. F., ed. Clark C. Heritage Memorial Series on Wood, Vol 3. Educational Modules for Materials Science and Engineering (EMMSE). University Park, PA: Pennsylvania State University; 1984.

Gillespie, R. H. Accelerated aging of adhesives in plywood-type joints. Madison, WI: Forest Products Journal. 15(9): 369−378; 1965.

Gillespie, R. H. Wood composites. In: Oliver, J. F., ed. Adhesion in cellulose and wood-based composites. New York: Plenum Press; 1981.

Gillespie, R. H.; Countryman, D.; Blomquist, R. F. Adhesives in building construction. Agric. Handb. 516. Washington, DC: U.S. Department of Agriculture; 1978.

Gillespie, R. H.; Page, W. D. Quality control of bonding in building construction. Report for Department of Housing and Urban Development, Office of Policy Development and Research, HUD/Res.−1215. Madison, WI: U.S. Department of Agriculture, Forest Service, Forest Products Laboratory; 1977. (Available from National Technical Information Service, Springfield, VA.)

Hardwood Plywood Manufacturers Association. American national standard for hardwood and decorative plywood. ANSI/ HPMA HP. New York: American National Standards Institute, Inc.; (see current edition).

Hoyle, R. J. Designing wood structures bonded with elastomeric adhesives. Madison, WI: Forest Products Journal. 26(3): 28−34; 1976.

Jarvi, R. A. Exterior glues for plywood. Madison, WI: Forest Products Journal. 17(1): 37−42; 1967.

Jokerst, R. W. Finger-jointed wood products. Res. Pap. FPL 382. Madison, WI: U.S. Department of Agriculture, Forest Service, Forest Products Laboratory; 1981.

Krueger, G. P. Design methodology for adhesives based on safety and durability. In: Blomquist, R. F., ed. Adhesive bonding of wood and other structural materials. Clark C. Heritage Memorial Series on Wood, Vol. 3. Educational Modules for Materials Science and Engineering (EMMSE). University Park, PA: Pennsylvania State University; 1984.

Krueger, G. P.; Sandberg, L. B. Durability of structural adhesives for use in the manufacture of mobile homes. Report for Department of Housing and Urban Development, Office of Policy Development and Research. NTIS PB81−107534. Houghton, MI: Michigan Tech. University, Institute of Wood Research; 1979. (Available from National Technical Information Service, Springfield, VA.)

Lambuth, A. L. Bonding tropical hardwoods with phenolic adhesives. In: Proceedings, IUFRO meeting on processing of tropical hardwoods; Laboratorio Nacional de Productos Forestales, Merida, Venezuela; Oct. 1977.

Marra, A. Application and needs for wood bonding. In: Blomquist, R. F. ed. Adhesive bonding of wood and other structural materials. Clark C. Heritage Memorial Series on Wood, Vol. 3. Educational Modules for Materials Science and Engineering (EMMSE). University Park, PA: Pennsylvania State University; 1984.

McGee, W. D.; Hoyle, R. J. Design method for elastomeric adhesive bonded wood joist-deck systems. Wood and Fiber. 6(2): 144–155; 1974.

Midwest Plan Service. Designs for glued trusses. Ames, IA: Iowa State University; 1981: 82.

Millett, M. A.; Gillespie, R. H. Precision of the rate-process method for predicting bondline durability. Report for Department of Housing and Urban Development, Office of Policy Development and Research. Madison, WI: U.S. Department of Agriculture, Forest Service, Forest Products Laboratory; 1978. (Available from National Technical Information Service, Springfield, VA.)

Millett, M. A.; Gillespie, R. H.; River, B. H. Evaluating wood adhesives and adhesive bonds: performance requirements, bonding variables, bond evaluation, procedural requirements. Report for Department of Housing and Urban Development, Office of Policy Development and Research. Madison, WI: U.S. Department of Agriculture, Forest Service, Forest Products Laboratory; 1977. (Available from National Technical Information Service, Springfield, VA.)

National Particleboard Association. Mat-formed wood particleboard. ANSI A208.1–79. New York: American National Standards Institute; 1979.

Rice, J. T. The bonding process. In: R. F. Blomquist, ed. Adhesive bonding of wood and other structural materials. Clark C. Heritage Memorial Series on Wood, Vol. 3. Educational Modules for Materials Science and Engineering (EMMSE). University Park, PA: Pennsylvania State University; 1984.

River, B. H. Mastic construction adhesives in fire exposure. Res. Pap. FPL 198. Madison, WI: U.S. Department of Agriculture, Forest Service, Forest Products Laboratory; 1973.

River, B. H. Strength and shear modulus of several construction adhesives as influenced by environment and loading conditions. Report for Department of Housing and Urban Development, Office of Policy Development and Research. Madison, WI: U.S. Department of Agriculture, Forest Service, Forest Products Laboratory; 1978. (Available from National Technical Information Service, Springfield, VA.)

Schneberger, G. L. Metals, plastics and inorganic bonding: practices and trends. In: Blomquist, R. F., ed. Adhesive bonding of wood and other structural materials. Clark C. Heritage Memorial Series on Wood, Vol. 3. Educational Modules for Materials Science and Engineering. (EMMSE). University Park, PA: Pennsylvania State University; 1984.

Selbo, M. L. Adhesive bonding of wood. Tech. Bull. 1512. Washington, DC: U.S. Department of Agriculture, Forest Service; 1975.

Skeist, I.; Miron, J. Handbook of adhesives, 2d ed. Skeist, I., ed. New York: Van Nostrand Reinhold; 1977.

Snogren, R. C. Handbook of surface preparation. New York: Palmerton Publishing; 1974.

Subramanian, R. V. The adhesive system. In: Blomquist, R. F., ed. Adhesive bonding of wood and other structural materials. Clark C. Heritage Memorial Series on Wood, Vol. 3. Educational Modules for Materials Science and Engineering (EMMSE). University Park, PA: Pennsylvania State University; 1984.

Vick, C. B. Elastomeric adhesives for field-gluing plywood floors. Madison, WI: Forest Products Journal. 21(8): 34–42; 1971.

Vick, C. B. Structural bonding of CCA-treated wood for foundation systems. Madison, WI: Forest Products Journal. 30(9): 25–32; 1980.

Wellons, J. D. Adhesion to wood substrates. In: Goldstein, I. S., ed. Wood technology: chemical aspects. ACS Symposium Series No. 43. Washington, DC: American Chemical Society; 1977.

Wellons, J. D. The adherends and their preparation for bonding. In: Blomquist, R. F., ed. Adhesive bonding of wood and other structural materials. Clark C. Heritage Memorial Series on Wood, Vol. 3. Educational Modules for Materials Science and Engineering (EMMSE). University Park, PA: Pennsylvania State University; 1984.

Chapter 10

Glued Structural Members

Glued Structural Members*

Glued-Laminated Timbers (Glulam)

Glued-laminated timbers (fig. 10−1) or glulam refer to two or more layers of wood glued together with the grain of all layers or laminations approximately parallel. The laminations may vary as to species, number, size, shape, and thickness. The individual laminations cannot exceed 2 inches in thickness and are typically made from nominal 1-inch or 2-inch sawn lumber. Glulam timber is an engineered, stress-rated product of a timber laminating plant, comprising assemblies of suitably selected and prepared wood laminations securely bonded together with approved adhesives.

Any species can be used for laminating provided its mechanical and physical properties are suited for the purpose. Standards already exist for many softwood and hardwood species. The most commonly used species groups for glulam members in the United States are Douglas fir-larch, southern pine, and hem-fir.

Uses and Advantages

Laminated wood was first used in the United States for furniture parts, cores of veneer panels, and sporting goods, but is now widely used for structural timbers.

The first use of glulam timbers was in Europe, where as early as 1893 laminated arches (probably glued with casein glue) were erected for an auditorium in Basel, Switzerland. Improvements in casein glue during World War I aroused further interest in the manufacture of glulam structural members, at first for aircraft and later as framing members of buildings.

In the United States one of the early examples of glulam arches designed according to engineering principles is in a building erected in 1934 at the Forest Products Laboratory. This installation was followed by many others in gymnasiums, churches, halls, factories, hangars, and barns. The development of durable synthetic-resin adhesives during World War II permitted the use of glulam members in bridges and marine construction where a high degree of resistance to severe service conditions is required. With growing public acceptance of glulam construction, its use has increased steadily until it now forms an important segment of the construction industry. Today glulam is used as load-carrying structural framing for roofs and other structural portions of buildings, and for other construction, such as bridges, towers, and marine installations.

Glulam timbers may be straight or curved. Curved arches have been used to span more than 300 feet in structures. Domes, like the Tacoma, WA, dome shown in figure 10−2,

have been built spanning more than 500 feet. Straight members spanning up to 100 feet are not uncommon, and some span up to 140 feet. Sections deeper than 7 feet have been used.

Straight beams can be designed and manufactured as horizontally laminated timbers (designed to resist bending loads applied perpendicular to the wide faces of the laminations), or vertically laminated timbers (designed to resist bending loads applied parallel to the wide faces of the laminations). Horizontally laminated timbers are the most widely used. Curved members are horizontally laminated to permit bending of laminations during gluing.

The advantages of glulam timber construction are many and significant. They include the following:

1. Ease of manufacturing large structural elements from standard commercial sizes of lumber, which is increasingly adaptable to future timber economy. Smaller trees result in more lumber in smaller sizes.

2. Achievement of excellent architectural effects and the possibility of individualistic decorative styling in interiors, as nearly unlimited curved shapes are possible.

3. Minimization of checking or other seasoning defects associated with large one-piece wood members because the laminations are thin enough to be readily seasoned before manufacture of members.

4. The opportunity to design on the basis of the strength of seasoned wood for dry service conditions, again because the individual laminations can be dried to provide members thoroughly seasoned throughout.

5. The opportunity to design structural elements that vary in cross section along their length in accordance with strength requirements.

6. The use of lower grade material for less highly stressed laminations without adversely affecting the structural integrity of the member.

Certain factors involved in the production of glulam timbers are not encountered in producing solid timbers:

1. Preparation of lumber for gluing and the gluing operation usually raises the cost of the final laminated product above that of solid sawn timbers in sizes that are reasonably available.

2. Because the strength of a laminated product depends upon the quality of the glue joints, the laminating process requires special equipment, plant facilities, and manufacturing skills not needed to produce solid sawn timbers.

3. Because several extra manufacturing operations are involved in manufacturing laminated members as compared with solid members, greater care must be exercised in each operation to ensure a product of high quality.

4. Large curved members are awkward to handle and ship by the usual carriers.

* Revision by Catherine M. Marx, General Engineer; Theodore L. Laufenberg, General Engineer; and Terry D. Gerhardt, General Engineer.

Figure 10 – 1—Glued-laminated timber member.

Figure 10 – 2—At 530 feet in diameter and 157 feet tall, this dome at Tacoma, WA, is both the largest wood-dome structure in the world and the longest clear-span wood roof structure in the world. (Courtesy of Western Wood Structures, Inc.)   (MC84 9025)

Standards and Specifications

Manufacture

ANSI A 190.1 of the American National Standards Institute (see Selected References) contains requirements for the production, testing, and certification of structural glulam timber in the United States. Further details and commentary on the requirements specified in ANSI A 190.1 are provided in AITC 200 (Inspection Manual of the American Institute of Timber Construction) which is part of ANSI A 190.1 by reference. A standard for glulam poles, ANSI 0 5.2, has been prepared which addresses special requirements for utility uses.

Requirements for the manufacture of structural glulam timbers in Canada are given in CSA Standard 0 122 of the Canadian Standards Association.

Deriving Design Values

ASTM D 3737 of the American Society for Testing and Materials covers the procedures for establishing design values for structural glulam timber in the United States. Properties considered include bending, tension and compression parallel to grain, modulus of elasticity, horizontal shear, radial tension, and compression perpendicular to grain.

Design Values and Procedures

The main design value reference for softwoods is AITC 117, "Standard Specifications for Structural Glued Laminated Timber of Softwood Species;" this specification contains data relating to design values and modification of stresses for the design of glulam timber members in the United States. The equivalent specification for hardwoods is AITC 119, "Standard Specifications for Hardwood Glued-Laminated Timber." The National Design Specification (NDS) summarizes the design information in AITC 117 and AITC 119 and defines the practice to be followed in structural design of glulam timbers. For additional design information see the Timber Construction Manual developed by the American Institute of Timber Construction.

In Canada, CAN3−0 86, code for engineering design in wood, provides design criteria for structural glulam timbers.

Manufacture

Some discussion in this section will be similar to parts of chapter 9, but here it will apply specifically to the manufacture of structural glulam timbers.

Glulam manufacture must conform to recognized standards (ANSI A 190.1) and procedures to justify design values developed as described later in this chapter. Quality control plays a very important role in the production of reliable structural glulam members. There are two main aspects of good quality control: initial qualification testing and on-going, daily testing.

Qualification tests are required before a laminator can be certified to be in compliance with ANSI A 190.1. All daily quality control tests and production line checks used by the plant, as well as tests related specifically to qualification of the manufacturer to produce glulam, are required for qualification.

Daily testing of samples or checking of manufacturing procedures is required during the production process and on specimens selected from production. Daily tests for glueline strength and durability are performed on samples taken from production to evaluate the adequacy of face, edge (where applicable), and end joint assembling and bonding procedures. The finished product is visually inspected to assure conformance to the requirements of ANSI A 190.1 and the job specifications. ANSI A 190.1 also requires that laminating plants be visited periodically by a qualified inspection agency to check their compliance with that standard.

The following is a brief overview of the manufacturing process for structural glulam timbers. Additional details are given in ANSI A 190.1, AITC 200, or U.S. Department of Agriculture Technical Bulletin 1069.

Lumber Drying and Grading

Lumber must be properly dried and graded for use in glulam timbers. Limits are placed on moisture content because it is important to avoid the development of appreciable internal stresses. Shrinking and swelling associated with moisture gradients within a timber are the fundamental causes of internal stresses that can result in checking.

The possibility of checking can be minimized by using lumber with (1) a relatively narrow range of moisture content and (2) an average moisture content near that which the timber will attain in service. A range in moisture content no greater than 5 percent among the laminations in the same assembly may be permitted without significant effect on serviceability.

A slight increase in average moisture content for the finished timber generally causes no harm but rapid drying could result in checking. Thus, the average moisture content should ideally be the same or slightly lower than that which the timber will attain in service, but that is difficult to achieve.

Structural laminating grades describe the characteristics used to segregate lumber for structural glulam timbers. Each grade has restrictions on the percentage of the lumber cross-section that strength-reducing characteristics may occupy. Allowable properties are not assigned separately to the laminating grades.

There are two kinds of graded sawn lumber: visually graded and E-rated material. Visually graded softwood material is graded according to one of three sets of rules. The first set is the rules for the American Softwood Lumber Standard with additional requirements for laminating. The second set is the laminating grades typically used for visually graded western species and includes three basic categories called L1, L2, and L3. The third set includes special requirements for tension members and outer tension laminations on bending members as given in AITC 117. Laminating plants pay special attention to grading tension laminations because they are the most critical in terms of controlling beam strength.

The visual grades for softwoods have provisions for dense, close-grain, medium-grain, or coarse-grain lumber as well as for strength-reducing characteristics. For hardwoods, knot size, slope of grain, and other requirements are given in AITC 119.

The E-rated grades are categorized by a combination of visual grading and lumber stiffness criteria. These grades are expressed in terms of the maximum allowable edge characteristic, along with the long-span modulus of elasticity (E) required. For example, the E-rating grade 1/6-2.2E allows up to one-sixth of the cross section to be occupied by edge characteristics (which include knots, knot holes, burls, distorted grain, or decay partially or wholly at edges of wide faces), while providing an E of 2.2×10^6 psi.

Other wood products such as manufactured lumber and laminated veneer lumber may be used as substitutes for sawn lumber in glulam members provided they meet the requirements in AITC 200.

End Jointing

For most glulam timbers, pieces of lumber must be joined end-to-end to provide laminations of sufficient length. The most commonly used end joint for structural glulam timbers is a finger joint approximately 1.1 inches long (fig. 10−3). Other configurations and types of joints, such as the scarf joint, are also used. Finger joints have the advantage over scarf joints of allowing a continuous production process and they occupy only a short length, thereby reducing the amount of lumber required to manufacture the joint. Melamine-urea or phenol-resorcinol are the two most common adhesives used for end jointing. The adhesive may be cured as the end joint travels through a continuous radio frequency (RF) tunnel.

Well-made end joints are necessary to ensure adequate beam performance, and can be achieved only through a rigor-

Figure 10—3—A typical finger type end joint used in the manufacture of glulam timber. (M84 0326-12)

Figure 10—4—Clamped assembly of laminated members. (M138 312)

ous quality control program. Many laminators have installed proof-loaders that have been shown to be effective in improving quality control. The proof-loading device applies a known bending or tension load to check the strength of end joints at a stress level high enough to detect and reject low-strength end joints for the most critical laminations, yet not so high as to cause damage in laminations accepted.

Face Gluing and Finishing

The best procedure to follow is to plane the end-jointed laminations just prior to face gluing to provide a fresh surface for bonding. Actual procedures vary depending upon plant practices and species characteristics. Some laminating plants have proven through testing that they can achieve adequate adhesive bonds by planing the lumber of some species prior to end jointing rather than after end jointing.

A glue extruder is normally used to apply an adequate amount of adhesive over the face of the laminations. Phenol-resorcinol is the most commonly used adhesive for face gluing, but other adhesives that have been adequately tested and proven to perform satisfactorily may be used.

Pressure must be applied while the adhesive is still tacky to achieve adequate bonding. The most common means of applying gluing pressure to laminated assemblies are clamping beds with the pressure supplied with screw or hydraulic presses (fig. 10—4). Common curing times under room temperature conditions may range from 6 to 24 hours.

After the glulam members have been adequately cured, the sides are surfaced and the members are sent to the finishing room. Here the laminator cuts, drills, etc. to meet job specifications, as well as finishes the members to meet the appearance grade specified. If required, members are then marked to comply with ANSI A 190.1 and prepared for shipping.

Protection During Transit, Storage, and Erection

End sealers, surface sealers, primer coats, and wrappings may be applied at the laminating plant to help protect structural glulam members. The protection specified should depend upon end use and final finish of the member.

Special precautions must also be taken during handling, storage, and erection to protect structural members. Padded or nonmarring slings should be used during handling and erection to prevent damage to finished surfaces. Cable slings or chokers should not be used to handle laminated materials unless adequate blocking is provided between the cable and the wood member to prevent damage. Glulam members should be stored at the job site with care to provide protection against moisture. See AITC 111 for additional details on protection during transit, storage, and erection.

Development of Design Values

Procedures for establishing design values for structural glulam timbers in the United States are given in ASTM D 3737. In general that procedure involves the determination of clear wood design stresses (CWDS) and then application of appropriate strength ratios (SR) to account for such strength-reducing characteristics as knots and slope of grain.

Clear Wood Design Stresses

Specific requirements are given in ASTM D 3737 for the

establishment of CWDS values for the properties of bending, tension and compression parallel to grain, modulus of elasticity, horizontal shear, compression perpendicular to grain, and radial tension for both visually graded and E-rated material.

Two different procedures have been used to determine CWDS values for bending: (1) Lower 5 percent exclusion limit values are calculated based on methods in ASTM D 2555 and information on strength properties for small, clear, straight-grained specimens of green lumber; then appropriate adjustment factors for duration of load, factor of safety, moisture content, size, and loading conditions are applied to obtain CWDS values. (2) Data on full-size beam tests have been used to justify CWDS values in bending for Douglas fir-larch, southern pine, and hem-fir beams.

CWDS values for some of the other properties such as compression parallel to grain and horizontal shear have been developed in a similar manner to procedure (1) for bending. See ASTM D 3737 for further details on CWDS values for those and other properties.

Strength Ratios

Bending and Axially Stressed Members—Knots and slope of grain are the two principal characteristics that influence bending and axially stressed members and therefore must be accounted for. These two factors are not cumulative; the lower of the two determines the strength ratio.

Knots—Test data have shown that several important properties depend upon the size and frequency of knots. However, the relationship between knots and strength varies by property.

The effect of knots on the bending strength of horizontally laminated timbers depends on the number, size, and position of the knots with respect to the neutral axis of the member. Traditionally this has been described using the I_K/I_G ratio, where I_K is the sum of the moments of inertia of cross-sectional areas occupied by knots within 6 inches of either side of a critical section, and I_G is the moment of inertia of the gross cross section of the member. The empirical relationship between bending strength and the I_K/I_G ratio is shown in Figure 10—5. Procedures for calculating I_K/I_G ratios are given in ASTM D 3737 and U.S. Department of Agriculture Technical Bulletin 1069.

Various research studies in the last 15 years have shown that additional grading restrictions must be used for laminations in the outer tension zone of horizontally laminated beams to justify stresses predicted by the I_K/I_G concept. This finding led to development of the special tension lamination grades given in AITC 117. Procedures for development of those grades are given in ASTM D 3737. If beams are fabricated without those special tension laminations, the allowable design bending strength is obtained by multiplying the design value predicted by the I_K/I_G theory by 0.85 if the depth of the beam

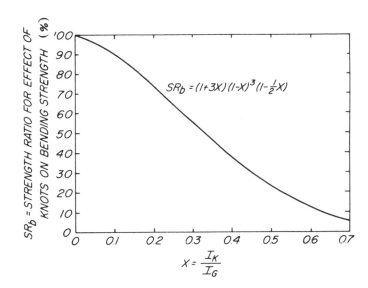

Figure 10—5—Relationship of bending strength ratio to moment of inertia of areas occupied by knots. (ML84 5801)

is 15 inches or less, or by 0.75 if the depth exceeds 15 inches.

The effect of knots on vertically laminated beams can be determined by calculating a strength ratio by the empirical relationship given in ASTM D 3737.

Tests have shown that axial compressive strength of short glulam compression members is related to the percentage of the cross section occupied by the largest knot in the lamination. Figure 10—6 shows an empirically derived relationship between SR_c, the compressive strength ratio, and Y_1, the near-maximum size knot (99.5 percentile) expressed as a decimal fraction of the dressed width of lumber used for the lamination. An extra adjustment is necessary to compensate for additional bending stresses in compression members with grades of lumber placed unsymmetrically.

Available test results have been used to derive the linear relationship between SR_t, the strength ratio in tension, and Y_2, the maximum edge knot size permitted in the grade and expressed as a decimal fraction of the dressed width of the wide face of the piece of lumber used for the lamination (fig. 10—7).

Slope of Grain—As discussed earlier, a strength ratio is equal to the lower of the two strength ratios determined by considering the effect of knots and the effect of slope of grain. The effect of slope of grain on tensile and compressive stresses is given in table 10—1. These values apply also to the tension and compression zones in horizontally laminated members, while the tension values are also used for vertically laminated members.

For horizontally laminated timbers, it is possible to vary the slope-of-grain limitations at different points in the depth

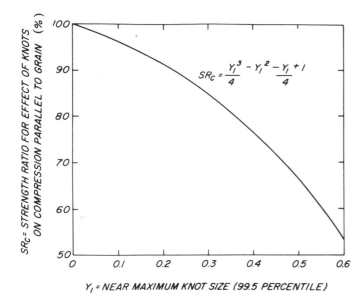

$$SR_c = \frac{Y_i^3}{4} - Y_i^2 - \frac{Y_i}{4} + 1$$

Y_i = NEAR MAXIMUM KNOT SIZE (99.5 PERCENTILE)

Figure 10–6—Relationship of compressive strength (ML84 5795) ratio to size of knots in laminations of laminated short columns.

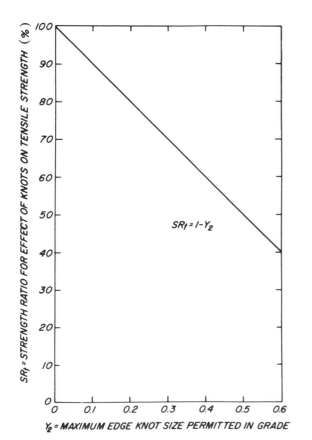

$$SR_t = 1 - Y_2$$

Y_2 = MAXIMUM EDGE KNOT SIZE PERMITTED IN GRADE

Figure 10–7—Relationship of tensile strength ratio (ML84 5796) to permitted knot size (expressed as fraction of dressed width) in laminations of tension members.

of the beam in accordance with the stress requirements. That is, steeper slope of grain may be permitted in laminations in the interior of the timber than in the laminations at and near the outside. The permitted variation should be based on the assumption of linear variation of strain across the depth and the modulus of elasticity of each lamination.

Modulus of Elasticity

The modulus of elasticity (E) of a glulam member directly depends on the long-span E's of the laminations used in its manufacture. By definition, the long-span E's of the laminations are determined using an approximate 100:1 span-to-depth ratio under center-point loading with properly applied adjustments, or by using values already determined by surveys of laminating grades. For axially stressed glulam members, the E of the member is equal to the weighted average of the E's of the laminations. For vertically laminated bending members, the E of the member is equal to the weighted average of the lamination E times an adjustment factor of 0.95. For horizontally laminated bending members, the E of the member may be calculated from formula (10–1).

$$E_h = 0.95 \frac{\sum_{i=1}^{N} E_i I_i}{I} \qquad (10\text{-}1)$$

where E_h = modulus of elasticity of the horizontally laminated member; N = number of laminations in the member; $E_i I_i$ = the product of the modulus of elasticity and the moment of inertia about the centroid of the member for each lamination; and I = moment of inertia of the gross cross section.

Table 10–1—*Strength ratios parallel to grain for designing glulam combinations*

Slope of grain	Strength ratio	
	Tension	Compression
1:4	0.27	0.46
1:6	.40	.56
1:8	.53	.66
1:10	.61	.74
1:12	.69	.82
1:14	.74	.87
1:16	.80	1.00
1:18	.85	1.00
1:20	1.00	1.00

Other Considerations

Effect of End Joints on Strength—Both finger joints and scarf joints can be manufactured with adequate strength for use in structural glulam timbers. The adequacy is determined by physical testing procedures and requirements in ANSI A 190.1.

Joints should be well scattered in portions of structural glulam timbers highly stressed in tension. Required spacings of end joints are given in ANSI A 190.1.

End joints of two qualities may be used in a glulam member, depending upon strength requirements at various depths of the cross section. Usually, however, laminators use the same joint throughout the members for ease in manufacture.

The highest strength values are obtained with well-made plain scarf joints; the lowest values are obtained with butt joints. This is because scarf joints with flat slopes have essentially side-grain surfaces that can be well bonded to develop high strength, while butt joints have end-grain surfaces that cannot be bonded effectively. Structural finger joints (either vertical or horizontal) are a compromise between scarf and butt joints, and their strength varies with joint design (see ch. 9).

The joint efficiency or joint factor, which is the joint strength as a percentage of the strength of clear straight-grain material, is useful in evaluating end joint designs. The joint factor for plain scarf joints in the tension portion of bending members or in tension members has been determined to be the following for accurately fitted and well-glued joints:

Scarf slope	Joint factor
	(Percent)
1 in 12 or flatter	90
1 in 10	85
1 in 8	80
1 in 5	65

For production scarf joints, the joint factor can vary from the values given above.

No statement can be made regarding the specific joint factor of finger joints because finger joint strength depends on the type and configuration of joint and the manufacturing process. However, the joint factor of commonly used finger joints in high-quality lumber used for laminating may be about 75 percent. High-strength finger joints can be made when the design is such that the fingers have relatively flat slopes and sharp tips. Tips are essentially a series of butt joints, which reduce the effectiveness of finger joints as well as being sources of stress concentration.

Butt joints generally can transmit no tensile stress and can transmit compressive stress only after considerable deformation or if a metal bearing plate is tightly fitted between the abutting ends. In normal assembly operations such fitting would not be done, and it is therefore necessary to assume that butt joints are ineffective in transmitting both tensile and compressive stresses. Because of this ineffectiveness, and because butt joints cause concentration of both shear stress and longitudinal stress, butt joints are not recommended for use in structural glued-laminated timbers.

Other types of end joints, such as a folded scarf joint, may be used in laminated members if they can be shown to meet strength and durability requirements.

Effect of Edge Joints on Strength—It is sometimes necessary to place laminations edge-to-edge to provide glulam members of sufficient width. Because of difficulties in fabrication, structural edge joint bonding may not be readily available and the designer should investigate the availability prior to specifying.

For tension, compression, and horizontally laminated bending members, the strength of edge joints is of little importance to the overall strength of the member. Therefore, from the standpoint of strength alone, it is unnecessary that edge joints be glued if they are not in the same location in adjacent laminations. However, for maximum strength, edge joints should be glued where torsional loading is involved. Other considerations, such as the appearance of face laminations or the possibility that water will enter the unglued joints and promote decay, also may dictate that edge joints should be glued.

For vertically laminated beams of depths in excess of a single lamination width, sufficient laminations must be edge glued to provide adequate shear resistance in the beam. Not only is adequate initial strength required of such joints, but they must be durable enough to retain that strength under the conditions to which the beam is exposed in service. If edge joints in vertically laminated beams are not glued, the shear strength can be determined by engineering analysis. Using standard laminating procedures with staggered edge joints, the shear strength of vertically laminated beams with unglued edge joints is approximately one-half of those with glued edge joints.

Effect of Shake, Checks and Splits on Shear Strength—In general, checks and splits have little effect on the shear strength of laminated timbers. Shake generally occurs infrequently but should be excluded from material for laminations. Most laminated timbers are made from laminations that are thin enough to season readily without developing significant checks and splits.

Future Trends

In the future, computer simulation models based primarily upon lumber properties will gradually replace some of the procedures previously discussed and will be used to predict the strength and stiffness distributions of glulam members.

Design Considerations

In general, the engineering formulas given in chapter 8 apply also to glued-laminated beams and columns. Moisture content adjustments for glulam differ somewhat from solid sawn design, and special considerations are necessary for both curved and deep members. For sharply curved members, ordinary engineering formulas may introduce appreciable error in the analysis, and special stress analysis methods available in mechanics textbooks should be used.

Some of the items requiring consideration when designing glulam structural members include:

Moisture Content

Design values tabulated in AITC 117 were developed based on an average moisture content of 12 percent and a maximum moisture content of less than 16 percent. Thus, those dry conditions-of-use design values are applicable when the moisture content in service is less than 16 percent, as in most covered structures. Wet conditions-of-use design values are applicable when the moisture content in service is 16 percent or more, as may occur in members not covered or in covered locations of high relative humidity. The factors in table 10−2 are commonly applied to dry-use values to obtain wet-use values when tabular information is not available.

Size Factor and Loading Conditions

Published glulam design stress values apply to 12-inch-deep members that are simply supported with a span-to-depth ratio (L/d) of 21:1 and a uniformly distributed load.

The bending strength of wood beams has been shown to decrease as the size of beams increases. Laminated members can be of considerable size, and the effect of size on strength should be considered in design. Therefore, when the depth of a rectangular beam exceeds 12 inches, the tabulated design stress values must be reduced by multiplying by the size factor, C_F, as determined by the following formula:

$$C_F = \left(\frac{12}{d}\right)^{1/9} \qquad (10-2)$$

where d is the depth of the member in inches.

For loading conditions other than L/d of 21:1 and a uniformly distributed load, greater accuracy may be obtained by applying the percentage changes given in table 10−3 directly to the size factor, C_F. Straight line interpolation may be used between the L/d ratios given in the table.

Curvature Factor

Stress is induced when laminations are bent to curved forms, such as arches and curved rafters. While much of this stress is quickly relieved, some remains and tends to reduce the strength of a curved member. For the curved portion of members, the design stress values must be reduced by multiplying by the following curvature factor, C_c:

$$C_c = 1 - 2{,}000\left(\frac{t}{R}\right)^2 \qquad (10-3)$$

where t is the thickness of a lamination in inches and R is the radius of curvature of the laminations in inches. The ratio t/R should not exceed 1/100 for hardwoods and southern pine or 1/125 for other softwoods.

No curvature factor is applied to the design stress values in the straight portion of an assembly, regardless of curvature elsewhere.

Radial Tension or Compression

When curved members are subjected to bending moments, stresses are set up in a direction parallel to the radius of curvature (perpendicular to grain). If the moment increases the radius (makes the member straighter), the stress is tension; if it decreases the radius (makes the member more sharply

Table 10−2—Factors for modification of design stresses for wet conditions

Type of stress	Wet-use factor
Bending and tension parallel to grain	0.80
Compression parallel to grain	.73
Compression perpendicular to grain	.53
Shear	.88
Modulus of elasticity	.83

Table 10−3—Factors for modification of the size factor to account for span-depth ratio and loading conditions

Span-to-depth ratio, L/d	Change	Loading conditions for simply supported beams	Change
	Pct		Pct
7	+6.2	Single concentrated load	+7.8
14	+2.3		
21	0	Uniform load	0
28	−1.6	Third-point load	−3.2
35	−2.8		

curved), the stress is compression. For members of constant depth, the stress is a maximum at the neutral axis and is approximately

$$f_R = \frac{3\,M}{2\,Rbh} \qquad (10-4)$$

where M is the bending moment in inch-pounds, R is the radius of curvature at the centerline of the member in inches, and b and h are, respectively, the width and height of the cross section in inches.

Values of f_R should be limited to those given in AITC 117, AITC 119, and the National Design Specification. If the values for radial tension are exceeded, mechanical reinforcement sufficient to resist all radial tension stresses is required, but the calculated radial tension stress still cannot exceed one-third of the design stress in horizontal shear.

For procedures to calculate radial stresses in curved beams with varying cross section, refer to the National Design Specification or contact the American Institute of Timber Construction.

Lateral Stability

Design stresses are applicable to members that are adequately braced. When deep, slender members are not adequately braced, a reduction must be applied based on computation of the slenderness factor of the member. The slenderness factor should be applied in design as shown in the AITC Timber Construction Manual. Lateral stability is discussed also in chapter 8.

Deflection

Deflection will often govern in members with long spans or in members subjected to light live loads. In general the deflection characteristics of glued-laminated timbers can be computed with formulas given in chapter 8.

The design modulus of elasticity values given in AITC 117 are effective values that include the effects of the grade and placement of the laminations used in the glulam member. Deflections calculated using those values will be only immediate values. Where deflection is critical, consideration must be given to the added deflection that occurs under long-time loading and that due to shear (see chs. 6 and 8). In addition, use of a reduced modulus of elasticity value may be deemed appropriate by the designer in certain applications where deflection may be critical. In addition to a sloped roof, camber is recommended for horizontally laminated beams to minimize ponding conditions. Camber will also help avoid the impression of excessive deflection. The recommended magnitude of camber is usually equal to 1.5 times the dead load deflection.

Duration of Load

Design stresses are based on normal load duration which implies stressing a member with the full design load for a duration of approximately 10 years, either continuously or cumulatively. Modification of glulam design stresses for other than normal load duration is the same as for lumber as discussed in chapter 6.

Treatment

If the glulam timber is to be used under conditions that raise its moisture content to more than 20 percent, the wood should be treated with approved preservative chemicals and only wet-use adhesives should be used. Experience has shown that some oil-borne preservatives, besides providing protection from fungi and insects, also retard moisture changes at the surface of the wood, and thus inhibit checking. If the size and shape of the timbers permit, laminated timbers can be treated with some preservatives after gluing, but penetration perpendicular to the planes of the glue joints will be distinctly retarded at the first glueline. Treatment of large glulam timbers with waterborne preservatives after gluing is not recommended because of difficulties in subsequent drying and associated checking.

Glulam timbers treated with oil-borne preservatives after gluing and fabrication have given excellent service in bridges and similar installations. Laminations also can be treated and then glued if suitable precautions are observed. The treated laminations should be conditioned and must be resurfaced just before gluing. Not all adhesives can be used to glue preservative-treated wood, but, if compatible adhesives and treatments are selected and the gluing is carefully done, laminated timbers can be produced that are entirely serviceable under moist, warm conditions that favor decay.

Although glulam has been considered as heavy timber and proven to be fire resistant, fire-retardant treatments have infrequently been used to meet certain flame spread requirements for some uses. The design values for structural glulam members pressure-impregnated with fire-retardant chemicals depend on the species and treatment combinations involved. The effect on strength must be determined for each treatment. The manufacturer of the treatment should be contacted for specific information on fire-retardant adjustments for all recommended design values. Fire-retardant-treated glulam members are not readily available and the availability should be checked prior to specifying.

Fastenings

Design loads or stresses for fastenings that are applicable to solid wood members (ch. 7) are also applicable to laminated timbers. Because greater depths are practical with laminated timbers than with solid timbers, design of fastenings for deep glulam timbers requires special consideration. It is desirable to have the moisture content of the timber as near as possible to that which it will attain in service. If moisture content is higher than the equilibrium moisture content in service, considerable shrinkage may occur between widely separated bolts; if the bolts are held in position by metal shoes or angles, large splitting forces may be set up. Slotting of the bolt holes in the metal fitting will tend to relieve the splitting stresses. Cross bolts will assist in preventing separation if splitting does occur; they will not, however, prevent splits.

For additional information on good fastening design practice see the AITC Timber Construction Manual.

Laminated Veneer Lumber (LVL)

Material made by parallel lamination of veneers into thicknesses common to solid sawn lumber (3/4 to 2-1/2 in) is called Laminated Veneer Lumber (LVL).

Work in the 1940's on this material concept was targeted for production of high-strength aircraft structures. Later research studies of LVL were aimed at defining the effects of processing variables and included veneer up to 1/2 inch thick. The industry presently uses veneers 1/8 to 1/10 inch in thickness, which are hot-pressed with phenol-formaldehyde adhesive into lengths from 8 to 60 feet or more. Joints between individual veneers are staggered along the LVL to avoid gross strength-reducing defects. Common practice in the United States is for these to be butt joints, or the veneer ends may overlap for some distance to provide load transfer. Some foreign producers are scarf jointing veneers. LVL may be made in 8-foot lengths having no end joints in the veneer. A scarf or finger joint can be used to join these lengths to one another or to sawn lumber.

A number of methods may be used to grade the veneer which make up the LVL. Visual grading, used in most plywood specifications, may be adequate for producing some structural products. The lack of a slope-of-grain requirement for the visual grading of veneer in PS 1−74 limits the effectiveness of this method in assessing strength characteristics. A recently utilized technology in veneer grading is that of ultrasonics. Veneers can be segregated into various stiffness classes as determined by an ultrasonic test and used to produce LVL with reduced variability in mechanical properties. Additional visual restrictions on the veneer quality beyond the requirements of PS 1−74 may also be used to upgrade the LVL properties.

Applications and Uses

Some of the first applications of LVL were inspired by rising costs and shortages of high-grade solid sawn lumber for use in parallel-chord trusses and as scaffold planks. Truss manufacturers have found the LVL concept capable of opening new markets. Strength-reducing defects are virtually

Figure 10−8—I−beams with laminated veneer lumber (LVL) flanges and panel-type webs. Two experimental products have a hardboard web A, and flakeboard web B, along with a commercially available product with a plywood web C. (M84 0468)

nonexistent, making possible the feasibility of light trusses and I−sections (fig. 10−8). The I−sections have a web composed of a structural panel product such as plywood or hardboard glued into a machined groove in the LVL flanges. Current markets for these I−section beams are as joists and rafters in light-frame construction.

Scaffold planks are an important LVL product. Uniformity of properties and an extended service life due mainly to splitting resistance are the primary reasons for the success of LVL in this market. This splitting resistance is attributed to the stress relief afforded by lathe checks in the veneer. Another industry that has embraced LVL is that of manufactured housing. LVL is used in manufactured housing components because of major weight and material savings on the fabricated components. LVL is considered to be cost-effective in secondary manufacturing operations because of reduced occurrence of such typical lumber defects as twist, crook, warp, splits, and strength-reducing knots.

LVL represents a new technology in wood utilization; thus the processes and uses for this material are evolving. Uses such as truck decking with abrasion-resistant species such as keruing (apitong) on wearing surfaces, box beams, kiln stickers, and door rails capitalize on the versatility of the LVL process and the consistency of mechanical and dimensional properties exhibited by this material.

Specifications for LVL

Two basic types of LVL material may be distinguished by the method of manufacture. These are: (1) The LVL produced on a continuous press in lengths greater than 8 feet and (2) that produced in traditional plywood presses at a nominal length of 8 feet (244 cm). Continuous-length LVL has joints in each veneer lamination at 8-foot intervals. Veneer joints may allow some load transfer between veneers as with finger, scarf, or lap joints. Butt joints made in the interior of continuous-length LVL have little effect on member strength due to the number of thin veneers in LVL. However, butt-jointed outer plies have been shown to seriously degrade member strengths.

Panel-length LVL material is typically made in presses suitable for making plywood. The 8-foot long, 3/4- to 2-1/2-inch-thick LVL material may be finger- or scarf-jointed together into longer lengths.

Model building codes have granted approvals to some manufacturers of LVL. Allowable design stresses of code-approved LVL products are listed below:

Design stress	Range
	Lb/in^2
Flexure	2,200−4,200
Tension parallel	1,600−2,800
Compression parallel	2,400−3,200
Compression perpendicular to the grain:	
Perpendicular to glueline	400−600
Parallel to glueline	400−800
Horizontal shear:	
Perpendicular to glueline	200−300
Parallel to glueline	100−200
Modulus of elasticity	$1.8 \times 10^6 - 2.8 \times 10^6$

These properties are for dry conditions of use and are subject to adjustment for duration of load and repetitive member use as provided for in building codes. In addition, the allowable loads for nails and bolts in the Douglas-fir LVL equal those allowed for nondense Douglas-fir lumber. Due to the reconstituted nature of the LVL material, fasteners that may work well in solid wood members should be used with caution in LVL as the veneer lathe checks seriously reduce its fracture toughness. One continuous-length product (up to 80 ft (24 m)) has three grades; 1.8 E, 2.0 E, and 2.2 E, corres-ponding to the modulus of elasticity (in million psi) associated with each grade. Thickness of 3/4 to 2-1/2 inches (20 to 64 mm) and member depths from 2-1/2 to 24 inches (64 mm to 61 cm) are allowed by this approval.

LVL manufactured in panel lengths may be end-jointed to provide longer stock and is approved in two grades: (1) with end joints between panel lengths and (2) without end joints. Allowable flexural stresses of 2,600 psi are allowed for the end-jointed material and 3,000 psi for the 8-foot unjointed material. The end-jointed material is specifically allowed for use in manufacturing glulam timber as a substitute for tension ply grade lumber.

A standard is currently being drafted by the American Plywood Association and LVL manufacturers for the grade-marking of LVL. Methods of determining design allowables, assuring certain levels of environmental durability, monitoring of quality levels, and marking of LVL products may be standardized under this policy.

Design Considerations

LVL is comprised of many layers of veneers; thus its properties are generally more uniform than for solid wood products. This is shown in figure 10−9 as frequency plots of bending strength for Douglas-fir solid sawn visually graded glulam press-lam, an LVL product. Coefficients of variation for the mechanical properties of LVL have been reported to be from 8 to 18 percent in comparison with values of 25 to 35 percent reported for visually graded lumber.

Limited experience with this material indicates that processing is an important factor in determining the durability of LVL products. Other characteristics that may be of importance in design are dimensional stability and treatability. The numerous veneer laminations allow sufficient dispersion of cross grain and compression wood in the veneers; thus LVL has minimum amounts of warp, twist, and crook. Splitting, which is commonly associated with the residual stresses induced during the drying of wood members, is minimized in LVL material. Treatment of LVL with preservatives and fire retardants is enhanced due to the lathe checks produced in the veneers by the peeling operation and by any butt joints that allow access to the veneer ends. These butt joints and lathe checks allow difficult-to-treat species, such as Douglas-fir heartwood, to be treated faster and more thoroughly than solid lumber.

Wood-Plywood Glued Structural Members

Highly efficient structural components can be produced by combining wood with plywood or other panel products through gluing. These components, including box beams, I−beams,

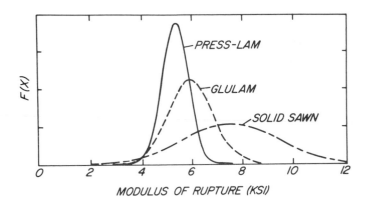

Figure 10−9—Comparison of bending strength distributions for Douglas-fir visually graded solid sawn lumber, glulam, and press-lam, an LVL product.　　(ML84 5799)

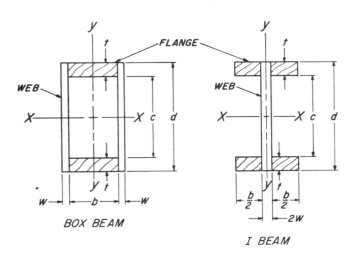

BOX BEAM

I BEAM

Figure 10−10—Beams with plywood webs.　　(M132 343)

"stressed-skin" panels, and folded plate roofs, are discussed in detail in technical publications of the American Plywood Association. Details on structural design will be given in the following portion of this chapter for beams with plywood webs and for stressed-skin panels wherein the parts are glued together with a rigid, durable adhesive. Many of the methods can probably be adopted for use with other panel materials such as flakeboard and hardboard. Recent publications that propose design methods for I−beams with hardboard webs are listed in the Selected References.

These highly efficient designs, while adequate structurally, may suffer from lack of resistance to fire and decay unless treatment or protection is provided. The rather thin portions of the cross section, the panel materials, are more vulnerable to fire damage than heavy, solid cross sections.

Beams With Plywood Webs

Box beams and I−beams with lumber or laminated flanges and glued plywood webs can be designed to provide desired stiffness, bending moment resistance, and shear resistance. The flanges resist bending moment, and the webs provide primary shear resistance. Proper design requires that the webs must not buckle under design loads. If lateral stability is a problem, the box beam design should be chosen because it is stiffer in lateral bending and in torsion than the I−beam. On the other hand, the I−beam should be chosen if buckling of the plywood web is of concern because its single web, double the thickness of that of a box beam, will offer greater buckling resistance.

Design details are presented for beam cross sections shown in figure 10−10. Both flanges are the same thickness in these beams because a construction symmetrical about the neutral plane provides the greatest moment of inertia for the amount of material employed. The following formulas were derived by basic principles of engineering mechanics. These methods can be extended to derive designs for unsymmetrical constructions if necessary.

Beam Deflections

Beam deflections can be computed with formula (8−2) (ch. 8). The bending stiffness $(EI)_x$ and shear stiffness (GA') for the box and I−beam shown in figure 10−10 are given by the following formulas. The bending stiffness is:

$$(EI)_x = \frac{1}{12}[E(d^3 - c^3)b + 2E_wWd^3] \qquad (10-5)$$

where E is flange modulus of elasticity and E_w is web modulus of elasticity. Values of E_w for the appropriate plywood construction and grain direction can be computed from formulas (11−2), (11−3), and (11−4) using the edgewise compression modulus of elasticity of plywood (ch. 11). Dimensions are as noted by symbols of figure 10−10.

An approximate expression for the shear stiffness is:

$$(GA') = 2WcG \qquad (10-6)$$

where G is plywood shear modulus for appropriate grain direction (ch. 11). An improvement in shear stiffness can be made if the grain direction of the plywood webs is placed at 45° to the beam neutral plane rather than at 0° or 90° to the neutral plane. Data in chapter 11 on shear modulus given by formulas (11−8) to (11−11) should be used. Formula (10−6) is conservative since it ignores the shear stiffness of the flange. This contribution can be included by use of American Plywood Association design methods which are based on FPL−0210.

Flange Compressive and Tensile Stresses

Flange compressive and tensile stresses at outer beam fibers are given by the formula:

$$f_x = \frac{6M}{(d^3 - c^3)\dfrac{b}{d} + \dfrac{2E_w W d^2}{E}} \qquad (10-7)$$

where M is bending moment and other symbols are as defined previously in this chapter.

Web Shear Stress

Web shear stress at the beam neutral plane is given by the formula:

$$f_{xy} = \frac{3V}{4W}\left[\frac{E(d^2 - c^2)b + 2E_w W d^2}{E(d^3 - c^3)b + 2E_w W d^3}\right] \qquad (10-8)$$

where V is shear load and other symbols are as defined previously in this chapter. The shear stress must not exceed values given by formulas in chapter 11 for plywood shear strength or the critical shear buckling stress, F_{scr}, given by formula (11-17) of chapter 11. To avoid web buckling, either the web should be increased in thickness or the clear length of web should be broken by stiffeners glued to the webs.

Web edgewise bending stresses at the inside of the flanges can be computed by the formula:

$$f_{xw} = \frac{6M}{\dfrac{E}{E_w}(d^3 - c^3)\dfrac{b}{c} + 2\dfrac{d^3}{c}W} \qquad (10-9)$$

where the symbols are as previously defined. Although it is not very likely, the web may buckle due to bending stresses. Thus the stresses given by formula (10-9) should not exceed those given by F_{bcr} of chapter 11, formula (11-18). Should buckling due to edgewise bending appear possible, the interaction of shear and edgewise bending buckling can be checked approximately by means of an interaction formula such as:

$$\left(\frac{f_{xy}}{F_{scr}}\right)^2 + \left(\frac{f_{xw}}{F_{bcr}}\right)^2 = 1. \qquad (10-10)$$

Web-to-flange glue shear stresses, f_{gl}, can be computed by the approximate formula:

$$f_{gl} = \frac{3V}{2}\left[\frac{Eb(d + c)}{E(d^3 - c^3)b + 2E_w W d^3}\right] \qquad (10-11)$$

where the symbols are as previously defined in this chapter.

The stresses computed by formula (10-11) should be less than those given for glued joints (ch. 9). They should also be less than the rolling shear stresses for solid wood, because the thin plies of the plywood web allow the glue shear stresses to be transmitted to adjacent plies and could cause rolling shear failure in the wood.

Possible Lateral Buckling

Possible lateral buckling of the entire beam should be checked by calculating the critical bending stress as shown in chapter 8 under the heading of Lateral-Torsional Buckling. The slenderness factor, ρ, required to calculate this stress includes terms for lateral flexural rigidity, EL_2, and torsional rigidity, GK, that are defined as follows:

$$(EL_2)_y = \frac{1}{12}E(d - c)b^3 + E_w \qquad (10-12)$$
$$[(b + 2W)^3 - b^3]\ d$$

$$(GK) = \left[\frac{(d + c)(d^2 - c^2)(b + W)^2 W}{(d^2 - c^2) + 4(b + W)W}\right]G \qquad (10-13)$$

For I-beams:

$$(EL_2)_y = \frac{1}{12}\left\{E\ [(b + 2W)^3 - (2W)^3\ ](d - c)\right.$$
$$\left. + E_{fw}(2W)^3 d\right\} \qquad (10-14)$$

$$(GK) = \frac{1}{3}\ [\tfrac{1}{4}(d - c)^3 b + d(2W)^3]\ G \qquad (10-15)$$

where E_{fw} is flexural elastic modulus of the plywood web as computed by formulas (11-20) or (11-21) of chapter 11.

In formulas (10-13) and (10-15) the shear modulus G can be assumed without great error to be about one-sixteenth of the flange modulus of elasticity, E_L. The resultant torsional stiffness GK will be slightly low if beam webs have plywood grain at 45° to the neutral axis. The lateral buckling of I-beams will also be slightly conservative because bending rigidity of the flange has been neglected in writing the formulas given here. If buckling of the I-beam seems possible at design loads, the more accurate analysis of Forest Products Laboratory Report 1318–B should be used before redesigning.

Stiffeners and Load Blocks

A determination of the number and sizes of stiffeners and load blocks needed in a particular construction does not lend itself to a rational procedure, but certain general rules can be given that will help the designer of a wood-plywood structure obtain a satisfactory structural member.

Stiffeners serve a dual purpose in a structural member of this type. One function is to limit the size of the unsupported panel in the plywood web, and the other is to restrain the flanges from moving toward each other as the beam is stressed.

Stiffeners should be glued to the webs and should be in contact with both flanges. No rational way of determining how thick the stiffener should be is available, but it appears, from tests of box beams made at the Forest Products Laboratory, that a thickness of at least six times the thickness of the plywood web is sufficient. Because stiffeners must also resist the tendency of the flanges to move toward each other, the stiffeners should be as wide as (extend to the edge of) the flanges.

For plywood webs containing plies with the grain of the wood oriented both parallel and perpendicular to the axis of the member, the spacing of the stiffeners is relatively unimportant for the web shear stresses that are allowed. Maximum allowable stresses are below those that will produce buckling.

A clear distance between stiffeners equal to or less than two times the clear distance between flanges is adequate.

Load blocks are special stiffeners placed along a wood-plywood structural member at points of concentrated load. Load blocks should be designed so that stresses caused by a load that bears against the side-grain material in the flanges do not exceed the design allowables for the flange material in compression perpendicular to grain.

Stressed-Skin Panels

Constructions consisting of plywood "skins" glued to wood stringers are often called stressed-skin panels. These panels offer efficient structural constructions for building floor, wall, and roof components. They can be designed to provide desired stiffness, bending moment resistance, and shear resistance. The plywood skins resist bending moment and the wood stringers provide shear resistance.

Details of design are given for a panel cross section shown in figure 10−11. The formulas given were derived by basic principles of engineering mechanics. A more rigorous design procedure that includes the effects of shear lag and requires a programmable calculator to implement is available in FPL 251.

Panel deflections can be computed with formula (8−2) chapter 8. The bending stiffness (EI) and shear stiffness (GA') are given by the following formulas for the stressed-skin panel shown in figure 10−11.

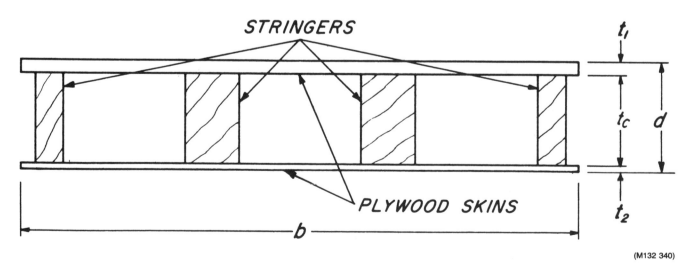

(M132 340)

Figure 10−11—Stressed-skin panel cross section.

$$(EI) = \frac{b}{(E_1t_1 + E_2t_2 + Et_c\frac{s}{b})} \left\{ E_1t_1E_2t_2[(t_1 \right.$$

$$+ t_c) + (t_2 + t_c)]^2 + E_1t_1Et_c$$

$$\left. \frac{s}{b}(t_1 + t_c)^2 + E_2t_2Et_c\frac{s}{b}(t_2 + t_c)^2 \right\} \qquad (10-16)$$

$$+ \frac{b}{12}[E_{f1}t_1^3 + E_{f2}t_2^3 + Et_c^3\frac{s}{b}]$$

where E_1 and E_2 are modulus of elasticity values for skins 1 and 2 (for compression values for plywood—see ch. 11), E_{f1} and E_{f2} are flexural modulus of elasticity values for skins 1 and 2 (also see ch. 11), E is stringer modulus of elasticity, and s is total width of all stringers in a panel. Other dimensions are as noted by symbols of figure 10–11. An approximate expression for the shear stiffness is:

$$(GA') = Gst_c \qquad (10-17)$$

where G is stringer shear modulus.

Skin Stresses

Skin tensile and compressive stresses are given by the formulas:

$$f_{x1} = \frac{ME_1y_1}{(EI)}$$
$$\qquad (10-18)$$
$$f_{x2} = \frac{ME_2y_2}{(EI)}$$

where (EI) is given by formula (10–16); M is bending moment; other symbols are as defined previously in this chapter; and

$$y_1 =$$

$$\frac{E_2t_2[(t_1 + t_c) + (t_2 + t_c)] + Et_c\frac{s}{b}(t_1 + t_c)}{2(E_1t_1 + E_2t_2 + Et_c\frac{s}{b})}$$

$$y_2 =$$

$$\frac{E_1t_1[(t_1 + t_c) + (t_2 + t_c)] + Et_c\frac{s}{b}(t_2 + t_c)}{2(E_1t_1 + E_2t_2 + Et_c\frac{s}{b})}$$

Either the skins should be thick enough or the stringers spaced closely enough so that buckling does not occur in the compression skin. The buckling stress is given by F_{ccr} (formula (11–7) in ch. 11). The plywood tensile and compressive strength (see ch. 11) should not be exceeded.

Stringer Bending Stress

The stringer bending stress is the larger value given by the formulas:

$$f_{sx1} = \frac{ME(y_1 - \frac{t_1}{2})}{(EI)}$$
$$\qquad (10-19)$$
$$f_{sx2} = \frac{ME(y_2 - \frac{t_2}{2})}{(EI)}$$

and these should not exceed appropriate values for the species.

The stringer shear stress is given by the formula:

$$f_{sxy} = \frac{V(EQ)}{s(EI)} \qquad (10-20)$$

where $(EQ) = (E_1t_1b + Es\frac{y_1}{2})\ y_1$. This also should not exceed appropriate values for the species.

Glue Shear Stress

Glue shear stress in the joint between the skins and stringers is given by the formula:

$$f_{gl} = \frac{V(EQ)}{s(EI)} \qquad (10-21)$$

where $(EQ) = E_1t_1by_1$. This stress should not exceed values for the glue and species. It should also not exceed the wood stress, f_{TR} ("rolling" shear) for solid wood because the thin plywood plies allow the glue shear stresses to be transmitted to adjacent plies and could cause rolling shear failure in the wood.

Buckling

Buckling of the stressed-skin panel of unsupported length, L, under end load applied in a direction parallel to the length of the stringers can be computed with the formula:

$$P_{cr} = \frac{\pi^2(EI)}{L^2} \qquad (10-22)$$

where L is unsupported panel length and *(EI)* is bending stiffness given by formula (10–16).

Compressive stress in the skins is given by the formula:

$$f_{xc1} = \frac{PE_1}{(EA)}$$

$$(10-23)$$

$$f_{xc2} = \frac{PE_2}{(EA)}$$

and in the stringers is given by the formula:

$$f_{sxc} = \frac{PE}{(EA)} \qquad (10-24)$$

where $(EA) = E_1 t_1 b + E_2 t_2 b + E t_c s$. These compressive stresses should not exceed stress values for plywood or stringer material. The plywood stress should also be lower than the buckling stress given by F_{ccr} (formula (11–7), ch. 11).

Loadings

Normally directed loads on the panel skins produce bending deflections and stresses in the plywood skins. Bending deflections and stresses from uniformly distributed or concentrated normal loads can be computed by formulas for plywood panels given in chapter 11.

Racking loads on panels shear the skins, and buckling of the plywood should not occur due to the shear stress caused by racking. The buckling stress can be computed as F_{scr} from formula (11–17) of chapter 11.

Selected References

Glulam

American Institute of Timber Construction. Timber construction manual. New York: John Wiley & Sons; (see current edition).

American Institute of Timber Construction. Recommended practice for protection of structural glued-laminated timber during transit, storage, and erection. AITC 111. Englewood, CO: AITC; (see current edition).

American Institute of Timber Construction. Standard specifications for structural glued-laminated timber of softwood species. AITC 117. Englewood, CO: AITC; (see current edition).

American Institute of Timber Construction. Standard specifications for hardwood glued-laminated timber. AITC 119. Englewood, CO: AITC; (see current edition).

American Institute of Timber Construction. Inspection manual. AITC 200. Englewood, CO: AITC; (see current edition).

American National Standards Institute. Structural glued laminated timber. ANSI A 190.1. New York: ANSI; (see current edition).

American National Standards Institute. Structural glued laminated timber for utility structures. ANSI 0 5.2. New York: ANSI; (see current edition).

American Society of Civil Engineers. Wood structures, a design guide and commentary. New York: ASCE; 1975.

American Society of Civil Engineers. Evaluation, maintenance and upgrading of wood structures. New York: ASCE; 1982.

American Society for Testing and Materials. Standard method for establishing stresses for structural glued-laminated timber (glulam). ASTM D 3737. Philadelphia, PA: ASTM; (see current edition).

Bohannan, B. Flexural behavior of large glued-laminated beams. Res. Pap. FPL 72. Madison, WI: U.S. Department of Agriculture, Forest Service, Forest Products Laboratory; 1966.

Canadian Standards Association. Structural glued-laminated timbers. CSA Standard 0122. Rexdale, ON: CSA; (see current edition).

Canadian Standards Association. Code for the engineering design of wood. CAN3-086. Rexdale, ON: CSA; (see current edition).

Freas, A. D.; Selbo, M. L. Fabrication and design of glued laminated wood structural members. U.S. Department of Agriculture Tech. Bull. 1069; 1954. Available from American Institute of Timber Construction, Englewood, CO.

Jokerst, R. W. Finger-jointed wood products. Res. Pap. FPL 382. Madison, WI: U.S. Department of Agriculture, Forest Service, Forest Products Laboratory; 1981.

Marx, C. M.; Moody, R. C. Strength and stiffness of small glued-laminated beams with different qualities of tension laminations. Res. Pap. FPL 381. Madison, WI: U.S. Department of Agriculture, Forest Service, Forest Products Laboratory; 1981.

Moody, R. C. Improved utilization of lumber in glued laminated beams. Res. Pap. FPL 292. Madison, WI: U.S. Department of Agriculture, Forest Service, Forest Products Laboratory; 1977.

National Forest Products Association. National design specification for wood construction. Washington, DC; (see current edition).

Peterson, J.; Madson, G.; Moody, R. C. Tensile strength of one-, two-, and three-ply glulam members of 2 by 6 Douglas-fir. Madison, WI: Forest Products Journal. 31(1): 42–48; 1981.

Selbo, M. L.; Knauss, A. C.; Worth, H. E. After two decades of service, glulam timbers show good performance. Madison WI: Forest Products Journal. 15(11): 466–472; 1965.

Wolfe, R. W.; Moody, R. C. Bending strength of vertically glued laminated beams with one to five plies. Res. Pap. FPL 333. Madison, WI: U.S. Department of Agriculture, Forest Service, Forest Products Laboratory; 1979.

Wolfe, R. W.; Moody, R. C. A summary of modulus of elasticity and knot size surveys for laminating grades of lumber. Gen. Tech Rep. FPL–31. Madison, WI: U.S. Department of Agriculture, Forest Service, Forest Products Laboratory; 1981.

Laminated Veneer Lumber

American Institute of Timber Construction. Standard specifications for structural glued laminated timber of softwood species. AITC 117. Englewood, CO: AITC; (see current edition).

Bohlen, J. C. Laminated veneer lumber—development and economics. Madison, WI: Forest Products Journal. 22(1): 18–26; 1972.

Jung, J. Properties of parallel-laminated veneer from stress-wave-tested veneers. Madison, WI: Forest Products Journal. 32(7): 30–35; 1982.

Kunesh, R. H. MICRO=LAM: Structural laminated veneer lumber. Madison, WI: Forest Products Journal. 28(7). 41–44. 1978.

Laufenberg, T. Parallel laminated veneer: Processing and performance research review. Madison, WI: Forest Products Journal. 33(9):21–28; 1983.

Moody, R. C. Tensile strength of lumber laminated from 1/8-inch thick veneers. Res. Pap. FPL 181. Madison, WI: U.S. Department of Agriculture, Forest Service, Forest Products Laboratory; 1972.

Press-lam Team. Feasibility of producing a high-yield laminated structural product. Res. Pap. FPL 175. Madison, WI: U.S. Department of Agriculture, Forest Service, Forest Products Laboratory; 1972.

Wood-Plywood

American Plywood Association. Plywood Design Specification. Tacoma, WA: APA; 1980:

    Supp. 1, Design and fabrication of plywood curved panels

    Supp. 2, Design and fabrication of plywood-lumber beams

    Supp. 3, Design and fabrication of plywood stressed-skin panels

    Supp. 4, Design and fabrication of plywood sandwich panels.

American Society for Testing and Materials. Standard methods of testing veneer, plywood, and other glued veneer constructions. ASTM D 805. Philadelphia, PA: ASTM; (see current edition).

Forest Products Laboratory. Design of plywood webs for box beams. FPL Rep. 1318. Madison, WI: U.S. Department of Agriculture, Forest Service, Forest Products Laboratory; 1943.

Chan, W.W.L. Design guide: The structural use of tempered hardboard. London: Fibre Building Board; 1980.

Kuenzi, E. W.; Zahn, J. J. Stressed-skin panel deflection and stresses. Res. Pap. FPL 251. Madison, WI: U.S. Department of Agriculture, Forest Service, Forest Products Laboratory; 1975.

Lewis, W. C.; Heebink, T. B.; Cottingham, W. S.; Dawley, E. R. Buckling in shear webs of box and I-beams and their effect upon design criteria. FPL Rep. 1318B. Madison, WI: U.S. Department of Agriculture, Forest Service, Forest Products Laboratory; 1943.

Lewis, W. C.; Heebink, T. B.; Cottingham, W. S. Effects of certain defects and stress concentrating factors on the strength of tension flanges of box beams. FPL Rep. 1513. Madison, WI: U.S. Department of Agriculture, Forest Service, Forest Products Laboratory; 1944.

Lewis, W. C.; Heebink, T. B.; Cottingham, W. S. Effect of increased moisture on the shear strength at glue lines of box beams and on the glue-shear and glue-tension strengths of small specimens. FPL Rep. 1551. Madison, WI: U.S. Department of Agriculture, Forest Service, Forest Products Laboratory; 1945.

Markwardt, L. J.; Freas, A. D. Approximate methods of calculating the strength of plywood. FPL Rep. 1630. Madison, WI: U.S. Department of Agriculture, Forest Service, Forest Products Laboratory; 1950.

McNatt, J. D. Hardboard-webbed beams: Research and application. Madison, WI: Forest. Products Journal. 30(10): 57−64; 1980.

Orosz, I. Simplified method for calculating shear deflections of beams. Res. Note FPL-0210. Madison, WI: U.S. Department of Agriculture, Forest Service, Forest Products Laboratory; 1970.

Superfesky, M. J.; Ramaker, T. J. Hardboard-webbed I−beams subjected to short-term loading. Res. Pap. FPL 264. Madison, WI: U.S. Department of Agriculture, Forest Service, Forest Products Laboratory; 1976.

Superfesky, M. J.; Ramaker, T. J. Hardboard-webbed I−beams: Effects of long-term loading and loading environment. Res. Pap. FPL 306. Madison, WI: U.S. Department of Agriculture, Forest Service, Forest Products Laboratory; 1978.

Timoshenko, S. Theory of elastic stability. 2nd Ed. New York: McGraw-Hill; 1961.

Chapter 11

Plywood

Plywood*

Plywood is a glued wood panel made up of relatively thin layers with the grain of adjacent layers at an angle, usually 90°. Each layer consists of a single thin sheet, or ply, or of two or more plies laminated together with grain direction parallel. The usual constructions have an odd number of layers. The outside plies are called faces or face and back plies, the inner plies are called cores or centers, and the plies with grain perpendicular to that of the face and back are called crossbands. The core may be veneer, lumber, or particleboard, with the total panel thickness typically not less than 1/16 inch or more than 3 inches. The plies may vary as to number, thickness, species, and grade of wood.

As compared with solid wood, the chief advantages of plywood are its approach to having properties along the length more nearly equal to properties along the width of the panel, its greater resistance to splitting, and its form, which permits many useful applications where large sheets are desirable. Use of plywood may result in improved utilization of wood, because it covers large areas with a minimum amount of wood fiber. This is because it is permissible to use plywood thinner than sawn lumber in some applications.

The properties of plywood depend on the quality of the different layers of veneer, the order of layer placement in the panel, the adhesive used, and the control of gluing conditions in the gluing process. The grade of the panel depends upon the quality of the veneers used, particularly of the face and back. The type of panel refers to glueline durability and depends upon the glue joint, particularly its water resistance, and upon veneer grades used. Generally, face veneers with figured grain that are used in panels where appearance is important have numerous short, or otherwise deformed, wood fibers. These may significantly reduce strength and stiffness of the panels. On the other hand, face veneers and other plies may contain certain sizes and distributions of knots, splits, or growth characteristics that have no undesirable effects on strength properties for specific uses. Such uses include structural applications such as sheathing for walls, roofs, or floors.

The plywood industry continues to develop new products. Hence, the reader should always refer directly to current specifications on plywood and its use for specific details.

Types of Plywood

Broadly speaking, two classes of plywood are available, covered by separate standards: Construction and industrial; hardwood and decorative. Construction and industrial plywood has traditionally been made from softwoods such as Douglas-fir, southern pine, white fir, larch, western hemlock, and redwood. However, the current standard lists a large number of hardwoods as qualifying for use. At the same time, the standard for hardwood and decorative plywood covers certain decorative softwood species for nonconstruction use.

Most construction and industrial plywood used in the United States is produced domestically, and U.S. manufacturers export some material. Generally speaking, the bulk of construction and industrial plywood is used where strength, stiffness, and construction convenience are more important than appearance. However, some grades of construction and industrial plywood are made with faces selected primarily for appearance and are used either with clear natural finishes or with pigmented finishes.

Hardwood and decorative plywood is made of many different species, both in the United States and overseas. Well over half of all such panels used in the United States are imported. Hardwood plywood is normally used in such applications as decorative wall panels and for furniture and cabinet panels where appearance is more important than strength. Most of the production is intended for interior or protected uses, although a very small proportion is made with adhesives suitable for exterior service, such as in marine applications. A significant portion of all hardwood plywood is available completely finished.

The adhesives used in the manufacture of the two classes of plywood are quite different, but each type is selected to provide the necessary performance required by the appropriate specifications.

Construction and industrial plywood covered by Product Standard PS 1 is classified by exposure capability (type) and grade. The two exposure capabilities are exterior and interior. Exterior plywood is bonded with exterior (waterproof) glue, and veneers used in manufacture cannot be less than "C" grade as defined in PS 1. Interior-type plywood may be bonded with interior, intermediate, or exterior (waterproof) glue. "D"-grade veneer is allowed as inner and back plies of certain interior-type grades. Gluebond performance requirements are specified in PS 1.

The four types of hardwood and decorative plywood in decreasing order of resistance to water are: Technical (Exterior), Type I (Exterior), Type II (Interior), and Type III (Interior); gluebond requirements for these are specified in ANSI/HPMA HP.

Product and Performance Specifications for Plywood

The most commonly used product specifications for plywood are the product standards established by the industry under the guidelines of nationally recognized procedures for standards development. The specifications for construction

---

* Revision by J. Dobbin McNatt, Forest Products Technologist.

and industrial plywood (Product Standard PS 1) are available from the National Bureau of Standards or the American Plywood Association (APA). The specifications for hardwood and decorative plywood (ANSI/HPMA HP) may be obtained from the Hardwood Plywood Manufacturers' Association. These specifications cover such items as the grading of veneer, the construction of panels, the glueline performance requirements, and recommendations for use.

Some plywood and composite panels with particleboard core are recognized for use as sheathing and combination subfloor-underlayment based on performance testing rather than on prescriptive standards. These are marketed under a quality program designed to assure that structural properties do not fall below those of panels successfully passing performance testing. Such a program also establishes minimum criteria for physical properties and durability.

Imported plywood is generally not produced in conformance with U.S. product specifications. However, some countries have their own specifications for plywood manufactured for export to the United States, and these follow the requirements of the domestic product standards.

Grademarking

All panels of construction and industrial plywood represented as complying with Product Standard PS 1 are identified with the mark of a qualified inspection and testing agency which gives the following information:

1. Species group classification, or class number, or span rating, depending on grade.

2. Either Interior or Exterior.

3. Grade name or grade of face and back veneers.

4. The symbol PS 1 signifying conformance with the standard.

5. Manufactured thickness—if nonstandard.

Examples of grade stamps for construction plywood conforming to PS 1 are shown in figure 11−1. A panel with one of these marks would have a C-grade veneer face and D-grade veneer back. The span rating, 32/16, refers to a maximum allowable roof support spacing of 32 inches and a maximum floor joist spacing of 16 inches. The panel would be interior type but would be bonded with exterior glue. Numbers where "zeros" are shown would identify the mill where the panel was manufactured. "NRB" signifies that the product is approved by the National Research Board and constitutes recognition and approval by the three code bodies which jointly sponsor the NRB.[1]

---

[1] Building Officials and Code Administrators International, Inc. (BOCA), Chicago, IL; International Conference of Building Officials (ICBO), Whittier, CA; and Southern Building Code Congress (SBCC), Birmingham, AL.

Figure 11−1—Typical grade marks for sheathing plywood conforming to PS 1−83.

(ML84 5771)

All hardwood plywood represented as conforming to American National Standard ANSI/HPMA HP-1983 is identified by one of two methods: (a) Each panel is marked with the symbol of the standard, the name or recognized identification of the producer, the species and grade of the face veneer, and the gluebond type (fig. 11−2) or (b) a written statement containing this information accompanies the order or shipment.

Special Plywood

In addition to the all-veneer panels, a variety of special plywood panels are made.

Composite Plywood

Composite plywood is a type of construction panel manufactured with a structural core of reconstituted wood and with veneer faces (fig. 11−3). For many applications it is interchangeable with certain grades of all-veneer plywood. Its intended uses include roof and wall sheathing, subflooring, and single-layer floors.

Overlaid Plywood

Paper, plastic, or metal layers are combined with wood veneers, usually as the face layer, to provide special panel characteristics such as improved surface properties (ch. 23).

Special resin-treated papers called "overlays" can be bonded to plywood panels, either on one or on both sides. These overlays are either of the "high-density" or the "medium-density" types, and are intended to provide improved resistance to abrasion or wear, better surfaces when appearance is important or for concrete-form use, or better paint-holding properties.

Specific plywood grades with paper-face overlays are available commercially. These are of three types—high density, medium density, and special overlay. Although they are designed for either exterior or interior service, all commercial overlay plywood conforming to the product standards is made in the exterior type.

By specification, the high-density type is one in which the surface on the finished product is hard, smooth, and of such character that further finishing by paint or varnish is not required. It consists of a cellulose fiber sheet or sheets, in which not less than 45 percent of the weight of the laminate is a thermosetting resin of the phenolic or melamine type.

By specification also, the medium-density type must present a smooth uniform surface suitable for high-quality paint finishes. It consists of a cellulose-fiber sheet or sheets in which not less than 17 percent resin solids by weight of the laminate for a beater-loaded sheet (or 22 percent by weight if impregnated) is a thermosetting resin of either the phenolic or melamine type.

The main difference between the two kinds of paper-face overlays for plywood is that the medium-density overlay face is opaque (of solid color) and not translucent like the high-density one. Some evidence of the underlying grain may appear, but compared to the high-density surface, there is no consistent showthrough.

Special overlays are those surfacing materials with special characteristics that do not fit the exact description of high- or medium-density overlay types but otherwise meet the test requirements for overlaid plywood. Some panels are also overlaid, usually on the face only, with high-density paper-base decorative laminates, hardboards, or metal sheets. A backing sheet having the same properties (modulus of elasticity, dimensional stability, and vapor transmission rate) as the decorative face sheet must be used to provide a panel free from warp and twist as moisture changes occur. Overlays may be applied in the original layup or they may be applied to plywood after the panels have been sanded. The two-step method permits a closer thickness tolerance. Requirements for certain types of overlaid plywood are included in the aforementioned product standards.

Other Special Products

Embossed, grooved, and other textured plywood panels are used primarily as interior paneling and exterior siding.

Prefinished plywood, particularly hardwood plywood, is available in wide variety. The finishes are normally applied in the plywood plants as clear or pigmented liquid finishes. Various printed film patterns are sometimes applied to plywood. Printed panels are also available using liquid finishing systems, and three-dimensional finishes can be achieved by passing the panels under an embossing roller. By these techniques, the appearance for such uses as furniture and wall paneling is improved. Some use is also made of clear, printed, and pigmented plastic films bonded to plywood for the same purpose.

Plywood may be purchased that has been specially treated for protection against fire or decay (see chs. 15 and 18). It is technically feasible to treat the veneers with chemical solutions and then glue them into plywood. A more general practice is to treat the plywood after gluing, with either fire-retardant or wood-preservative solutions, and then redry the panels. Panels must be glued with durable glues of the exterior type to withstand such treatment.

Large-size panels for special purposes are made by end jointing standard-size panels with scarf joints or finger joints. This is done mainly with softwood panels for structural use as in boats or trailers. Requirements for joints are given in the product standards for conventional plywood.

Curved plywood is sometimes used, particularly in certain furniture items, as a specialty product. Much curved plywood is made by gluing the individual veneers to the desired shape and curvature in special jigs or presses. Flat plywood can also be bent to simple curvatures after gluing using techniques similar to those for bending solid wood (see ch. 13).

Factors Affecting Dimensional Stability of Plywood

Arrangement of Plies

The tendency of crossbanded products to warp, as the result of uneven shrinking and swelling caused by moisture changes, is largely eliminated by balanced construction. Balanced construction involves plies arranged in similar pairs, a ply on each side of the core. Similar plies have the same thickness, kind of wood with particular reference to shrinkage and density, moisture content at the time of gluing, and grain direction. The importance of having the grain direction of similar plies parallel cannot be overemphasized. A face ply with its grain at a slight angle to the grain of the back ply may result in a panel that will twist excessively with moisture content changes.

An odd number of plies permits an arrangement that gives a substantially balanced effect; that is, when three plies are glued together with the grain of the outer two identical plies at right angles to the grain of the core, the panel is balanced and tends to remain flat through moisture content changes. With five, seven, or some other uneven number of plies, panels may be similarly balanced. Four- or six-ply panels made with the two center plies having grain parallel may also be balanced panels (fig. 11−4).

Balanced construction is highly important in panels that must remain flat. Conversely, in certain curved members, the natural cupping tendency of an even number of plies may be used to advantage.

Because the outer or face plies of a crossbanded construction are restrained on only one side, changes in moisture content induce relatively large stresses in the outer glue joints. The magnitude of the stresses depends upon such factors as thickness of outer plies, density of the veneer involved, and the rate and amount of change in moisture content. In general, the thinner the face veneer the less problem with face checking.

Quality of Plies

In thin plywood where dimensional stability is important, all the plies affect the shape and permanence of form of the panel. All plies should be straight grained, smoothly cut, and of sound wood.

In thick five-ply lumber-core panels the crossbands, in

GLUE BOND
TYPE II
ANSI/HPMA
HP-1983

FLAME SPREAD
200 OR LESS
ASTM E84

SIMULATED
DECORATIVE
FINISH ON
3.6MM THICK

MILL 00
SPECIALTY GRADE

Figure 11−2—Grade stamp for specialty hardwood plywood conforming to HP-1983. (M84 5316)

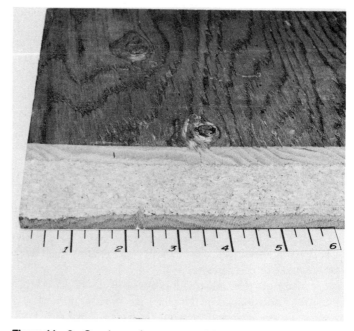

Figure 11−3—Specimen of a veneer-particle panel. (M150 063) The edge has been bevel-cut to accentuate the wood-particle core. (Numbered scale in inches.)

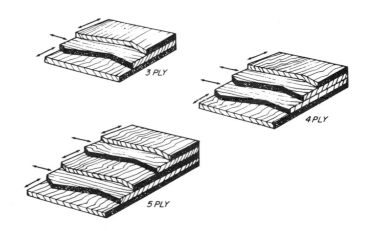

Figure 11−4—Typical 3-, 4-, and 5-ply construction. (ML84 5821)

11-5

particular, affect the quality and stability of the panel. In such panels thin and dissimilar face and back plies can be used without upsetting the stability of the panel. Imperfections in the crossbands, such as marked differences in the texture of the wood or irregularities in the surface, are easily seen through thin surface veneers. Cross grain that runs sharply through the crossband veneer from one face to the other causes the panels to cup. Cross grain that runs diagonally across the face of the crossband veneer causes the panel to twist unless the two crossbands are laid with their grain parallel. Failure to observe this simple precaution accounts for much warping in crossbanded construction.

In many hardwood and decorative plywood uses, both appearance and dimensional stability are important. The best woods for cores of such high-grade panels are those of low density and low shrinkage, of slight contrast between earlywood and latewood, and of species that are easily glued. Edge-grained cores are better than flat-grained cores because they shrink less in width. In softwoods with pronounced latewood, moreover, edge-grained cores are better because the hard bands of latewood are less likely to show through thin veneer, and the panels show fewer irregularities in the surfaces. In most species, a core made entirely of either quartersawed or plainsawed material remains more uniform in thickness during moisture content changes than one in which the two types of material are combined.

Distinct distortion of surfaces has been noted, particularly in softwoods, when the core boards were neither distinctly flat-grained nor edge-grained.

For many uses of construction plywood, as in sheathing, strength characteristics of the panel are most important. By contrast, appearance, moderate tendencies to warp, and small dimensional changes are of minor importance. Strength and stiffness in bending are particularly important. In such panels, veneers are selected mainly to provide strength properties. This selection often permits controlled amounts of knots, splits, and other irregularities that might be considered objectionable from an appearance standpoint.

Moisture Content

The tendency of plywood panels to warp is affected by changes in moisture content as a result of changes in atmospheric moisture conditions or wetting of the surface by free water. Surface appearance may also be affected. The response of plywood to moisture content changes in service may depend on whether it was manufactured by hot or cold pressing because of differences in the state of internal stresses built into the panels.

Most plywood is made in hot presses, but other panels are cold-pressed.

Hot-pressed panels come out of the press quite dry. The original moisture content of the veneer and the amount of water added by the adhesive must be kept low to avoid blister problems in hot-pressing the panels. In addition, water is lost from the adhesive and the wood during heating. On the other hand, cold-pressed panels are generally fairly high in moisture content when removed from the press—the actual values depending on the original moisture content of the veneer, the amount of water in the adhesive, and the amount of adhesive spread. Such panels may lose considerable moisture while reaching equilibrium in service.

Either type of panel may be used satisfactorily if it is properly designed for the service condition. To minimize dimensional changes and problems associated with them, panels should be conditioned at a relative humidity as close as possible to service conditions.

Expansion or Contraction

The dimensional stability of plywood, associated with moisture and thermal changes, involves not only cupping, twisting, and bowing but includes expansion or contraction. The usual swelling and shrinking of the wood is effectively reduced because grain directions of adjacent layers are placed at right angles. The low dimensional change parallel to the grain in one layer restrains the normal swelling and shrinking across the grain in the layer glued to it. An additional restraint results from the fact that modulus of elasticity parallel to the grain is about 20 times that across the grain (see ch. 4). The expansion or contraction of plywood can be closely approximated by the formula

$$\epsilon = \frac{\Sigma m_i E_i t_i}{\Sigma E_i t_i} \qquad (11-1)$$

where ϵ is the expansion or contraction of the plywood, m_i is the coefficient of expansion of the ith ply over the range of moisture or thermal change anticipated, E_i is the modulus of elasticity of the ith layer, and t_i is the thickness of the ith layer. The units of ϵ correspond to the units of m. If m is percent, ϵ is percent; if m is total expansion, ϵ is total expansion. Values of m can be obtained from data given in chapter 3 in the sections Shrinkage and Thermal Properties. Values of E are obtainable from data given in chapter 4. Plywood expansion or contraction as given by formula $(11-1)$ will be about equal to the parallel-to-grain movement of the wood species in the veneers.

For all practical purposes, the dimensional change of plywood in thickness does not differ from that of solid wood.

Thermal Properties

The thermal conductivity (k) of plywood has essentially the same value as that of solid wood of the same species and density. Average k values for plywood made from the four species groups as defined in PS 1 at 12 percent moisture content are:

Species Group	Value of k
	(Btu·in/hr·°F)
1	1.02
2	.89
3	.86
4	.76

Structural Design of Plywood

The stiffness and strength of plywood can be computed by formulas relating the plywood properties to the construction of the plywood and to the properties of particular wood species in the component plies. Testing all of the many possible combinations of layer thickness, species, number of layers, and variety of structural components is impractical. The various formulas developed mathematically and presented here were checked by tests to verify their applicability. Some approximate properties of sheathing-grade plywood are shown in table 11−1.

Plywood may be used under loading conditions that require the addition of stiffeners to prevent it from buckling (ch. 10). It may also be used in the form of cylinders or curved plates, which are beyond the scope of this handbook but are discussed in U.S. Department of Defense Bulletin ANC−18.

It is obvious from its construction that a strip of plywood cannot be so strong in tension, compression, or bending as a strip of solid wood of the same size. Those layers having their grain direction oriented at 90° to the direction of stress can contribute only a fraction of the strength contributed by the corresponding areas of a solid strip, because they are stressed perpendicular to the grain. Strength properties in the length and width directions tend to be equalized in plywood because adjacent layers are oriented at an angle of 90° to each other.

Table 11 − 1—*General property values for sheathing-grade plywood[1]*

Properties	Values	ASTM Test Method[2] (where applicable)
Linear hygroscopic expansion (30−90 pct RH)	<.020 pct	
Linear thermal expansion	0.0000034 in/in/°F	
Flexure (psi):		
Modulus of rupture	7,000−10,000	D 3043
Modulus of elasticity	1,200,000−1,500,000	
Tensile strength (psi)	4,000−5,000	D 3500
Compressive strength (psi)	4,500−6,000	D 3501
Shear through the thickness (edgewise shear) (psi):		
Shear strength	800−1,000	D 2719
Shear modulus	85,000−110,000	D 3044
Shear in the plane of the plies (rolling shear) (psi):		D 2718
Shear strength	250−300	
Shear modulus	20,000−30,000	

[1] All mechanical properties are on the basis of total panel thickness, parallel to the grain direction of the face plies where applicable. Note: The data presented are general average values, accumulated from a number of sources, and are not to be used in developing allowable design values. Information on engineering design methods for plywood is available from the American Plywood Association, Tacoma, WA.

[2] Standard methods of testing strength and elastic properties of plywood are given in ASTM Standards (see Selected References).

The formulas given in this chapter may be used, in general, for calculating the stiffness of plywood, stresses at proportional limit or ultimate, or for estimating working stresses, depending upon the veneer or wood species property that is substituted in the formulas. Values of the wood properties are given in tables in chapter 4. Chapter 4 also gives values of property coefficients of variation to indicate the variability and reliability of the data as an aid in selection of allowable property values. Modulus of elasticity values given in table 4−2 (ch. 4) should be increased by 10 percent before being used to predict plywood properties; values in table 4−2 represent test data wherein the measured deflection was attributed to bending stiffness and did not consider shear deflection, which effectively increases by about 10 percent the deflection of wood beams having a span-depth ratio of 14:1 as tested.

Properties in Edgewise Compression

Modulus of Elasticity

The modulus of elasticity of plywood in compression parallel to or perpendicular to the face-grain direction is equal to the weighted average of the moduli of elasticity of all plies parallel to the applied load. That is,

$$E_w \text{ or } E_x = \frac{1}{h} \sum_{i=1}^{i=n} E_i h_i \qquad (11-2)$$

where E_w is the modulus of elasticity of plywood in compression parallel to the face grain; E_x, the modulus of elasticity of plywood in compression perpendicular to the face grain; E_i, the modulus of elasticity parallel to the applied load of the veneer in layer i; h_i, the thickness of the veneer in layer i; h, the thickness of the plywood; and n, the number of layers.

When all layers are of the same thickness and wood species, the formula reduces to

$$E_w = \frac{1}{2n} \left[(E_L + E_T)n + (E_L - E_T) \right]$$

$$E_x = \frac{1}{2n} \left[(E_L + E_T)n - (E_L - E_T) \right]$$

$(11-3)$

where n is the number of layers (n is odd), E_L is the modulus of elasticity of the veneer parallel to the grain, and E_T is the modulus of elasticity of the veneer in the tangential direction (for rotary-cut veneer). For quarter-sliced veneer, use the modulus of elasticity of the veneer in the radial direction, E_R, instead of E_T.

The modulus of elasticity in compression at angles to the face-grain direction other than 0° or 90° is given approximately by:

$$(11-4)$$

$$\frac{1}{E_\theta} = \frac{1}{E_w} \cos^4 \theta + \frac{1}{E_x} \sin^4 \theta + \frac{1}{G_{wx}} \sin^2 \theta \cos^2 \theta$$

where E_θ is the modulus of elasticity of a plywood strip in compression at an angle θ to the face grain; G_{wx} is the modulus of rigidity associated with plywood distortion under edgewise shearing forces along axes w (parallel to face grain) and x (perpendicular to face grain); and the other terms are as defined in formula (11−3). Formulas for computing values of G_{wx} are given under Properties in Edgewise Shear.

Strength

The compressive strength of plywood subjected to edgewise forces is given by:

$$F_{cw} = \frac{E_w}{E_{cL}} F_{cL}$$

$$(11-5)$$

$$F_{cx} = \frac{E_x}{E_{cL}} F_{cL}$$

where F_{cw} is the compressive strength of plywood parallel to the face grain; F_{cx}, the compressive strength of plywood perpendicular to the face grain; F_{cL}, the compressive strength of the veneer parallel to the grain; and E_{cL}, the modulus of elasticity of the veneer parallel to the grain. If more than one species is used in the longitudinal plies, values for the species having the lowest ratio of F_{cL}/E_{cL} should be used in the formulas given.

When plywood is loaded at an angle to the face grain, its compressive strength may be computed from:

$$(11-6)$$

$$F_{c\theta} = \frac{1}{\sqrt{\frac{\cos^4\theta}{F_{cw}^2} + \frac{\sin^4\theta}{F_{cx}^2} + \left(\frac{1}{F_{swx}^2} - \frac{1}{F_{cw}F_{cx}}\right) \sin^2\theta \cos^2\theta}}$$

where $F_{c\theta}$ is the compressive strength of plywood at an angle θ to the face grain, and F_{swx} is the shear strength of plywood under edgewise shearing forces along axes w (parallel to face grain) and x (perpendicular to face grain) (formula (11−12)); and the other terms are as previously defined.

If the plywood is a thin panel, compressive edge loads can cause buckling and consequently reduction of load-carrying capacity. Plywood panels in stressed-skin constructions must be designed to preclude buckling under edge compression loads or bending loads causing edgewise compression on one facing. The critical compressive buckling stress for a plywood panel with face grain parallel to edges and simply supported at four edges is given approximately by the formula:

$$F_{crr} = \frac{\pi^2}{6}\left(\sqrt{E_{fa}E_{fb}} + 0.17E_L\right)\frac{h^2}{b^2} \qquad (11-7)$$

$$\text{for } a \geq b\left(\frac{E_{fa}}{E_{fb}}\right)^{1/4}$$

where F_{ccr} is the critical compressive buckling stress; h is plywood thickness; b is width of the loaded edge of the plywood panel; a is plywood panel length; E_{fa} and E_{fb} are flexural moduli of elasticity of the plywood in the a and b directions, respectively; and E_L is the modulus of elasticity of the plywood species. Formulas for computing E_f values are elsewhere in this section.

Properties in Edgewise Tension

Modulus of Elasticity

Values of modulus of elasticity in tension are the same as those in compression.

Strength

The strength of a plywood strip in tension parallel or perpendicular to the face grain may be taken as the sum of the strength values of the plies having their grain direction parallel to the applied load.

The tensile strength parallel to the face grain will be designated as F_{tw} and the tensile strength perpendicular to the face grain as F_{tx}.

The tensile strength at an angle to the face grain may be computed from formula (11-6) by substituting the subscript "t" for "c" wherever it appears.

Properties in Edgewise Shear

Modulus of Rigidity

The modulus of rigidity of plywood may be calculated from:

$$G_{wx} = \frac{1}{h}\sum_{i=1}^{i=n} G_i h_i \qquad (11-8)$$

where G_{wx} is the modulus of rigidity of plywood under edge-wise shear; G_i is the modulus of rigidity of the ith layer; h_i is the thickness of the ith layer; and h is the plywood thickness.

When the plywood is made of a single species of wood:

$G_{wx} = G_{LT}$ for rotary-cut veneers
$G_{wx} = G_{LR}$ for quarter-sliced veneer

Values of G_{LT} and G_{LR} are given in terms of the modulus of elasticity parallel to grain (E_L) in table 4-2 of chapter 4.

The modulus of rigidity at an angle to the face grain may be computed from:

$$\frac{1}{G_\theta} = \frac{1}{G_{wx}}\cos^2 2\theta + \left[\frac{1}{E_w} + \frac{1}{E_x}\right]\sin^2 2\theta \qquad (11-9)$$

This formula gives a maximum value for G_θ when $\theta = 45°$; thus shear deflections of constructions such as box- and I-beams, wherein plywood webs offer principal resistance to shear, can be reduced by orienting the plywood face grain at 45° to the beam axis. The formula for the special case of $\theta = 45°$ reduces to

$$G_{45°} = \frac{E_w E_x}{E_w + E_x} \qquad (11-10)$$

This formula has a maximum value for plywood arranged to have the same area of parallel grain layers in the two principal directions to produce $E_w = E_x = \frac{1}{2}(E_L + E_T)$. This maximum 45° shear modulus is then

$$\max G_{45°} = \frac{1}{4}(E_L + E_T) \qquad (11-11)$$

For quarter-sliced veneer, E_R (radial direction) is to be substituted for E_T (tangential direction).

Strength

The ultimate strength of plywood elements in shear, with the shearing forces parallel and perpendicular to the face-grain direction, is given by the empirical formula:

$$F_{swx} = 55\frac{n-1}{h} + \frac{9}{16h}\sum_{i=1}^{i=n} F_{swxi}h_i \qquad (11-12)$$

where n is the number of layers and F_{swxi} is the shear strength of the ith layer.

In using this formula, the factor $(n-1)/h$ should not be assigned a value greater than 35.

In some commercial grades of plywood, gaps in the core or crossbands are permitted. These gaps reduce the shear strength of plywood, and the formula just given should be corrected to account for this effect. This may be done approximately by subtracting from the number of layers (n) in the first term twice the number of layers containing openings at any one section, and omitting from the summation in the second term all layers containing openings at any one section. Since the first term represents the contribution of the glue layers to shear, twice the number of layers containing openings at any one section is subtracted to account for the lack of glue on each side of the opening. The modification for the effect of core gaps just outlined represents a logically derived procedure not confirmed by test.

When the plywood is stressed in shear at an angle to the face grain, ultimate shear strength with face grain in tension or compression is given by the following formulas:

$$(11-13)$$

$$F_{s\theta t} = \frac{1}{\sqrt{\left(\dfrac{1}{F_{tw}^2} + \dfrac{1}{F_{tw}F_{cx}} + \dfrac{1}{F_{cx}^2}\right)\sin^2 2\theta + \dfrac{\cos^2 2\theta}{F_{swx}^2}}}$$

$$(11-14)$$

$$F_{x\theta c} = \frac{1}{\sqrt{\left(\dfrac{1}{F_{cw}^2} + \dfrac{1}{F_{cw}F_{tx}} + \dfrac{1}{F_{tx}^2}\right)\sin^2 2\theta + \dfrac{\cos^2 2\theta}{F_{swx}^2}}}$$

These formulas have maximum values for $\theta = 45°$ as did the modulus of rigidity formula. For $\theta = 45°$, the formulas reduce to:

$$F_{s45t} = \frac{1}{\sqrt{\dfrac{1}{F_{tw}^2} + \dfrac{1}{F_{tw}F_{cx}} + \dfrac{1}{F_{cx}^2}}} \qquad (11-15)$$

$$F_{s45c} = \frac{1}{\sqrt{\dfrac{1}{F_{cw}^2} + \dfrac{1}{F_{cw}F_{tx}} + \dfrac{1}{F_{tx}^2}}} \qquad (11-16)$$

If the plywood is a thin panel, edgewise shearing loads can cause buckling and subsequent reduction of load-carrying capacity. Plywood panels in structures, such as webs of I− or box−beams or walls subjected to racking, must be so designed as to preclude buckling due to shearing loads. The critical shear buckling stress for a plywood panel with face grain parallel to edges and simply supported at four edges is given approximately by the formula:

$$F_{scr} = \frac{K_s}{3}\left(E_{fa}E_{fb}^3\right)^{1/4} \frac{h^2}{b^2} \qquad (11-17)$$

where F_{scr} is the critical shear buckling stress; h is plywood thickness; E_{fa} and E_{fb} are flexural moduli of elasticity of the plywood in the a and b directions, respectively; K_s is a buckling coefficient given by figure 11−5; and a and b panel dimensions are chosen so that the abscissa quantity in figure 11−5 is <1.0. If the shear buckling stress is too low for the intended use, the buckling stress can be increased considerably by placing the plywood in shear so that the face grain is in compression and at 45° to the panel edge. Details of design for grain directions other than parallel or perpendicular to panel edges are given in ANC−18 Bulletin.

Properties in Edgewise Bending

For the occasional use where plywood is subjected to edgewise bending, such as in plywood box− and I−beam webs, the values of modulus of elasticity, modulus of rigidity, and strength are the same as those for plywood in compression, tension, or shear, whichever loading is appropriate in the design. If the plywood is a thin panel, edgewise bending can cause buckling, which reduces load-carrying capacity. The critical buckling stress for a simply supported plywood panel under pure edgewise bending is approximately equal to six times the compression buckling stress; thus

$$F_{bcr} \approx 6F_{ccr} \qquad (11-18)$$

$$\text{for } a \geq 0.7b\left(\frac{E_{fa}}{E_{fb}}\right)^{1/4}$$

Properties in Flexure

The following material pertains to flexure of plywood that causes curvature of the plane of the plywood sheet.

Modulus of Elasticity

The modulus of elasticity in flexure is equal to the average of the moduli of elasticity parallel to the span of the various plies weighted according to their moment of inertia about the neutral plane. That is,

$$E_{fw} \text{ or } E_{fx} = \frac{1}{I}\sum_{i=1}^{i=n} E_i I_i \qquad (11-19)$$

where E_{fw} is the modulus of elasticity of plywood in bending when the face grain is parallel to the span; E_{fx}, the modulus of elasticity of plywood in bending when the face grain is perpendicular to the span; E_i, the modulus of elasticity of the ith layer in the span direction; I_i, the moment of inertia of the ith layer about the neutral plane of the plywood; and I, the moment of inertia of the total cross section about its centerline. When all layers are of the same thickness and wood species, the formula reduces to

$$E_{fw} = \frac{1}{2n^3}[(E_L + E_T)\, n^3 + (E_L - E_T)\,(3n^2 - 2)]$$

$$(11-20)$$

$$E_{fx} = \frac{1}{2n^3}[(E_L + E_T)\, n^3 - (E_L - E_T)\,(3n^2 - 2)]$$

where n is the number of layers (n is odd), E_L is the modulus of elasticity of the veneer parallel to the grain, and E_T is the modulus of elasticity of the veneer in the tangential direction (see ch. 4). For quarter-sliced veneer, the modulus of elasticity in the radial direction, E_R, should be substituted for E_T.

The effective moduli of elasticity E_{fw} and E_{fx} are useful in computing the deflections of plywood strips that are subjected primarily to bending on a long span. Deflections due to shear are low for strips on long spans but become important for short spans; they can be computed by analyses given in references by March in the Selected References for this chapter.

The deflection of a plywood plate simply supported on all four edges depends also on plywood bending stiffness and plate aspect ratio. The center deflection of a plywood plate of width b and length a under a uniformly distributed load of intensity p is given by:

$$w = 0.155K \frac{pb^4}{E_{fb}h^3} \qquad (11-21)$$

where K is given in figure 11−6 and h is plywood thickness. The center deflection of a plywood plate of width b and length a under a center concentrated load P is given by:

$$w = 0.252K \left(\frac{E_{fb}}{E_{fa}}\right)^{1/4} \frac{Pb^2}{E_{fb}h^3} \qquad (11-22)$$

where K is given in figure 11−6.

Strength

The resisting moment of plywood strips having face grain parallel to the span is given by:

$$M = 0.85\left(\frac{E_{fw}}{E_L}\right)\left(\frac{F_b I}{c}\right) \qquad (11-23)$$

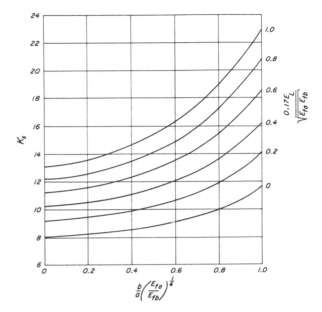

Figure 11−5—Shear buckling coefficient for plywood panels having face grain parallel to panel edges. (M135 060)

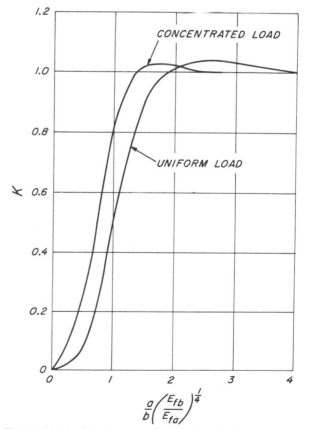

Figure 11−6—Deflection coefficients for simply supported plywood plates under normal load. (M135 059)

11-11

For face grain perpendicular to the span,

$$M = 1.15 \left(\frac{E'_{fx}}{E_L}\right)\left(\frac{F_b I}{c}\right) \text{for three-layer plywood} \quad (11-24)$$

$$M = \left(\frac{E'_{fx}}{E_L}\right)\left(\frac{F_b I}{c}\right) \begin{array}{l} \text{for plywood having} \\ \text{five or more layers} \end{array} \quad (11-25)$$

where M is the resisting moment of the plywood; F_b, the strength of the outermost longitudinal layer; c, the distance from the neutral plane to the outer fiber of the outermost longitudinal layer; and E'_{fx} is the same as E_{fx}, except that the outermost layer in tension is neglected, and the other terms are as defined previously.

For plywood having five or more layers, the use of E_{fx} in place of E'_{fx} in calculating the resisting moment will result in negligible error. It should be noted that E'_{fx} is used only in strength calculations and is not to be used in deflection calculations.

Other Design Considerations

Plywood of thin, cross-laminated layers is very resistant to splitting and therefore nails and screws can be placed close together and close to the edges of panels (ch. 7).

Highly efficient, rigid joints can, of course, be obtained by gluing plywood to itself or to heavier wood members such as are needed in box-beams and stressed-skin panels. Glued joints should not be designed primarily to transmit load in tension normal to the plane of the plywood sheet because of the rather low tensile strength of wood in a direction perpendicular to the grain. Glued joints should be arranged to transmit loads through shear. It must be recognized that shear strength across the grain of wood (often called rolling shear strength because of the tendency to roll the wood fibers) is only 20 to 30 percent of that parallel to the grain (see ch. 4). Thus sufficient area must be provided between plywood and flange members of box-beams and between plywood and stringers of stressed-skin construction to avoid shearing failure perpendicular to the grain in the face veneer, in the crossband veneer next to the face veneer, or in the wood member. Various details of design are given in chapter 10.

Selected References

American Plywood Association. Performance standards and policies for structural-use panels. Tacoma, WA: APA; 1982.

American Society for Testing and Materials. Standard methods of testing plywood. ASTM D 2718, D 2719, 3043, 3044, 3499, 3500, and 3501. Philadelphia, PA: ASTM; (see current edition).

American Society for Testing and Materials. Definitions of terms relating to veneer and plywood. ASTM D 1038. Philadelphia, PA: ASTM; (see current edition).

Baldwin, R. F. Plywood manufacturing practices. San Francisco, CA: Miller Freeman Publications, Inc.; 1975.

Carll, C. G.; Dickerhoof, H. E.; Youngquist, J. A. U.S. wood-based panel industry: Part II. Standards for panel products. Forest Products Journal. 32(7): 12-15; 1982.

Forest Products Laboratory. Manufacture and general characteristics of flat plywood. Res. Note FPL-064. Madison, WI: U.S. Department of Agriculture, Forest Service, Forest Products Laboratory; 1964.

Forest Products Laboratory. Bending strength and stiffness of plywood. Res. Note FPL-059. Madison, WI: U.S. Department of Agriculture, Forest Service, Forest Products Laboratory; 1964.

Forest Products Laboratory. Some causes of warping in plywood and veneered products. Res. Note FPL-0136. Madison, WI: U.S. Department of Agriculture, Forest Service, Forest Products Laboratory; 1966.

Hardwood Plywood Manufacturers Association. American national standard for hardwood and decorative plywood. ANSI/HPMA HP. New York: American National Standards Institute, Inc.; (see current edition).

Heebink B. G. Fluid-pressure molding of plywood. FPL Rep. 1624. Madison, WI: U.S. Department of Agriculture, Forest Service, Forest Products Laboratory; 1959.

Heebink, B. G. Importance of balanced construction in plastic-faced wood panels. Res. Note FPL-021. Madison, WI: U.S. Department of Agriculture, Forest Service, Forest Products Laboratory; 1963.

Heebink, B. G.; Kuenzi, E. W.; Maki, A. C. Linear movement of plywood and flakeboards as related to the longitudinal movement of wood. Res. Note FPL-073. Madison, WI: U.S. Department of Agriculture, Forest Service, Forest Products Laboratory; 1964.

Liska, J. A. Methods of calculating the strength and modulus of elasticity of plywood in compression. FPL Rep. 1315, rev. Madison, WI: U.S. Department of Agriculture, Forest Service, Forest Products Laboratory; 1955.

March, H. W. Bending of a centrally-loaded rectangular strip of plywood. Physics. 7(1): 32-41; 1936.

March, H. W.; Smith, C. B. Buckling of flat sandwich panels in compression. FPL Rep. 1525. Madison, WI: U.S. Department of Agriculture, Forest Service, Forest Products Laboratory; 1945.

Norris, C. B. Strength of orthotropic materials subjected to combined stresses. FPL Rep. 1816. Madison, WI: U.S. Department of Agriculture, Forest Service, Forest Products Laboratory; 1950.

Norris, C. B.; Werren, F.; McKinnon, P. F. The effect of veneer thickness and grain direction on the shear strength of plywood. FPL Rep. 1801. Madison, WI: U.S. Department of Agriculture, Forest Service, Forest Products Laboratory; 1948.

Superfesky, M. J. Investigating methods to evaluate impact behavior of sheathing materials. Res. Pap. FPL 260. Madison, WI: U.S. Department of Agriculture, Forest Service, Forest Products Laboratory; 1975.

Superfesky, M. J.; Montrey, H. M.; Ramaker, T. J. Floor and roof sheathing subjected to static loads. Wood Science 10(1). 31-41; 1977.

U.S. Department of Commerce. Construction and industrial plywood. Prod. Stand. PS 1-83. Washington, DC: USDC; 1983.

U.S. Department of Defense. Design of wood aircraft structures. ANC-18 Bull. (Issued by Subcommittee on Air Force-Navy-Civil Aircraft Design Criteria, Aircraft Comm.) 2nd ed. Washington, DC: Munitions Board; 1951.

Zahn, J. J.; Romstad, K. M. Buckling of simply supported plywood plates under combined edgewise bending and compression. Res. Pap. FPL 50. Madison, WI: U.S. Department of Agriculture, Forest Service, Forest Products Laboratory; 1965.

Chapter 12

Structural Sandwich Construction

Structural Sandwich Construction *

Structural sandwich construction is a layered construction formed by bonding two thin facings to a thick core (fig. 12–1). The thin facings are usually of a strong and dense material, since they resist nearly all the applied edgewise loads and flatwise bending moments. The core, which is of a weak and low-density material, separates and stabilizes the thin facings and provides most of the shear rigidity of the sandwich construction. By proper choice of materials for facings and core, constructions with high ratios of stiffness to weight can be achieved. As a crude guide to the material proportions, an efficient sandwich is obtained when the weight of the core is roughly equal to the total weight of the facings. Sandwich construction is also economical, for the relatively expensive facing materials are used in much smaller quantity than the usually inexpensive core materials. The materials are positioned so that each is used to its best advantage.

Specific nonstructural advantages can be incorporated in a sandwich construction by proper selection of facing and core materials. An impermeable facing can act as a moisture barrier for a wall or roof panel in a house; an abrasion-resistant facing can be used for the top facing of a floor panel; and decorative effects can be obtained by using panels with plywood or plastic facings for walls, doors, tables, and other furnishings. Core material can be chosen to provide thermal insulation, fire resistance, and decay resistance. Because of the light weight of the construction, sound transmission problems must also be considered in choosing sandwich component parts.

Methods of joining sandwich panels to each other and to other structures must be planned so that the joints function properly and allow for possible dimensional change due to temperature and moisture variations. Both structural and non-structural advantages need to be analyzed in light of the strength and service requirements for the sandwich construction. Moisture-resistant facings, cores, and adhesives should be employed if the construction is to be exposed to adverse moisture conditions. Similarly, heat-resistant or decay-resistant facings, cores, and adhesives should be used if exposure to elevated temperatures or decay organisms is expected.

Fabrication of Sandwich Panels

Facing Materials

One of the advantages of sandwich construction is the great latitude it provides in choice of facings and the opportunity to use thin sheet materials because of the nearly continuous support by the core. The stiffness, stability, and, to a

Figure 12–1—A cutaway section of sandwich construction with plywood facings and a paper honeycomb core. (M93 157)

large extent, the strength of the sandwich are determined by the characteristics of the facings. Some of the different facing materials used include plywood, single veneers or plywood overlaid with a resin-treated paper, fiberboard, particleboard, glass-fiber-reinforced polymers or laminates, veneer bonded to metal, and such metals as aluminum, enameled steel, stainless steel, magnesium, or titanium.

Core Materials

Many lightweight materials, such as balsa wood, rubber foam, resin-impregnated paper, reinforced plastics, perforated chipboard, several kinds of expanded plastics, foamed glass, lightweight concrete and clay products, and formed sheets of cloth, metal, or paper, have been used as core for sandwich construction. New materials and new combinations of old materials are constantly being proposed and used. Cores of formed sheet materials are often called honeycomb cores. By varying the sheet material, sheet thickness, cell size, and cell shape, cores of a wide range in density can be produced. Various core configurations are shown in figures 12–2 and 12–3. The core cell configurations shown in figure 12–2 can be formed to moderate amounts of single curvature, but cores shown in figure 12–3 as configurations A, B, and C can be formed to severe single curvature and mild compound curvature (spherical).

---

* Revision by Jen Y. Liu, General Engineer.

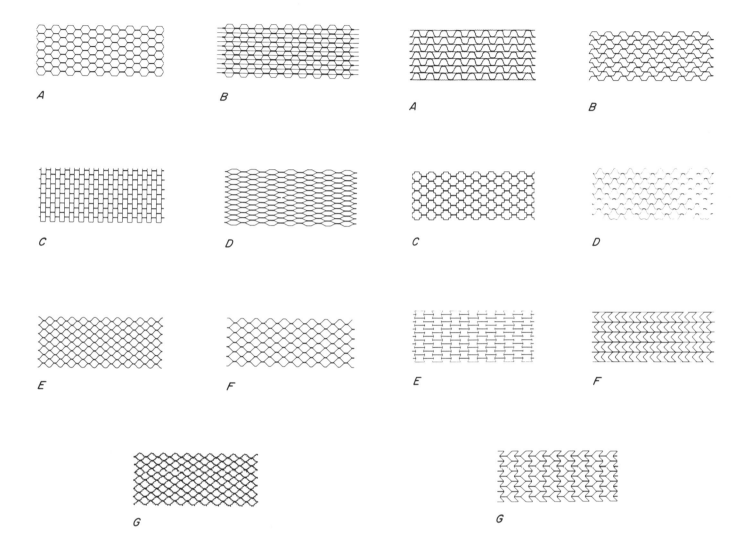

Figure 12–2—Honeycomb core cell configurations. (M134 056)

Figure 12–3—Cell configurations for formable paper honeycomb cores. (M134 055)

Four types of readily formable cores are shown as configurations *D*, *E*, *F*, and *G* in figure 12–3. The type *D* and *F* cores form to cylindrical shape, the type *D* and *E* cores to spherical shape, and the type *D* and *G* cores to various compound curvatures.

If the sandwich panels are likely to be subjected to damp or wet conditions, a core of paper honeycomb should contain a synthetic resin. Paper with 15 percent phenolic resin provides good strength when wet, decay resistance, and desirable handling characteristics during fabrication. Resin amounts in excess of about 15 percent do not seem to produce a gain in strength commensurate with the increased quantity of resin required. Smaller amounts of resin may be combined with fungicides to offer primary protection against decay.

Manufacturing Operations

The principal operation in the manufacture of sandwich panels is bonding the facings to the core. Special presses are needed for sandwich panel manufacture to avoid crushing lightweight cores, because the pressures required are usually lower than can be obtained in the range of good pressure control on presses ordinarily used for plywood or plastic products. Because pressure requirements are low, simple and perhaps less costly presses could be used. Continuous roller presses or hydraulic pressure equipment may also be suitable. In the pressing of sandwich panels, certain special problems may arise, but the manufacturing process is basically not complicated.

Adhesives must be selected and applied to provide the necessary joint strength and permanence. The facing materials, especially if metallic, may need special surface treatment before the adhesive is applied.

In certain sandwich panels, loading rails or edgings are placed between the facings at the time of assembly. Special fittings or equipment, such as heating coils, plumbing, or electrical wiring conduit, can easily be installed in the panel before its components are fitted together.

Some of the most persistent difficulties in the use of sandwich panels are caused by the necessity to introduce edges, inserts, and connectors in them. In some cases, the problem involves tying together thin facing materials without causing severe stress concentrations; in other cases, such as furniture manufacture, the problem is "showthrough" of core or inserts through decorative facings. These difficulties are minimized by a choice of materials in which the rate and degree of differential dimensional movement between core and insert are at a minimum.

Structural Design of Sandwich Construction

The structural design of sandwich construction may be compared to the design of an I—beam. The facings and core of the sandwich are analogous to the flanges and web of the I—beam, respectively. The two thin and stiff facings, separated by a thick and light core, are to carry the bending loads. The functions of the core are to support the facings against lateral wrinkling caused by in-plane compressive loads, and to carry, through the bonding adhesive, shearing loads. When the strength requirements for the facings and core in a particular design are met, the construction should also be checked for possible buckling, as for a column or panel in compression, and for possible wrinkling of the facings.

The contribution of the core material to the stiffness of the sandwich construction can be neglected because of its low modulus of elasticity; therefore, the shear stress may be assumed constant over the depth of the core. The facing moduli of elasticity are usually more than 100 times as great as the core modulus of elasticity. The core material may also have a small shear modulus. This small shear modulus causes increased deflections of sandwich constructions subjected to bending and decreased buckling loads of columns and edge-loaded panels, compared to constructions in which the core shear modulus is large. The effect of this low shear modulus is greater for short beams and columns and small panels than it is for long beams and columns and large panels.

The bending stiffness of sandwich beams having facings of equal or unequal thickness is given by:

$$D = \frac{h^2 t_1 t_2 (E_1 t_2 + E_2 t_1)}{(t_1 + t_2)^2} + \frac{1}{12} (E_1 t_1^3 + E_2 t_2^3) \qquad (12-1)$$

where D is the stiffness per unit width of sandwich construction (product of modulus of elasticity and moment of inertia of the cross section); E_1, E_2 are the moduli of elasticity of facings 1 and 2; t_1, t_2 are the facing thicknesses; and h is the distance between facing centroids.

The shear stiffness per unit width is given by:

$$U = \frac{h^2}{t_c} G_c \qquad (12-2)$$

where G_c is the core shear modulus associated with distortion of the plane perpendicular to the facings and parallel to the sandwich length; and t_c is the thickness of the core.

The bending stiffness, D, and shear stiffness, U, are used to compute deflections and buckling loads of sandwich beams.

The general expression for the deflection of flat sandwich beams is given by:

$$\frac{d^2 y}{dx^2} = -\frac{M_x}{D} + \frac{1}{U} \left(\frac{dS_x}{dx} \right) \qquad (12-3)$$

where y is deflection; x is distance along the beam; M_x is the bending moment per unit width at point x; and S_x is shearing force per unit width at point x. Integration of this equation leads to the following general expression for deflection of a sandwich beam:

$$y = \frac{k_b P a^3}{D} + \frac{k_s P a}{U} \qquad (12-4)$$

where P is total load per unit width of beam; a is span; and k_b and k_s are constants dependent upon the loading condition. The first term in the right side of this formula gives the bending deflection and the second term the shear deflection. Values of k_b and k_s for several loadings are given in table 12—1.

For sandwich panels supported on all four edges, the theory of plates must be applied to obtain analytical solutions. A comprehensive treatment of sandwich plates under various loading conditions can be found in the Selected References at the end of this chapter.

The buckling load per unit width of a sandwich column is given by:

$$N = \frac{N_E}{1 + \dfrac{N_E}{U}} \qquad (12-5)$$

where the Euler load, N_E, is

$$N_E = \frac{\pi^2 n^2 D}{a^2} \qquad (12-6)$$

in which n is the number of half waves into which the column buckles and a is the panel length. For a simply supported column, $n = 1$. This buckling form is often called "general buckling" and is illustrated in sketch A of figure 12−4.

The buckling load, N, is often expressed in the equivalent form:

$$\frac{1}{N} = \frac{1}{N_E} + \frac{1}{U} \qquad (12-7)$$

When U is finite, N is less than the Euler load, N_E; when U is infinite, $N = N_E$; and when N_E is infinite, i.e. $n = \infty$ in the formula for N_E, $N = U$, which is often called the "shear instability" limit. The appearance of this buckling failure resembles a crimp as illustrated in sketch B of figure 12−4. "Shear instability" or "crimping" failure is always possible for edge-loaded sandwich and is a limit for general instability and not a localized failure.

For a sandwich panel under edge load and with edge members, inserts perhaps, the edge members will carry a load proportional to their transformed area (area multiplied by ratio of edge member modulus of elasticity to the facing modulus of elasticity). Edge members will also raise the overall

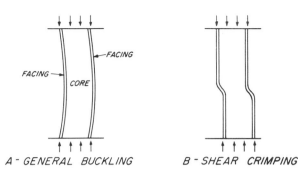

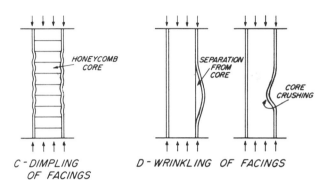

Figure 12−4—Modes of failure of sandwich construction under edgewise loads. A, general buckling; B, shear crimping; C, dimpling of facings; D, wrinkling of facings either away from or into core.　(M117 844)

panel buckling load because of restraints at edges. Buckling criteria for flat rectangular sandwich panels under edgewise compressive load, shear load, bending load, and combined loads have all been investigated by various researchers. Details can be found in Military Handbook 23A of U.S. Department of Defense and some other publications listed in the Selected References.

Table 12−1—*Values of k_b and k_s for several beam loadings*

Loading	Beam ends	Deflection at—	k_b	k_s
Uniformly distributed	Both simply supported	Midspan	5/384	1/8
	Both clamped	Midspan	1/384	1/8
Concentrated at midspan	Both simply supported	Midspan	1/48	1/4
	Both clamped	Midspan	1/192	1/4
Concentrated at outer quarter points	Both simply supported	Midspan	11/768	1/8
	Both simply supported	Load point	1/96	1/8
Uniformly distributed	Cantilever, 1 free, 1 clamped	Free end	1/8	1/2
Concentrated at free end	Cantilever, 1 free, 1 clamped	Free end	1/3	1

Buckling of sandwich walls of cylinders under torsion, axial compression or bending, and combined loads has also been presented in Military Handbook 23A.

Buckling of sandwich components has been emphasized because it causes complete failure, usually producing severe shear crimping at the edges of the buckles. Another important factor is the necessity that the facing stress be no more than its allowable value at the design load. The facing stress is obtained by dividing the load by the facing area under load. For an edgewise compressive load per unit width, N, the facing stress is given by:

$$ f = \frac{N}{t_1 + t_2} \qquad (12-8) $$

In a strip of sandwich construction subjected to bending moments, the mean facing stresses are given by:

$$ f_i = \frac{M}{t_i h}, \; i = 1,2 \qquad (12-9) $$

where f_i is the mean compressive or tensile stress per unit width in facing i, and M is the bending moment per unit width of sandwich. If the strip is subjected to shear loads, the shear stress in the core is given by:

$$ f_{cs} = \frac{S}{h} \qquad (12-10) $$

where S is the applied shear load per unit width of sandwich.

Localized failure of sandwich must be avoided. Such failure is shown as dimpling of the facings in sketch C and wrinkling of the facings in sketch D of figure 12-4. The stress at which dimpling of the facing into a honeycomb core begins is given by the empirical formula:

$$ f_d = 2E \left(\frac{t_f}{s} \right)^2 \qquad (12-11) $$

where f_d is facing stress at dimpling, E is effective facing modulus of elasticity at stress f_d, t_f is facing thickness, and s is cell size of honeycomb core (diameter of inscribed circle). Increase in dimpling stress can be attained by decreasing the cell size. Wrinkling of the sandwich facings can occur because they are thin and supported by a lightweight core which functions as their elastic foundation. It can also occur because of poor facing-to-core bond, resulting in a separation of facing from core (fig. 12-4,D). Increase in bond strength should produce wrinkling by core crushing. Thus a convenient rule

of thumb is to require that the sandwich flatwise tensile strength (bond strength) is no less than flatwise compressive core strength. Wrinkling and other forms of local instability are described in detail in Military Handbook 23A and in the book by Allen. Localized failure is not accurately predictable and designs should be checked by ASTM tests of small specimens.

Because sandwich constructions are composed of several materials, it is often of interest to attempt to design a construction of minimum weight for a particular component. One introduction to the problem of optimum design is presented by Kuenzi. For a sandwich with similar facings having a required bending stiffness, D, the dimensions for the minimum weight design are given by:

$$ h = 2 \left(\frac{Dw}{Ew_c} \right)^{1/3} \qquad (12-12) $$

$$ t = \frac{w_c}{4w} h $$

where h is distance between facing centroids, t is facing thickness, E is facing modulus of elasticity, w is facing density, and w_c is core density. The resulting construction will have very thin facings on a very thick core and will be proportioned so that the total core weight is two-thirds the total sandwich weight minus bond weight. However, such a construction may be impracticable as the required facings may be too thin to be practical.

Many detailed design procedures necessary for rapid design of sandwich components for aircraft are summarized in Military Handbook 23A. The principles contained therein and in some other publications in the Selected References are broad and can be applied to sandwich components of all structures.

Dimensional Stability, Durability, and Bowing of Sandwich Panels

In a sandwich panel any dimensional movement of one facing with respect to the other due to changes in moisture content and temperature causes bowing of an unrestrained panel. Thus, although the use of dissimilar facings is often desirable from an economic or decorative standpoint, the dimensional instability of the facings during panel manufacture or exposure may rule out possible benefits. If dimensional change of both facings is equal, the length and width of the panel will increase or decrease but bowing will not result.

The problem of dimensional stability is chiefly related to the facings, because the core is not stiff enough either to cause bowing of the panel or to cause it to remain flat. The magnitude of the bowing effect, however, depends on the thickness of the core.

It is possible to calculate mathematically the bowing of a sandwich construction if the percentage of expansion of each facing is known. The maximum deflection is given approximately by:

$$\Delta = \frac{ka^2}{800h} \qquad (12-13)$$

where k is the percentage of expansion of one facing as compared to the opposite facing; a, the length of the panel; and h, the distance between facing centroids.

In conventional construction, vapor barriers are often installed to block migration of vapor to the cold side of a wall. Various methods have been tried or suggested for reducing vapor movement through sandwich panels, which causes a moisture differential with resultant bowing of the panels. These include bonding metal foil within the sandwich construction; blending aluminum flakes with the resin bonding adhesives; and using plastic vapor barriers between veneers, overlay papers, special finishes, or metal or plastic facings. Because added cost is likely, some of these methods should not be resorted to unless need for them has been demonstrated.

A large test unit simulating use of sandwich panels in houses was constructed at the Forest Products Laboratory. The panels used, consisting of a variety of facing materials including plywood, aluminum, particleboard, fiberboard, paperboard, and cement asbestos, with cores of paper honeycomb, polyurethane or extruded polystyrene, were evaluated for bowing and general performance after various lengths of service between 1947 and 1978. The experimental assembly shown in figure 12−5 represents the type of construction used in the test unit. The major conclusions were that bowing was least for aluminum-faced panels, and was greater for plywood-faced panels with polyurethane or polystyrene cores than for plywood-faced panels with paper cores, and that with proper combinations of facings, core, and adhesives, satisfactory sandwich panels can be assured by careful fabrication techniques.

Thermal Insulation of Sandwich Panels

Satisfactory thermal insulation can best be obtained with sandwich panels by using cores having low thermal conductivity, although the use of reflective layers on the facings is of some value. Paper honeycomb cores have thermal conductivity values (k values) ranging from 0.30 to 0.65 British thermal units per hour per 1 °F per square foot per inch of

Figure 12−5—A cut-away to show details of sandwich construction in an experimental structure.

(M76 939)

thickness, depending on the particular core construction. The k value does not vary linearly with core thickness for a true honeycomb core because of direct radiation through the core cell opening from one facing to the other. Honeycomb with open cells can also have greater conductivity if cells are large enough (larger than about 3/8 in) to allow convection currents to develop.

An improvement in the insulation value can be realized by filling the honeycomb core with fill insulation or a foamed-in-place resin.

Fire Resistance of Sandwich Panels

In tests at the Forest Products Laboratory, the fire resistance of wood-faced sandwich panels was appreciably higher than that of hollow panels faced with the same thickness of plywood. Fire resistance was greatly increased when coatings that intumesce on exposure to heat were applied to the core material. The spread of fire through the honeycomb core depended to a large extent on the alignment of the flutes in the core. In panels having flutes perpendicular to the facings, only slight spread of flame occurred. In cores in which flutes were parallel to the length of the panel, the spread of flame occurred in the vertical direction along open channels. Resistance to flame spread could be improved by placing a barrier sheet at the top of the panel or at intervals in the panel height, or, if strength requirements permit, by simply turning the length of the core blocks at 90° to the vertical direction.

Selected References

Allen, H. G. Analysis and design of structural sandwich panels. Oxford, England: Pergamon; 1969.

American Society for Testing and Materials. Methods of test for structural sandwich constructions. ASTM Section 15, Vol. 15.03. Philadelphia, PA: ASTM; (see current edition).

Cheng, S. Torsion of sandwich plates of trapezoidal cross section. Journal of Applied Mechanics, American Society of Mechanical Engineering Transactions. 28(3): 363−366; 1961.

Fazio, P.; Rizzo, S. Nonlinear elastic analysis of panelized shear sandwich walls. American Society of Civil Engineering, Journal of Structural Division. 105(6): 1187−1203; 1979.

Ha, K. H.; Hussein, R.; Fazio, P. Analytic solution for continuous sandwich plates. Proceedings, American Society of Civil Engineers, Engineering Mechanics Division. 108(EM2): 228−241; 1982.

Holmes, Carlton A.; Eickner, Herbert W.; Brenden, John J.; Peters, Curtis C.; White, Robert H. Fire development and wall endurance in sandwich and wood-frame structures. Res. Pap. FPL 364. Madison, WI: U.S. Department of Agriculture, Forest Service, Forest Products Laboratory; 1980.

Jones, Robert E. Field sound insulation of load-bearing sandwich panels for housing. Noise Control Engineering. 16(2): 90−105; 1981.

Kimel, W. R. Elastic buckling of a simply supported rectangular sandwich panel subjected to combined edgewise bending and compression—results for panels with facings of either equal or unequal thickness and with orthotropic cores. FPL Rep. 1857−A. Madison, WI: U.S. Department of Agriculture, Forest Service, Forest Products Laboratory; 1956.

Kuenzi, E. W. Minimum weight structural sandwich. Res. Note FPL−086. Madison, WI: U.S. Department of Agriculture, Forest Service, Forest Products Laboratory; 1970.

Kuenzi, E. W.; Bohannan, B.; Stevens, G. H. Buckling coefficients for sandwich cylinders of finite length under uniform external lateral pressure. Res. Note FPL−0104. Madison, WI: U.S. Department of Agriculture, Forest Service, Forest Products Laboratory; 1965.

Kuenzi, E. W.; Ericksen, W. S.; Zahn, J. J. Shear stability of flat panels of sandwich construction. FPL Rep. 1560. Madison, WI: U.S. Department of Agriculture, Forest Service, Forest Products Laboratory; rev. 1962.

Lewis, W. C. Thermal insulation from wood for buildings: Effects of moisture control. Res. Pap. FPL 86. Madison, WI: U.S. Department of Agriculture, Forest Service, Forest Products Laboratory; 1968.

Miyairi, H.; Nagai, M.; Muramatsu, A. Flexural fracture and mechanical properties of nonsymmetrical sandwich construction. Bulletin of Japanese Society of Mechanical Engineering. 21(161): 1588−1594; 1978.

Murase, K.; Nishimura, T. Analysis of sandwich plates by finite element method and its applications. Bulletin of Japanese Society of Mechanical Engineering. 20(144): 680−687; 1977.

Palms, J.; Sherwood, G. E. Structural sandwich performance after 31 years of service. Res. Pap. FPL 342. Madison, WI: U.S. Department of Agriculture, Forest Service, Forest Products Laboratory; 1979.

Plantema, F. J. Sandwich construction. New York: John Wiley; 1966.

Rizzo, S.; Fazio, P. Comprehensive strength analysis of framed sandwich panels. Canadian Journal of Civil Engineering. 6(4): 514−522; 1979.

Seidl, R. J. Paper honeycomb cores for structural sandwich panels. FPL Rep. R1918. Madison, WI: U.S. Department of Agriculture, Forest Service, Forest Products Laboratory; 1952.

Sherwood, G. E. Longtime performance of sandwich panels in Forest Products Laboratory experimental unit. Res. Pap. FPL 144. Madison, WI: U.S. Department of Agriculture, Forest Service, Forest Products Laboratory; 1970.

Smolenski, C.; Krokosky, E. Optimal multi-factor design procedure for sandwich panels. American Society of Civil Engineers Journal of Structural Division. 96(ST4): 823−837; 1970.

U.S. Department of Defense. Structural sandwich composites. Military Handbook 23A. Washington, DC: Superintendent of Documents; 1968.

Zahn, J. J.; Cheng, S. Edgewise compressive buckling of flat sandwich panels: Loaded ends simply supported and sides supported by beams. Res. Note FPL−019. Madison, WI: U.S. Department of Agriculture, Forest Service, Forest Products Laboratory; 1964.

Chapter 13

Bent Wood Members

Bent Wood Members *

Bending can provide a variety of functional and esthetically pleasing wood members, ranging from large curved arches to small furniture components. Bent wood may be formed with or without softening or plasticizing treatments and with or without end pressure. The curvature of the bend, size of the member, and intended use of the product determine the production method.

Laminated Members

In the United States, curved pieces of wood were once laminated chiefly to produce such small items as parts for furniture and pianos. However, the principle was extended to the manufacture of arches for roof supports in farm, industrial, and public buildings (fig. 13−1) and other types of structural members.

Both softwoods and hardwoods are suitable for laminated bent structural members, and thin material of any species can be bent satisfactorily for such purposes. The choice of species and adhesive depends primarily on the cost, required strength, and demands of the application (see chs. 9 and 10).

Laminated curved members are produced from dry stock in a single bending and gluing operation. This process has several advantages over bending single-piece members:

(1) Bending thin laminations to the required radius involves only moderate stress and deformation of the wood fibers, eliminating the need for treatment with steam or hot water and associated drying and conditioning of the finished product. Also, the moderate stresses involved in curving laminated members result in stronger members when compared to curved single-piece members.

(2) The tendency of laminated members to change shape with changes in moisture content is less than that of single-piece bent members.

(3) Ratios of thickness of member to radius of curvature that are impossible to obtain by bending single pieces can be attained readily by laminating.

(4) Curved members of any desired length can be produced.

Design criteria for glued-laminated timbers are discussed in chapter 10.

Straight laminated members can be steamed and bent after they are glued. However, this type of procedure requires an adhesive that will not be affected by the steaming or boiling treatment and complicates conditioning of the finished product.

Curved Plywood

Curved plywood is produced (1) by bending and gluing the plies in one operation, or (2) by bending previously glued flat plywood. Plywood curved by bending and gluing simultaneously is more stable in curvature than plywood curved by bending previously glued material.

Plywood Bent and Glued Simultaneously

In bending and gluing plywood in a single operation, glue-coated pieces of veneer are assembled and pressed over or between curved forms; pressure and sometimes heat are applied through steam or electrically heated forms until the glue sets and holds the assembly to the desired curvature. Some of the laminations are at an angle, usually 90°, to other laminations, as in the manufacture of flat plywood. The grain direction of the thicker laminations is normally parallel to the axis of the bend to facilitate bending.

A high degree of compound curvature can be obtained in an assembly comprising a considerable number of thin veneers. First, for both the face and back of the assembly, the two outer plies are bonded at 90° to each other in a flat press. The remaining veneers are then glue-coated and assembled at any desired angle to each other. The entire assembly is hot-pressed to the desired curvature.

Bonding the two outer plies before molding allows a higher degree of compound curvature without cracking the face plies than could otherwise be obtained. Where a high degree of compound curvature is required, the veneer should be relatively thin, 1/32 inch or less, with a moisture content of about 12 percent.

The molding of plywood with fluid pressure applied by flexible bags of some impermeable material produces plywood parts of various degrees of compound curvature. In "bag molding," fluid pressure is applied through a rubber bag by air, steam, or water. The veneer may be wrapped around a form and the whole assembly enclosed in a bag and subjected to pressure in an autoclave, the pressure in the bag being "bled." Or the veneer may be inserted inside a metal form and, after the ends have been attached and sealed, pressure applied by inflating a rubber bag. The form may be heated electrically or by steam.

The advantages of bending and gluing plywood simultaneously to form a curved shape are similar to those for curved laminated members, and in addition, the cross plies give the curved members properties characteristic of cross-banded plywood. Curved plywood shells for furniture manufacture are examples of these bent veneer and glued products.

*Revision by William T. Simpson, Forest Products Technologist; Ronald W. Jokerst, Forest Products Technologist; and Sidney R. Boone, Forest Products Technologist.

Figure 13–1—Curved laminated arch provides pleasing lines in this building. The laminations are bent without end pressure against a form and glued together.

(M139 409)

Plywood Bent After Gluing

After the plies are glued together, flat plywood is often bent by methods that are somewhat similar to those used in bending solid wood. To bend plywood properly to shape, it must be plasticized by some means, usually moisture or heat, or a combination of both. The amount of curvature that can be introduced into a flat piece of plywood depends on numerous variables, such as moisture content, direction of grain, thickness and number of plies, species and quality of veneer, and the technique applied in producing the bend. Plywood is normally bent over a form or a bending mandrel.

Flat plywood glued with a waterproof adhesive can be bent to compound curvatures after gluing. No simple criterion, however, is available for predetermining whether a specific compound curvature can be imparted to flat plywood. Soaking the plywood prior to bending, and the use of heat during forming are aids in manipulation. Normally the plywood to be postformed is first thoroughly soaked in hot water and then dried between heated male and female dies attached to a hydraulic press. If the use of postforming for bending flat plywood to compound curvatures is contemplated, exploratory trials to determine the practicability and the best procedure are recommended. It should be remembered that in postforming plywood to compound curvatures, all of the deformation must be by compression or shear, as plywood cannot be stretched. Hardwood species, such as birch, poplar, and gum, are normally used in plywood that is to be postformed.

13-3

Veneered Curved Members

Veneered curved members are usually produced by gluing veneer to one or both faces of a curved solid wood base. The bases are ordinarily bandsawed to the desired shape or bent from a piece grooved with saw kerfs on the concave side at right angles to the directions of bend. Pieces bent by making saw kerfs on the concave side are commonly reinforced and kept to the required curvature by gluing splines, veneer, or other pieces to the curved base.

Veneering over curved solid wood finds use mainly in furniture. The grain of the veneer is commonly laid in the same general direction as the grain of the curved wood base. The use of crossband veneers, that is, veneers laid with the grain at right angles to the grain of the back and face veneer, reduces the tendency of the member to split.

Bending Solid Members

Principles of Bending

Wood of certain species steamed or soaked in boiling water can be compressed as much as 25 to 30 percent parallel to the grain. The same wood can be stretched only 1 to 2 percent. Because of the relation between attainable tensile and compressive deformations, it is necessary, if bending involves severe deformation, that most of the deformation be forced to take place as compression. The inner or concave side must assume the maximum amount of compression and the outer or convex side, zero strain or a slight tension. To accomplish this, a metal strap equipped with end fittings is customarily used. The strap makes contact with the outer or convex side and, acting through the end fittings, places the whole piece of wood in compression. The tensile stress that would normally develop in the outer side of the piece of wood during bending is borne by the metal strap. An example of a bending form is shown in figure 13−2.

Selection of Stock

In general, hardwoods possess better bending quality than softwoods, and certain hardwoods surpass others in this quality. The species commonly used to produce bent members are: White oak, red oak, elm, hickory, ash, beech, birch, maple, walnut, sweetgum, and mahogany. Softwoods have poor bending quality, as a rule, and are not often used in bending operations. Yew and Alaska-cedar are probable exceptions to this rule. Douglas-fir, southern pine, northern and Atlantic white-cedar, and redwood are used for ship and boat planking, for which purpose they are often bent to moderate curvature after being steamed or soaked.

Bending stock should be free from serious cross grain and distorted grain, such as may occur near knots. The slope of cross grain should not be steeper than about 1 to 15. Decay, knots, shake, pith, surface checks, and exceptionally light or brashy wood should be avoided. Such irregularities tend to cause the pieces of wood to fail during bending, particularly if they are on the face that is to be the inner or concave side at the time of bending and if the piece is to be bent to a sharp curvature.

Moisture Content of Bending Stock

Although green wood can be bent to produce many curved members, difficulties are encountered in drying and fixing the bend. Another disadvantage with green stock is that hydrostatic pressure may be developed during bending. Hydrostatic pressure may cause compression failures on the concave side if the wood is compressed by an amount greater than the air space in the cells of the green wood.

Bending stock that has been dried to a low moisture content requires a long steaming or soaking process to increase its moisture content to the point where it can be made sufficiently plastic for successful bending. For most chair and furniture parts, the moisture content of the bending stock should be 12 to 20 percent before it is steamed. The preferred moisture content varies with the severity of the curvature to which the wood is bent and the method used in drying and fixing the bend member. For example, chair-back slats, which have a slight curvature and are subjected to severe drying conditions between steam-heated platens, can be produced successfully from stock at a moisture content of 12 percent. For furniture parts with a more severe bend where the part must be bent over a form, a moisture content of 15 to 20 percent is recommended.

Plasticizing of Bending Stock

Heat and moisture make certain species of wood sufficiently plastic for bending operations. Steaming at atmospheric or a low gage pressure or soaking in boiling or nearly boiling water are satisfactory methods of plasticizing wood. Wood at 20 to 25 percent moisture content needs to be heated without losing moisture; at lower moisture content, heat and moisture must be added. As a consequence, the recommended plasticizing processes are steaming or boiling for about 1/2 hour per inch of thickness for wood at 20 to 25 percent moisture content and steaming or boiling for about 1 hour per inch of thickness for wood at lower moisture content values. Steaming at high pressures causes wood to become plastic, but wood treated with high-pressure steam generally does not bend as successfully as wood treated at atmospheric or low pressure.

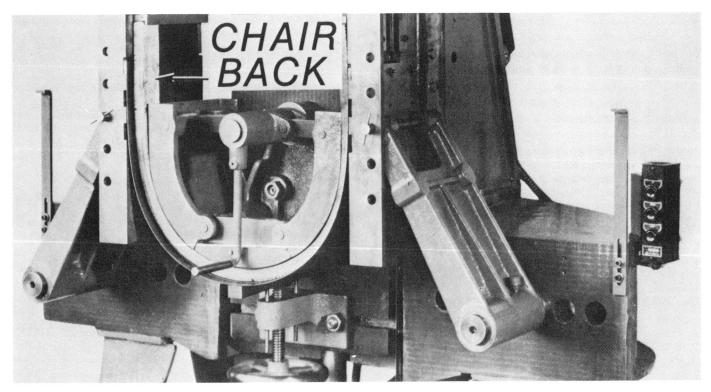

Figure 13–2—Chair back being bent through an arc of 180° in a bending machine.

(M105 257)

Wood can be plasticized by a great variety of chemicals. Such chemicals behave like water, in that they are adsorbed and cause swelling. Common chemicals that plasticize wood include urea, dimethylol urea, low-molecular-weight phenol-formaldehyde resin, dimethyl sulfoxide, and liquid ammonia. Urea and dimethylol urea have received limited commercial attention, and a free-bending process using liquid ammonia has been patented. Wood members can be readily molded or shaped after immersion in liquid ammonia or treatment under pressure with ammonia in the gas phase. As the ammonia evaporates, the wood stiffens and retains its new shape.

Bending Operation and Apparatus

After being plasticized, the stock should be quickly placed in the bending apparatus and bent to shape. The bending apparatus consists essentially of a form (or forms) and a means of forcing the piece of steamed wood against the form. If the curvature to be obtained demands a difference of much more than 3 percent between lengths of the outer and inner surfaces of the pieces, then the apparatus should include a device for applying end pressure. This generally takes the form of a metal strap or pan provided with end blocks, end bars, or clamps.

Fixing the Bend

After being bent, the piece should be cooled and dried while held in its curved shape. One method is to dry the piece in the bending machine between the plates of a hot-plate press. Another method is to secure the bent piece to the form and place both the piece and the form in a drying room. Still another is to keep the bent piece in a minor strap with tie rods or stays, so that it can be removed from the form and placed in a drying room. When the bent member has dried to a moisture content suited for its intended use, the restraining devices can be removed and the piece will hold its curved shape.

Characteristics of Bent Wood

After a bent piece of wood is dried, the curvature will be maintained unless the wood undergoes changes in moisture content. An increase in moisture content causes the piece to lose some of its curvature. A decrease in moisture content causes the curve to become sharper, although repeated changes in moisture content bring about a gradual straightening. These changes are caused primarily by lengthwise swelling or shrinking of the inner (concave) face, the fibers of which were

13-5

wrinkled or folded during the bending operation.

A bent piece of wood possesses less strength than a similar but unbent piece. The reduction in strength brought about by bending, however, is seldom serious enough to affect the utility value of the member.

Selected References

Clark, W. M. Veneering and wood bending in the furniture industry. New York: Pergamon Press; 1965.

Davidson, R. W. Plasticizing wood with anhydrous ammonia. Syracuse, NY: New York State College of Forestry; n.d.

Heebink, B. G. Fluid-pressure molding of plywood. FPL Rep. 1624. Madison, WI: U.S. Department of Agriculture, Forest Service, Forest Products Laboratory; 1959.

Hoadley, R. B. Understanding wood: a craftsman's guide to wood technology. Newtown, CT: The Taunton Press; 1980.

Hurst, K. Plywood bending. Australian Timber Journal.; June 1962.

Jorgensen, R. N. Furniture wood bending, Part I. Furniture Design and Manufacturing; Dec. 1965.

Jorgensen, R. N. Furniture wood bending, Part II. Furniture Design and Manufacturing; Jan. 1966.

McKean, H. B.; Blumenstein, R. R.; Finnorn, W. F. Laminating and steam bending of treated and untreated oak for ship timbers. Southern Lumberman. 185: 2321; 1952.

Peck, E. C. Bending solid wood to form. Agric. Handb. 125. Washington, DC: U.S. Department of Agriculture; 1957.

Perry, T. D. Curves from flat plywood. Wood Products. 56(4); 1951.

Schuerch, C. Principles and potential of wood plasticization. Madison, WI: Forest Products Journal. 14(9): 377−381; 1964.

Stevens, W. C.; Turner, N. Wood bending handbook. London, England: Her Majesty's Stationery Office; 1970.

Chapter 14

Control of Moisture Content and Dimensional Changes

Control of Moisture Content and Dimensional Changes*

Correct drying, handling, and storage of wood will minimize moisture content changes that might occur in service. If moisture content is controlled within reasonable limits by such methods, major problems from dimensional changes will be avoided. Wood is subject naturally to dimensional changes. In the living tree, wood contains large quantities of water. As green wood dries, most of this water is removed. The moisture remaining in the wood tends to come to equilibrium with the relative humidity of the air. Also, when the moisture content is reduced below the fiber saturation point, shrinkage starts to occur.

This discussion is concerned with moisture content determination, recommended moisture content values, drying methods, methods of calculating dimensional changes, design factors affecting such changes in structures, and moisture content control during transit, storage, and construction. Data on green moisture content, fiber saturation point, shrinkage, and equilibrium moisture content are given with information on other physical properties in chapter 3.

Wood in service is virtually always undergoing at least slight changes in moisture content. The changes in response to daily humidity changes are small and usually of no consequence. Changes due to seasonal variation, although gradual, tend to be of more concern. Protective coatings retard changes but do not prevent them.

Generally, no significant dimensional changes will occur if wood is fabricated or installed at a moisture content corresponding to the average atmospheric conditions to which it will be exposed. When incompletely dried material is used in construction, some minor changes can be tolerated if the proper design is used.

Determination of Moisture Content

The amount of moisture in wood is ordinarily expressed as a percentage of the weight of the wood when ovendry. Four methods of determining moisture content are covered by ASTM D 2016. Two of these, the ovendrying and the electrical method, are described here.

Ovendrying has been the most universally accepted method for determining moisture content, but it is slow and necessitates cutting the wood. In addition it gives values slightly higher than true moisture content with woods containing volatile extractives. The electrical method is rapid, does not require cutting the wood, and can be used on wood in place in a structure. However, considerable care must be taken to use and interpret the results correctly. Generally, use of the elec-

trical method is limited to moisture content values below 30 percent.

Ovendrying Method

In the ovendrying method, specimens are taken from representative boards or pieces of a quantity of lumber or other wood units. With lumber, the specimens should be obtained at least 20 inches from the ends of the pieces. They should be free from knots and other irregularities, such as bark and pitch pockets. Specimens from lumber should be full cross sections 1 inch in length. Specimens from larger items may be representative sectors of such sections or subdivided increment borer or auger chip samples. Convenient amounts of chips and particles can be selected at random from larger batches, with care being taken to ensure that the sample is representative of the batch. Samples of veneer should be selected from four or five locations in a sheet to ensure that the sample average will accurately indicate the average of the sheet.

Each specimen should be weighed immediately, before any drying or reabsorption of moisture has taken place. If the specimen cannot be weighed immediately after it is taken, it should be placed in a plastic bag or tightly wrapped in metal foil to protect it from moisture change until it can be weighed. After weighing, the specimen is placed in an oven heated to 214 to 221 °F (101 to 105 °C) and kept there until constant weight is reached. A lumber section will reach a constant weight in 12 to 48 hours. Smaller specimens will take less time.

The constant or ovendry weight and the weight of the specimen when cut are used to determine the percentage moisture content with the following formula:

$$\text{Percent moisture content} = \frac{\text{Weight when cut} - \text{Ovendry weight}}{\text{Ovendry Weight}} \times 100 \quad (14-1)$$

Electrical Method

The electrical method of determining the moisture content of wood uses the relationships between moisture content and measurable electrical properties of wood such as conductivity (or its inverse, resistivity), dielectric constant, or power-loss factor. These properties vary in a definite and predictable way with changing moisture content, but correlations are not perfect. Moisture determinations using electric methods therefore always are subject to some uncertainty.

Electric moisture meters are available commercially that are based on each of these properties, and are identified by

---

*Revision by William T. Simpson, Forest Products Technologist; William L. James, Physicist; and J. Dobbin McNatt, Forest Products Technologist.

the property measured. Conductance-type (or resistance) meters measure moisture content in terms of the direct current conductance of the specimen. Dielectric-type meters are of two types: Those based principally on dielectric constant are called capacitance or capacitive admittance meters, and those based on loss factor are called power-loss meters.

The principal advantages of the electrical method over the ovendrying method are its speed and convenience. Only a few seconds are required for the determination, and the piece of wood being tested is not cut or damaged, except for driving electrode needle points into the wood when using the resistance-type meters. Thus, the electrical method is adaptable to rapid sorting of lumber on the basis of moisture content, measuring the moisture content of wood installed in a building, or, when used in accordance with ASTM D 2016, establishing the moisture content of a quantity of lumber or other wood items.

For resistance meters, needle electrodes 5/16- to 7/16-inch long are appropriate for wood that has been in use for 6 months or longer, or for lumber up to 1-1/2 inches thick with a normal drying moisture gradient. For wood with normal moisture gradients, the pins should be driven to a depth of one-fifth to one-fourth of the wood thickness. For thicker specimens, electrodes with longer pins are available. If other than normal drying gradients are present, best accuracy can be obtained by exploring the gradient through readings made at various penetration depths.

Dielectric-type meters are fitted with surface contact electrodes designed for the type of specimen material being tested. The electric field from these electrodes penetrates well into the specimen, but with a strength that decreases rapidly with depth of penetration. For this reason the readings of dielectric meters are influenced predominantly by the surface layers of the specimen, and material near midthickness may not be adequately represented in the meter reading.

Ordinarily, moisture meters should not be used on lumber with wet or damp surfaces, because the wet surface will cause inaccurate readings. A resistance meter with insulated-pin electrodes can be used, with caution, on such stock.

Although some meters have scales that go up to 120 percent, the range of moisture content that can be measured reliably is 0 to about 30 percent for dielectric meters and about 6 to 30 percent for resistance meters. The precision of the individual meter readings decreases near the limits of these ranges. Any readings above 30 percent must be considered only qualitative. When the meter is properly used on a quantity of lumber dried to a reasonably constant moisture content below fiber saturation, the average moisture content from the corrected meter readings should be within 1 percent of the true average.

To obtain accurate moisture content values, each instrument should be used in accordance with its manufacturer's instructions. The electrodes should be appropriate for the material being tested and properly oriented according to manufacturers instructions. The readings should be carefully taken as soon as possible after inserting the electrode. A species correction supplied with the instrument should be applied when appropriate. Temperature corrections then should be made if the temperature of the wood differs considerably from the temperature of calibration used by the manufacturer. Approximate corrections for conductance-type (resistance) meters are to add or subtract about 0.5 percent for each 10 °F the wood differs from the calibration temperature; the correction factors are added to the readings for temperatures below the calibration temperature and subtracted from the readings for temperatures above this temperature.

Temperature corrections for dielectric meters are rather complex, and are best made from charts such as are published in report FPL−6, or in special calibration charts such as published by Bramhall.

Conditioning Wood Products

Wood and wood products are often conditioned to various levels of constant moisture content. In production it is often desirable or necessary to store solid or reconstituted wood products at optimum moisture content conditions while awaiting further fabrication into final products, or to store the final products themselves. In certain stages of many drying operations, it is necessary to control temperature and relative humidity so that conditions of constant moisture content are maintained. In research and testing, moisture content must often be a controlled variable because many properties of wood depend on moisture content.

Relative humidity and temperature are the principal variables that affect the equilibrium moisture content of wood. To hold a conditioning environment with any given tolerance of equilibrium moisture content, it is necessary to know the effect of relative humidity and temperature on equilibrium moisture content. This information is well known for wood and to a somewhat lesser extent for reconstituted wood products. However, the quantitative information is not available in a form that one can quickly and easily use to determine the necessary control of the variables to control equilibrium moisture conditions.

The required relative humidity control necessary to hold equilibrium moisture content of solid wood within prescribed limits is shown in figure 14−1. For example, to maintain control of equilibrium moisture control within ±1 percent moisture content at 80 °F and 30 percent relative humidity, it is necessary to control relative humidity within approximately ±7 percent relative humidity (i.e., between 23 and 37 percent relative humidity). To control equilibrium moisture con-

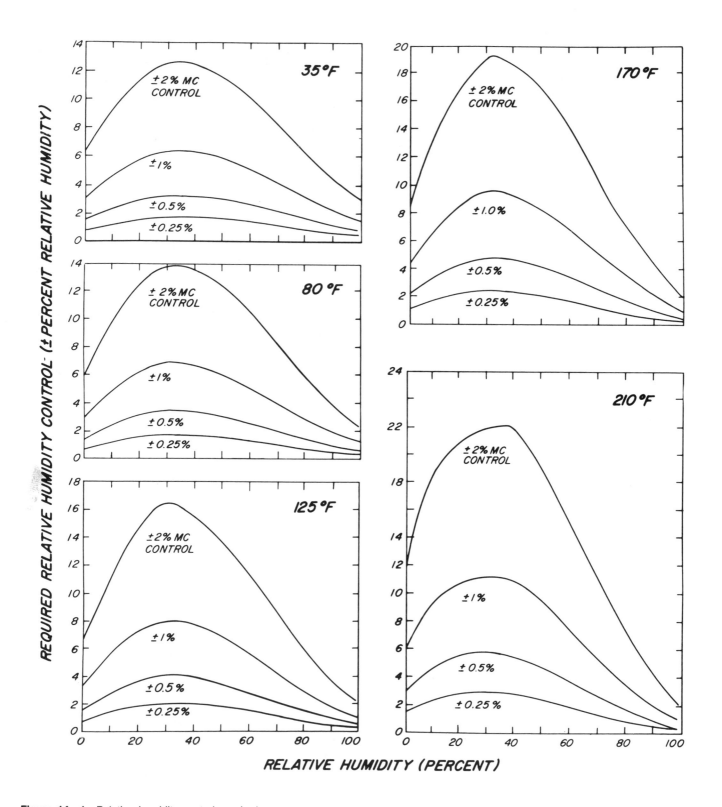

Figure 14–1—Relative humidity control required to maintain equilibrium moisture content of solid wood within required limits at various temperatures.

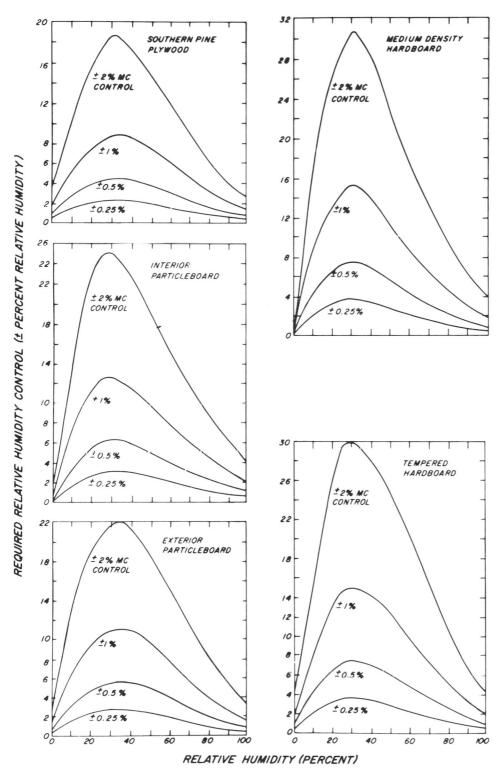

Figure 14−2—Relative humidity control required at 75 °F to maintain EMC within required limits for various composite wood products.

(ML84 5197)

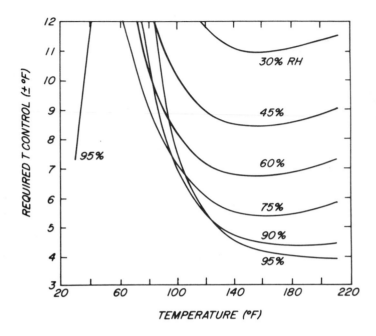

Figure 14—3—Temperature control required to maintain EMC of solid and composite wood products within ±0.25 percent moisture content.

(ML84 5818)

tent within these same limits at 90 percent relative humidity, it is necessary to maintain ±2 percent relative humidity control. At high levels of relative humidity, control must improve to maintain a given equilibrium moisture content tolerance. Similar information for several composite wood products is shown in figure 14—2.

The effect of temperature on equilibrium moisture content is less than the effect of relative humidity, and is shown in figure 14—3. For example, at 75 percent relative humidity and 120 °F, temperature control need only be ±6 °F to maintain equilibrium moisture content control within ±0.25 percent moisture content.

Recommended Moisture Content

Installation of wood at the moisture content percentages recommended here for different environments will reduce future changes in moisture content, thus minimizing dimensional changes after the wood is placed in service. The service condition to which the wood will be exposed—outdoors, in unheated buildings, or in heated and air-conditioned buildings—should be considered in determining moisture content requirements.

Timbers

Ideally, solid timbers should be dried to the average moisture content they will reach in service. While this optimum is possible with lumber less than 3 inches thick, it is seldom practical to obtain fully dried timbers, thick joists, and planks. When thick solid members are used, some shrinkage of the assembly should be expected. In the case of built-up assemblies such as roof trusses, it may be necessary to tighten the bolts or other fastenings from time to time to maintain full bearing of the connectors as the members shrink.

Lumber

The moisture content requirements are more exacting for finish lumber and for wood products used inside heated and air-conditioned buildings than those for lumber used outdoors or in unheated buildings. For general areas of the United States, the recommended moisture content values for wood used inside heated buildings are shown in figure 14—4. Values and tolerances both for interior and exterior use of wood in various forms are given in table 14—1. If the average moisture content value is within 1 percent of that recommended and all pieces fall within the individual limits, the entire lot is probably satisfactory.

General commercial practice is to kiln-dry wood for some products such as flooring and furniture, to a slightly lower moisture content than service conditions demand, anticipating a moderate increase in moisture content during processing and construction. This practice is intended to assure uniform distribution of moisture among the individual pieces. Common grades of softwood lumber and softwood dimension are not normally dried to the moisture content values indicated in table 14—1. When they are not, shrinkage effects should be considered in the structural design and construction methods.

Dry lumber as defined in the American Softwood Lumber Standard has a maximum moisture content of 19 percent. Much softwood dimension lumber meets this requirement. Some industry grading rules provide for even lower maximums. For example, to be grademarked KD (kiln dry) the maximum moisture content permitted is generally 15 percent.

Glued Wood Products

When veneers are bonded together with cold-setting glues to make plywood, they absorb comparatively large quantities of moisture. To keep the final moisture content low and to minimize redrying of the plywood, the initial moisture con-

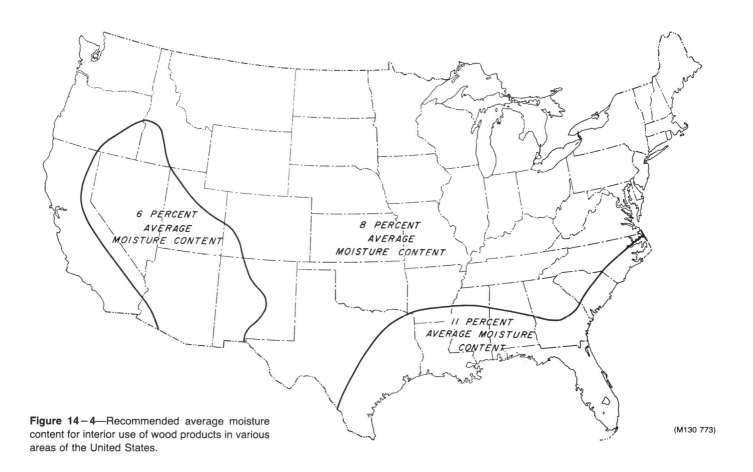

Figure 14–4—Recommended average moisture content for interior use of wood products in various areas of the United States.

(M130 773)

Table 14–1—*Recommended moisture content values for various wood items at time of installation*

Use of wood	Most areas of United States		Dry southwestern area[1]		Damp, warm coastal areas[1]	
	Average[2]	Individual pieces	Average[2]	Individual pieces	Average[2]	Individual pieces
	- Percent -					
Interior: Woodwork, flooring, furniture, wood trim, laminated timbers, cold-press plywood	8	6–10	6	4–9	11	8–13
Exterior: Siding, wood trim, framing, sheathing, laminated timbers	12	9–14	9	7–12	12	9–14

[1] Major areas are indicated in fig. 14–4.

[2] To obtain a realistic average, test at least 10 percent of each item. If the amount of a given item is small, several tests should be made. For example, in an ordinary dwelling having about 60 floor joists, at least 10 tests should be made on joists selected at random.

tent of the veneer should be as low as practical. Very dry veneer, however, is difficult to handle without damage, so the minimum practical moisture content is about 4 percent. Freshly glued plywood intended for interior service should be dried to the moisture content values given in table 14–1.

Hot-pressed plywood and other board products, such as particleboard and hardboard, often do not arrive at the same equilibrium moisture content values as those given for lumber. The high temperatures used in hot presses cause these products to assume a lower moisture content for a given relative humidity. Because this lower equilibrium moisture content varies widely, depending on the specific type of hot-pressed product, it is recommended that such products be conditioned at 40 to 50 percent relative humidity for interior use and 65 percent for exterior use.

Lumber used in the manufacture of large laminated members should be dried to a moisture content slightly below the moisture content expected in service so that moisture absorbed from the glue will not cause the moisture content of the product to exceed the service value. The range of moisture content between laminations assembled into a single member should not exceed 5 percentage points. Although laminated members are often massive and respond rather slowly to changes in environmental conditions, it is desirable to follow the recommendations in table 14–1 for moisture content at time of installation.

Drying of Wood

Well-developed techniques have been established for removing the large amounts of moisture normally present in green wood (ch. 3). In addition to drying, some end uses require equalizing and conditioning treatments to improve moisture uniformity within and between pieces, and to relieve residual stresses and sets. Careful techniques are necessary, especially during the drying phase, to protect the wood from stain and decay and from excessive drying stresses that cause defects and degrade. The established drying methods are air drying, accelerated air drying, and kiln drying. Other methods, such as high-frequency dielectric heating, vapor drying, and solvent drying, have been developed for special uses.

Drying reduces the weight of wood, with a resulting decrease in shipping costs; reduces or eliminates shrinkage, checking, and warping in service; increases strength and nail-holding power; decreases susceptibility to infection by blue stain and other fungi; reduces chance of attack by insects; and improves the capacity of wood to take preservative and fire-retardant treatment and to hold paint.

Sawmill Practice

It is common practice at most softwood sawmills to kiln dry all upper grade lumber intended for finish, flooring, and cut stock. Lower grade boards are often air dried. Dimension lumber is air dried or kiln dried, although some mills ship certain species without drying. Timbers are generally not held long enough to be considered dry, but some drying may take place between sawing and shipment or while they are held at a wholesale or distributing yard. Sawmills cutting hardwoods commonly classify the lumber for size and grade at the time of sawing. Some mills send all freshly sawed stock to the air-drying yard or an accelerated air-drying operation. Others kiln dry directly from the green condition. Air-dried stock may be kiln dried at the sawmill, at a custom drying operation during transit, or at the remanufacturing plant before being made up into such finished products as furniture, cabinet work, interior finish, and flooring.

Air Drying

Air drying is not a complete drying process, except as preparation for uses for which the recommended moisture content is not more than 5 percent below that of the air-dry stock. Even when air-drying conditions are mild, air-dry stock used without kiln drying may have some residual stress and set that can cause distortions after nonuniform surfacing or machining. On the other hand, rapid air drying accomplished by low relative humidities produces a large amount of set that will assist in reducing warp during the final kiln drying. Rapid surface drying also greatly decreases the incidence of chemical and sticker stain, blue stain, and decay.

Air drying is an economical method when carried out (1) in a well-designed yard or shed, (2) with proper piling practices, and (3) in favorable drying weather. In cold or humid weather, air drying is slow and cannot readily reduce wood moisture to levels suitable for rapid kiln drying or for use.

Accelerated air drying involves the use of fans to force the air through the lumber piles in a shed or under other protection from the weather. Sometimes small amounts of heat are used to reduce relative humidity and slightly increase temperature. Accelerated air drying to moisture content levels between 20 and 30 percent may take only one-half to one-fourth as long as ordinary air drying. Moisture content in the stock dried with such acceleration may vary somewhat more than that of stock air dried under natural conditions to the same average moisture level.

Kiln Drying

In kiln drying, higher temperatures and fast air circulation are used to increase the drying rate considerably. Average moisture content can be reduced to any desired value. Specific schedules are used to control the temperature and humidity in accordance with the moisture and stress situation within the wood, thus minimizing shrinkage-caused defects. For some purposes, equalizing and conditioning treatments are used to improve moisture content uniformity and relieve stresses and set at the end of drying, so that the material will not warp when resawed or machined to smaller sizes or irregular shapes. Further advantages of kiln drying are the setting of pitch in resinous woods, the killing of staining or decay fungi or insects in the wood, and reductions in weight greater than those achieved by air drying. At the end of kiln drying, moisture-monitoring equipment is sometimes used to sort out moist stock for redrying and to ensure that the material ready for shipment meets moisture content specifications.

Temperatures of ordinary kiln drying generally are between 110 and 180 °F. Elevated-temperature (180 to 212 °F) and high-temperature (above 212 °F) kilns are becoming increasingly common, although some strength loss is possible with higher temperatures (see ch. 4).

Drying Degrade

Lumber grading associations specify the types and amounts of defects permitted in the various grades of dimension stock. Drying defects and other degrading factors are considered. The higher grades are practically free of such defects, but high-quality material may not be needed for all uses. The defects permitted in the other approved grades have no detrimental effect on the wood's utility in many applications.

Drying defects that cause degrade may be classified into three main groups: (1) Those caused by unequal shrinkage (surface checks (fig. 14—5), end checks (fig. 14—6), honeycomb (fig. 14—7), warp (fig. 14—8), loosening of knots (fig. 14—9), and collapse (fig. 14—10)); (2) those associated with the action of fungi (molds and stains) (fig. 14—11); and (3) those associated with soluble wood constituents (brown stain and sticker stain). Collapse and warp affect appearance and ease of application. Checking and honeycombing may, in addition, reduce strength. Defects caused by fungi may affect both appearance and strength (see ch. 17).

Brown stain occurs in softwoods. It is a yellow to dark-brown discoloration apparently caused by concentration or chemical transformation of water-soluble materials in the wood. Sticker stain may occur in the air drying of both softwoods and hardwoods. It is caused by color changes of water-soluble materials in the wood under the stickers. Conversely, the chemicals through the rest of the board may undergo color changes while those under the stickers do not, causing light-colored "sticker marking." General blue-gray or gray-brown chemical stain of hardwoods may also be a problem. Although chemical stains cause considerable degrade because of appearance, none of them lower the strength.

Figure 14—5—Surface checking in Douglas-fir dimension lumber. (M22523)

Figure 14—6—End checks in oak. (M3510)

14-9

Figure 14–7—Honeycomb in oak. (M11293)

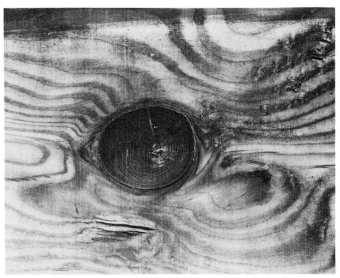

Figure 14–9—Loose knot in southern pine. (M16 268)

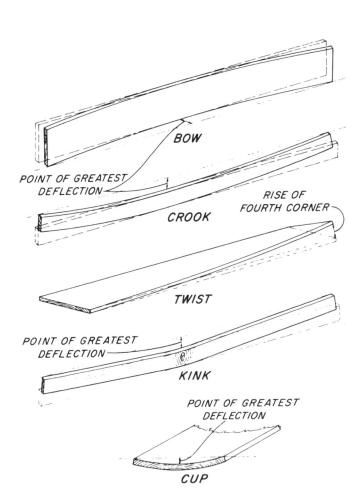

Figure 14–8—Various types of warp. (M133 580)

Drying defects can be largely eliminated by good practice in either air drying or kiln drying. The period immediately after sawing is the most critical. Lumber should be piled on stickers under good drying conditions within 2 or 3 days after sawing. Rapid surface drying decreases the incidence of mold, stain, and decay but sometimes additional measures, such as dipping in chemicals, are required (ch. 17). Too-rapid surface drying, however, may cause checking and splitting. Honeycombing and collapse are more likely to occur in hardwoods than in softwoods. These defects are more likely to occur under improper kiln drying than under air drying, although very severe air-drying conditions can cause them to occur.

Moisture-resistant coatings are sometimes applied to the end-grain surfaces of green lumber to retard end drying and minimize the formation of end checks and end splits. To be effective, the coatings must be applied to the freshly trimmed green lumber before any checking has started. Sprayable wax emulsions are sometimes used on the ends of lumber subject to considerable loss by end checking.

Moisture Content of Dried Lumber

The trade terms "shipping dry," "air dry," and "kiln dry," although widely used, may not have identical meanings as to moisture content in the different producing regions. Despite the wide variations in the use of these terms, they are sometimes used to describe dried lumber. The following statements, which are not exact definitions, outline the categories:

Shipping-Dry Lumber

Lumber that is partially dried to prevent stain or mold in

Figure 14–10—Severe collapse in western red-cedar.   (M111 997)

Figure 14–11—Sap stain in southern pine. Color   (M92 499)
ranges from bluish gray to black.

brief periods of transit, preferably with the outer 1/8 inch dried to 25 percent moisture content or below.

Air-Dry Lumber

Lumber that has been dried by exposure to the air outdoors or in a shed, or by forced circulation of unhumidified air that has not been heated above 120 °F. Commercial air-dry stock generally will have an average moisture content low enough for rapid kiln drying or for rough construction use. These values generally would be in the range of 20 to 25 percent for dense hardwoods and 15 to 20 percent for softwoods and low-density hardwoods. Extended exposure can bring 1- and 2-inch lumber within a percentage point or two of the average exterior equilibrium moisture content of the region. For much of the United States, the minimum moisture content of thoroughly air-dried lumber is 12 to 15 percent.

Kiln-Dry Lumber

Lumber that has been dried in a kiln or by some special drying method to an average moisture content specified or understood to be suitable for a certain use. The average should have upper and lower tolerance limits, and all values should fall within the limits. Kiln-dry lumber generally has an average moisture content of 12 percent or below and can be specified to be free of drying stresses.

The importance of suitable moisture content values is recognized, and provisions covering them are now incorporated in some standards and grading rules. Moisture content values in the general grading rules may or may not be suitable for a specific use; if not, a special moisture content specification should be made.

Moisture Control During Transit and Storage

Lumber and other wood items may change in moisture content and dimension while awaiting shipment, during fabrication, or in transit, as well as when stored in a wholesale or retail yard.

When 1-inch dry softwood lumber is shipped in tightly closed boxcars or trucks, or in packages with complete and intact wrappers, average moisture content changes for a package can generally be held to 0.2 percent per month or less. In holds or between decks of ships, dry material absorbs usually about 1.5 percent moisture during normal shipping periods. If green material is included in the cargo, the moisture regain of the dry lumber may be doubled. On the top deck, the moisture regain may be as much as 7 percent.

When 1-inch softwood lumber, kiln dried to 8 percent or less, is piled solid under a good pile roof in a yard in humid weather, average moisture content of a pile can increase at the rate of about 2 percent per month during the first 45 days. An absorption rate of about 1 percent per month then may be sustained throughout a humid season. Comparable initial and sustaining absorption rates are about 1 percent per month in open (roofed) sheds and 0.3 percent per month in closed sheds. Stock piled in an open shed in a western location increased 2.7 percent on the inside of solid piles and 3.5 percent on the outside of the pile in a year.

All stock on which any manufacturing has been done should be protected from precipitation and spray, because water that gets into a solid pile tends to be absorbed by the wood instead of evaporating. The extent to which additional control of the storage environment is required depends upon the use to which the wood will be put and the corresponding moisture content recommendations. The moisture content of all stock should be determined when it is received. If moisture content is not as specified or required, stickered storage in an appropriate condition could ultimately bring the stock within the desired moisture content range. If the degree of moisture change required is large the stock must be redried.

Sheathing and Structural Items

Green or only partially dried lumber and timbers should be open piled on stickers and protected from sunshine and precipitation by a tight roof. Framing lumber and plywood with 20 percent or less moisture content can be solid piled in a shed that provides good protection against sunshine and direct or wind-driven precipitation. However, a better practice for stock above 12 percent moisture content is stickered piling to bring moisture content more in line with the moisture content in use. Dry lumber can be piled solid in the open for relatively short periods, but at least a minimum pile cover of waterproofed paper should be used whenever possible. Protective treatments containing a fungicide and water repellents reduce moisture absorption about 50 percent under exposure to intermittent short-term wetting, but do not protect against absorption when exposure to water is prolonged. Because it is difficult to keep rain out completely, long storage of solid-piled lumber in the open is not recommended. If framing lumber must be stored in the open for a long time, the lumber should be piled on stickers over good supports, and the piles should be roofed. Solid-piled material that has become wet again should also be repiled on stickers.

Finish and Factory Lumber

Such kiln-dried items as exterior finish, siding, and exterior millwork should be stored in a closed but unheated shed. They should be placed on supports raised above the floor, at least 6 inches if the floor is paved, 12 inches if not paved.

Interior trim, flooring, cabinet work, and lumber for processing into furniture should be stored in a room or closed shed that is heated or dehumidified. Kiln-dried and machined hardwood dimension or softwood cut stock also should be stored under controlled humidity conditions. Under uncontrolled conditions, the ends of such stock may come to a higher moisture content than the balance of the length; then when the stock is straight-line ripped or jointed before edge gluing, subsequent shrinkage will cause splitting or open glue joints.

The simplest way to reduce relative humidity in storage areas of all sizes is to heat the space to a temperature slightly above that of the outside air. Dehumidifiers can be used in small, well-enclosed spaces. If the heating method is used, and there is no source of moisture except that contained in the air, the equilibrium moisture content can be maintained by raising the temperature of the storage area above the outside temperature by the amount in table 14−2.

When a dehumidifier is used, the average temperature in the storage space should be known or controlled and table 3−4 should be used to select the proper relative humidity to give the desired average moisture content.

Wood in a factory awaiting or following manufacture can become too dry if the area is heated to 70 °F or higher when there is a low outdoor temperature. This often occurs in the northern United States during the winter. Under such circumstances, exposed ends and surfaces of boards or cut pieces will tend to dry to the low equilibrium moisture content condition, causing shrinkage and warping. Also an equilibrium moisture content of 4 percent or more below the moisture content of the core of freshly crosscut boards may cause end checking. Simple remedies are to cover piles of partially manufactured items with plastic film and to use properly lowered shop temperatures during nonwork hours. More precise control can be obtained in critical shop and storage areas by humidification. In warm weather, cooling may increase relative humidity, and dehumidification may be necessary.

Table 14 − 2—*Amount the temperature of storage area must be raised above outside temperature to maintain equilibrium moisture content*

Outside relative humidity	Desired equilibrium moisture content—percent						
	6	7	8	9	10	11	12
Percent	- - - - - - - - - - - - -			°F	- - - - - - - - - - - - -		
90	33	29	23	18	15	11	9
80	30	25	19	14	11	8	6
70	25	20	15	10	7	4	3
60	20	15	9	6	3	—	—
50	15	10	5	1	—	—	—

Dimensional Changes in Wood Items

Dry wood undergoes small changes in dimension with normal changes in relative humidity. More humid air will cause slight swelling, and drier air will cause slight shrinkage. These changes are considerably smaller than those involved with shrinkage from the green condition. Approximate changes in dimension can be estimated by a simple formula involving a dimensional change coefficient when moisture content remains within the range of normal use.

Estimate Using Dimensional Change Coefficient

The change in dimension within the moisture content limits of 6 to 14 percent can be estimated satisfactorily by using a dimensional change coefficient based on the dimension at 10 percent moisture content:

$$\Delta D = D_I [C_T(M_F - M_I)] \qquad (14-2)$$

where:

ΔD = change in dimension,

D_I = dimension in inches or other units at start of change,

C_T = dimensional change coefficient, tangential direction (for radial direction, use C_R),

M_F = moisture content (percent) at end of change,

M_I = moisture content (percent) at start of change.

Values for C_T and C_R, derived from total shrinkage values, are given in table 14-3. When M_F is less than M_I, the quantity $(M_F - M_I)$ will be negative, indicating a decrease in dimension; when greater, it will be positive, showing an increase in dimension.

As an example, assuming the width of a flat-grained white fir board is 9.15 inches at 8 percent moisture content, its change in width at 11 percent moisture content is estimated as:

$$
\begin{aligned}
\Delta D &= 9.15[0.00245(11-8)] \\
&= 9.15[0.00735] \\
&= 0.06725 \text{ or } 0.067 \text{ inch}
\end{aligned}
$$

Then dimension at end of change = $D_I + \Delta D$
$$
\begin{aligned}
&= 9.15 + 0.067 \\
&= 9.217 \text{ inches}
\end{aligned}
$$

The thickness at 11 percent moisture content of the same board can be estimated by using the coefficient $C_R = 0.00112$.

Because commercial lumber is often not perfectly flatsawn or quartersawn, this procedure will probably overestimate width shrinkage and underestimate thickness shrinkage.

Calculation Based on Green Dimensions

Approximate dimensional changes associated with moisture content changes larger than 6 to 14 percent, or when one moisture value is outside of those limits, can be calculated by:

$$\Delta D = \frac{D_I (M_F - M_I)}{\dfrac{30(100)}{S_T} - 30 + M_I} \qquad (14-3)$$

where: ΔD = change in dimension, D_I = dimension in inches or other units at start of change, M_F = moisture content (percent) at end of change, M_I = moisture content (percent) at start of change, S_T = tangential shrinkage (percent) from green to ovendry (tables 3-5 and 3-6) (use radial shrinkage S_R when appropriate).

Neither M_I nor M_F should exceed 30, the assumed moisture content value when shrinkage starts for most species.

Design Factors Affecting Dimensional Change in a Structure

Framing Lumber in House Construction

Ideally, house framing lumber should be dried to the moisture content it will reach in use, thus minimizing future dimensional changes due to frame shrinkage. This ideal condition is difficult to achieve, but some shrinkage of the frame may take place without being visible or causing serious defects after the house is completed. If, at the time the wall and ceiling finish is applied, the moisture content of the framing lumber is not more than about 5 percent above that which it will reach in service (table 14-1), there will be little or no evidence of defects caused by shrinkage of the frame. In heated houses in cold climates, joists over heated basements, studs, and ceiling joists may reach a moisture content as low as 6 to 7 percent. In mild climates the minimum moisture content will be higher.

The most common evidences of excessive shrinkage are cracks in plastered walls, open joints and nail pops in drywall construction, distortion of door openings, uneven floors, or loosening of joints and fastenings. The extent of vertical shrinkage after the house is completed is proportional to the depths of wood used as supports in a horizontal position, such as girders, floor joists, and plates. After all, shrinkage occurs primarily in the width of members, not the length.

Thorough consideration should be given to the type of framing best suited to the whole building structure. Methods should be selected that will minimize or balance the use of

Table 14-3—Coefficients for dimensional change due to shrinkage or swelling within moisture content limits of 6 to 14 percent

HARDWOODS

Species	Dimensional change coefficient [1] Radial C_R	Dimensional change coefficient [1] Tangential C_T
Alder, red	0.00151	0.00256
Apple	.00205	.00376
Ash:		
Black	.00172	.00274
Oregon	.00141	.00285
Pumpkin	.00126	.00219
White, green	.00169	.00274
Aspen, quaking	.00119	.00234
Basswood, American	.00230	.00330
Beech, American	.00190	.00431
Birch:		
Paper	.00219	.00304
River	.00162	.00327
Yellow, sweet	.00256	.00338
Buckeye, yellow	.00123	.00285
Butternut	.00116	.00223
Catalpa, northern	.00085	.00169
Cherry, black	.00126	.00248
Chestnut, American	.00116	.00234
Cottonwood:		
Black	.00123	.00304
Eastern	.00133	.00327
Elm:		
American	.00144	.00338
Rock	.00165	.00285
Slippery	.00169	.00315
Winged, cedar	.00183	.00419
Hackberry	.00165	.00315
Hickory:		
Pecan	.00169	.00315
True hickory	.00259	.00411
Holly, American	.00165	.00353
Honeylocust	.00144	.00230

Species	Dimensional change coefficient [1] Radial C_R	Dimensional change coefficient [1] Tangential C_T
Locust, black	.00158	.00252
Madrone, Pacific	.00194	.00451
Magnolia:		
Cucumbertree	.00180	.00312
Southern	.00187	.00230
Sweetbay	.00162	.00293
Maple:		
Bigleaf	.00126	.00248
Red	.00137	.00289
Silver	.00102	.00252
Sugar, black	.00165	.00353
Red oak:		
Commercial red	.00158	.00369
California black	.00123	.00230
Water, laurel, willow	.00151	.00350
White oak:		
Commercial white	.00180	.00365
Live	.00230	.00338
Oregon white	.00144	.00327
Overcup	.00183	.00462
Persimmon, common	.00278	.00403
Sassafras	.00137	.00216
Sweetgum	.00183	.00365
Sycamore, American	.00172	.00296
Tanoak	.00169	.00423
Tupelo:		
Black	.00176	.00308
Water	.00144	.00267
Walnut, black	.00190	.00274
Willow:		
Black	.00112	.00308
Pacific	.00099	.00319
Yellow-poplar	.00158	.00289

Table 14 – 3—Coefficients for dimensional change due to shrinkage or swelling within moisture content limits of 6 to 14 percent—Continued

SOFTWOODS

Species	Dimensional change coefficient [1]		Species	Dimensional change coefficient [1]	
	Radial C_R	Tangential C_T		Radial C_R	Tangential C_T
Baldcypress	.00130	.00216	Larch, western	.00155	.00323
Cedar:			Pine:		
Alaska-	.00095	.00208	Eastern white	.00071	.00212
Atlantic white-	.00099	.00187	Jack	.00126	.00230
Eastern redcedar	.00106	.00162	Loblolly, pond	.00165	.00259
Incense-	.00112	.00180	Lodgepole, Jeffrey	.00148	.00234
Northern white-[2]	.00101	.00229	Longleaf	.00176	.00263
Port-Orford-	.00158	.00241	Ponderosa	.00133	.00216
Western redcedar [2]	.00111	.00234	Red	.00130	.00252
Douglas-fir:			Shortleaf	.00158	.00271
Coast-type	.00165	.00267	Slash	.00187	.00267
Interior north	.00130	.00241	Sugar	.00099	.00194
Interior west	.00165	.00263	Virginia	.00144	.00252
Fir:			Western white	.00141	.00259
Balsam	.00099	.00241	Redwood:		
California red	.00155	.00278	Old-growth [2]	.00120	.00205
Noble	.00148	.00293	Second-growth [2]	.00101	.00229
Pacific silver	.00151	.00327	Spruce:		
Subalpine	.00088	.00259	Black	.00141	.00237
White, grand	.00112	.00245	Engelmann	.00130	.00248
Hemlock:			Red, white	.00130	.00274
Eastern	.00102	.00237	Sitka	.00148	.00263
Western	.00144	.00274	Tamarack	.00126	.00259

14-15

Table 14–3—*Coefficients for dimensional change due to shrinkage or swelling within moisture content limits of 6 to 14 percent*—Continued

Species	Dimensional change coefficient [1]		Species	Dimensional change coefficient [1]	
	Radial C_R	Tangential C_T		Radial C_R	Tangential C_T
IMPORTED WOODS					
Andiroba, crabwood	.00137	.00274	Light red 'Philippine mahogany"	.00126	.00241
Angelique	.00180	.00312	Limba	.00151	.00187
Apitong, keruing [2] (All *Dipterocarpus* spp.)	.00243	.00527	Mahogany [2]	.00172	.00238
			Meranti	.00126	.00289
Avodire	.00126	.00226	Obeche	.00106	.00183
Balsa	.00102	.00267	Okoume	.00194	.00212
Banak	.00158	.00312	Parana pine	.00137	.00278
Cativo	.00078	.00183	Pau marfim	.00158	.00312
Cuangare	.00183	.00342	Primavera	.00106	.00180
Greenheart [2]	.00390	.00430	Ramin	.00133	.00308
Iroko [2]	.00153	.00205	Santa Maria	.00187	.00278
Khaya	.00141	.00201	Spanish-cedar	.00141	.00219
Kokrodua [2]	.00148	.00297	Teak [2]	.00101	.00186
Lauans:					
Dark red "Philippine mahogany"	.00133	.00267			

[1] Per 1 percent change in moisture content, based on dimension at 10 percent moisture content and a straight-line relationship between the moisture content at which shrinkage starts and total shrinkage. (Shrinkage assumed to start at 30 percent for all species except those indicated by footnote 2.)

[2] Shrinkage assumed to start at 22 percent moisture content.

wood across the grain in vertical supports. These involve variations in floor, wall, and ceiling framing. The factors involved and details of construction are covered extensively in "Wood-Frame House Construction," USDA Agriculture Handbook 73.

Heavy Timber Construction

In heavy timber construction, a certain amount of shrinkage is to be expected. A column that bears directly on a wood girder may result in a structure settling due to the perpendicular-to-grain shrinkage of the girder. If not provided for in the design, shrinkage may cause weakening of the joints or uneven floors or both. One means of eliminating part of the shrinkage in mill buildings and similar structures is with metal post caps, separating the upper column from the lower column only by the metal in the post cap. The same thing is accomplished by supporting the upper column on the lower column with wood corbels bolted to the side of the lower column to support the girders.

Where joist hangers are used, the top of the joist, when installed, should be above the top of the girder; otherwise, when the joist shrinks in the stirrup, the floor over the girder will be higher than that bearing upon the joist.

Heavy planking used for flooring should be near 12 percent in moisture content to minimize openings between boards as they approach moisture equilibrium. When 2- or 3-inch joists are nailed together to provide a laminated floor of greater depth for heavy design loads, the joist material should be somewhat below 12 percent moisture content if the building is to be heated.

Interior Finish

The normal seasonal changes in the moisture content of interior finish are not enough to cause serious dimensional change if the woodwork is carefully designed. Large members, such as ornamental beams, cornices, newel posts, stair stringers, and handrails, should be built up from comparatively small pieces. Wide door and window trim and base should be hollow-backed. Backband trim, if mitered at the corners, should be glued and splined before erection; otherwise butt joints should be used for the wide faces. Large, solid pieces, such as wood paneling, should be so designed and installed that the panels are free to move across the grain. Narrow widths are preferable.

Flooring

Flooring is usually dried to the moisture content expected in service so that shrinkage and swelling are minimized and buckling or large gaps between boards does not occur. When used in basement, large hall, or gymnasium floors, however, enough space should be left around the edges to allow for some expansion.

Wood Care and Scheduling During Construction

Lumber and Sheathing

Lumber and sheathing received at the building site should be protected from wetting and other damage. Construction lumber in place in a structure before it is enclosed may be wet during a storm, but the wetting is mostly on the exposed surface, and the lumber can dry out quickly. Dry lumber may be solid piled at the site, but the piles should be at least 6 inches off the ground and covered with canvas or waterproof paper laid to shed water from the top, sides, and ends of the pile.

Lumber that is green or nearly green, and lumber or plywood that has been used for concrete forms, should be piled on stickers under a roof for more thorough drying before it is built into the structure. The same procedure is required for preservative-treated lumber that has not been fully redried.

If framing lumber has higher moisture content when installed than that recommended in table 14−1, some shrinkage may be expected. Framing lumber, even thoroughly air-dried stock, will generally have a moisture content higher than that recommended when it is delivered to the building site. If carelessly handled in storage at the site, it may take up more moisture. Builders may schedule their work so an appreciable amount of drying can take place during the early stages of construction. This minimizes the effects of further drying and shrinkage after completion.

When the house has been framed, sheathed, and roofed, the framing is so exposed that in time it can dry to a lower moisture content than would ordinarily be expected in yard-dried lumber. The application of the wall and ceiling finish is delayed while wiring and plumbing are installed. If the delay is for about 30 days in warm, dry weather, framing lumber should lose enough moisture so that any further drying in place will be relatively unimportant. In cool, damp weather, or if wet lumber is used, the period of exposure should be extended. Checking moisture content of door and window headers and floor and ceiling joists at this time with an electric moisture meter is good practice. When these members approach an average of 12 percent moisture content, interior finish and trim can normally be installed. Closing the house and using the heating system will hasten the rate of drying.

Before wall finish is applied, the frame should be examined and any defects that may have developed during drying, such as warped or distorted studs, shrinkage of lintels over openings, or loosened joints, should be corrected.

Exterior Trim and Millwork

Exterior trim such as cornice and rake mouldings, fascia boards, and soffit material is normally installed before the shingles are laid. Trim, siding, and window and door frames should be protected on the site by storing in the house or garage if they are received some time before the contractor can use them. While items such as window frames and sash are usually treated with some type of water-repellent preservative to resist absorption of water, they should be stored in a protected area if they cannot be installed soon after delivery. Wood siding is often received in packaged form and can ordinarily remain in the package until it is applied.

Finish Floor

Cracks develop in flooring if it absorbs moisture either before or after it is laid and then shrinks when the building is heated. Such cracks can be greatly reduced by observing the following practices: (1) Specify flooring manufactured according to association rules and sold by dealers that protect it properly during storage and delivery; (2) do not allow the flooring to be delivered before the masonry and plastering are completed and fully dry, unless a dry storage space is available; (3) have the heating plant installed before the flooring is delivered; (4) break open the flooring bundles and expose all sides of the flooring to the atmosphere inside the structure; (5) close up the house at night and raise the temperature about 15 °F above the outdoor temperature for about 3 days before laying the floor; (6) if the house is not occupied immediately after the floor is laid, keep the house closed at night or during damp weather and supply some heat if necessary.

Better and smoother sanding and finishing can be done when the house is warm and the wood has been kept dry.

Interior Finish

In a building under construction, the relative humidity will average higher than it will in an occupied house because of the moisture that evaporates from wet concrete, brickwork, and plaster, and even from the structural wood members. The average temperature will be lower, because workmen prefer a lower temperature than is common in an occupied house. Under such conditions the finish tends to have a higher moisture content during construction than it will have during occupancy.

Before any interior finish is delivered, the outside doors and windows should be hung in place so that they may be kept closed at night; in this way conditions of the interior can be held as close as possible to the higher temperature and lower humidity that ordinarily prevail during the day. Such protection may be sufficient during dry summer weather, but during damp or cool weather it is highly desirable that some heat be maintained in the house, particularly at night. Whenever possible, the heating plant should be placed in the house before the interior trim goes in, to be available for supplying the necessary heat. Portable heaters also may be used. The temperatures during the night should be maintained about 15 °F above outside temperatures but should not be allowed to drop below about 70 °F during the summer or 62 °F when outside temperatures are below freezing.

After buildings have thoroughly dried, there is less need for heat, but unoccupied houses, new or old, should not be allowed to stand without some heat during the winter. A temperature of about 15 °F above outside temperatures and above freezing at all times will keep the woodwork, finish, and other parts of the house from being affected by dampness or frost.

Plastering

During a plastering operation in a moderate-sized six-room house approximately 1,000 pounds of water are used, all of which must be dissipated before the house is ready for the interior finish. Adequate ventilation to remove the evaporated moisture will avoid that moisture being absorbed by the framework. In houses plastered in cold weather the excess moisture may also cause paint to blister on exterior finish and siding. During warm, dry, summer weather with the windows wide open, the moisture will be gone within a week after the final coat of plaster is applied. During damp, cold weather, the heating system or portable heaters are used to prevent freezing of plaster and to hasten its drying. Adequate ventilation should be provided at all times of the year, because a large volume of air is required to carry away the amount of water involved. Even in the coldest weather, the windows on the side of the house away from the prevailing winds should be opened 2 or 3 inches, preferably from the top.

Selected References

American Society for Testing and Materials. Standard methods of test for moisture content of wood. ASTM D 2016. Philadelphia, PA: ASTM; (see current edition).

Anderson, L. O. Wood-frame house construction. Agr. Handb. 73. Washington, DC: U.S. Department of Agriculture, Forest Service; rev. 1970.

Bramhall, G.; Salamon, M. Combined species-temperature correction tables for moisture meters. Inf. Rep. VP−X−103. Vancouver, BC: Canadian Forestry Service, Western Forest Products Laboratory; 1972.

Forest Products Laboratory. Wood floors for dwellings. Agric. Handb. 204. Washington, DC: U.S. Department of Agriculture; 1961.

Forest Products Laboratory. Methods of controlling humidity in woodworking plants. Res. Note FPL−0218. Madison, WI: U.S. Department of Agriculture, Forest Service, Forest Products Laboratory; 1972.

James, W. L. Electric moisture meters for wood. Tech. Rep. FPL−6. Madison, WI: U.S. Department of Agriculture, Forest Service, Forest Products Laboratory; 1975.

James, W. L. Effect of temperature on readings of electric moisture meters. Madison, WI: Forest Products Journal 18(10). 23−31; 1968.

McMillen, J. M.; Wengert, E. M. Drying eastern hardwood lumber. Agric. Handb. 528. Washington, DC: U.S. Department of Agriculture; 1978.

Rasmussen, E. F. Dry kiln operator's manual. Agric. Handb. 188. Washington, DC: U.S. Department of Agriculture; 1961.

Rietz, R. C. Storage of lumber. Agric. Handb. 531. Washington, DC: U.S. Department of Agriculture; 1978.

Rietz, R. C.; Page, R. H. Air drying of lumber: A guide to industry practices. Agric. Handb. 402. Washington, DC: U.S. Department of Agriculture; 1971.

Simpson, W. T. Drying wood: a review. Drying Technology, An International Journal, Part 1. 2(2):235−265, Part 2, 2(3):353−368.

U.S. Department of Commerce. American softwood lumber standard. NBS Voluntary Prod. Stand. PS 20−70; Washington, DC: USDC; 1970.

Chapter 15

Fire Safety in Wood Construction

Fire Safety in Wood Construction *

Fire safety is an important concern in all types of construction. Information on the fire-safe use of wood in construction is covered in this chapter. This includes fire performance characteristics, such as ignition, charring, flame spread, heat release, and smoke. When evaluating fire safety, basic data are needed on performance characteristics of building materials. Even more important than the performance of these materials is the design of the building. Therefore, methods are discussed for improving fire safety through design and fire-retardant treatments that can improve the fire performance of wood.

Major building codes generally recognize five classifications of construction based on types of materials and required fire resistance ratings. Of the five, wood is permitted in three of the classifications. These three types of construction have traditionally been referred to as heavy timber, ordinary, and light-frame. Heavy timber construction has wood columns, floors, roofs and interior partitions of certain minimum dimensions. For example, beams and girders may be not less than nominal 6 inches in width and not less than nominal 10 inches in depth. Ordinary construction has smaller size wood members, such as nominal 2-inch-thick wood joists. In both heavy timber and ordinary construction, the exterior walls are of noncombustible materials. In light-frame construction, the walls, floors, and roofs may be nominal 2-inch-thick wood framing and the exterior walls may be of combustible materials. The fire resistance of light-frame and heavy timber construction will be discussed later. While the other two classifications, fire-resistive and noncombustible constructions, basically restrict the construction to noncombustible materials, fire-retardant-treated wood is permitted in limited applications.

The high level of national concern for fire safety is reflected in limitations and design requirements in the building codes. The codes provide the minimum statutory requirements for fire safety. Adherence to codes will result in an improved level of fire safety. Code officials should be consulted early in the design of a building, because the codes offer alternatives. For example, floor areas can be increased with the addition of automatic sprinkler systems. Code officials have the option to approve alternative materials and methods of construction and modify provisions of the codes when equivalent fire protection and structural integrity is documented. Insurance rating bureaus and fire insurance engineers are available to help lower insurance costs. As a supplement to the building codes the National Fire Protection Association's "Life Safety Code" provides guidelines for life safety from fire in buildings and structures. As with the model building codes, provisions of the life safety code are statutory requirements when adopted by local or state authorities.

Fire Performance Characteristics of Wood

Wood will burn when exposed to heat and air. Several characteristics, discussed in this section, can be used to quantify this behavior.

Ignition

Wood products ignite when subjected to certain conditions of high temperature in surroundings that provide oxygen for combustion. The wood typically responds to these external exposures by decomposing, or pyrolyzing, into volatiles and a char residue. After pyrolysis, the char may burn in place or disintegrate by glowing or smoldering; the volatiles mix with oxygen in the air and may undergo flaming combustion. Smoldering can proceed with or without glowing, so that the only evidence of thermal degradation may be a color change accompanied by a weight loss over a matter of a few minutes in the affected region. Sound-deadening board has been known to exhibit this mode of smoldering. Ignition occurs in one of two ways (modes)—piloted or nonpiloted. Piloted ignition refers to the presence of a flame that serves as an ignition source for the volatiles resulting from thermal decomposition. Mode of ignition is a key concept in understanding the ignition phenomena.

The question of ignitability is also governed by the fire exposure in terms of heat flows and time. High heat flows cause high temperatures, which are associated with shorter times to ignition.

Because many factors affect the accumulation of heat by fire-exposed wood assemblies, there is no characteristic ignition temperature. At the present time, it is not possible to give specific ignition data that apply to a broad range of cases. For radiant heating of cellulosic solids, nonpiloted (spontaneous) transient ignition is reported at 1,112 °F (600 °C), with piloted transient ignition at 572 to 770 °F (300 to 410 °C). Persistent flaming ignition has been obtained at temperatures greater than about 608 °F (320 °C). With convective heating of wood, nonpiloted ignition is reported as low as 518 °F (270 °C) and as high as 878 °F (470 °C).

A frequent concern is the "maximum safe-working temperature" for wood exposed over long periods (i.e. the maximum temperature that will not lead to ignition). A temperature of 212 °F (100 °C) is often used, although a lower figure of 170 °F (77 °C) is sometimes specified to allow a margin of safety. The Selected References include one by Schaffer for conservatively estimating the initiation of smoldering in cases where the volatile degradation products cannot accumulate and lead to flaming ignition.

---

* Revision by Robert H. White, Forest Products Technologist; Susan LeVan, Chemical Engineer; and John J. Brenden, Chemical Engineer.

Building codes do not generally consider ignition. As a result, general design criteria have not been developed. Rather, this subject is considered in conjunction with limits on combustibility. Combustibility may be defined either in terms of flame spread or in terms of a standard test method (usually ASTM E 136 or its equivalent). Fire-retardant treatments can be used to alter the thermal degradation products from wood, usually resulting in reduced ignitability and flame spread.

An emerging area of concern about ignition phenomena has resulted from fires in attics where cellulosic-based insulation has contacted electrical fixtures (see Combustible Insulation).

Charring and Fire Resistance

As noted before, wood exposed to high temperatures will decompose to provide an insulating layer of char that retards further degradation of the wood. The load-carrying capacity of a structural wood member depends upon its cross-sectional dimensions. Thus, the charring rate is the major factor in the fire endurance of structural wood members.

Thermal degradation of wood occurs in stages. The degradation process and the exact products of thermal degradation depend upon the rate of heating as well as the temperatures. At temperatures up to 302 °F (150 °C), wood becomes dehydrated and evolves water vapor. In the range of temperatures from 212 to 482 °F (100 to 250 °C), slow degradation occurs and the wood will eventually become charred. The gaseous products given off during slow degradation are mostly noncombustible. Active or fast degradation occurs at temperatures of 536 to 932 °F (280 to 500 °C). Combustible gases are evolved during this stage. Flaming is the combustion of volatile organic products of thermal degradation.

The standard test method for determining the ability of structural assemblies to withstand a fire is ASTM E 119. When wood is first exposed to fire, the wood chars and eventually flames. Ignition occurs in about 2 minutes under the ASTM E 119 fire-test exposures. Charring then proceeds at a rate of approximately 1/30 inch per minute for the next 8 minutes. Thereafter, the char layer has an insulating effect, and the rate decreases to 1/40 inch per minute. Considering the initial ignition delay, the fast initial charring, and then the slowing down to a constant rate, the average constant-charring rate is about 1/40 inch per minute or 1-1/2 inches per hour (Douglas-fir, 7 percent moisture content).

The rate of char penetration is inversely related to the wood's density and moisture content. The permeability of the wood may also be a factor in the charring rates. Burn-through rates for vertically fire-exposed 1-inch boards of American species (table 15−1) are obtained under ASTM E 119 fire exposure. Empirical equations, relating charring rate under

ASTM E 119 fire exposure to density and moisture content, are available for Douglas-fir, southern pine, and white oak. These equations for rates transverse to the grain are:

$R = 1/[(57.4 + 1.16M)\rho + 8.4]$ for Douglas-fir
$R = 1/[(11.7 + 0.24M)\rho + 25.7]$ for southern pine (15−1)
$R = 1/[(40.1 + 0.81M)\rho + 15.0]$ for white oak

where

R = char rate (inches per minute)
M = moisture content (percent)
ρ = dry specific gravity (dimensionless)

Charring in the longitudinal direction is reportedly double that in the transverse direction. These charring rates can also be affected by the severity of the fire exposure.

Table 15−1—_Burn-through rate for vertically fire-exposed 1-inch boards under ASTM E 119 fire exposure_

Species	Specific gravity[1]	Rate[2]
		In/hr
Baldcypress	0.44	1.7
Basswood	.42	2.4
Birch, yellow	.63	2.0
Chestnut	.45	1.7
Douglas-fir	.45	1.6
Hemlock, eastern	.40	1.6
Maple, sugar	.64	2.1
Oak:		
Northern red	.61	1.8
White	.67	1.5
Pine:		
Eastern white	.39	1.6
Ponderosa	.42	2.1
Southern	.55	2.2
Sugar	.32	2.0
Redwood	.38	1.6
Spruce, Sitka	.43	1.7
Sweetgum:	.53	1.8
Sapwood	.55	2.4
Heartwood	.52	1.6
Yellow-poplar	.44	2.1

[1] Specific gravity is based on weight when ovendry and volume at 6 percent moisture content.

[2] Moisture content of 6 to 7 percent.

The temperature at the innermost zone of the char layer is approximately 550 °F (288 °C). Due to the low thermal conductivity of wood, the wood 1/4 inch inward from the base of the char layer is a maximum of 360 °F (182 °C). This steep temperature gradient means the remaining uncharred cross-sectional area of a large wood member remains at a low temperature and can continue to carry a load. Moisture is driven into the wood as charring progresses. A moisture content peak is created inward from the char base. The peak moisture content occurs where the temperature of the wood is about 212 °F (100 °C) at about 1/2 inch from the char base.

Panel products provide a thermal (or fire) barrier for structural members and combustible insulation. The fire-resistance test criteria for thermal barriers generally is the time until there is a 250 °F (139 °C) average temperature rise or 325 °F (181 °C) localized temperature rise on the unexposed side of the panel product when the other side is subjected to ASTM E 119 fire exposure. Increased density, moisture content, and thickness of a panel product will significantly increase the time to achieve the 250/325 °F (139/181 °C) temperature rise. The protected element or other materials immediately behind the thermal barrier affects the performance of the thermal barrier. Insulative materials behind a thermal barrier reduce the time for the critical temperature rise because they conduct little heat away from the barrier. In small-scale tests of the times to reach 250 °F (139 °C) temperature rise, the performances of 5/8-inch-thick exterior plywood and 1/2-inch-thick regular gypsum wallboard were equivalent.

Flame Spread

Code authorities attempt to eliminate hazardous materials and improve life safety in buildings by regulating the flame spread of building materials. Thus, flame spread is one of the most tested fire-performance properties of a material. Numerous flame-spread tests are used, but the most common one cited by building codes is ASTM E 84, the 25-foot tunnel test. This test method involves the use of a 20-inch by 25-foot specimen exposed horizontally to a furnace operating under forced draft conditions. The operator positions the specimen on the bottom of the furnace cover. The flame impinges on the underside of the specimen at one end. The operator records the distance and time of maximum flame front travel in a 10-minute period.

For regulatory purposes, interior finish materials are classified according to their flame spread. The classes are 0−25 for Class A or I, 26−75 for Class B or II, and 76−200 for Class C or III (see "Interior Finish").

In the past, red oak flooring was used as a standard and was given a flame-spread index of 100. Today, red oak flooring still has an index around 100 but is no longer used in the calculation of the ASTM E 84 flame-spread index. Calibration of the ASTM E 84 test is now based on operational specifications measured on a standard noncombustible material. Most wood species have flame-spread index values from 90 to 160 by the ASTM E 84 method. A few species have flame-spread index values slightly less than 75 (western redcedar, redwood) and qualify for Class B applications (see "Interior Finish"). Flame-spread indexes reported in the literature for several species of nominal 1-inch-thick lumber are listed in table 15−2. Values may be higher for thinner thicknesses.

Several versions of the corner wall test have been used to measure flame spread. In these tests, the test material is placed on two walls and a ceiling forming a corner. An ignition source is placed on the floor of the corner. Rate of flame spread is expressed as the time required for the flame to reach the ceiling.

Heat Release and Heat of Combustion

The total heat of combustion of ovendry wood varies from about 8,000 to about 12,000 Btu's per pound of original wood, depending on species, resin content, moisture, and other factors. The contribution to fire growth from this total depends on the circumstances of the fire exposure, the completeness of combustion, and, to a critical extent, on the rate at which the heat is released. In recent years, the concept of heat release rate has become a more important criterion than total heat available. Additionally, rate-of-heat-release information is required for input into mathematical models of fires and fire exposures.

Initially, research efforts in this area were directed toward quantitative measurements under a variety of fire exposure conditions. Results of the early work showed that heat-release-rate values depended strongly on the exposure conditions and on the experimental apparatus used. Eventually the procedure known as ASTM E 906 became the most widely accepted means of making these determinations. Results of this test method have been used in the development of fire models for compartments.

At present, this area of fire research is changing rapidly due to the emergence of a technique known as the "oxygen depletion" method for measuring heat release rates. The method is based on the experimental observation that the heats of combustion, *per unit of oxygen consumed*, are approximately constant for a wide variety of organic substances. Because the concept of heat release rates is new, this phenomenon is not usually covered in design methodologies or building codes.

Table 15—2—ASTM E 84 flame-spread indexes for various wood species of 1-inch nominal solid lumber as reported in the literature

Species[1]	Flame-spread index	Source
SOFTWOODS		
Baldcypress (cypress)	145–150	UL
Cedar:		
Eastern redcedar	[2]110	HUD/FHA
Alaska		
(Pacific Coast yellow cedar)	78	CWC
Western redcedar	70	HPMA
Douglas-fir	70–100	UL
Hemlock, western (West Coast)	60–75	UL
Pine:		
Western white		
(western white, Idaho white)	[3]75,72	UL,HPMA
Eastern white (eastern		
white, northern white)	85, [3]120–215	CWC,UL
Lodgepole	65–110	CWC
Ponderosa	[4]105–230	UL
Red	142	CWC
Southern (southern yellow)	[3]130–195	UL
Redwood	70	UL
Spruce:		
Eastern (white, northern)	65	CWC,UL
Sitka (western)	100	UL
HARDWOODS		
Birch, yellow	105–110	UL
Cottonwood	115	UL
Maple (maple flooring)	104	CWC
Oak (red or white)	100	UL
Sweetgum (gum, red)	140–155	UL
Walnut	130–140	UL
Yellow-poplar (poplar)	170–185	UL

Sources:
CWC — Canadian Wood Council. Fire protective design—flame-spread rating. Data file FP–6, 1973.
HPMA — Hardwood Plywood Manufacturers Assoc., Tests 596 and 592.
HUD/FHA — U.S. Department of Housing and Urban Development. Manual of acceptable practices to the HUD minimum property standards, 1973.
UL — Underwriters Laboratories, Inc. Wood-fire hazard classification. Card Data Service, Serial No. UL 527, 1971.

[1] In cases where the name given in the source did not conform to the official nomenclature of the Forest Service, the likely official nomenclature name is given. The name given by the source is given in parentheses.

[2] Thickness of 1/2 in.

[3] UL footnote—due to wide variations in the different species of the pine family, and local connotations of their popular names, exact identification of the types of pine tested was not possible. The effects of differing climatic and soil conditions on the burning characteristics of a given species have not been determined.

[4] UL footnote—In 18 tests of ponderosa pine, 3 had values over 200 and the average of all tests is 154.

Smoke and Toxic Gases

One of the most important problems associated with fires is the smoke they produce. The term ''smoke'' is frequently used in an all inclusive sense to mean the mixture of pyrolysis products and air that is present near the fire site. In this context, smoke contains gases, solid particles, and droplets of liquid. Smoke presents potential hazards because it interacts with light to obscure vision and because it contains noxious and toxic substances.

Generally, two approaches are used to deal with the smoke problem: First, limit smoke production; and second, control the smoke that has been produced. The control of smoke flow is most often a factor in the design and construction of large or tall buildings. In these buildings, combustion products may have serious effects in areas remote from the actual fire site. Several references relating to smoke flow rates are given at the end of this chapter.

Currently, several laboratory-scale test methods provide comparative smoke yield information on materials and assemblies. Each method has entirely different exposure conditions; none are generally correlated to full-scale fire conditions or experience. Up until the middle 1970's, smoke yield restrictions in building codes were almost always based on data from ASTM E 84. However, the method of ASTM E 662 has recently gained increasing recognition and use because it can be applied to a variety of fire exposure situations.

Toxicity of combustion products is an area of emerging concern. Something on the order of 75 or 80 percent of fire victims are not touched by the flame, but die as a result of exposure to smoke, to toxic gases, or as the result of oxygen depletion. These life-threatening conditions can result from burning contents, such as furnishings, as well as from the structural materials involved. The toxicity resulting from the thermal decomposition of wood and cellulosic substances is not well understood. Part of the reason is due to the wide variety of compounds found in wood smoke. Their individual concentration depends on such factors as the fire exposure, the oxygen and moisture present, the species of wood, any treatments or finishes that may have been applied, and other considerations. One approach used to estimate the toxicity hazard of wood smoke is to find the toxicology data for individual smoke components. This, of course, neglects possible synergistic effects. The other approach is to measure the response of laboratory animals to the products of combustion. Improved toxicity data are being gathered and should be available in the near future.

Improving Fire Safety Through Design

Methods for improving fire safety can be grouped into three categories: (1) prevention, (2) containment, and (3) detection and evacuation. Fire prevention basically means preventing the ignition of combustible materials by controlling either the source of heat or the combustible materials. This generally involves proper design, installation/construction and maintenance of the building and its contents. Topics in this category include wood-fueled heaters and chimneys, combustible insulation, and wood roof coverings.

Design deficiencies are often responsible for spread of heat and smoke in a fire. Fire containment depends upon the use of design methods which limit fire growth and spread within a compartment and those that limit the spread of the fire outside the compartment of origin. Topics related to fire growth within a compartment include interior finish, area and height limitations, and automatic sprinklers. Topics related to spread of a fire out of a compartment include fire resistance of assemblies, firestops and draftstops, doors and stairways.

The ability to escape from a fire often is a critical factor in life safety. Topics related to fire evacuation from a building include fire detectors and exits.

Each of the above topics is discussed in this section. Proper fire safety measures depend upon the occupancy or processes taking place in the building. Future trends in fire safety include systems approaches to design. Methodologies for systematically evaluating fire risks include fire safety decision trees and network diagrams. In addition to being integrated approaches to fire safety, systems methodologies permit greater flexibility in design and a better evaluation of the actual degree of risk.

Wood-Fueled Heaters and Chimneys

Home heating systems are one of the leading causes of fires. As the use of solid-fuel heating equipment has increased, fires associated with heating systems have also increased. Improper installation is often the cause for fires involving heating equipment using wood as a fuel.

Manufacturer's recommendations and local building codes should be consulted whenever any home heating system is installed. Wood construction can be protected against ignition by heat from heaters, fireplaces, and chimneys.

Proper clearances for the heater, chimney connector, and chimney are necessary to protect nearby combustible materials. Clearance-related requirements for heaters depend upon the type, size, temperature characteristics of heater, and the clearance itself. When wall clearance is small, a non-combustible shield is required to be placed between the wall and the heater. Spacers are used to allow ventilation between the wall and the noncombustible shield.

The chimney connector or smoke pipe connects the heater to the chimney. It should be as short and straight as possible. The proper size ventilating thimble should be used when it is necessary to pass through an interior wall. Smoke pipes should not pass through an exterior wall. Smoke pipes should not pass through floors or ceilings but should join the chimney on the same floor where they originate.

The chimney must be the proper type and of adequate size. Only one heating unit should be connected to one chimney flue. With multiple venting, it would not be possible to suffocate a fire burning out of control. Also, improper venting from the heaters can result with multiple venting. Wood beams, joists, or rafters should be separated from any chimney by at least a 2-inch space and this space should be filled with noncombustible materials. There are minimum requirements for the extension of the chimney beyond the roof surface in the codes.

Combustible Insulation

In addition to heating with wood, there is a great interest in insulation to reduce home heating or cooling losses. Two highly effective types of insulation are cellulosic insulation and foam plastics. While these are combustible materials, proper precautions allow for their safe use.

Cellulosic fibers provide an economical method of insulating homes and properly treated fibers significantly reduce any potential fire hazard. Reports of fire incidences involving cellulosic insulation indicate that the fires are initiated by smoldering combustion caused by the overheating of recessed light fixtures and other electrical devices improperly covered with insulation. Barriers should be built around the top of recessed light fixtures to separate them from the insulation and to allow ventilation. Cellulosic insulation with approved fire-retardant treatment should be specified.

Requirements for the safe use of foam plastics include maximum flame-spread and smoke-density ratings and the installation of a thermal barrier over the interior surface of the foam plastic. The thermal barrier (see Charring and Fire Resistance) is designed to protect the foam plastics from fire exposure from the interior of the building. Building codes require a 15-minute thermal barrier in most applications. Generally, 1/2-inch-thick gypsum board is accepted as being a 15-minute thermal barrier.

Wood Roof Coverings

Many builders consider wood shingles to be the ultimate in roof covering materials because of their esthetic appearance, natural durability, and practicality. In geographical areas subject to high winds, wood shingles are more resistant than asphalt shingles to windstorm damage.

Safer heating systems, increased separation between structures, and improved fire-resistant materials have reduced the number of fires attributed to sparks on all roof types. Sometimes code ordinances require special fire-retardant-treated wood shingles and shakes for structures in dry and brushy areas, areas where fire protection is difficult to supply, or within certain fire zones. These shingles and shakes are treated with a leach-resistant fire retardant to improve the fire performance. Installation of shingles or shakes over fire-resistant underlayment will further improve the fire performance.

Roof-covering materials are designated either Class A, B, or C according to their performance in ASTM E 108. Class A is the most fire-resistant and Class C the least under this system or rating. This standard includes intermittent flame exposure, spread of flame, burning brand, flying brand, and rain tests. Leach-resistant fire-retardant-treated shingles are available that carry a Class B or C fire rating. Some testing laboratories publish lists that identify products which meet these ratings (i.e., UL Building Material Directory). Class A can only be met by fully noncombustible roof coverings. For Class B and C ratings, factory-treated redcedar shingles and shakes must be laid in accordance with manufacturers' instructions and are limited to roof decks capable of receiving and retaining nails and to inclines sufficient to permit drainage. In addition to the above requirements, Class B shingles and shakes are limited to use on 1/2-inch minimum thickness plywood decks covered with a layer of 0.002-inch polyethylene coated steel foil.

Interior Finish

Interior finish commonly used in building construction consists of the exposed interior surface of walls and ceilings. Interior floor finishes mean the exposed floor surfaces of buildings, including floor coverings such as carpets and floor tiles. Many codes exclude trim and incidental finish, as well as decorations and furnishings that are not affixed to the structure, from the more rigid requirements for walls and ceilings. Many local building regulations have been expanded to include provisions regulating floors and floor coverings, either by including them as interior finish or by requiring them to meet other test criteria, for example, a critical radiant flux (ASTM E 648).

Materials are classified into groups based on their ASTM E 84 flame-spread index ($0-25$ is Class A or I, $26-75$ is Class B or II, and $76-200$ is Class C or III). Codes generally specify maximum flame-spread indexes for interior finish based on building occupancy, location within the building, and availability of automatic sprinkler protection. The more restrictive requirements, A or B, for surface flammability of interior finish are generally prescribed for stairways and corridors providing access to exits. In general, the next classification indicative of greater flammability is permitted for the interior finish used in other areas of the building that are not considered exitways and for materials that are protected by automatic sprinkler devices. Where interior finish is restricted for flame spread, Class C finish is usually the least restrictive requirement in buildings.

Wallpaper and paint may be exempt from these requirements unless they are judged to be unusual fire hazards. Most wood species have flame-spread index values of 90 to 160 by ASTM E 84 (see Flame Spread). As a result, unfinished wood, 3/8 inch or thicker, is generally acceptable for interior finish applications requiring a Class C rating. Fire retardant treatments for wood are usually necessary when Class A and B flame spread is required. A few species (see "Flame Spread") can meet the Class B flame-spread performance and only require fire-retardant treatments to meet a Class A rating. Most common paints and varnishes have negligible effects on the flame-spread ratings of wood (see "Coatings").

Area and Height Limitations

Fire safety in structures is improved by limiting building areas and heights of compartments. A compartment is the area within surrounding exterior walls and fire walls. Building codes limit building areas and heights primarily upon the type of building construction and occupancy. In addition to breaking up an area with fire walls, fire safety can be improved by adding automatic sprinklers, increasing property line setbacks, or specifying a more fire-resistant construction. Building codes recognize improved fire safety resulting from application of these factors by increasing the allowable areas and heights. Thus, proper site planning and building design may result in a desired building area classification being achieved with wood construction.

Automatic Sprinklers

Properly installed and maintained automatic sprinklers will significantly improve the fire safety of a building. In addition to improving fire safety, there are economic incentives. Where sprinklers are installed, building code requirements for flame spread and fire resistance may be less restrictive. As mentioned before, area and height limitations may be less restrictive for sprinklered buildings. Reduced insurance premiums are still another advantage.

Automatic sprinklers are common in industrial and commercial buildings but are presently rarely used in single-family residences. Efforts are being made to make sprinklers more practical for dwellings and other small buildings. In addition to its standard on sprinklers (NFPA 13), the National Fire Protection Association has a standard for the installation of sprinkler systems in one- and two-family dwellings and mobile homes (NFPA 13D).

Since design features of a building may affect the effectiveness of the sprinkler system, the sprinkler system should be included in the initial planning of a building. Sprinklers provide the initial attack on a fire. To be most effective in extinguishing or controlling the spread of a fire, sprinklers should be installed in all portions of a building. To allow the water to reach a fire and put it out, precautions should be taken so the water from the sprinklers will not be obstructed by either part of the building or its contents before reaching any areas the sprinklers are meant to protect. Proper site planning includes considerations of adequate access and water supply for fire department operations.

Fire Resistance

A fire starting in one room or compartment of a building will be confined to that room for a variable period of time, depending on the amount and distribution of combustible contents in the room, the amount of ventilation, and the fire resistance of walls, doors, ceilings, and floors. Fire resistance is the ability of materials or their assemblies to prevent or retard the passage of excessive heat, hot gases, or flames while continuing to support their structural loads. The standard test for fire resistance is ASTM E 119.

The self-insulating qualities of wood, particularly in the large wood sections in heavy timber construction, are an important factor in providing a good degree of fire resistance. Light wood-frame construction can be provided with a high degree of fire resistance through use of conventional gypsum board interior finish.

Heavy Timber Construction

The low thermal conductivity and slow penetration of fire by charring allow heavy timber members to maintain a high percentage of their original strength during a fire. Heavy timber construction is generally defined in the building codes and standards by minimum sizes for the various members or portions of a building. For example, columns may not be less than 8 inches in any dimension. The acceptance of heavy timber construction is based on experience with its performance in actual fires, the lack of concealed spaces, and the high fire resistance of walls in this type of construction. Heavy timber construction simplifies fire-fighting operations because there are few concealed spaces in which fire can begin and spread unnoticed.

The fire resistance of glued-laminated structural members, such as arches, beams, and columns, is approximately equivalent to the fire resistance of solid members of similar size. Available information indicates that laminated members glued with phenol, resorcinol, or melamine adhesives are at least equal in their fire resistance to a one-piece member of the same size. Laminated members glued with casein have only slightly less fire resistance.

Proper heavy timber construction includes using approved fastenings, avoiding concealed spaces under floors or roofs, and providing required fire resistance in the interior and exterior walls. The dimensions of the various members or portions of a building should meet the minimum size requirements for heavy timber.

The lack of an ASTM E 119 fire-resistance rating for large timber members has limited their use in buildings not classified as heavy timber construction. Methodologies for calculating the fire resistance of timber beams are described in building codes. As a result, large timber members will be permitted in a wide range of applications on which a 1-hour fire-resistance rating is required.

Light-Frame Construction

Frame construction is generally subdivided into two parts, protected and unprotected. For protected frame construction, most structural elements have a 1-hour fire-resistance rating. In the standard fire-resistance test (ASTM E 119), there are three failure criteria: element collapse, passage of flames, or temperature rise on the nonfire-exposed surface exceeding 250 °F (139 °C) average or 325 °F (181 °C) maximum.

Traditional constructions of gypsum wallboard or lath and plaster over wood joists and studs have fire-resistance ratings of 15 to 30 minutes and appear to be sufficient in most cases. Many recognized assemblies involving wood-framed walls, floors, and roofs provide a 1-hour or a 2-hour fire resistance. As an example, a wall with one layer of 5/8-inch-thick type X gypsum wallboard on each side of the stud has a 1-hour rating. Fire-resistance ratings of various assemblies are listed in several of the references, including the Fire Protection Handbook.

Fire-resistance ratings are usually obtained by conducting standard fire tests. Efforts have also been made to develop procedures for predicting the fire resistance of an assembly. Currently, fire-resistance design methods do not consider the variability in performance of an assembly. Evaluation of an assembly is based on a single test result. To develop more rational levels of safety, reliability-based design methods have been proposed for unprotected wood joist and wood truss floor systems. These methods will include the variability in fire-severity and fire-resistance performance.

While fire-resistance ratings are for the entire assembly, the fire resistance of a wall or floor can be viewed as the sum of the resistance of the interior finish and the resistance of the framing members. The interior finish provides the initial fire resistance. The method of fastening the interior finish to the framing members and the treatment of the joints are significant factors in the fire resistance of an assembly. The type and quantity of any insulation may also affect the fire resistance of an assembly.

Gypsum board provides an effective protective interior finish. Type X gypsum board has textile glass filaments and other ingredients that help to keep the gypsum core intact during a fire. Type X gypsum board, by definition, is a gypsum board that provides a 1-hour fire-resistance rating for 5/8-inch thickness or a 3/4-hour fire-resistance rating for 1/2-inch thickness when applied in a single layer and properly fastened to each face of load-bearing wood-framing members.

The relatively fine structural behavior of a traditional wood member in a fire test results from the fact that strength is generally uniform through the mass of the piece. Thus the unburned fraction of the member retains high strength and its load carrying capacity is diminished only in proportion to its loss of cross section. Innovative designs for structural wood members often reduce the mass of the member and locate the principal load-carrying components at the outer edges where they are most vulnerable to fire, as in structural sandwich panels. With high-strength facings attached to a low-strength core, unprotected load-bearing sandwich panels have failed to support their load in less than 15 minutes. If a sandwich panel is to be used as a load-bearing assembly, it should be protected with gypsum wallboard or some other thermal barrier.

Quality of workmanship is important in achieving adequate fire resistance. Inadequate nailing and less than required thickness of the interior finish can reduce the fire resistance of an assembly. Electrical receptacle outlets, pipe chases, and other poke-throughs that are not adequately firestopped can affect the fire resistance.

Unprotected light-frame wood buildings do not have the natural fire resistance achieved with heavier wood members. In these, as in all buildings, attention to good construction details is important to minimize fire hazards. In addition to the design of walls, ceilings, floors, and roofs for fire resistance, stairways, doors and firestops are of particular importance.

Firestops and Draftstops

Fire in buildings spreads by the movement of high-temperature air and gases through open channels. Firestops and draftstops in concealed air spaces are designed to interfere with the passage of flames up or across a building. In addition to halls, stairways, and other large spaces, heated

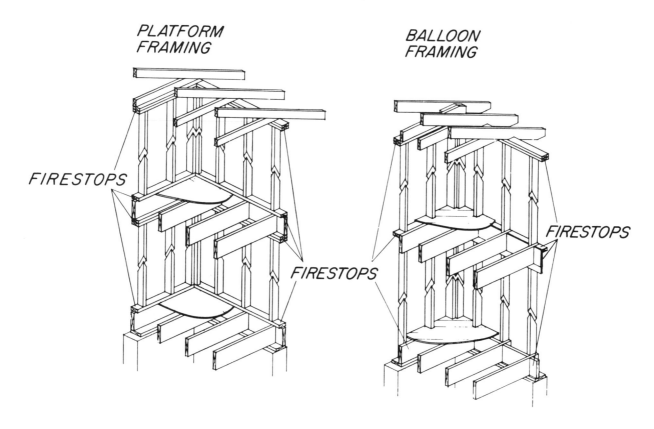

PLATFORM FRAMING

BALLOON FRAMING

FIRESTOPS

FIRESTOPS

FIRESTOPS

FIRESTOPS

Figure 15 – 1—Typical firestopping in concealed spaces of stud walls and partitions, including furred spaces at ceiling and floor levels for platform framing and balloon framing (NFPA).

(M149 464-1)

gases also follow the concealed spaces between floor joists, and between studs in partitions and walls of frame construction. Obstruction of these hidden channels provides an effective means of restricting fire from spreading to other parts of the structure.

Firestops are obstructions in relatively small concealed passages in building components such as floors, walls, and stairs (fig. 15–1 and 15–2). Effective firestops include two thicknesses of 1-inch nominal lumber with broken lap joint, one thickness of 3/4-inch plywood with joints backed by 3/4-inch plywood, 2-inch nominal lumber with tight joints and some noncombustible materials. Good practice includes the use of: (1) firestops in exterior walls at each floor level, and at the level where the roof connects with the wall; (2) firestops at each floor level in partitions that are continuous through two or more stories; (3) firestops at all interconnections between concealed vertical and horizontal spaces such as occur at soffits, drop ceilings, and cove ceilings; (4) headers at the top and bottom of the space between stair carriages; (5) mineral wool or equivalent noncombustible material packed tightly around pipes or ducts that pass through a floor or a firestop; and (6) self-closing doors on vertical shafts such as clothes chutes.

Draftstops are barriers in large concealed passages. New design and construction techniques such as suspended or dropped ceilings and parallel chord trusses have resulted in new draftstop requirements. Draftstopping materials include 1/2-inch gypsum board and 3/8-inch plywood. Two locations where draftstops should be used to break up a large area are floor-ceiling assemblies in which the ceilings are either suspended below solid wood joists or the joists are open-web trusses, and attics and other concealed roof spaces such as mansards and overhangs.

Some construction practices increase the risk of a fire spreading to the concealed spaces. Installing cabinets, shower stalls, and other fixtures without an interior wall lining on the studs allows easier penetration into the wall cavities. A built-in bathtub provides interconnections between two walls and the floor. A thin plywood cover over a trapdoor allows a fire to spread easily to the attic or other concealed space.

Doors and Stairways

Doors can be critical in preventing the spread of fires. Doors left open or doors with little fire resistance can easily defeat the purpose of a fire-rated wall.

The standard methods of fire testing door assemblies are given in ASTM E 152. Ratings for doors are generally 1/3, 1/2, 3/4, 1, 1-1/2, and 3 hours. Listings of fire-rated doors, frames, and accessories are provided by various testing agencies. When selecting a fire-rated door, details about which type of door, mounting, hardware, and closing mechanism are acceptable for any given location should be obtained from authorities having jurisdiction.

Some solid wood core or particleboard core wood flush doors have 20- or 30-minute ratings. Hollow core flush doors offer less resistance to fire penetration. Various wood-covered composite doors with an insulating core have 3/4-, 1-, or 1-1/2-hour ratings. Metal doors are available with 3-hour ratings.

The enclosing of stairways retards rapid spread of fire from floor to floor. If the interior design calls for an open stairway below, it can often be closed at the top with a door. The location of stairways should be planned with emergency evacuation in mind. Stairways should have no dead ends and should have doors at each floor. Walls of the stairway enclosure should have adequate fire resistance and an interior finish with a low flame spread.

Fire Detectors

The ability to escape from or extinguish a fire depends largely upon early detection. Heat and smoke detectors facilitate discovery. Smoke detectors have the advantage of responding to smoldering fires and giving quicker response to flaming fires.

The low cost of battery-operated smoke detectors has led to their wide public acceptance in residences. Available means of escape should be considered in the placement of smoke detectors. It is generally recommended that, at a minimum, there be one smoke detector outside each sleeping area and one on each habitable story, and basement, of the home. Reports on fire incidents confirm that smoke detecors save lives. Smoke detectors need to be continuously maintained.

Exits

Once people are aware of a fire, the number and accessibility of exits becomes important. Thus, fire safety can be improved by using favorable spatial arrangement and proper exit design. There should be at least two means of escape from all living and sleeping rooms. People should be able to

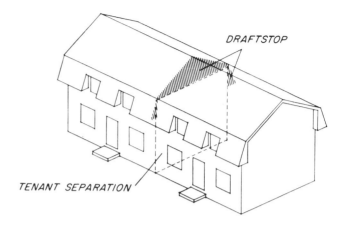

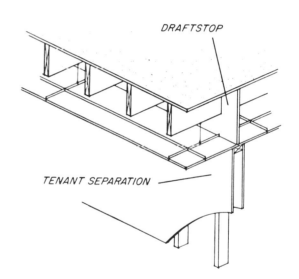

Figure 15–2—Draftstops in multifamily buildings. (M149 458-1) Top, in the floor-ceiling assemblies and bottom, in attics, mansards, overhang, or other concealed roof space above and in line with tenant separation when tenant separation walls do not extend to the roof sheathing above (NFPA).

easily climb out of a window when it is considered an exit. Thus, it should be easy to open, of adequate size, and not too high from the floor. Hallways should not have portions that are dead ends. Doors should swing in the direction of emergency exit and be readily opened. The effect of accessibility of exits should be considered when improving building security.

Fire-Retardant Treatments

Two general application methods are available for improving the fire performance of wood with fire-retardant chemicals. One method consists of pressure impregnating the wood with water- or organic solvent-borne chemicals. The second method consists of applying fire-retardant chemical coatings to the wood surface. The impregnation method is usually more effective and longer lasting. For wood in existing constructions, surface application of fire-retardant paints or other finishes offers a practical method to reduce flame spread.

Impregnation

In the impregnation treatments, wood is pressure-impregnated with chemical solutions using full-cell pressure processes similar to those used for chemical preservative treatments. Retentions of the chemicals must be fairly high to be effective.

Full-cell pressure impregnation provides the most effective method for getting chemicals into the wood at the high retention levels needed for reduced flame spread. Standards C20 and C27 of the American Wood-Preservers' Association recommend the treating conditions for lumber and plywood. The wood is usually treated in the air-dried or kiln-dried condition, but certain species may be treated green if the wood is first given a steam treatment for periods of up to 4 hours.

The penetration of the chemicals into the wood depends on the species, wood structure, and moisture content. Since some species are difficult to treat, the degree of impregnation required to obtain a Class A category may not be possible. Certain wood species are incised prior to treatment to improve the depth of penetration. Knife checks and end grain at panel edges improve the ease of impregnation on sheets of plywood, thus eliminating the need for incising. With water-soluble impregnation, only exterior-grade plywood should be used, to prevent the plies from delaminating.

After wood is removed from the treating solution, it must be carefully dried and, in certain cases, cured under the proper conditions. Various laboratories perform fire performance rating tests on these treated materials and maintain lists of products that meet certain standards.

Coatings

Many commercial coating products are available to provide varying degrees of protection to wood against fire. These coatings generally have low surface flammability characteristics and "intumesce" to form an expanded low-density film upon exposure to fire. This film insulates the wood surface below from high temperatures. Also, coatings have ingredients that restrict the flaming of any released combustible vapors. These formulations may contain chemicals that promote the rapid decomposition of the wood surface to charcoal and water rather than forming intermediate volatile flammable products.

Chemicals

Several different kinds of chemicals are used in current fire-retardant formulations and range from inexpensive inorganic salts to more complex and expensive chemicals.

Inorganic salts are the most commonly used fire retardant for interior wood products, and their characteristics have been known for over 50 years. These salts include monoammonium and diammonium phosphate, ammonium sulfate, zinc chloride, sodium tetraborate, and boric acid. These salts are combined in formulations to develop optimum fire performance yet still retain acceptable property characteristics such as hygroscopicity, strength, corrosivity, machinability, surface appearance, gluability, and paintability. Cost also must be considered in these formulations. Many commercial formulations are available.

Water-soluble organic fire retardants have recently been developed to meet the need for nonleachable systems. This type of compound falls into two categories: Resins polymerized after impregnation into wood, and graft polymer fire retardants attached directly to cellulose.

The amino resin system is the most commonly used type of the first category. It is based on urea, melamine, dicyandiamide, and related compounds. The process is simple, chemicals are inexpensive, and the polymer is insoluble. Several different formulations of this system are commercially available.

Graft polymer fire retardants for wood resulted from work in fire-retardant treatment of textiles. However, the work is still developmental and not commercially available for wood products.

Besides the water-soluble fire retardants mentioned above, oil-soluble fire retardants can be used for exterior application such as railroad timbers and ties. Triaryl and tricresyl phosphates have been used in conjunction with creosote for treatment of timbers. However, high retentions of oil-type preservatives can create a greater fire hazard than do low retentions, and require increased percentages of fire retardants.

Fire-retardant coatings include those based on water-soluble silicates, urea resins, carbohydrates and alginates, polyvinyl emulsions and oil-base alkyd, and pigmented paints. In many of the water-soluble paints, manufacturers use ammonium phosphate or sodium borate to obtain fire-retardant characteristics. The oil-base paints frequently make use of chlorinated paraffins and alkyds plus antimony trioxide to limit the flammability of any pyrolysis products produced. Inert materials, such as zinc borate, mica, kaolin, and inorganic pigments, are also included in these formulations. The natural characteristics of some of the ingredients, such as isano oil, may assist intumescence.

Application

Fire-retardant treatment of wood improves the fire performance by reducing the amount of flammable volatiles released during fire exposure, consequently reducing the rate flames spread over the surface. Treatment also reduces the amount of heat released by the volatiles liberated during the initial stages of fire. The wood may then self-extinguish when the primary heat source is removed (fig. 15–3).

Fire-retardant treatment of wood does not prevent the wood from decomposing and charring under fire exposure—the rate of fire penetration through treated wood approximates the rate through untreated wood. One can obtain slight improvement in the fire endurance of doors and walls when using fire-retardant-treated material. Most of this improvement is associated with the reduction in surface flammability, rather than any changes in charring rates.

For most rating purposes, commercial laboratories use the ASTM E 84 test (see Flame Spread) for evaluating the surface flame-spread characteristics of interior materials. Effective fire-retardant treatment can reduce the flame-spread index of lumber and most wood products to 25 or less. Some treatments qualify for a special "FR–S" rating. This rating is assigned to products which pass an extended 30-minute test instead of the usual 10-minute test. This rating indicates a flame-spread classification of not over 25 and no evidence of significant progressive combustion in this extended 30-minute ASTM E 84 test.

Fire-retardant-treated wood and plywood is currently being used for interior finish and trim in rooms, auditoriums, and corridors where codes require materials with low surface flammability. While fire-retardant-treated wood is not considered a noncombustible material, many codes have accepted the use of fire-retardant-treated wood and plywood in fire-resistive and noncombustible constructions for the framing of nonload-bearing walls, roof assemblies, and decking. Fire-retardant-treated wood is also used for such special purposes as wood scaffolding and for the frame, rails, and stiles of wood fire doors.

Durability

The chemicals used as fire retardants are thermally stable to temperatures up to 330 °F; therefore, fire-retardant-treated wood remains durable and effective under normal conditions.

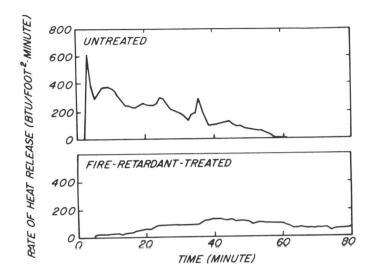

Figure 15–3—In addition to reducing flame spread, (ML84 5807) fire-retardant treatment reduces the rate at which heat is released from wood during a fire.

Treated wood which has been in service for over 40 years has demonstrated this durability.

Inorganic salt fire retardants are water soluble and are leached out in exterior applications or with repeated washings. Inorganic salts can also make the wood more hygroscopic than untreated wood. Therefore, hygroscopic salts are recommended only for those applications where the relative humidity never exceeds 80 percent.

For exterior applications, such as wood shingles or shakes, one can use leach-resistant types of fire retardants. These types of formulations maintain their effectiveness with exterior weathering conditions, and do not increase the hygroscopicity of the product. Materials are tested for durability in an accelerated weathering apparatus. Exterior fire-retardant treatments should be specified whenever the wood is exposed to exterior weathering conditions.

Strength

Fire-retardant treatment results in some slight reduction in the strength properties of wood. Tests indicate that some current treatments decrease the modulus of elasticity and modulus of rupture when both treated and untreated samples are conditioned at the same relative humidity conditions. Fire-retardant-treated wood is more brash than untreated wood. While this reduced resistance to impact is not usually considered in design, the work to maximum load, which measures brashness, may be decreased.

Because evidence indicates some reduction in the strength properties of pressure-impregnated fire-retardant-treated lumber, design values for the allowable unit stress are reduced compared to untreated wood. Design values, including fastener design loads, for lumber and structural glued laminated timber pressure impregnated with fire retardant chemicals can be obtained from the company providing the treating and redrying service.

Hygroscopicity

Wood treated with inorganic fire-retardant salts is usually more hygroscopic than untreated wood, particularly at high relative humidities. Increases in equilibrium moisture content of treated wood will depend upon the type of chemical, level of chemical retention, and size and species of wood involved. The increase in equilibrium moisture content at 80 °F and 30 to 50 percent relative humidity is negligible; at 80 °F and 65 percent relative humidity, it is 2 to 8 percent; at 80 °F and 80 percent relative humidity, it is 5 to 15 percent and may result in the exudation of chemicals from the wood.

Currently, commercial manufacturers are marketing new types of interior fire retardants that are effective for applications where the relative humidity exceeds 80 percent. This treatment is nonhygroscopic, is suitable up to relative humidities of 95 percent, and is not corrosive to metal fasteners.

Corrosion

Corrosion of fasteners can occur naturally in untreated wood. However this corrosion can be accelerated under conditions of high humidity and in the presence of fire-retardant salts as evidenced by the problems of corrosion occurring with truss plates. Fire-retardant treatments that are hygroscopic display a greater propensity to corrode certain metals than nonhygroscopic treatments. Therefore, fire-retardant-treated wood in areas of high humidity or for exterior purposes requires the appropriate fire-retardant treatment to reduce the corrosion of fasteners. Manufacturers of truss plates also recommend particular types of truss plates to be used with fire retardant treated wood.

For fire-retardant treatments containing inorganic salts, the type of metal and chemical in contact with each other greatly affects the rate of corrosion. For example, monoammonium phosphate is very corrosive on iron, steel, and copper, somewhat corrosive on brass, and only slightly corrosive on zinc. Some chemicals such as borax are noncorrosive on brass, steel, or zinc. Sodium dichromate can inhibit corrosion in most cases; but in some cases it can accelerate corrosion (i.e., ammonium phosphates with respect to zinc, monoammonium phosphate-boric acid mixture with respect to zinc,

and zinc chloride with respect to steel). To prevent corrosion, exterior or interior treatments should be selected that are noncorrosive in adverse moisture conditions.

Machinability

The presence of salt crystals in wood has an abrasive effect on cutting tools. Increased tool life can be obtained by using cutting and shaping tools tipped with tungsten carbide or similar abrasion-resistant alloys. Regular high-speed steel tools are practical only for cutting of no more than a few hundred feet of fire-retardant-treated wood. The usual practice in preparing fire-retardant-treated wood for use in trim and moldings is to cut the material to approximate finish size before treatment so that a minimum of machining is necessary after treatment.

Gluing Characteristics

Certain phases of the gluing of fire-retardant-treated woods still remain a problem. However, untreated veneer facings can be satisfactorily glued over treated plywood cores with the conventional hot-press phenolic adhesives. For assembly gluing of fire-retardant-treated wood for nonstructural purposes, one can use adhesives such as casein, urea, and resorcinol types. The major problem is in the structural bonding of fire-retardant-treated wood to provide bonds, which under both interior and exterior exposures are equivalent to those obtainable for the untreated wood. Special resorcinol-resin adhesives, which employ a high formaldehyde content hardener, have proven capabilities for gluing fire-retardant-treated wood. This type of adhesive improves bonding when cured at temperatures of 150° F or higher.

Paintability

The fire-retardant treatment of wood does not generally interfere with the adhesion of decorative paint coatings unless the treated wood has an increased moisture content. For woods treated with hygroscopic inorganic salts, one should reduce the moisture content to 12 percent or less at the time of coating application. Crystals may appear on the surface of paint coatings applied over wood treated with hygroscopic salts. This usually occurs when the wood is exposed to high relative humidity.

Natural finishes can be used on certain fire-retardant treatments. However, in general, fire-retardant treatment and subsequent drying often causes darkening and irregular staining. Manufacturers usually prepare decorative fire-retardant plywoods by treating the plywood core and then bonding a thin, untreated decorative veneer facing to these cores. This eliminates the stained surfaces, which may be difficult to finish properly to a natural wood appearance.

Many of the commercial fire-retardant finishes have been tested according to ASTM E 84 (25-ft tunnel furnace) when the coatings are applied over a substrate of Douglas-fir lumber. These coatings, when properly applied to lumber and wood products, can reduce the surface flame-spread index to 25 or less. To obtain this reduction in surface flammability, users must apply coatings in thicknesses greater than generally used for conventional decorative finishes. However, many of these coatings do not have as good brushing characteristics because of the added ingredients.

Most of the fire-retardant coatings are intended for interior use, although some products on the market can be used on the exterior of a structure. Some manufacturers recommend an application of thin coatings of conventional paint products over the fire-retardant coatings to improve their durability. Most conventional decorative paint coating products will slightly reduce the flammability of wood products when applied in conventional film thicknesses. A limited number of clear fire-retardant finishes are available.

Selected References

General

American Plywood Association. Construction for fire protection. Tacoma, WA. n.d.

Canadian Wood Council. Fire protective design. Data files 1−20. Ottawa, ON, Canada: CWC; (see current edition).

Egan, M. David. Concepts in building firesafety. New York: John Wiley and Sons, Inc.; 1978.

Gage-Babcock & Associates, Inc. Fire safety in housing. Washington, DC: U.S. Department of Housing and Urban Development; 1975.

National Fire Protection Association. Fire protection handbook. Quincy, MA: NFPA; (see current edition).

Underwriters Laboratories, Inc. Building materials directory. Northbrook, IL: UL; (see current edition).

U.S. Department of Housing and Urban Development. Manual of acceptable practices to the HUD minimum property standards. Washington, DC; 1973.

Codes

Building Officials and Code Administrators International, Inc. The BOCA basic building code. Homewood, IL: BOCA; (see current edition).

Council of American Building Officials. One and two family dwelling code. Published jointly by BOCA, ICBO and SBCC; (see current edition).

International Conference of Building Officials. Uniform building code. Whittier, CA: ICBO; (see current edition.)

National Fire Protection Association. Code for safety to life from fire in buildings and structures (Life Safety Code). NFPA 101. Quincy, MA: NFPA; (see current edition).

Southern Building Code Congress International, Inc. Standard building code. Birmingham, AL: SBCC; (see current edition).

U.S. Department of Housing and Urban Development. Minimum property standards. Washington, DC: HUD; (see current edition).

Ignition

American Society for Testing and Materials. Standard test method for behavior of materials in a vertical tube furnace at 750 °C. ASTM E 136. Philadelphia, PA: ASTM; (see current edition).

Schaffer, E. L. Smoldering in cellulosics under prolonged low-level heating. Fire Technology. 16(1): 22−28; 1980.

Charring and Fire Resistance

Schaffer, E. L. Review of information related to the charring rate of wood. Res. Note FPL-145. Madison, WI: U.S. Department of Agriculture, Forest Service, Forest Products Laboratory; 1966, rev. 1980.

White, Robert H. Wood-based paneling as thermal barriers. Res. Pap. FPL 408. Madison, WI: U.S. Department of Agriculture, Forest Service, Forest Products Laboratory; 1982.

Flame Spread

American Society for Testing and Materials. Standard test method for critical radiant flux of floor-covering systems using a radiant heat energy source. ASTM E 648. Philadelphia, PA: ASTM; (see current edition).

American Society for Testing and Materials. Test for surface burning characteristics of building materials. ASTM E 84. Philadelphia, PA: ASTM; (see current edition).

Heat Release and Heat of Combustion

American Society for Testing and Materials. Standard method of test for heat and visible smoke release rates from materials and products. ASTM E 906. Philadelphia, PA: ASTM; (see current edition).

Chamberlain, D. L. Rate of heat release—tool for the evaluation of the fire performance of materials. Presented at the joint meeting of the Central States and Western States section of the Combustion Institute; 1975 April 21−22; San Antonio, TX.

Smoke and Toxic Gases

American Society for Testing and Materials. Test for specific optical density of smoke generated by solid materials. ASTM E 662. Philadelphia, PA: ASTM; (see current edition).

Fathergill, J. W. Computer-aided design technology for smoke control and removal systems. Fire Technology. 14(2); 1978.

Heselden, A.J.M.; Baldwin, R. The movement and control of smoke on escape routes in buildings. Fire Technology. 14(3); 1978.

Wood Heaters and Chimneys

Building Officials and Code Administrators International, Inc. Burning solid fuel safely. Homewood, IL: BOCA; 1982.

International Conference of Building Officials. Installation and operation of solid-fuel-burning appliances. Whittier, CA: ICBO; 1981.

Combustible Insulation

Consumer Product Safety Commission. Standard 16 CFR, Part 1209: Cellulose insulation—interim safety standard; and Part 1404: Cellulose insulation. Washington, DC: CPSC; (see current edition).

Nosse, John H. Enforcement of foam plastic regulation. Building Safety. (special combined issue of The Building Official and Code Administrator, Building Standards, and Southern Building); Mar./Apr. 1979: 14−20.

The Society of the Plastics Industry, Inc. Model code provisions pertaining to rigid foam plastics insulation. PICC-402. New York: SPI; (see current edition).

Wood Roof Coverings

American Society for Testing and Materials. Standard methods of fire tests of roof coverings. ASTM E 108. Philadelphia, PA: ASTM; (see current edition).

Holmes, C. A. Methods of evaluating fire-retardant treatments for wood shingles. Madison, WI: Forest Products Journal. 22(3): 45−50; 1972.

Moore, Howard E. Protecting residences from wildfires: a guide for homeowners, lawmakers, and planners. Gen. Tech. Rep. PSW−50. Berkeley, CA: U.S. Department of Agriculture, Forest Service, Pacific Southwest Forest and Range Experiment Station; 1981.

Area and Height Limitations

National Forest Products Association. Code conforming wood design, allowable heights and areas. Washington, DC: NFPA; (see current edition.)

Automatic Sprinklers

National Fire Protection Association. Standard for the installation of sprinkler systems. NFPA 13. Quincy, MA: NFPA; (see current edition).

National Fire Protection Association. Standard for sprinkler systems in one- and two-family dwellings and mobile homes. NFPA 13D. Quincy, MA: NFPA; (see current edition.)

Fire Resistance

American Institute of Timber Construction. Timber construction manual. New York: John Wiley and Sons, Inc.; (see current edition).

American Insurance Association. Fire resistance ratings. New York: AIA; (see current edition).

American Society for Testing and Materials. Standard methods of fire tests of building construction and materials. ASTM E 119. Philadelphia, PA: ASTM; (see current edition).

Gypsum Association. Fire resistance design manual. Evanston, IL: GA; (see current edition).

Schaffer, Erwin. State of structural timber fire endurance. Wood and Fiber. 9(2): 145−170; 1977.

Underwriters Laboratories, Inc. Fire resistance directory. Northbrook, IL: UL; (see current edition).

U.S. Department of Housing and Urban Development. Guideline on fire ratings of archaic materials and assemblies. Rehabilitation Guidelines, Part 8. Washington, DC: Superintendent of Documents; 1980.

U.S. National Bureau of Standards. Fire resistance classifications of building constructions. Building Materials and Structures Rep. 92; 1942. (Available from National Technical Information Service, Springfield, VA.)

Firestops and Draftstops

National Forest Products Association. Improved fire safety: design of firestopping for concealed spaces. Washington, DC: NFPA; (see current edition).

Doors and Stairways

American Society for Testing and Materials. Standard methods of fire tests of door assemblies. ASTM E 152. Philadelphia, PA: ASTM; (see current edition).

Briber, A. A., Sr. Construction, testing, and use of composite fire doors. Forest Prod. Journal. 16(3): 62−63; 1966.

Degenkolb, John. The 20-minute door—and other considerations. Southern Building. Aug.-Sept. 1975: 7−12.

Warnock Hersey International, Inc. Fire rating services—building materials and equipment. Antioch, CA: WHI; (see current edition).

Fire-Retardant Treatments

American Society for Testing and Materials. Standard method of tests for accelerated weathering of fire-retardant treated wood for fire testing. ASTM D 2898. Philadelphia, PA: ASTM; (see current edition).

American Wood-Preservers' Association. Structural lumber, fire-retardant treatment by pressure processes. Standard C20. Washington, DC: AWPA; (see current edition).

American Wood-Preservers' Association. Plywood, fire-retardant treatment by pressure processes. Standard C27. Washington, DC: AWPA; (see current edition).

Holmes, C. A. Effect of fire-retardant treatments on performance properties of wood. In: Goldstein, I. S., ed. Wood technology: chemical aspects. ACS Symposium Series 43. Washington, DC: American Chemical Society; 1977.

Holmes, C. A.; Knispel, R. 0. Exterior weathering durability of some leach-resistant fire-retardant treatments for wood shingles: A five-year report. Res. Pap. FPL 403. Madison, WI: U.S. Department of Agriculture, Forest Service, Forest Products Laboratory; 1981.

LeVan, Susan L. Chemistry of fire retardancy. In: Rowell, Roger M., ed. The chemistry of solid wood. Advances in chemistry series 207. Washington, DC: American Chemical Society; 1984: Chapter 14.

Chapter 16

Finishing of Wood

Finishing of Wood *

The primary function of any wood finish (paint, varnish, wax, stain, oil, etc.) is to protect the wood surface, help maintain appearance, and provide cleanability. Unfinished wood can be used both outdoors and indoors without further protection. However, wood surfaces exposed to the weather without any finish change color, are roughened by photo-degradation and surface checking, and erode slowly. Wood surfaces exposed indoors may change color and accumulate dirt and grease if left unprotected without some finish.

Wood and wood-based products in a variety of species, grain patterns, textures, and colors can be finished effectively by many different methods. Selection of any finish will depend on the appearance and degree of protection desired, and on the substrates used. Because different finishes give varying degrees of protection, the type of finish, its quality, quantity, and the application method must be considered in selecting and planning the finishing or refinishing of wood and wood products.

Factors Affecting Finish Performance

Satisfactory performance of wood finishes is achieved when full consideration is given to the many factors that affect these finishes. These factors include the effect of the wood substrate, the properties of the finishing material, details of application, and severity of exposure. Some of the more important considerations are reviewed in this chapter. Sources of more detailed information are given in the Bibliography at the end of this chapter.

Wood Properties

Wood surfaces that shrink and swell the least are best for painting. For this reason, vertical- or edge-grained surfaces (fig. 16—1) are far better than flat-grained surfaces of any species, especially for exterior use where wide ranges of relative humidity and periodic wetting can produce equally wide ranges of swelling and shrinking.

Also, because the swelling of wood is directly proportional to density, low-density species are preferred over high-density species. However, even high-swelling and dense wood surfaces with flat grain can be stabilized with a resin-treated paper overlay (overlaid exterior plywood and lumber) to provide excellent surfaces for painting. Medium-density, stabilized fiberboard products with a uniform, low-density surface or paper overlay are also good substrates for exterior use. However, vertical-grained heartwood of western redcedar and redwood are the species most widely used as exterior siding

and trim when painting is desired. These species are classified in group I, those woods easiest to keep painted (table 16—1). Vertical-grain surfaces of all species actually are considered excellent for painting, but most species are generally available only as flat-grain lumber.

Species that are normally cut as flat-grained lumber, that are high in density and swelling, or have defects such as knots or pitch, are classified in groups II through V, depending upon their general paint-holding characteristics. Many species in groups II through IV are commonly painted, particularly the pines, Douglas-fir, and spruce; but these species generally require more care and attention than the vertical-grain surfaces of group I. Exterior paint will be more durable on vertical-grain boards than on flat-grain boards for any species with marked differences in density between earlywood and latewood, even if the species are rated in group I (fig. 16—2). Flat-grain boards that are to be painted should be installed in areas protected from rain and sun.

Plywood for exterior use nearly always has a flat-grain surface. In addition, cycles of wetting and drying with subsequent swelling and shrinking tend to check the face veneer of plywood much more than lumber. This checking sometimes extends through paint coatings to detract from their appearance and durability. Plywood with a resin-treated paper overlay does not check. It has excellent paintability and would be equal to or better than vertical-grain lumber of group I.

Types of Wood Products Used Outdoors

Three general categories of wood products are commonly used in exterior construction: (1) lumber, (2) plywood, and (3) reconstituted wood products such as hardboard and particleboard. Each product has unique characteristics that will affect the durability of any finish applied to it.

Lumber

Lumber is being used less and less as exterior siding, but was once the most common wood material used in construction. Many older homes have wood siding. The ability of lumber to retain and hold a finish is affected by species, by ring direction with respect to the surface (or how the piece was sawn from the log), and by smoothness.

The weight of wood varies tremendously between species. Some common construction woods such as southern pine are dense and heavy compared with the lighter weight ones such as redwood and cedar. The weight of wood is important because heavy woods shrink and swell more than light ones. This dimensional change in lumber occurs as the wood gains or loses moisture. Excessive dimensional change in wood constantly stresses a paint film and may result in early failure.

Some species have wide bands of earlywood and latewood (fig. 16—3). Wide, prominent bands of latewood are charac-

*Revision by William C. Feist, Chemist.

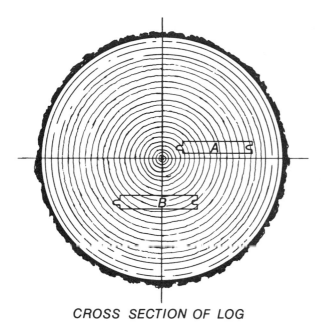

CROSS SECTION OF LOG

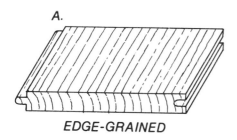

A.

EDGE-GRAINED

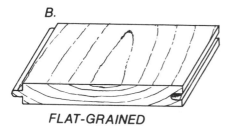

B.

FLAT-GRAINED

Figure 16 – 1—Edge-grained (or vertical-grained or quartersawed) board *A*, and flat-grained (or slash-grained or plainsawed) board *B*, cut from a log. (M148 631)

Figure 16 – 2—Paint applied over edge-grained boards (top and bottom) performs better than that applied to flat-grained boards (middle). (M147 211-12)

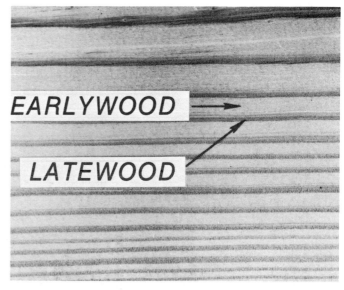

EARLYWOOD

LATEWOOD

Figure 16 – 3—Earlywood and latewood bands in southern pine. These distinct bands often lead to early paint failure. (M83 0023)

16-3

teristic of southern pine and most Douglas-fir, and paint will not hold well on these species. By contrast redwood and cedar do not have wide latewood bands, and these species are preferred when paint will be used.

Ring direction also affects paint-holding characteristics and is determined at the time lumber is cut from a log (fig. 16—2). Most standard grades of lumber contain a high percentage of flat grain. Lumber used for board and batten siding, drop siding, or shiplap is frequently flat-grained. Bevel siding is commonly produced in several grades. In some cases, the highest grade is required to be vertical-grained and all-heart over most of the width for greater paint durability. Other grades may be flat-grained, vertical-grained, or mixed-grain and without requirements as to heartwood.

Plywood

Exterior plywood with a rough-sawn surface is commonly used for siding. Smooth-sanded plywood is not recommended for siding, but it is often used in soffits. Both sanded and rough-sawn plywood will develop surface checks, especially when exposed to moisture and sunlight. These surface checks can lead to early paint failure with oil or alkyd paint systems (fig. 16—4). Quality acrylic latex primer and topcoat paint

Table 16—1—*Characteristics of woods for painting and weathering (omissions in the table indicate inadequate data for classification)*

Wood	Ease of keeping well painted; I—easiest, V—most exacting[1]	Weathering		Appearance	
		Resistance to cupping; 1—best, 4—worst	Conspicuousness of checking; 1—least, 2—most	Color of heartwood (sapwood is always light)	Degree of figure on flat-grained surface

SOFTWOODS

Wood					
Cedar:					
Alaska-	I	1	1	Yellow	Faint
California incense-	I	—	—	Brown	Do.
Port-Orford-	I	—	1	Cream	Do.
Western redcedar	I	1	1	Brown	Distinct
White-	I		—	Light brown	Do.
Cypress	I	1	1	do.	Strong
Redwood	I	1	1	Dark brown	Distinct
Products[2] overlaid with resin-treated paper	I	—	1	—	—
Pine:					
Eastern white	II	2	2	Cream	Faint
Sugar	II	2	2	do.	Do.
Western white	II	2	2	do.	Do.
Ponderosa	III	2	2	do.	Distinct
Fir, commercial white	III	2	2	White	Faint
Hemlock	III	2	2	Pale brown	Do.
Spruce	III	2	2	White	Do.
Douglas-fir (lumber and plywood)	IV	2	2	Pale red	Strong
Larch	IV	2	2	Brown	Do.
Lauan (plywood)	IV	2	2	do.	Faint
Pine:					
Norway	IV	2	2	Light brown	Distinct
Southern (lumber and plywood)	IV	2	2	do.	Strong
Tamarack	IV	2	2	Brown	Do.

systems generally perform better. The flat-grained pattern present in nearly all plywood can also contribute to early paint failure. Therefore, if smooth or rough-sawn plywood is to be painted, special precautions should be exercised. Penetrating stains are often more appropriate for rough-sawn exterior plywood surfaces but quality acrylic latex paints also perform very well.

Reconstituted Wood Products

Reconstituted wood products are those made by forming small pieces of wood into large sheets, usually 4 by 8 feet or as required for a specialized use such as beveled siding.

Figure 16 – 4—Early paint failure on plywood because of penetration of moisture into surface checks of plywood. (M83 0013-6)

Table 16 – 1—*Characteristics of woods for painting and weathering (omissions in the table indicate inadequate data for classification)*—**Continued**

| Wood | Ease of keeping well painted; I—easiest, V—most exacting[1] | Weathering | | Appearance | |
		Resistance to cupping; 1—best, 4—worst	Conspicuousness of checking; 1—least, 2—most	Color of heartwood (sapwood is always light)	Degree of figure on flat-grained surface
HARDWOODS					
Alder	III	—	—	Pale brown	Faint
Aspen	III	2	1	do.	Do.
Basswood	III	2	2	Cream	Do.
Cottonwood	III	4	2	White	Do.
Magnolia	III	2	—	Pale brown	Do.
Yellow-poplar	III	2	1	do.	Do.
Beech	IV	4	2	do.	Do.
Birch	IV	4	2	Light brown	Do.
Cherry	IV	—	—	Brown	Do.
Gum	IV	4	2	Brown	Do.
Maple	IV	4	2	Light Brown	Do.
Sycamore	IV	—	—	Pale brown	Do.
Ash	V or III	4	2	Light brown	Distinct
Butternut	V or III	—	—	do.	Faint
Chestnut	V or III	3	2	Light brown	Distinct
Walnut	V or III	3	2	Dark brown	Do.
Elm	V or IV	4	2	Brown	Do.
Hickory	V or IV	4	2	Light brown	Do.
Oak, white	V or IV	4	2	Brown	Do.
Oak, red	V or IV	4	2	do.	Do.

[1] Woods ranked in group V for *ease of keeping well painted* are hardwoods with large pores that need filling with wood filler for durable painting. When so filled before painting, the second classification recorded in the table applies.

[2] Plywood, lumber, and fiberboard with overlay or low-density surface.

These products may be classified as fiberboard or particleboard, depending upon the nature of the basic wood component.

Fiberboards are produced from mechanical pulps. Hardboard is a relatively heavy type of fiberboard, and its tempered or treated form designed for outdoor exposure is used for exterior siding. It is often sold in 4- by 8-foot sheets as a substitute for solid wood beveled siding.

Particleboards are manufactured from whole wood in the form of splinters, chips, flakes, strands, or shavings. Waferboard and flakeboard are two types of particleboard made from relatively large flakes or shavings.

Some fiberboards and particleboards are manufactured for exterior use. Film-forming finishes such as paints and solid color stains will give the most protection to these reconstituted wood products. Some reconstituted wood products may be factory primed with paint, and some may even have a factory-applied top coat. Also, some may be overlaid with a resin-treated cellulose fiber sheet to provide a superior surface for paint.

Treated Wood

Wood is sometimes used in severe outdoor situations where special treatments and finishes are required for proper protection and best service. These situations involve the need for protection against decay (rot), insects, fire, and harsh exposures such as marine environments.

Although not generally classified as wood finishes, preservatives in wood do protect against weathering (in addition to decay), and a great quantity of preservative-treated wood is exposed outdoors without any additional finish. There are three main types of preservatives (see ch. 18): (1) the preservative oils (e.g., coal-tar creosote), (2) the organic solvent solutions (e.g., pentachlorophenol), and (3) waterborne salts (e.g., chromated copper arsenate). These preservatives can be applied in several ways, but pressure treatment generally gives the greatest protection against decay. Higher preservative content of pressure-treated wood generally results in greater resistance to weathering and improved surface durability. The chromium-containing preservatives also protect against ultraviolet degradation, an important factor in the weathering process.

Water-repellent preservatives introduced into wood by a vacuum-pressure or dipping process (NWMA Industry Standard 4–81) are paintable. Coal-tar creosote or other dark oily preservatives tend to stain through paint, especially light-colored paint, unless the treated wood has weathered for many years before painting.

The fire-retardant treatment of wood does not generally interfere with adhesion of decorative paint coatings, unless the treated wood has an extremely high moisture content because of its increased hygroscopicity. It is most important that only those fire-retardant treatments specifically prepared and recommended for outdoor exposure be used for that purpose. These treated woods are generally painted according to recommendations of the manufacturer rather than being left unfinished because the treatment and subsequent drying often darken and irregularly stain the wood.

Extractives in Wood

Water-soluble colored extractives occur naturally in the heartwood of such species as western redcedar, cypress, and redwood. It is to these substances that the heartwood of these species owes its attractive color, good stability, and natural decay resistance. However, discoloration of paint may occur when the extractives are dissolved and leached from the wood by water. When the solution of extractives reaches the painted surface, the water evaporates, leaving the extractives as a yellow to reddish-brown stain. The water that gets behind the paint and causes moisture blisters also causes migration of extractives. The discoloration produced when water wets siding from the back frequently forms a rundown or streaked pattern.

Moisture-Excluding Effectiveness of Finishes

Shrinking and swelling and the accompanying stresses in wood that cause warping, checking, and contribute to weathering are brought about by changes in the moisture content. Such changes occur whenever wood is exposed to varying atmospheric conditions. Effective protection against fluctuating atmospheric conditions is furnished by coatings of various moisture-retardant finishes, provided that the coating is applied to all surfaces of wood through which moisture might gain access. No coating is entirely moistureproof, however, and there is as yet no way of completely keeping moisture out of wood that is exposed to dampness constantly or for prolonged periods.

The protection afforded by coatings in excluding moisture from wood depends on a great number of variables. Among them are coating film thickness, defects and voids in the film, type of pigment, chemical composition of the vehicle, volume ratio of pigment to vehicle, vapor-pressure gradient across the film, and length of exposure period.

The relative effectiveness of several typical treating and finishing systems for wood in retarding adsorption of water vapor at 90 percent relative humidity is compared in table 16–2. Perfect protection, or no adsorption of water, would be represented by 100 percent effectiveness; complete lack of protection (as with unfinished wood) by 0 percent.

Values in table 16–2 are only representative and indicate the range in protection against moisture in vapor form for

Table 16−2—*Moisture-excluding effectiveness of various finishes. Ponderosa pine sapwood was initially finished and conditioned to 80 °F and 30 percent relative humidity and then exposed to 80 °F and 90 percent relative humidity*

Finish type	No. of coats	Moisture-excluding effectiveness		
		1 day	7 days	14 days
		- - - - - - Percent - - - - - -		
Linseed oil sealer (50 pct)	1	7	0	0
	2	15	1	0
	3	18	2	0
Linseed oil	1	12	0	0
	2	22	0	0
	3	33	2	0
Tung oil	1	34	0	0
	2	46	2	0
	3	52	6	2
Paste furniture wax	1	6	0	0
	2	11	0	0
	3	17	0	0
Water repellent	1	12	0	0
	2	46	2	0
	3	78	27	11
Latex flat wall paint (vinyl acrylic resin)	1	5	0	0
	2	11	0	0
	3	22	0	0
Latex enamel wall paint (epoxy)	1	45	10	3
	2	66	17	5
	3	73	26	10
Latex primer wall paint (butadiene-styrene resin)	1	78	37	20
	2	86	47	27
	3	88	55	33
Alkyd flat wall paint (soya alkyd)	1	9	1	0
	2	21	2	0
	3	37	5	0
Acrylic latex house primer paint	1	43	6	1
	2	66	14	2
	3	72	20	4
Acrylic latex flat house paint	1	52	12	5
	2	77	28	11
	3	84	39	16
Solid color latex stain (acrylic resin)	1	5	0	0
	2	38	4	0
	3	50	6	0

Table 16 – 2—Moisture-excluding effectiveness of various finishes. Ponderosa pine sapwood was initially finished and conditioned to 80 °F and 30 percent relative humidity and then exposed to 80 °F and 90 percent relative humidity—Continued

Finish type	No. of coats	Moisture-excluding effectiveness		
		1 day	7 days	14 days
		- - - - - - Percent - - - - - -		
Solid color oil-based stain	1	45	7	1
(linseed oil)	2	84	48	26
	3	90	64	42
FPL natural finish	1	62	14	3
(linseed oil-based semi-	2	70	21	6
transparent stain)	3	76	30	11
Semitransparent oil-based stain	1	7	0	0
(commercial)	2	13	0	0
	3	21	1	0
Marine enamel—gloss	1	79	38	18
(soya alkyd)	2	91	66	46
	3	93	74	57
Marine enamel—flat	1	22	1	0
	2	76	28	10
	3	89	57	32
Aluminum paint	1	80	35	13
(linseed/phenolic/menhaden)	2	97	87	76
	3	98	91	82
Alkyd house primer paint	1	85	46	24
(tall maleic alkyd resin)	2	93	70	49
	3	95	78	60
Oil-based house paint	1	72	23	8
(tall/soya alkyds)	2	86	52	29
	3	90	63	41
Enamel paint—satin	1	93	69	50
(soya/tung/alkyd;	2	96	83	70
interior/exterior)	3	97	86	80
	4	98	92	85
	5	98	93	88
	6	98	94	89
Enamel paint—gloss	1	83	45	25
(soya alkyd; interior)	2	91	64	43
	3	94	76	59
	4	95	80	65
	5	96	84	72
	6	96	85	74

Table 16−2—*Moisture-excluding effectiveness of various finishes. Ponderosa pine sapwood was initially finished and conditioned to 80 °F and 30 percent relative humidity and then exposed to 80 °F and 90 percent relative humidity—*Continued

Finish type	No. of coats	Moisture-excluding effectiveness		
		1 day	7 days	14 days
		- - - - - - Percent - - - - - -		
Floor and deck enamel	1	80	31	18
(phenolic alkyd)	2	89	53	35
	3	92	63	46
Polysilicone enamel	1	75	30	12
(silicone alkyd)	2	88	59	36
	3	91	69	48
Latex pigmented shellac	1	30	3	0
	2	38	8	1
	3	44	8	1
Shellac	1	65	10	3
	2	84	43	20
	3	91	64	42
	4	93	75	58
	5	94	81	67
	6	95	85	73
Pigmented shellac	1	91	67	44
	2	95	81	65
	3	96	85	73
	4	96	88	79
	5	97	89	81
	6	97	90	83
Nitrocellulose lacquer	1	40	4	1
	2	70	22	8
	3	79	37	19
Floor seal	1	31	1	0
(phenolic resin/tung oil)	2	80	37	18
	3	88	56	35
Gym seal (linseed oil/	1	53	9	1
phenolic resin/tung oil)	2	87	53	28
	3	91	66	44
Spar varnish (soya alkyd)	1	48	6	0
	2	80	36	15
	3	87	53	30

Table 16−2—*Moisture-excluding effectiveness of various finishes. Ponderosa pine sapwood was initially finished and conditioned to 80 °F and 30 percent relative humidity and then exposed to 80 °F and 90 percent relative humidity*—Continued

Finish type	No. of coats	Moisture-excluding effectiveness		
		1 day	7 days	14 days
		- - - - - - Percent - - - - - -		
Urethane varnish (oil-modified)	1	55	10	2
	2	83	43	23
	3	90	64	44
	4	91	68	51
	5	93	72	57
	6	93	76	62
Aluminum flake pigmented urethane varnish (oil-modified)	1	90	61	41
	2	97	87	77
	3	98	91	84
	4	98	93	87
	5	98	94	89
	6	99	95	90
Polyurethane finish—clear (2-component)	1	48	6	0
	2	90	66	46
	3	94	81	66
Polyurethane paint—gloss (2-component)	1	91	66	44
	2	94	79	62
	3	96	86	74
Epoxy finish—clear (2-component)	1	93	73	54
	2	98	93	88
	3	98	95	91
Epoxy paint—gloss (2-component)	1	93	77	53
	2	98	90	82
	3	98	93	87
Epoxy paint—gloss (1-component)	1	92	37	18
	2	93	69	49
	3	94	76	59
Epoxy varnish—gloss (1-component)	1	54	11	3
	2	87	54	34
	3	92	69	50
Paraffin wax—brushed	1	97	82	69
Paraffin wax—dipped	1	100	97	95

some conventional finish systems when exposed to continuous high humidity. The degree of protection provided also depends on the kind of exposure. For example, the water-repellent treatment, which may have 0 percent effectiveness against water vapor after 2 weeks at 80 °F and 90 percent relative humidity, would have an effectiveness of over 60 percent when tested after immersion in water for 30 minutes. The high degree of protection provided by water repellents and water-repellent preservatives to short periods of wetting by liquid water is the major reason they are recommended for exterior finishing.

Porous paints, such as the latex paints and low-luster (flat) or breather type oil-base paints formulated at a pigment volume concentration usually above 40 percent, afford little protection against moisture. These paints permit rapid entry of water vapor and water from dew and rain unless applied over a nonporous primer.

The moisture-excluding effectiveness of many coatings improves slightly with age. Good exterior coatings either retain their maximum effectiveness for a considerable time or lose effectiveness slowly. As long as the original appearance and integrity of the coatings are retained, most of the effectiveness remains. Paint that is slowly fading or chalking remains effective if vigorous rubbing will remove the chalk and disclose a glossy film underneath. Deep chalking, checking, or cracking indicates serious impairment of the effectiveness.

The numerical values for percentage effectiveness in table 16−2 should be considered relative rather than absolute because percentage effectiveness varies substantially with the conditions under which exposure to moisture takes place. The values for effective coatings (60 percent or over) are reliable in the sense that they can be reproduced closely on repeating the test; values for ineffective coatings (less than 20 percent) must be regarded as rough approximations only. The percentages shown are based on average amounts of moisture absorbed per unit surface area by newly coated and by uncoated wood panels.

Finishing Wood Exposed Outdoors

Weathering

The simplest finish for wood is that created in the weathering process. Without paint or treatment of any kind, wood surfaces gradually change in color and texture and then may stay almost unaltered for a long time if the wood does not decay. Generally, the dark-colored woods become lighter and the light-colored woods become darker. As weathering continues, all woods become gray, accompanied by photodegradation and gradual loss of wood cells at the surface. As a result, exposed unfinished wood will slowly wear away in a process called erosion (fig. 16−5).

The weathering process is a surface phenomenon and is so slow that softwoods erode at an average rate of about 1/4 inch per century. Dense hardwoods will erode at a rate of only 1/8 inch per century. Low-density softwoods such as western redcedar may erode at a rate as high as 1/2 inch per century. In cold northern climates, erosion values as low as 1/32 inch per century have been reported.

The physical loss of wood substance from the wood surface (erosion) during weathering depends not only on wood species and density, but also on growth rate, ring orientation, amount of irradiation, rain action, wind, and degree of exposure. It also occurs most rapidly in thin-walled fibers of earlywood in softwoods and, at a slower rate, in dense latewood. Accompanying this loss of wood substance are the swelling and shrinking stresses set up by fluctuations in moisture content. All this results in surface roughening, grain raising, differential swelling of earlywood and latewood bands, and the formation of many small parallel checks and cracks. Larger and deeper cracks may also develop and warping frequently occurs.

The weathering process is usually accompanied by the growth of dark-colored spores and mycelia of fungi or mildew on the surface, which give the wood a dark gray, blotchy, and unsightly appearance. In addition, highly colored wood extractives in such species as western redcedar and redwood add to the variable color of weathered wood. The dark brown color of extractives may persist for a long time in areas not exposed to the sun and where the extractives are not removed by the washing action of rain.

Natural Wood Finishes

Some wood finishes often find application as so-called natural finishes for wood. Each finish system offers various advantages and disadvantages for this use. These systems can be classified as film-forming or penetrating finishes. The penetrating finishes can be subdivided further into transparent, semitransparent, and waterborne salts. The simplest natural finish is supplied by nature itself in the weathering process. The unpigmented natural finishes provide the least protection of all the wood finishes for outdoor exposure.

Film-Forming

Varnishes are the primary transparent film-forming materials used for natural wood finishes outdoors, and their use greatly enhances the natural beauty and figure of wood. Varnishes lack exterior permanence unless protected from direct exposure to sunlight, and varnish finishes on wood exposed outdoors without protection will generally require refinishing every 1 to 2 years.

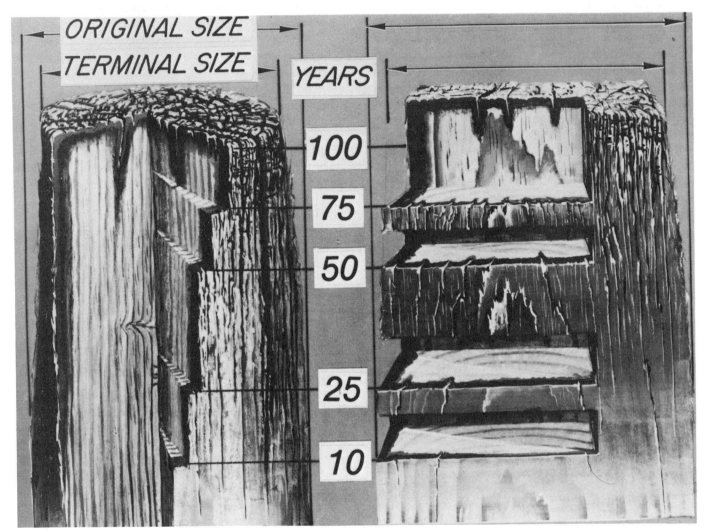

ORIGINAL SIZE
TERMINAL SIZE

YEARS

100
75
50
25
10

(M146 221)

Figure 16 – 5—Artist's rendition showing the weathering process of round and square timbers. Cutaway shows that interior wood below the surface is relatively unchanged.

Penetrating

The penetrating finishes are the second broad classification of natural wood finishes. These finishes do not form a film on the wood surface and are further divided into (1) transparent or clear systems, (2) pigmented or semitransparent systems, and (3) waterborne salts.

Water-repellent preservatives are the most important of the transparent penetrating natural finish systems. Treating wood surfaces with a water-repellent preservative will protect wood exposed outdoors with little change in appearance. A clean,

golden-tan color can be achieved with most wood species. The treatment reduces warping and cracking, prevents water staining at edges and ends of wood siding, and helps control mildew growth. The first application of a water-repellent preservative may protect exposed wood surfaces for only 1 to 2 years, but subsequent reapplications may last 2 to 4 years because the weathered boards absorb more of the finish. Drying oils (linseed, tung, etc.) are also sometimes used by themselves as natural finishes.

The semitransparent stains are the second group of the

16-12

penetrating natural wood finishes. These stain finishes provide a less natural appearance because they contain pigment that partially hides the original grain and color of the wood. They are generally much more durable than are varnishes or water-repellent preservatives and provide more protection against weathering. With these stain systems, weathering is slowed by retarding the alternate wetting and drying of wood. The presence of pigment particles on the wood surface minimizes the degrading effects of sunlight. The amount of pigment in the semitransparent stains can vary considerably, thus providing different degrees of protection against ultraviolet degradation and masking of the appearance of the original wood surface. Higher pigment loading yields greater protection against weathering.

Waterborne inorganic salts containing chromium are a special group of penetrating finishes. These surface treatments result in a natural finish related to the semitransparent penetrating finishes, because they change the color of the wood and leave a thin surface deposit of material similar to the pigment found in semitransparent stains.

Types of Exterior Wood Finishes

The outdoor finishes described in this section, their properties, treatment, and maintenance are summarized in table 16-3. The suitability and expected life of the most commonly used finishes on several wood and wood-based products is summarized in table 16-4. Information in tables 16-3 and 16-4 should be considered as general guidelines only. Many factors affect the performance and lifetime of wood finishes as described earlier.

Paint

Paints are coatings commonly used on wood and provide the most protection. They come in a wide range of colors and may be either oil-or latex-based. Latex-based paints and stains are waterborne, and oil or alkyd paints are organic solventborne. Paints are used for esthetic purposes, to protect the wood surface from weathering and to conceal certain defects.

Paints are applied to the wood surface and do not penetrate it deeply. The wood grain is completely obscured, and a surface film is formed. Paints perform best on smooth, edge-grained lumber of lightweight species. Paints are the only way to achieve a bright white finish. This surface film can blister or peel if the wood is wetted or if inside water vapor moves through the house wall and wood siding because of the absence of a vapor-retarding material.

Latex paints are generally easier to use because water is used in cleanup. They are also porous and, thus, will allow some moisture movement. In comparison, oil-based paints require organic solvents for cleanup, and some are resistant to moisture movement.

Of all the finishes, paints provide the most protection for wood against surface erosion and offer the widest selection of colors. A nonporous paint film retards penetration of moisture and reduces the problem of discoloration by wood extractives, paint peeling, and checking and warping of the wood. PAINT IS NOT A PRESERVATIVE. IT WILL NOT PREVENT DECAY IF CONDITIONS ARE FAVORABLE FOR FUNGAL GROWTH. Original and maintenance costs are often higher for a paint finish than for a water-repellent preservative or penetrating stain finish.

Solid Color Stains

Solid color stains are opaque finishes (also called hiding, or heavybodied, or blocking) which come in a wide range of colors and are made with a much higher concentration of pigment than the semitransparent penetrating stains. As a result they will totally obscure the natural wood color and grain. Oil-based solid color stains tend to form a film much like paint and as a result can also peel loose from the substrate. Latex-based solid color stains are also available and form a film as do the oil-based solid color stains. Both these stains are similar to thinned paints and can usually be applied over old paint or stains.

Semitransparent Penetrating Stains

Semitransparent penetrating stains are only moderately pigmented and, thus, do not totally hide the wood grain. These stains penetrate the wood surface, are porous, and do not form a surface film like paints. As a result, they will not blister or peel even if moisture moves through the wood. Penetrating stains are alkyd- or oil-based, and some may contain a fungicide as well as a water repellent. Moderately pigmented latex-based (waterborne) stains are also available, but they do not penetrate the wood surface as do the oil-based stains.

Stains are most effective on rough lumber or rough-sawn plywood surfaces, but they also provide satisfactory performance on smooth surfaces. They are available in a variety of colors and are especially popular in the brown or red earth tones since they give a "natural or rustic wood appearance." These stains are not available in white. They are an excellent finish for weathered wood. Semitransparent stains are not effective when applied over a solid color stain or over old paint.

An effective stain of this type is the Forest Products Laboratory natural finish. The finish has a linseed oil vehicle; a fungicide, pentachlorophenol (see Caution), to protect the oil and wood from mildew; and a water repellent, paraffin wax, to protect the wood from excessive penetration of water. Durable red and brown iron oxide pigments simulate the natural colors of redwood and cedar. A variety of other colors (except pure white) can also be achieved with this type of finish.

Table 16–3—*Exterior wood finishes: Types, treatment, and maintenance*[1]

Finish	Initial treatment	Appearance of wood	Cost of initial treatment	Maintenance procedure	Maintenance period of surface finish	Maintenance cost
Preservative oils (creosotes)	Pressure, hot and cold tank steeping	Grain visible. Brown to black in color, fading slightly with age	Medium	Brush down to remove surface dirt	5–10 yr only if original color is to be renewed; otherwise no maintenance is required	Nil to low
	Brushing	do.	Low	do.	3–5 yr	Low
Waterborne preservatives	Pressure	Grain visible. Greenish in color, fading with age	Medium	Brush down to remove surface dirt	None, unless stained, painted, or varnished as below	Nil, unless stains, varnishes, or paints are used. See below
	Diffusion plus paint	Grain and natural color obscured	Low to medium	Clean and repaint	7–10 yr	Medium
Organic solvents preservatives[2]	Pressure, steeping, dipping, brushing	Grain visible. Colored as desired	Low to medium	Brush down and re-apply	2–3 yr or when preferred	Do.
Water repellent[3]	One or two brush coats of clear material or, preferably, dip applied	Grain and natural color visible, becoming darker and rougher textured	Low	Clean and apply sufficient material	1–3 yr or when preferred	Low to medium
Stains	One or two brush coats	Grain visible. Color as desired	Low to medium	do.	3–6 yr or when preferred	Do.
Clear varnish	Four coats (minimum)	Grain and natural color unchanged if adequately maintained	High	Clean and stain bleached areas, and apply two more coats	2 yr or when breakdown begins	High

Finish	Initial treatment	Appearance of wood	Cost of initial treatment	Maintenance procedure	Maintenance period of surface finish	Maintenance cost
Paint	Water repellent, prime, and two topcoats	Grain and natural color obscured	Medium to high	Clean and apply topcoat; or remove and repeat initial treatment if damaged	7–10 yr[4]	Medium to high

[1] This table is a compilation of data from the observations of many researchers.
[2] Pentachlorophenol, bis(tri-n-butyltin oxide), copper naphthenate, copper-8-quinolinolate, and similar materials.
[3] With or without added preservatives. Addition of preservative helps control mildew growth and gives better performance.
[4] Using top-quality acrylic latex topcoats.

Water Repellents and Water-Repellent Preservatives

A water-repellent preservative may be used as a natural finish. It contains a fungicide (such as pentachlorophenol or copper naphthenate), a small amount of wax as a water repellent, a resin or drying oil, and a solvent such as turpentine or mineral spirits. Water-repellent preservatives do not contain any coloring pigments. Therefore, the resulting finish will vary in color depending upon the wood itself. The preservative also prevents wood from darkening (graying) through mildew growth.

The initial application to smooth surfaces is usually short lived. When a surface starts to show a blotchy discoloration due to extractives or mildew, it should be cleaned with liquid household bleach and detergent solution and re-treated after drying. During the first few years, the finish may have to be applied every year or so. After the wood has gradually weathered to a uniform color, the treatments are more durable and need refinishing only when the surface starts to become unevenly colored by fungi.

CAUTION: Because of the toxicity of pentachlorophenol, a commonly used fungicide in water-repellent preservative solutions and some semitransparent stains, care should be exercised to avoid excessive contact with the solution or its vapor or with the treated wood. Shrubs and plants should also be protected from accidental contamination.

Inorganic pigments also can be added to the water-repellent preservative solutions to provide special color effects, and the mixture is then classified as a pigmented penetrating stain. Two to six fluid ounces of colors-in-oil or tinting colors can be added to each gallon of treating solution. Colors that match the natural color of the wood and extractives are usually preferred. As with semitransparent stains, the addition of pigment to the finish helps stabilize the color and increase the durability of the finish.

Water-repellent preservatives may also be used as a treatment for bare wood before priming and painting or in areas where old paint has peeled, exposing bare wood, particularly around butt joints or in corners. This treatment keeps rain or dew from penetrating into the wood, especially at joints and end grain, and thus decreases the shrinking and swelling of wood. As a result, less stress is placed on the paint film, and its service life is extended (fig. 16–6). This stability is achieved by the small amount of wax present in water-repellent preservatives. The wax decreases the capillary movement or wicking of water up the back side of lap or drop siding. The fungicide inhibits surface decay.

Water repellents are also available. These are water-repellent preservatives with the preservative left out. Water repellents are not effective natural finishes by themselves but can be used as a stabilizing treatment before priming and painting.

Figure 16−6—A, Window sash and frame treated with a water-repellent preservative and then painted and B, window sash and frame not treated before painting. Both treatments were weathered for 5 years.

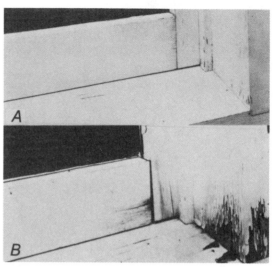

(MC84 9072)

Table 16−4—*Suitability of finishing methods for exterior wood surfaces[1]*

Type of exterior wood surfaces	Water-repellent preservative		Stains		Paints	
	Suitability	Expected life[2]	Suitability	Expected life[3]	Suitability	Expected life[4]
		Yr		*Yr*		*Yr*
Siding:						
Cedar and redwood						
Smooth (vertical grain)	High	1−2	Moderate	2−4	High	4−6
Rough sawn or weathered	High	2−3	Excellent	5−8	Moderate	3−5
Pine, fir, spruce, etc.						
Smooth (flat grain)	High	1−2	Low	2−3	Moderate	3−5
Rough (flat grain)	High	2−3	High	4−7	Moderate	3−5
Shingles						
Sawn	High	2−3	Excellent	4−8	Moderate	3−5
Split	High	1−2	Excellent	4−8	—	—
Plywood (Douglas-fir and southern pine)						
Sanded	Low	1−2	Moderate	2−4	Moderate	3−5
Rough sawn	Low	2−3	High	4−8	Moderate	3−5
Medium-density overlay[5]	—	—	—	—	Excellent	6−8
Plywood (cedar and redwood)						
Sanded	Low	1−2	Moderate	2−4	Moderate	3−5
Rough sawn	Low	2−3	Excellent	5−8	Moderate	3−5
Hardboard, medium density[6]						
Smooth						
Unfinished	—	—	—	—	High	4−6
Preprimed	—	—	—	—	High	4−6
Textured						
Unfinished	—	—	—	—	High	4−6
Preprimed	—	—	—	—	High	4−6

Transparent Coatings

Clear coatings of conventional spar or marine varnishes, which are film-forming finishes, are not generally recommended for exterior use on wood. Such coatings embrittle by exposure to sunlight and develop severe cracking and peeling, often in less than 2 years. Areas that are protected from direct sunlight by overhang or are on the north side of the structure can be finished with exterior-grade varnishes. Even in protected areas, a minimum of three coats of varnish is recommended, and the wood should be treated with water-repellent preservative before finishing. The use of pigmented stains and sealers as undercoats also will contribute to greater life of the clear finish. In marine exposures, six coats of varnish should be used for best performance.

Fire-Retardant Coatings

Many commercial fire-retardant coating products are available to provide varying degrees of protection of wood against fire. These paint coatings generally have low surface flammability characteristics and "intumesce" to form an expanded low-density film upon exposure to fire, thus insulating the wood surface below from heat and retarding pyrolysis reactions. The paints have added ingredients to restrict the flaming of any released combustible vapors. Chemicals may also be present in these paints to promote decomposition of the wood surface to charcoal and water rather than forming volatile flammable products.

Most fire-retardant coatings are intended for interior use, but some are available for exterior application. Conventional paints have been applied over the fire-retardant coatings to improve their durability. Most conventional decorative coatings will in themselves slightly reduce the flammability of wood products when applied in conventional film thicknesses.

Table 16 – 4—*Suitability of finishing methods for exterior wood surfaces[1]*—Continued

Type of exterior wood surfaces	Water-repellent preservative		Stains		Paints	
	Suitability	Expected life[2]	Suitability	Expected life[3]	Suitability	Expected life[4]
		Yr		Yr		Yr
Millwork (usually pine):						
Windows, shutters, doors, exterior trim	High[7]	—	Moderate	2–3	High	3–6
Decking:						
New (smooth)	High	1–2	Moderate	2–3	Low	2–3
Weathered (rough)	High	2–3	High	3–6	Low	2–3
Glued-laminated members:						
Smooth	High	1–2	Moderate	3–4	Moderate	3–4
Rough	High	2–3	High	6–8	Moderate	3–4
Waferboard	—	—	Low	1–3	Moderate	2–4

[1] This table is a compilation of data from the observations of many researchers. Expected life predictions are for an average continental U.S. location; expected life will vary in extreme climates or exposure (desert, seashore, deep woods, etc.).

[2] Development of mildew on the surface indicates a need for refinishing.

[3] Smooth, unweathered surfaces are generally finished with only one coat of stain, but rough-sawn or weathered surfaces, being more adsorptive, can be finished with two coats, with the second coat applied while the first coat is still wet.

[4] Expected life of two coats, one primer and one topcoat. Applying a second topcoat (three-coat job) will approximately double the life. Top-quality acrylic latex paints will have best durability.

[5] Medium-density overlay is generally painted.

[6] Semitransparent stains are not suitable for hardboard. Solid color stains (acrylic latex) will perform like paints. Paints are preferred.

[7] Exterior millwork, such as windows, should be factory treated according to Industry Standard IS4–81. Other trim should be liberally treated by brushing before painting.

Application of Wood Finishes

Paint

Proper surface care and preparation before applying paint to wood is essential for good performance. Wood and wood-based products should be protected from the weather and wetting on the jobsite and after they are installed. Surface contamination from dirt, oil, and other foreign substances must be eliminated. It is most important to paint wood surfaces within one week, weather permitting, after installation.

To achieve maximum paint life, follow these steps:

1. Wood siding and trim should be treated with a paintable water-repellent preservative or water repellent. Water repellents protect the wood against the entrance of rain and dew and thus help to minimize swelling and shrinking. They can be applied by brushing or dipping. Lap and butt joints and the edges of panel products such as plywood, hardboard, and particleboard should be especially well treated because paint normally fails in these areas first. Allow at least 2 warm, sunny days for adequate drying before painting the treated surface. If the wood has been dip treated, allow at least 1 week of favorable weather.

2. After the water-repellent preservative or water repellent has dried, the bare wood must be primed. As the primer coat forms a base for all succeeding paint coats, it is very important. For woods with water-soluble extractives such as redwood and cedar, the best primers are good quality oil-based and alkyd-based paints or stain-blocking acrylic latex-based paints. The primer seals in the extractives so that they will not bleed through the topcoat. A primer should be used whether the topcoat is an oil-base or latex-base paint. For species that are predominantly sapwood and free of extractives, such as pine, a high-quality acrylic latex topcoat paint may be used as both a primer and topcoat. Enough primer should be applied to obscure the wood grain. Do not spread the primer too thinly. Follow the application rates recommended by the manufacturer. A primer coat that is uniform and of the proper thickness will distribute the swelling stresses which develop in wood and thus help to prevent premature paint failure.

3. Two coats of a good-quality acrylic latex house paint should be applied over the primer. Other paints that are used include the oil-based, alkyd-based, and vinyl-acrylic. If it is not practical to apply two topcoats to the entire house, consider two topcoats for fully exposed areas on the south and west sides as a minimum for good protection. Areas fully exposed to sunshine and rain are the first to deteriorate and therefore should receive two coats. On those wood surfaces best suited for painting, one coat of a good house paint over a properly applied primer (a conventional two-coat paint system) should last 4 to 5 years, but two coats can last up to 10 years (table 16–4).

4. One gallon of paint will cover about 400 square feet of smooth wood surface area. However, coverage can vary with different paints, surface characteristics, and application procedures. Research has indicated that the optimum thickness for the total dry paint coat (primer and two topcoats) is 4 to 5 mils or about the thickness of a sheet of newspaper. The quality of paint is usually, but not always, related to price. Brush application is always superior to roller or spray application, especially for the first coat.

To avoid future separation between paint coats, the first topcoat should be applied within 2 weeks after the primer and the second coat within 2 weeks of the first. As certain paints weather, they can form a soaplike substance on their surface that may prevent proper adhesion of new paint coats. If more than 2 weeks elapse before applying another paint coat, scrub the old surface with water using a bristle brush or sponge. If necessary, to remove all dirt and deteriorated paint, use a mild detergent. Then rinse well with water, and allow the surfaces to dry before painting.

To avoid temperature blistering, oil-based paints should not be applied on a cool surface that will be heated by the sun within a few hours. Temperature blistering is most common with thick paint coats of dark colors applied in cool weather. The blisters usually show up in the last coat of paint and occur within a few hours or up to 1 or 2 days after painting. They do not contain water.

Oil-based paint may be applied when the temperature is 40 °F or above. A minimum of 50 °F is desired for applying latex-based waterborne paints. For proper curing of these latex paint films, the temperature should not drop below 50 °F for at least 24 hours after paint application. Low temperatures will result in poor coalescence of the paint film and early paint failure.

To avoid wrinkling, fading, or loss of gloss of oil-based paints and streaking of latex paints, the paint should not be applied in the evenings of cool spring and fall days when heavy dews form during the night before the surface of the paint has thoroughly dried. Serious water absorption problems and major finish failure can also occur with some latex paints when applied under these conditions.

Solid Color Stains

Solid color stains may be applied to a smooth surface by brush, spray, or roller application; but brush application is best. These stains act much like paint. One coat of solid color stain is considered adequate for siding, but two coats will provide significantly better protection and longer service. These stains are not generally recommended for horizontal wood surfaces such as decks and window sills.

Unlike paint, lap marks may form with a solid color stain. Latex-based stains are particularly fast-drying and are more likely to show lap marks than those with an oil base. To prevent lap marks, follow the procedures suggested under application of semitransparent penetrating stains.

Semitransparent Penetrating Stains

Semitransparent penetrating stains may be brushed, sprayed, or rolled on. Brushing will give the best penetration and performance. These stains are generally thin and runny, so application can be messy. Lap marks may form if stains are improperly applied. They can be prevented by staining only a small number of boards or one panel at a time. This method prevents the front edge of the stained area from drying out before a logical stopping place is reached. Working in the shade is desirable because the drying rate is slower. One gallon will usually cover about 200 to 400 square feet of smooth wood surface and from 100 to 200 square feet of rough or weathered surface.

For long life with penetrating oil-based stain on rough-sawn or weathered lumber, use two coats and apply the second coat before the first is dry. Apply the first coat to a panel or area in a manner to prevent lap marks. Then work on another area so that the first coat can soak into the wood for 20 to 60 minutes. Apply the second coat before the first coat has dried. (If the first coat dries completely, it may seal the wood surface so that the second coat cannot penetrate into the wood.) About an hour after applying the second coat, use a cloth, sponge, or dry brush lightly wetted with stain to wipe off the excess stain that has not penetrated into the wood. Otherwise areas of stain that did not penetrate may form an unsightly surface film and glossy spots. Avoid intermixing different brands or batches of stain. Stir stain occasionally and thoroughly during application to prevent settling and color change.

CAUTION: Sponges or cloths that are wet with oil-based stain are particularly susceptible to spontaneous combustion. To prevent fires, bury them, immerse them in water, or seal them in an airtight metal container immediately after use.

A two-coat system on rough wood may last as long as 10 years in certain exposures due to the large amount of stain absorbed. By comparison, if only one coat of penetrating stain is used on new smooth wood, its life expectancy is 2 to 4 years; however, succeeding coats will last longer (table 16–4).

Water Repellents

The most effective method of applying a water repellent or water-repellent preservative is to dip the entire board into the solution. However, brush treatment is also effective. When wood is treated in place, liberal amounts of the solution should be applied to all lap and butt joints, edges and ends of boards, and edges of panels where end grain occurs. Other areas especially vulnerable to moisture, such as the bottoms of doors and window frames, should not be overlooked. One gallon will cover about 250 square feet of smooth surface or 150 square feet of rough surface. The life expectancy is only

1 to 2 years as a natural finish, depending upon the wood and exposure. Treatments on rough surfaces are generally longer-lived than those on smooth surfaces. Repeated brush treatment to the point of refusal will enhance durability and performance. Treated wood that is painted will not need re-treating unless the protective paint layer weathers away (table 16–4).

Finishing Porches and Decks

Exposed flooring on porches and decks is sometimes painted. The recommended procedure of treating with water-repellent preservative and primer is the same as for wood siding. After the primer, an undercoat (first topcoat) and matching second topcoat of porch and deck enamel should be applied. These paints are especially formulated to resist abrasion and wear.

Many fully exposed decks are more effectively finished with only a water-repellent preservative or a penetrating-type semitransparent pigmented stain. These finishes will need more frequent refinishing than painted surfaces, but this is easily done because there is no need for laborious surface preparation as when painted surfaces start to peel. Solid color stains should not be used on any horizontal surface such as decks because early failure may occur.

Finishing Treated Wood

Wood pressure treated with waterborne chemicals, such as copper, chromium, and arsenic salts (CCA-treated wood) that react with the wood or form an insoluble residue, presents no major problem in finishing if the wood is properly redried and thoroughly cleaned after treating. Wood treated with solvent or oilborne preservative chemicals, such as pentachlorophenol, is not considered paintable until all the solvents have evaporated. Solvents such as methylene chloride or liquified petroleum gas evaporate readily. When heavy oil solvents with low volatility are used to treat wood under pressure, successful painting is usually impossible. Even special drying procedures for wood pressure treated with the water-repellent preservative formulas that employ highly volatile solvents do not restore complete paintability.

Woods that have been pressure treated for decay or fire resistance sometimes have special finishing requirements. All the common *pressure* preservative treatments (creosote, pentachlorophenol, water-repellent preservatives, and waterborne) will not significantly change the weathering characteristics of woods. Certain treatments such as water-borne treatments containing chromium reduce the degrading effects of weathering. Except for esthetic or visual reasons, there is generally no need to apply a finish to most preservative-treated woods. If needed, oil-base, semitransparent penetrating stains can be used but only after the preservative-treated wood has weathered for 1 to 2 years depending on exposure. The only preservative-treated woods that can be painted or stained immediately after treatment and without further exposure are CCA-treated woods. Manufacturers generally have specific recommendations for good painting and finishing practices for fire-retardant and preservative-treated woods.

Marine Uses

The marine environment is particularly harsh on wood. The earlier discussion on wood weathering indicated that the natural surface deterioration process occurs slowly. Marine environments speed up the natural weathering process to some degree, and wood is often finished with paint or varnish for protection. Certain antifouling paints are also used for protection against marine organisms on piers and ship hulls.

For best protection, wood exposed to marine environments above water and above ground should be treated with a water-repellent preservative, painted with a suitable paint primer, and topcoated (at least two coats) with quality exterior products. *Any wood in contact with water or the ground should be pressure treated to specifications recommended for inground or marine use.* Such treated woods are not always paintable. As indicated earlier, the CCA-treated woods are paintable when dry and clean.

Natural finishes (varnishes) for marine-exposed woods need regular and frequent care and refinishing. Varnishes should be specially formulated for harsh exposure and be applied in three- to six-coat thicknesses for best performance.

Refinishing Wood

Exterior wood surfaces need be repainted only when the old paint has worn thin and no longer protects the wood. In repainting with oil paint, one coat may be adequate if the old paint surface is in good condition. Dirty paint can often be freshened by washing with detergent. Too frequent repainting with oil-base systems produces an excessively thick film that is likely to crack abnormally across the grain of the wood. Complete paint removal and repainting is the only cure for cross-grain cracking (see later section).

Paint and Solid Color Stains

In refinishing an old paint coat (or solid color stain), proper surface preparation is essential if the new coat is to give the expected performance. First, scrape away all loose paint. Use sandpaper on any remaining paint to "feather" the edges smooth with the bare wood. Then scrub any remaining old paint with a brush or sponge and water. Rinse the scrubbed surface with clean water. Wipe the surface with your hand. If

the surface is still dirty or chalky, scrub it again using a detergent. Mildew should be removed with a dilute household bleach solution (see section on mildew). Rinse the cleaned surface thoroughly with fresh water and allow it to dry before repainting. Areas of exposed wood should be treated with a water-repellent preservative, or water repellent, and allowed to dry for at least 2 days, and then primed. Topcoats can then be applied.

It is particularly important to clean areas protected from sun and rain such as porches, soffits, and side walls protected by overhangs. These areas tend to collect dirt and water-soluble materials that interfere with adhesion of the new paint. It is probably adequate to repaint these protected areas every other time the house is painted.

Latex paint can be applied over freshly primed surfaces and on weathered paint surfaces if the old paint is clean and sound. Where old sound paint surfaces are to be repainted with latex paint, a simple test should be conducted first. After cleaning the surface, repaint a small, inconspicuous area with latex paint, and allow it to dry at least overnight. Then, to test for adhesion, firmly press one end of a Band-Aid type adhesive bandage onto the painted surface. Jerk it off with a snapping action. If the tape is free of paint, it tells you that the latex paint is well bonded and that the old surface does not need priming or additional cleaning. If the new latex paint adheres to the tape, the old surface is too chalky and needs more cleaning or the use of an oil base primer. If both the latex paint and the old paint coat adhere to the tape, the old paint is not well bonded to the wood and must be removed before repainting.

Semitransparent Penetrating Stains

Semitransparent penetrating stains are relatively easy to refinish. Heavy scraping and sanding are generally not required. Simply use a stiff-bristle brush to remove all surface dirt, dust, and loose wood fibers, and then apply a new coat of stain. The second coat of penetrating stain often lasts longer than the first since it penetrates into small surface checks that open up as wood weathers.

Water-Repellent Preservatives

Water-repellent preservatives used for natural finishes can be renewed by a simple cleaning of the old surface with a bristle brush and an application of a new coat of finish. To determine if a water-repellent preservative has lost its effectiveness, splash a small quantity of water against the wood surface. If the water beads up and runs off the surface, the treatment is still effective. If the water soaks in, the wood needs to be refinished. Refinishing is also required when the wood surface shows signs of graying. Gray discoloration can be removed by using liquid household bleach (see mildew).

NOTE: Steel wool and wire brushes should *not* be used to clean surfaces to be finished with semitransparent stains or water-repellent preservatives as small iron deposits may be left behind. The small iron deposits can react with certain water-soluble extractives in woods like western redcedar, redwood, Douglas-fir, and the oaks to yield dark blue-black stains on the surface (see section on iron stains). In addition, pentachlorophenol in any finish may cause iron remaining on the surface to corrode rapidly. The corrosion products may then react with the extractives to form a dark-blue, unsightly discoloration that becomes sealed beneath the new finishing system. Pentachlorophenol is present in some semitransparent penetrating stains and water-repellent preservatives.

Finish Failure or Discoloration Problems

Paint is probably the most common exterior finish in use on wood today. It appears somewhere on practically every house and on most commercial buildings. Even brick and aluminum-sided houses usually have some painted wood trim. When properly applied to the appropriate type of wood substrate, paint can give a service life of up to 10 years. All too often, however, problems may develop during the application of the paint or the paint coat fails to achieve the expected service life.

Paint properly applied and exposed under normal conditions is usually not affected by the first 2 to 3 years of exposure. Areas which deteriorate the fastest are those exposed to the greatest amount of sunshine and rain, usually on the south and west sides of a building. The normal deterioration process leads first to soiling or slight accumulation of dirt, and then to a flattening stage when the coating gradually starts to chalk and erode away. However, early paint failure may develop under certain conditions of service. Excess moisture, flat-grained wood, high porosity in the coating, and applying a new paint coat without proper preparation of the old surface can all contribute to early paint failure.

The most common cause of premature paint failure on wood is moisture. Paint on the outside walls of houses is subject to wetting from rain, dew, and frost. Equally serious is "unseen" moisture moving from inside the house to the outside. This is particularly true on houses without a proper vapor-retarding material in cold northern climates.

Temperature Blisters

Temperature blisters are bubble-like swellings that occur on the surface of the paint film as early as a few hours after painting or as long as one to two days later (fig. 16—7). They occur only in the last coat of paint. They are caused when a thin dry skin has formed on the outer surface of the fresh paint and the liquid thinner in the wet paint under the dry skin changes to vapor and cannot escape. A rapid rise in tempera-

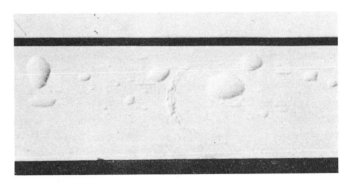

Figure 16—7—Temperature blisters can result when partially dried paint is suddenly heated by the direct rays of the sun. (M147 706-4)

ture, as when the rays of the sun fall directly on freshly painted wood, will cause the vapors to expand and produce blisters. Usually only oil-based paint blisters in this way. Dark colors that absorb heat and thick paint coats are more likely to blister than white paints or thin coats.

To prevent temperature blisters, avoid painting surfaces that will soon be heated. "Follow the sun around the house" for the best procedure. Thus, the north side of the building should be painted early in the morning, the east side late in the morning, the south side well into the afternoon, and the west side late in the afternoon.

If blistering does occur, allow the paint to dry for a few days. Scrape off the blisters, smooth the edges with sandpaper, and spot paint the area.

Moisture Blisters

Moisture blisters are also bubble-like swellings of the paint film on the wood surface. As the name implies, they usually contain moisture when they are formed. They may occur where outside moisture such as rain enters the wood through joints and other end-grain areas of boards and siding. Moisture may also enter from poor construction and maintenance practices. Paint blisters caused by outside water are usually concentrated around joints and the end grain of wood. Damage appears after spring rains and throughout the summer. Paint failure is most severe on the sides of buildings facing the prevailing winds and rain. Blisters may occur in both heated and unheated buildings.

Moisture blisters may also result from inside water moving to the outside. Plumbing leaks, overflow of sinks, bathtubs or shower spray, and improperly sealed walls are sources of inside water. Such damage is not seasonal and occurs when the faulty condition develops.

Moisture blisters usually include all paint coats down to the wood surface. After the blisters appear, they dry out and

Figure 16–8—Paint peeling from wood can result when excessive moisture moves through the house wall. Some cross-grain cracking is also evident on this older home. (MC83 9023)

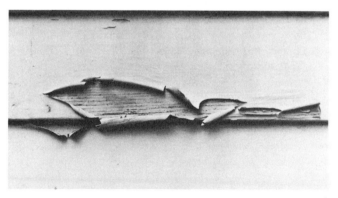

Figure 16–9—Intercoat peeling of paint is usually caused by poor preparation of the old paint surface. (MC83 9022)

collapse. Small blisters may disappear completely, fairly large ones may leave a rough spot, and in severe cases the paint will peel (fig. 16–8). Thin coatings of new oil-based paint are the most likely to blister. Old, thick coats are usually too rigid to swell and form blisters. Therefore, cracking and peeling will usually result.

Elimination of the moisture problem is the only practical way to prevent moisture blisters in paint. The moisture source should be identified and eliminated to avoid more serious problems such as wood decay (rot) and loss of insulating value.

Figure 16–10—Cross-grain cracking results from an excessive buildup of paint. (MC83 9025)

Intercoat Peeling

Intercoat peeling is the separation of the new paint film from the old paint coat, indicating a weak bond between the two (fig. 16–9). Intercoat peeling usually results from inadequate cleaning of the weathered paint and usually occurs within 1 year of repainting. This type of paint peeling can be prevented by following good painting practices.

Intercoat peeling can also result from allowing too much time between the primer coat and topcoat in a new paint job. If more than 2 weeks elapse between applying an oil-based primer and a topcoat, soap-like materials may form on the surface and interfere with the bonding of the next coat of paint. When the period between applications exceeds 2 weeks, scrub the surface before applying the second coat. Do not apply a primer coat in the fall and wait until spring to finish with the topcoat.

Cross-Grain Cracking

Cross-grain cracking occurs when oil-based or alkyd paint coatings become too thick (fig. 16–10). This problem often occurs on older homes that have been painted many times. Paint usually cracks in the direction it was brushed onto the wood. Cross-grain cracks run across the grain of the paint. Once cross-grain cracking has occurred the only solution is to completely remove the old paint and apply a new finishing system on the bare wood.

To prevent cross-grain cracking, follow the paint manufacturer's recommendations for spreading rates. Do not repaint unweathered, protected areas such as porch ceilings and roof overhangs as often as the rest of the house. If possible, repaint these areas only as they weather and require new paint. However, if repainting is required, be sure to scrub the areas with a sponge or bristle brush and detergent in water to remove any dirt and watersoluble materials that would otherwise interfere with adhesion of the new paint.

Chalking

Chalking results when a paint film gradually weathers or deteriorates, releasing the individual particles of pigment. These individual particles act like a fine powder on the paint surface. Most paints chalk to some extent. This phenomenon is desirable since it allows the paint surface to be self-cleaning. However, chalking is objectionable when it washes down over a surface with a different color or when it causes premature disappearance of the paint film through excess erosion.

Discoloration problems from chalking can be reduced by selection of a paint with slow chalking tendencies. The manner in which the paint is formulated usually determines how fast it chalks. Therefore, if chalking is likely to be a problem, select a paint that the manufacturer has indicated will chalk slowly.

When repainting surfaces that have chalked excessively, proper preparation of the old surface is essential if the new paint coat is expected to last. Scrub the old surface thoroughly with a detergent solution to remove all old deposits and dirt. Rinse thoroughly with clean water before repainting. The use of a top-quality oil-based primer or a stain-blocking acrylic latex primer may be necessary before latex topcoats are applied. Otherwise, the new paint coat may peel. Discoloration or chalk that has run down on a lower surface may be removed by vigorous scrubbing with a good detergent. This discoloration will also gradually weather away if the chalking problem on the painted surface above has been corrected.

Mildew

Mildew is probably the most common cause of house paint discoloration and gray discoloration of unfinished wood (fig. 16–11). Mildew is a form of stain fungi or microscopic plant life. The most common species are black, but some are red, green, or other colors. It grows most extensively in warm, humid climates but is also found in cold northern states. Mildew may be found anywhere on a building, but it is most common on walls behind trees or shrubs where air movement is restricted. Mildew may also be associated with the dew pattern of the house. Dew will form on those parts of the

Figure 16–11—Mildew is most common in shaded, moist, or protected areas. (M147 706-12)

house that are not heated and tend to cool rapidly, such as eaves and ceilings of carports and porches. This dew then provides a source of moisture for the mildew.

Mildew fungi can be distinguished from dirt by examination under a high-power magnifying glass. In the growing stage, when the surface is damp or wet, the fungus is characterized by its threadlike growth. In its dormant stage, when the surface is dry, it has numerous egg-shaped spores; by contrast, granular particles of dirt appear irregular in size and shape. A simple test for the presence of mildew on wood and paint can be made by applying a drop or two of liquid household bleach solution (5 percent sodium hypochlorite) to the stain. The dark color of mildew will usually bleach out in 1 or 2 minutes. A surface stain that does not bleach is probably dirt. It is important to use fresh bleach solution since it deteriorates upon standing and loses its potency.

In warm, damp climates where mildew occurs frequently, use a paint containing zinc oxide and a mildewcide for topcoats over a primer coat that also contains a mildewcide. For mild cases of mildew, use a paint containing a mildewcide.

Before repainting, the mildew must be killed or it will

grow through the new paint coat. To kill mildew on wood or on paint, and to clean an area for general appearance or for repainting, use a bristle brush or sponge to scrub the painted surface with the following solution:

1/3 cup household detergent

1 quart (5 percent) sodium hypochlorite (household bleach)

3 quarts warm water

WARNING: Do not mix bleach with ammonia or with any detergents or cleansers containing ammonia! Mixed together the two are a lethal combination, similar to mustard gas. In several instances people have died from breathing the fumes from such a mixture. Many household cleaners contain ammonia, so be extremely careful in selecting the type of cleaner you mix with bleach.

When the surface is clean, rinse it thoroughly with fresh water. Avoid splashing the cleaning solution on yourself or on shrubbery or grass as it may have harmful effects. Before the cleaned surface can become contaminated, repaint it with a paint containing a mildewcide. For unpainted wood finish with a water-repellent preservative.

Water-Soluble Extractives

In some species of wood the heartwood contains water-soluble extractives, while sapwood does not. These extractives can occur in both hardwoods and softwoods. Western redcedar and redwood are two common softwood species used in construction that contain large quantities of extractives. The extractives give these species their attractive color, good stability, and natural decay resistance, but they can also discolor paint. Woods such as Douglas-fir and southern yellow pine can also cause occasional extractive staining problems.

When extractives discolor paint, moisture is usually the culprit. The extractives are dissolved and leached from the wood by water. The water then moves to the paint surface, evaporates, and leaves the extractives behind as a reddish-brown stain (fig. 16−12).

Diffused discoloration from wood extractives is caused by water that comes from rain and dew that penetrates a porous or thin paint coat. It may also be caused by rain and dew that penetrate joints in the siding or by water from faulty roof drainage and gutters.

Diffused discoloration is best prevented by following good painting practices. Apply a water-repellent preservative or water repellent to the bare wood before priming. Use an oil-based, stain-resistant primer or a latex primer especially formulated for use over staining woods. Do not use porous paints such as flat alkyds and latex directly over the staining-type woods. If the wood is already painted, clean the surface, apply an oil-based or latex stain-resistant primer and then top

coat. Before priming and repainting, apply a water-repellent preservative or water repellent to any wood left bare from peeling paint.

A rundown or streaked type of discoloration can also occur when water soluble extractives are present. This discoloration results when the back of the siding is wetted, the extractives are dissolved, and then the colored water runs down the face of the adjacent painted boards from the lap joint.

Water which produces a rundown discoloration can result from water vapor within the house moving to the exterior walls and condensing during cold weather. Major sources of water vapor are humidifiers, unvented clothes dryers, showers, normal respiration, and moisture from cooking and dish-washing. Rundown discoloration may also be caused by water draining into exterior walls from roof leaks, faulty gutters, ice dams, and wind-driven rain and snow at louvers.

Rundown discoloration can be prevented by reducing condensation or the accumulation of moisture in the wall. New houses or those undergoing remodeling should have a vapor-retarding material on the inside of all exterior walls. If such a barrier is not practical, the inside of all exterior walls should be painted with a vapor-resistant paint. Water vapor in the house can be reduced by using exhaust fans vented to the outside in bathrooms and kitchens. Clothes dryers should be vented to the outside and not to the crawl space or attic. Avoid the use of humidifiers. If the house contains a crawl space, the soil should be covered with a vapor-retarding material to prevent migration of water vapor up into the living quarters.

Water from rain and snow can be prevented from entering the walls by proper maintenance of gutters and the roof. Ice dam formation can be prevented by installing adequate insulation in the attic and by providing proper ventilation.

If discoloration is to be stopped, moisture problems must be eliminated. The remaining rundown discoloration will usually weather away in a few months. However, discoloration in protected areas can become darker and more difficult to remove with time. In these cases, wash the discolored areas with a mild detergent soon after the problem develops. Paint cleaners are effective on darker stains.

Blue Stain

Blue stain is caused by microscopic fungi that commonly infect the sapwood of all woody species. Although microscopic, they collectively produce a blue black discoloration of the wood. Blue stain does not weaken wood structurally, but conditions that favor stain development are also ideal for serious wood decay and paint failure.

Wood in service may contain blue stain, and no detrimental effects will result so long as the moisture content is kept below 20 percent. (Wood in properly designed and well-maintained houses usually has a moisture content of 8 to 13 percent.) However, if the wood is exposed to moisture from sources such as rain, condensation, or leaking plumbing, the moisture content will increase and the blue-stain fungi will develop and become visible.

To prevent blue stain from discoloring paint, follow good construction and painting practices. First, do whatever is possible to keep the wood dry. Provide an adequate roof overhang, and properly maintain the shingles, gutters, and downspouts. Window and door casings should slope out from the house, thus allowing water to drain away rapidly. Use a vapor-resisting material on the interior side of all exterior walls to prevent condensation in the wall. Vent clothes dryers, showers, and cooking areas to the outside, and avoid the use of humidifiers. Untreated wood should be treated with a water-repellent preservative, then a nonporous mildew-resistant primer, and finally at least one topcoat also containing a mildewcide. If the wood has already been painted, remove the old paint and allow the wood to dry thoroughly. Apply a water-repellent preservative, and then repaint as described above.

A 5 percent sodium hypochlorite solution (ordinary liquid household bleach) may sometimes remove blue-stain discoloration, but it is not a permanent cure. Be sure to use fresh bleach since its effectiveness can diminish with age. The moisture problem must be corrected if a permanent cure is expected.

Iron Stain

Rust may be one type of staining problem associated with iron. When standard ferrous nails are used on exterior siding and then painted, a red-brown discoloration may occur through the paint in the immediate vicinity of the nailhead. To prevent rust stains, use corrosion-resistant nails. These include high-quality galvanized, stainless steel, and aluminum nails. Poor-quality galvanized nails can corrode easily and, like ferrous nails, can cause unsightly staining of the wood and paint. The galvanizing on the heads of the nails should not "chip loose" as they are driven into the wood. If rust is a serious problem on a painted surface, the nails should be countersunk, calked, and the area spot primed, and then topcoated.

Unsightly rust stains may also occur when standard ferrous nails are used in association with any of the other finishing systems such as solid color or opaque stains, semitransparent penetrating stains, and water-repellent preservatives. Rust stains can also result from screens and other metal objects or fasteners which are subject to corrosion and leaching.

A chemical reaction with iron resulting in an unsightly blue-black discoloration of wood can also occur. In this case,

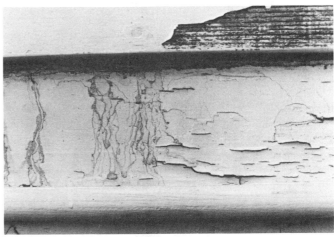

Figure 16–12—Water-soluble extractive discoloration can result from water wetting the back of the siding and then running down the front (top). Water causing discoloration also leads to paint failure (bottom). (M85 0013)

the iron reacts with certain wood extractives to form the discoloration. Ferrous nails are the most common source of iron for chemical staining, but problems have also been associated with traces of iron left from cleaning the wood surface

16-25

with steel wool or wire brushes. The discoloration can sometimes become sealed beneath a new finishing system.

Oxalic acid will remove the blue-black chemical discoloration from iron providing it is not already sealed beneath a finishing system. The stained surface should be given several applications of the solution containing at least 1 pound of oxalic acid per gallon of water, preferably hot. After the stains disappear, the surface should be thoroughly washed with warm fresh water to remove the oxalic acid and any traces of the chemical causing the stain. If all sources of iron are not removed or protected from corrosion, the staining problem may reoccur.

CAUTION: Extreme care should be exercised when using oxalic acid since this chemical is VERY TOXIC.

Brown Stain Over Knots

The knots in many softwood species, particularly pine, contain an abundance of resin. This resin can sometimes cause paint to peel or turn brown. In most cases, this resin is "set" or hardened by the high temperatures used in kiln drying construction lumber.

Good painting practices should eliminate or control brown stain over knots. Apply a good primer to the bare wood first. Then follow with two topcoats. Do not apply ordinary shellac or varnish to the knot area first as this may result in early paint failure in outdoor exposure. Some specially formulated exterior-grade shellacs are available for this purpose.

Finishing Interior Wood

Interior finishing differs from exterior finishing chiefly in that interior woodwork usually requires much less protection against moisture but more exacting standards of appearance and cleanability. Good finishes used indoors should last much longer than paint or other coatings on exterior surfaces. Veneered panels and plywood, however, present special finishing problems because of the tendency of these wood constructions to surface check.

Opaque Finishes

Interior surfaces may be easily painted by procedures similar to those for exterior surfaces. As a rule, however, smoother surfaces, better color, and a more lasting sheen are demanded for interior woodwork, especially wood trim; therefore, enamels or semigloss enamels are used rather than flat paints.

Before enameling, sand the wood surface extremely smooth and remove the surface dust with a tack cloth. Imperfections such as planer marks, hammer marks, and raised grain, are accentuated by enamel finish. Raised grain is especially trou-

blesome on flat-grained surfaces of the denser softwoods because the hard bands of latewood are sometimes crushed into the soft earlywood in planing, and later expand when the wood changes moisture content. For the smoothest surface, it is helpful to sponge softwoods with water, allow them to dry thoroughly, and then sand them lightly with new sandpaper before enameling. In new buildings, woodwork should be allowed adequate time to come to its equilibrium moisture content in the heated building before finishing.

To effectively finish hardwoods with large pores, such as oak and ash, the pores must be filled with wood filler (see section on fillers). After filling and sanding, successive applications of interior primer and sealer, undercoat, and enamel are used. Knots in the white pines, ponderosa pine, or southern pine should be sealed with shellac or a special knot sealer before priming. A coat of pigmented shellac or special knot sealer is also sometimes necessary over white pines and ponderosa pine to retard discoloration of light-colored enamels by colored matter present in the resin of the heartwood of these species.

One or two coats of enamel undercoat are next applied; this should completely hide the wood and also present a surface that easily can be sandpapered smooth. For best results, the surface should be sanded just before applying the finish enamel; however, this step is sometimes omitted. After the finishing enamel has been applied, it may be left with its natural gloss, or rubbed to a dull finish. When wood trim and paneling are finished with a flat paint, the surface preparation need not be as exacting.

Transparent Finishes

Transparent finishes are used on most hardwood and some softwood trim and paneling, according to personal preference. Most finishing consists of some combination of the fundamental operations of sanding, staining, filling, sealing, surface coating, or waxing. Before finishing, planer marks and other blemishes on the wood surface that would be accentuated by the finish should be removed.

Stains

Both softwoods and hardwoods are often finished without staining, especially if the wood has a pleasing and characteristic color. When stain is used, however, it often accentuates color differences in the wood surface because of unequal absorption into different parts of the grain pattern. With hardwoods, such emphasis of the grain is usually desirable; the best stains for the purpose are dyes dissolved either in water or solvent. The water stains give the most pleasing results, but raise the grain of the wood and require an extra sanding operation after the stain is dry.

The most commonly used stains are the "nongrain-raising"

ones in solvents which dry quickly, and often approach the water stains in clearness and uniformity of color. Stains on softwoods color the earlywood more strongly than the latewood, reversing the natural gradation in color unless the wood has been sealed first with a wash coat. Pigment-oil stains, which are essentially thin paints, are less subject to this problem and are therefore more suitable for softwoods. Alternatively, the softwood may be coated with penetrating clear sealer before applying any type of stain to give more nearly uniform coloring.

Fillers

In hardwoods with large pores, the pores must be filled, usually after staining and before varnish or lacquer is applied, if a smooth coating is desired. The filler may be transparent and without effect on the color of the finish, or it may be colored to contrast with the surrounding wood.

For finishing purposes, the hardwoods may be classified as follows:

Hardwoods with large pores	Hardwoods with small pores
Ash	Alder, red
Butternut	Aspen
Chestnut	Basswood
Elm	Beech
Hackberry	Cherry
Hickory	Cottonwood
Lauans	Gum
Mahogany	Magnolia
Mahogany, African	Maple
Oak	Sycamore
Sugarberry	Yellow-poplar
Walnut	

Birch has pores large enough to take wood filler effectively when desired, but small enough as a rule to be finished satisfactorily without filling.

Hardwoods with small pores may be finished with paints, enamels, and varnishes in exactly the same manner as softwoods.

A filler may be a paste or liquid, natural or colored. It is applied by brushing first across the grain and then brushing with the grain. Surplus filler must be removed immediately after the glossy wet appearance disappears. Wipe first across the grain to pack the filler into the pores; then complete the wiping with a few light strokes along the grain. Filler should be allowed to dry thoroughly and sanded lightly before the finish coats are applied.

Sealers

Sealers are thinned varnish or lacquer and are used to prevent absorption of surface coatings and also to prevent the bleeding of some stains and fillers into surface coatings, espe-cially lacquer coatings. Lacquer sealers have the advantage of being very fast drying.

Surface Coats

Transparent surface coatings over the sealer may be gloss varnish, semigloss varnish, shellac, nitrocellulose lacquer, or wax. Wax provides protection without forming a thick coating and without greatly enhancing the natural luster of the wood. Coatings of a more resinous nature, especially lacquer and varnish, accentuate the natural luster of some hardwoods and seem to permit the observer to look down into the wood. Shellac applied by the laborious process of French polishing probably achieves this impression of depth most fully, but the coating is expensive and easily marred by water. Rubbing varnishes made with resins of high refractive index for light (ability to bend light rays) are nearly as effective as shellac. Lacquers have the advantages of drying rapidly and forming a hard surface, but require more applications than varnish to build up a lustrous coating.

Varnish and lacquer usually dry with a highly glossy surface. To reduce the gloss, the surfaces may be rubbed with pumice stone and water or polishing oil. Waterproof sandpaper and water may be used instead of pumice stone. The final sheen varies with the fineness of the powdered pumice stone; coarse powders make a dull surface and fine powders produce a bright sheen. For very smooth surfaces with high polish, the final rubbing is done with rottenstone and oil. Varnish and lacquer made to dry to semigloss or satin finish are also available.

Flat oil finishes commonly called Danish oils are also very popular. This type of finish penetrates the wood and forms no noticeable film on the surface. Two or more coats of oil are usually applied, which may be followed with a paste wax. Such finishes are easily applied and maintained but are more subject to soiling than a film-forming type of finish. Simple boiled linseed oil or tung oil are also used extensively as wood finishes.

Finishes for Floors

Wood possesses a variety of properties that make it a highly desirable flooring material for homes and industrial and public structures. A variety of wood flooring products permits a wide selection of attractive and serviceable wood floors. Selection is available not only from a variety of different wood species and grain characteristics, but also from a considerable number of distinctive flooring types and patterns.

The natural color and grain of wood floors make them inherently attractive and beautiful. Floor finishes enhance the natural beauty of wood, protect it from excessive wear and abrasion, and make the floors easier to clean. A complete finishing process may consist of four steps: Sanding the surface, applying a filler for open-grain woods, applying a stain to achieve a desired color effect, and finally applying a finish. Detailed procedures and specified materials depend largely on the species of wood used and individual preference in type of finish.

Careful sanding to provide a smooth surface is essential for a good finish because any irregularities or roughness in the wood surface will be magnified by the finish. Development of a top-quality surface requires sanding in several steps with progressively finer sandpaper, usually with a machine unless the area is small. The final sanding is usually done with a 2/0 grade paper. When sanding is complete, *all* dust must be removed with a vacuum cleaner and then a tack rag. Steel wool should not be used on floors unprotected by finish because minute steel particles left in the wood may later cause staining or discoloration.

A filler is required for wood with large pores, such as oak and walnut, if a smooth, glossy, varnish finish is desired.

Stains are sometimes used to obtain a more nearly uniform color when individual boards vary too much in their natural color. Stains may also be used to accent the grain pattern. If the natural color of the wood is acceptable, staining is omitted. The stain should be an oil-based or a nongrain-raising type. Stains penetrate wood only slightly; therefore, the finish should be carefully maintained to prevent wearing through the stained layer. It is difficult to renew the stain at worn spots in a way that will match the color of the surrounding area.

Finishes commonly used for wood floors are classified either as sealers or varnishes. Sealers, which are usually thinned varnishes, are widely used in residential flooring. They penetrate the wood just enough to avoid formation of a surface coating of appreciable thickness. Wax is usually applied over the sealer; however, if greater gloss is desired, the sealed floor makes an excellent base for varnish. The thin surface coat of sealer and wax needs more frequent attention than varnished surfaces. However, rewaxing or resealing and waxing of high-traffic areas is a relatively simple maintenance procedure.

Varnish may be based on phenolic, alkyd, epoxy, or polyurethane resins. Varnish forms a distinct coating over the wood and gives a lustrous finish. The kind of service expected usually determines the type of varnish. Varnishes especially designed for homes, schools, gymnasiums, or other public buildings are available. Information on types of floor finishes can be obtained from the flooring associations or the individual flooring manufacturers.

Durability of floor finishes can be improved by keeping them waxed. Paste waxes generally give the best appearance and durability. Two coats are recommended and, if a liquid wax is used, additional coats may be necessary to get an adequate film for good performance.

Finishes for Wood Kitchen and Eating Utensils

Wood salad bowls, spoons, and forks used for food service need a finish that is resistant to abrasion, water, acids, and stains, with a surface that is easy to clean when soiled.

Many finishes are available such as varnishes, lacquers, shellac, and other miscellaneous types. Most of these are what could be called coating finishes. They resist many elements; but because they coat the surface, they may eventually chip, peel, alligator, or crack. They are generally not desirable finishes. They may also be toxic.

One of the most satisfactory finishes for wood surfaces is the penetrating wood sealer. A wood sealer, like an oil finish, sinks into the pores of the wood, fills the cavities of the wood cells, and saturates the surface. This prevents the absorption of moisture, makes the surface easy to clean when soiled, and makes it resistant to scratching.

Penetrating wood sealers are easy to apply, dry quickly, and require less skill than other finishes. Worn places in the finish may be patched without showing lapping around the edges, which ordinarily cannot be done with other types of finishes. Treated utensils should be allowed to dry thoroughly for several weeks before use.

Vegetable oils (olive, corn, etc.) are edible and are sometimes used to finish wood utensils. These are nondrying oils. They are applied heavily in several coats and can be refurbished easily. Occasionally, rancidity can develop if the utensils are not cleaned thoroughly. Walnut oil is an example of an edible slow-drying oil particularly suited for wood utensils. Mineral oil is a nondrying oil also used for a penetrating finish.

Penetrating drying oils (linseed, tung) can also be used as wood utensil finishes. These drying oils are applied heavily in repeated applications following manufacturer's instructions. All treated utensils should be allowed to dry several weeks before use.

One of the simplest treatments for wood utensils, especially cutting boards and butcher blocks, is the application of melted paraffin wax (the type used for home canning). The wax is melted in a double boiler over hot water and liberally brushed on the wood surface. Excess wax may be left on or scraped off as desired. Refinishing is simple and easy.

NOTE: Whatever finish is chosen for wood utensils used for storing, handling, or eating food, it is important to be sure that the finish is safe and not toxic (poisonous). For information on the safety and toxicity of any finish, check the label, contact the manufacturer, or check with your local extension home economics expert or county agent.

Selected References

American Plywood Association. Stains and paints on plywood. Pamphlet B407B. Tacoma, WA: APA; 1979.

Black, J. M.; Laughnan, D. F.; Mraz, E. A. Forest Products Laboratory natural finish. Res. Note FPL−046. Madison, WI: U.S. Department of Agriculture, Forest Service, Forest Products Laboratory; 1979.

Black, J. M.; Mraz, E. A. Inorganic surface treatments for weather-resistant natural finishes. Res. Pap. FPL−232. Madison, WI: U.S. Department of Agriculture, Forest Service, Forest Products Laboratory; 1974.

Browne, F. L. Wood properties and paint durability. Misc. Publ. 629. Madison, WI: U.S. Department of Agriculture; 1962.

Cassens, D. L.; Feist, W. C. Wood finishing: Finishing exterior plywood, hardboard and particleboard. North Central Region Extension Publ. 132. West Lafayette, IN: Purdue University, Cooperative Extension Service; 1980.

Cassens, D. L.; Feist, W. C. Wood finishing: Paint failure problems and their cure. North Central Region Extension Publ. 133. West Lafayette, IN: Purdue University, Cooperative Extension Service; 1980.

Cassens, D. L.; Feist, W. C. Wood finishing: Discoloration of house paint—causes and cures. North Central Region Extension Publ. 134. West Lafayette, IN: Purdue University, Cooperative Extension Service; 1980.

Cassens D. L.; Feist, W. C. Wood finishing: Selection and application of exterior finishes for wood. North Central Region Extension Publ. 135. West Lafayette, IN: Purdue University, Cooperative Extension Service; 1980.

Cassens, D. L.; Feist, W. C. Wood finishing: Finishing and maintaining wood floors. North Central Region Extension Publ. 136. West Lafayette, IN: Purdue University, Cooperative Extension Service; 1980.

Cassens, D. L.; Feist, W. C. Finishing wood exteriors: Selection, application, and maintenance. Agric. Handb. 647. Washington, DC; U.S. Department of Agriculture; 1986.

Feist, W. C. Protection of wood surfaces with chromium trioxide. Res. Pap. FPL−339. Madison, WI: U.S. Department of Agriculture, Forest Service, Forest Products Laboratory; 1982.

Feist, W. C. Weathering of wood in structural uses. In: R. W. Meyer and R. M. Kellogg, eds. Structural use of wood in adverse environments. New York: Van Nostrand Reinhold Co.; 1982: 156−178.

Feist, W. C. Weathering characteristics of finished wood-based panel products. Journal of Coating Tech. 54(686): 43−50; 1982.

Feist, W. C.; Hon, D. N.-S. Chemistry of weathering and protection. In: R. M. Rowell, ed. The chemistry of solid wood. ACS Advances in Chemistry Series No. 207. Washington, DC: American Chemical Society; 1984.

Feist, W. C.; Mraz, E. A. Wood finishing: Water repellents and water-repellent preservatives. Res. Note FPL−0124. Madison, WI: U.S. Department of Agriculture, Forest Service, Forest Products Laboratory; 1978.

Feist, W. C.; Mraz, E. A. Durability of exterior natural wood finishes in the Pacific Northwest. Res. Pap. FPL 366. Madison, WI: U.S. Department of Agriculture, Forest Service, Forest Products Laboratory; 1980.

Feist, W. C.; Mraz, E. A. Performance of mildewcides in a semitransparent stain wood finish. Madison, WI: Forest Products Journal. 30(5): 43−46; 1980.

Feist, W. C.; Oviatt, A. E. Wood siding—installing, finishing, maintaining. Home and Garden Bull. 203. Washington, DC: U.S. Department of Agriculture; rev. 1983.

Forest Products Laboratory. Wood finishing: Weathering of wood. Res. Note FPL−0135. Madison, WI: U.S. Department of Agriculture, Forest Service, Forest Products Laboratory; 1975.

Forest Products Laboratory. List of publications on wood finishing. 81−024. Madison, WI: U.S. Department of Agriculture, Forest Service, Forest Products Laboratory; 1981.

National Woodwork Manufacturers Association. Industry standard for water-repellent preservative treatment for millwork. IS4−81. Chicago, IL: NWMA; 1981.

Chapter 17

Protection From Organisms That Degrade Wood

Protection From Organisms That Degrade Wood *

Under proper conditions, wood will give centuries of service. Where conditions permit development of organisms that can degrade wood, however, protection must be provided in milling, merchandising, and building to ensure maximum service life of wood elements.

The principal organisms that can degrade wood are fungi, insects, bacteria, and marine borers.

Molds, most sapwood stains, and decay are caused by fungi, which are microscopic, threadlike plants that must have organic material to live. For some of them, wood offers the required food supply. The growth of fungi depends on suitably mild temperatures, moisture, and air (oxygen). Chemical stains, although they are not caused by organisms, are mentioned in this chapter because they resemble stains caused by fungi.

Insects also may damage wood, and in many situations must be considered in protective measures. Termites are the major insect enemy of wood, but, on a national scale, they are a less serious threat than fungi.

Bacteria in wood ordinarily are of little consequence, but some may make the wood excessively absorptive. Additionally, some may cause strength losses over long periods of exposure.

Marine borers are a fourth general type of wood-degrading organism. They can attack susceptible wood rapidly, and in salt-water harbors are the principal cause of damage to piles and other wood marine structures.

Wood degradation by organisms has been studied extensively, and many preventive measures are well known and widely practiced. By taking ordinary precautions with the finished product, the user can contribute substantially to ensuring a long service life.

Fungus Damage and Control

Fungus damage to wood may be traced to three general causes: (1) Lack of suitable protective measures when storing logs or bolts; (2) improper seasoning, storing, or handling of the raw material produced from the log; and (3) failure to take ordinary simple precautions in using the final product. The incidence and development of molds, decay, and stains caused by fungi depend heavily on temperature and moisture conditions.

Molds and Fungus Stains

Molds and fungus stains are confined largely to sapwood and are of various colors. The principal fungus stains are usually referred to as "sap stain" or "blue stain." The distinction between molding and staining is made largely on the basis of the depth of discoloration; with some molds and the lesser fungus stains there is no clear-cut differentiation. Typical sap stain or blue stain penetrates into the sapwood and cannot be removed by surfacing. Also, the discoloration as seen on a cross section of the wood often tends to exhibit some radial alinement corresponding to the direction of the wood rays (fig. 17−1). The discoloration may completely cover the sapwood or may occur as specks, spots, streaks, or patches of varying intensities of color. The so-called "blue" stains, which vary from bluish to bluish-black and gray to brown, are the most common, although various shades of yellow, orange, purple, and red are sometimes encountered. The exact color of the stain depends on the infecting organisms and the species and moisture condition of the wood. The fungal brown stain mentioned here should not be confused with chemical brown stain.

Mold discolorations usually first become noticeable as largely fuzzy or powdery surface growths, with colors ranging from light shades to black. Among the brighter colors, green and yellowish hues are common. On softwoods—though the fungus may penetrate deeply—the discoloring surface growth often can easily be brushed or surfaced off. On hardwoods, however, the wood beneath the surface growth is commonly stained too deeply to be surfaced off. The staining tends to occur in spots of varying concentration and size, depending on the kind and pattern of the superficial growth.

Under favorable moisture and temperature conditions, staining and molding fungi may become established and develop rapidly in the sapwood of logs shortly after they are cut. In addition, lumber and such products as veneer, furniture stock, and millwork may become infected at any stage of manufacture or use if they become sufficiently moist. Freshly cut or unseasoned stock that is piled during warm, humid weather may be noticeably discolored within 5 or 6 days. Recommended moisture control measures are given in chapter 14.

Ordinarily, stain and mold affect the strength of the wood only slightly; their greatest effect is usually confined to strength properties that determine shock resistance or toughness (ch. 4). They also increase the absorptivity of wood, render-

*Revision by Wallace E. Eslyn, Plant Pathologist, and Glenn R. Esenther, Entomologist.

[1] Mention of a chemical in this chapter does not constitute a recommendation; only those chemicals registered by the U.S. Environmental Protection Agency (EPA) may be recommended. Registration of preservatives is under constant review by EPA and the Department of Agriculture. Use only preservatives that bear an EPA Registration Number and carry directions for home and farm use. Preservatives such as creosote and pentachlorophenol should not be applied to the interior of dwellings which are occupied by humans. Because all preservatives are under constant review by EPA, a responsible State or Federal Agency should be consulted as to the current status of any preservative.

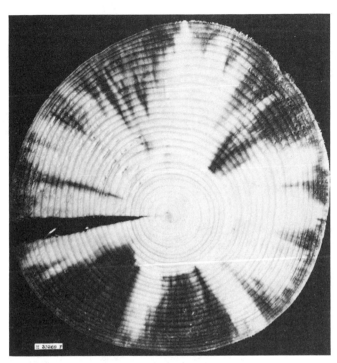

Figure 17–1—Typical radial penetration of log by blue stain. The pattern is a result of more rapid penetration by the fungus radially— through the wood rays—than tangentially.　(M33 269)

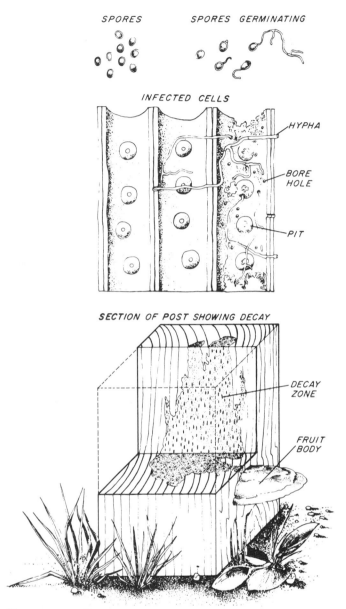

Figure 17–2—The decay cycle (top to bottom). Thousands of spores produced in a fruitbody are distributed by wind or insects. On contacting moist, susceptible wood they germinate to create new infections in the wood cells. In time serious decay develops that may be accompanied by formation of new fruitbodies.　(M124 755)

ing it more susceptible to attack by typical wood-decay fungi.

Stain- and mold-infected stock is practically unimpaired for many uses in which appearance is not a limiting factor and a small amount of stain may be permitted by standard grading rules. Stock with stain and mold may not be entirely satisfactory for siding, trim, and other exterior millwork because of its greater water absorptiveness. Also, incipient decay may be present—though inconspicuous—in the discolored areas. Both of these factors increase the possibility of decay in wood that is wet by rain unless the wood has been treated with a suitable preservative.

Chemical Stains

One type of stain in unseasoned sapwood may resemble blue or fungal brown stain but is not caused by a fungus. This is called chemical stain or sometimes oxidation stain because it is brought about by a reaction between oxygen in the air and certain constituents of the exposed wood. Chemical brown stain is the only one of this type that is particularly serious in softwoods; however, many hardwoods are degraded by a gray-appearing chemical stain. Chemical staining is largely a problem of seasoning; it can usually be prevented by rapid drying at low temperatures (ch. 14).

Decay

Decay-producing fungi may, under conditions that favor their growth, attack either heartwood or sapwood; the result is a condition variously designated as decay, rot, dote, or doze (fig. 17–2). Fresh surface growths of decay fungi may appear as fan-shaped patches (fig. 17–3), strands, or root-

17-3

Figure 17-3—Mycelial fans on a wood door.

(M52 236)

like structures, usually white or brown. Sometimes fruiting bodies are produced that take the form of toadstools, brackets, or crusts. The fungus, in the form of microscopic, threadlike strands, permeates the wood and uses parts of it as food. Some fungi live largely on the cellulose; others use the lignin as well as the cellulose.

Certain decay fungi attack the heartwood (causing "heartrot"), and rarely the sapwood of living trees, whereas others confine their activities to logs or manufactured products, such as sawed lumber, structural timbers, poles, and ties. Most of the tree-attacking groups cease their activities after the trees have been cut, as do the fungi causing brown pocket (peck) in baldcypress or white pocket in Douglas-fir. Relatively few continue their destruction after the trees have been cut and worked into products, and then only if conditions remain favorable for their growth.

Most decay can progress rapidly at temperatures that favor growth of plant life in general. For the most part decay is relatively slow at temperatures below 50 °F and much above 90 °F. Decay essentially ceases when the temperature drops as low as 35 °F or rises as high as 100 °F.

Serious decay occurs only when the moisture content of the wood is above the fiber saturation point (average 30 percent). Only when previously dried wood is contacted by water, such as provided by rain, condensation, or contact with wet ground, will the fiber saturation point be reached. The water vapor in humid air alone will not wet wood sufficiently to support significant decay, but it will permit development of some mold. Fully air-dry wood usually will have a moisture content not exceeding 20 percent, and should provide a reasonable margin of safety against fungus damage. Thus wood will not decay if it is kept air dry—and decay already present from infection incurred earlier will not progress.

Wood can be too wet for decay as well as too dry. If it is water soaked, there may be insufficient access of air to the interior of a piece to support development of typical decay fungi. For this reason, foundation piles buried beneath the water table and logs stored in a pond or under a suitable system of water sprays are not subject to decay by typical wood-decay fungi.

The early or incipient stages of decay are often accompanied by a discoloration of the wood, which is more evident on freshly exposed surfaces of unseasoned wood than on dry wood. Abnormal mottling of the wood color—with either unnatural brown or "bleached" areas—is often evidence of decay infection. Many fungi that cause heartrot in the standing tree produce incipient decay that differs only slightly from the normal color of the wood or gives a somewhat water-soaked appearance to the wood.

Typical or late stages of decay are easily recognized, because the wood has undergone definite changes in color and properties, the character of the changes depending on the organism and the substances it removes.

Two kinds of major decay are recognized—"brown rot" and "white rot." With brown rot, only the cellulose is extensively removed, the wood takes on a browner color, and it tends to crack across the grain and to shrink and collapse (fig. 17−4). With white rot, both lignin and cellulose usually are removed; the wood may lose color and appear "whiter" than normal, it does not crack across the grain, and until severely degraded it retains its outward dimensions and does not shrink or collapse.

Brown, crumbly rot, in the dry condition, is sometimes called "dry rot," but the term is incorrect because wood must be damp to decay, although it may become dry later. A few fungi, however, have water-conducting strands; such fungi are capable of carrying water (usually from the soil) into buildings or lumber piles, where they moisten and rot wood that would otherwise be dry. They are sometimes referred to technically as "dry rot fungi" or "water-conducting fungi." The latter term better describes the true situation as these fungi, like the others, must have water.

A third and generally less important kind of decay is known as soft rot. Soft rot is caused by fungi related to the molds rather than those responsible for brown and white rot. Soft rot typically is relatively shallow; the affected wood is greatly degraded and often soft when wet, but immediately beneath the zone of rot the wood may be firm (fig. 17−5). Because soft rot usually is rather shallow it is most likely to damage relatively thin pieces such as slats in cooling towers. It is favored by wet situations but is also prevalent on surfaces that have been alternately wet and dry over a substantial period. Heavily fissured surfaces—familiar to many as "weathered" wood—generally have been considerably degraded by soft rot fungi.

Decay Resistance of Wood

For a discussion of the natural resistance of wood to fungi and a grouping of species according to decay resistance, see chapter 3. Among decay-resistant domestic species, only the heartwood has significant resistance, because the natural preservative chemicals in wood that retard the growth of fungi are essentially restricted to the heartwood. Natural resistance

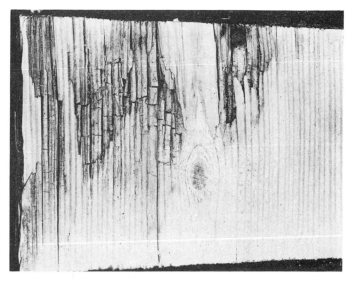

Figure 17−4—Brown rot in southern pine railroad tie. Note the darker color and the cubical checking in the wood. (M31 833)

M 116 113

Figure 17−5—Soft-rotted preservative-treated pine utility pole. Note the shallow depth of decay. (M116 113)

of species to fungi is important only where conditions conducive to decay exist or may develop. Where wood is subjected to severe decay conditions pressure-treated wood, rather than resistant heartwood, is generally prescribed.

Effect of Decay on Strength of Wood

Toughness, or the ability of wood to withstand impacts, is affected first by decay. This is generally followed by reductions in strength values related to static bending. Eventually all strength properties are seriously reduced.

Strength losses during early stages of decay can be considerable, depending to a great extent upon the fungus involved and, to a lesser extent, upon the type of wood undergoing decay. In laboratory tests, losses in toughness have ranged from 6 percent to more than 50 percent, by the time a 1 percent weight loss had occurred in the wood as a result of fungal attack. By the time weight losses due to decay have reached 10 percent most strength losses may be expected to exceed 50 percent. As decay is detectable, at such weight losses, only through microscopical observations, it may be assumed that wood with visually discernible decay has been greatly reduced in all strength values.

Prevention of Mold, Stain, and Decay

Logs, Poles, Piles, and Ties

The wood species, section of the country, and time of the year determine what precautions must be taken to avoid serious damage from fungi in poles, piles, ties, and similar thick products during seasoning or storage. In dry climates, rapid surface seasoning of poles and piles will retard development of mold, stains, and decay. First the bark is peeled from the pole and the peeled product is decked on high skids or piled on high, well-drained ground in the open to dry. In humid regions, such as the Gulf States, these products often do not air dry fast enough to avoid losses from fungi. Preseasoning treatments with approved preservative solutions can be helpful in these circumstances.

For logs, rapid conversion into lumber, or storage in water or under a water spray (fig. 17−6) is the surest way to avoid fungus damage. Preservative sprays promptly applied in the woods will protect most timber species during storage for 2 to 3 months. For longer storage, an end coating is needed to prevent seasoning checks, through which infection can enter the log.

Lumber

Growth of decay fungi can be prevented in lumber and other wood products by rapidly drying them to a moisture content of 20 percent or less and keeping them dry. Standard air-drying practices will usually dry the wood fast enough to protect it, particularly if the protection afforded by drying is supplemented by dip or spray treatment of the stock with an approved fungicidal solution. Successful control by this method depends not only upon immediate and adequate treatment but also upon the proper handling of the lumber after treatment. However, kiln drying is the most reliable method of rapidly reducing moisture content.

Air-drying yards should be kept as sanitary and as open as possible to air circulation (fig. 17−7). Recommended practice includes locating yards and sheds on well-drained ground; removing debris, which serves as a source of infection, and weeds, which reduce air circulation; and employing piling methods that permit rapid drying of the lumber and protect against wetting. Storage sheds should be constructed and maintained to prevent significant wetting of the stock; an ample roof overhang on open sheds is desirable. In areas where termites or water-conducting fungi may be troublesome, stock to be held for long periods should be set on foundations high enough so it can be inspected from beneath.

The user's best assurance of receiving lumber free from decay or other than light stain is to buy stock marked by a lumber association in a grade that eliminates or limits such quality-reducing features. Surface treatment for protection at the drying yard is only temporarily effective. Except for temporary structures, lumber to be used under conditions conducive to decay should be all heartwood of a naturally durable species or should be adequately treated with a wood preservative (ch. 18).

Buildings

The lasting qualities of properly constructed wood buildings are apparent in all parts of the country. Serious decay problems are almost always a sign of faulty design or construction, lack of reasonable care in the handling of the wood, or improper maintenance of the structure.

Construction principles that assure long service and avoid decay in buildings include: (1) Build with dry lumber, free of incipient decay and not exceeding the amounts of mold and blue stain permitted by standard grading rules; (2) use designs that will keep the wood dry and accelerate runoff; (3) for parts exposed to above-ground decay hazards, use wood treated with a preservative or heartwood of a decay-resistant species; and (4) for the high-hazard situation associated with ground contact, use pressure-treated wood.

A building site that is dry or for which drainage is provided will reduce the possibility of decay. Stumps, wood debris, stakes, or wood concrete forms frequently lead to decay if left under or near a building.

Unseasoned or infected wood should not be enclosed until it is thoroughly dried. Unseasoned wood may be infected because of improper handling at the sawmill or retail yard, or after delivery on the job.

Untreated wood parts of substructures should not be permitted to contact the soil. A minimum of 8 inches clearance between soil and framing and 6 inches between soil and siding is recommended. Where frequent hard rains occur a foundation height above grade of 12 to 18 inches is advocated. An exception may be made for certain temporary constructions. If contact with soil is unavoidable, the wood should be pressure treated (ch. 18).

Sill plates and other wood resting on a concrete-slab foundation generally should be pressure treated, and additionally

Figure 17–6—Water spraying logs protects against fungal stain and decay.

(M84 0367-0)

protected by installing beneath the slab a moisture-resistant membrane such as polyethylene. Girder and joist openings in masonry walls should be big enough to assure an air space around the ends of these wood members; if the members are below the outside soil level, moistureproofing of the outer face of the wall is essential.

In the crawl space of basementless buildings on damp ground, wetting of the wood by condensation during cold weather may result in serious decay damage. However, serious condensation leading to decay can be prevented by providing openings on opposite sides of the foundation walls for cross ventilation or by laying a barrier such as polyethylene on the soil; both provisions may be helpful in very wet

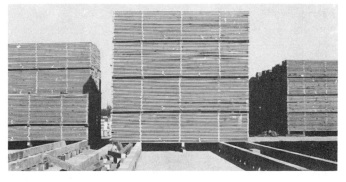

Figure 17–7—A sanitary, well-drained air-drying yard.

(M98 792)

situations. To facilitate inspection and ventilation of the crawl space, at least an 18-inch clearance should be left under wood joists.

Porches, exterior steps, and platforms present a decay hazard that cannot be fully avoided by construction practices. Therefore, in the wetter climates the use of preservative-treated wood (ch. 18) or heartwood of a durable species is advisable for such items.

Protection from entrance or retention of rainwater or condensation in walls and roofs will prevent the development of decay in these areas. A fairly wide roof overhang (2 ft) with gutters and downspouts that are never permitted to clog is very desirable. Sheathing papers under the siding should be of a "breathing" or vapor-permeable type (asphalt paper not exceeding 15-lb weight). Vapor-retarding materials should be near the warm face of walls and ceilings. Roofs must be kept tight, and cross ventilation in attics is recommended. The use of sound, dry lumber is important in all parts of buildings.

Where service conditions in a building are such that the wood cannot be kept dry, as in textile mills, pulp and paper mills, and cold-storage plants, lumber properly treated with an approved preservative or lumber containing all heartwood of a naturally decay-resistant species should be used.

In making repairs necessitated by decay, every effort should be made to correct the moisture condition leading to the damage. If the condition cannot be corrected, all infected parts should be replaced with treated wood or with all-heartwood lumber of a naturally decay-resistant wood species. If the sources of moisture that caused the decay are entirely eliminated, it is necessary only to replace the weakened wood with dry lumber.

Other Structures and Products

In general, the principles underlying the prevention of mold, stain, or decay, damage to veneer, plywood, containers, boats, and other wood products and structures are similar to those described for buildings—dry the wood rapidly and keep it dry or treat it with approved protective and preservative solutions. Interior grades of plywood should not be used where the plywood will be exposed to moisture; the adhesives, as well as the wood, may be damaged by fungi and bacteria as well as being degraded by moisture. With exterior-type panels, joint construction should be carefully designed to prevent the entrance of rainwater.

In treated bridge or wharf timbers checking may occur that exposes nontreated wood to fungal attack. Annual in-place treatment of these checks will provide protection from decay. Similarly, pile tops may be protected by treatment with a wood preservative followed by application of a suitable capping compound.

Wood boats present certain problems that are not encountered in other uses of wood. The parts especially subject to decay are the stem, knighthead, transom, and frameheads; these are reached by rainwater from above or condensation moisture from below. Faying surfaces are more likely to decay than exposed surfaces, and in salt-water service hull members just below the weather deck are more vulnerable than those below the waterline. Recommendations for avoiding decay include: (1) Use only heartwood of durable species, free of infection, and preferably below 20 percent in moisture content; (2) provide and maintain ventilation in the hull and all compartments; (3) keep water out as much as is practicable, especially fresh water; and (4) where it is necessary to use sapwood or nondurable heartwood, impregnate the wood with an approved preservative or treat the fully cut, shaped, and bored wood before installation by soaking it for a short time in preservative solution. Where such mild soaking treatment is used, the wood most subject to decay should also be flooded with an approved preservative at intervals of 2 or 3 years. When retreating, the wood should be dry so that joints are relatively loose.

Remedial Treatment of Internally Decayed Wood

Decay fungi in Douglas-fir utility poles and wharf timbers have been successfully eradicated through application of sodium methyl dithiocarbamate (vapam) and trichloronitromethane (chloropicrin). Both of these fumigants have been registered for use to arrest internal decay in poles and large timbers. Fumigants have been found to travel in poles as much as 8 feet above and below the treating level and to be effective for as long as 6 to 10 years, depending upon the fumigant used.

Bacteria

Most wood that has been wet for any considerable length of time probably will contain bacteria. The sour smell of logs that have been held under water for several months—or of lumber cut from them—manifests bacterial action. Usually bacteria have little effect on wood properties, except over long periods of time, but some may make the wood excessively absorptive. This can result in excessive pickup of preservatives during treatment or of moisture in use. This effect has been a problem in the sapwood of millwork cut from pine logs that have been stored in ponds. There also is evidence that bacteria developing in pine veneer bolts held under water or water spray may cause noticeable changes in the physical character of the veneer—including some strength loss. Additionally, mixtures of different bacteria, and probably fungi also, were found capable of accelerating decay of treated cooling tower slats and mine timbers.

Insect Damage and Control

The more common types of damage caused by wood-attacking insects are shown in table 17—1. Methods of controlling and preventing insect attack of wood are described in the following paragraphs.

Beetles

Bark beetles may damage the components of log and other rustic structures from which the bark has not been removed. They are reddish-brown to black and vary in length from about 1/16 to 1/4 inch. They bore through the outer bark to the soft inner part, where they make tunnels in which they lay their eggs. In making tunnels, bark beetles push out fine brownish-white sawdust-like particles. If many beetles are present, their extensive tunneling will loosen the bark and permit it to fall off in large patches, making the structure unsightly.

To avoid bark beetle damage, logs may be stored in water or under a water spray, or cut during the dormant season (October or November, for instance). If cut during this period, logs should immediately be piled off the ground where there will be good air movement, to promote rapid drying of the inner bark before the beetles begin to fly in the spring. Drying the bark will almost always prevent damage by insects that prefer freshly cut wood. Another protective measure is to thoroughly spray the logs with an approved insecticidal solution.

Ambrosia beetles, roundheaded and flatheaded borers, and some powder-post beetles that get into freshly cut timber can cause considerable damage to wood in rustic structures and some manufactured products. Certain beetles may complete development and emerge a year or more after the wood is dry, often raising a question as to the origin of the infestation. Proper cutting practices and spraying the material with an approved chemical solution, as recommended for bark beetles, will control these insects. Damage by ambrosia beetles can be prevented in freshly sawed lumber by dipping the product in a chemical solution. The addition of one of the sap-stain preventives approved for controlling molds, stains, and decay will keep the lumber bright.

Powder-post beetles attack both hardwoods and softwoods, and both freshly cut and seasoned lumber and timber. The powder-post beetles that cause most damage to dry hardwood lumber belong to the *Lyctus* species. They attack the sapwood of ash, hickory, oak, and other large-pored hardwoods as it begins to season. Eggs are laid in pores of the wood, and the larvae burrow through the wood, making tunnels from 1/16 to 1/12 inch in diameter, which they leave packed with a fine powder. Powder-post damage is indicated by holes left in the surface of the wood by the winged adults as they emerge and by the fine powder that may fall from the wood.

Susceptible hardwood lumber used for manufacturing purposes should be protected from powder-post beetle attack as soon as it is sawed and also when it arrives at the plant. An approved insecticide applied in water emulsion to the green lumber will provide protection. Such treatment may be effective even after the lumber is kiln dried—until it is surfaced.

Good plant sanitation is extremely important in alleviating the problem of infestations. Proper sanitation measures can often eliminate the necessity for other preventative steps. Damage to manufactured items frequently is traceable to infestations that occur before the products are placed on the market, particularly if a finish is not applied to the surface of the items until they are sold. Once wood is infested, the larvae will continue to work, even though the surface is subsequently painted, oiled, waxed, or varnished.

When selecting hardwood lumber for building or manufacturing purposes, any evidence of powder-post infestation should not be overlooked, for the beetles may continue to be active long after the wood is put to use. Sterilization of green wood with steam at 130 °F or sterilization of wood with a lower moisture content at 180 °F under controlled conditions of relative humidity for about 2 hours is effective for checking infestation or preventing attack of 1-inch lumber. Thicker material requires a longer time. A 3-minute soaking in a petroleum oil solution containing an insecticide is also effective for checking infestation or preventing attack of lumber up to 1 inch thick. Small dimension stock also can be protected by brushing or spraying with approved chemicals. For infested furniture or finished woodwork in a building, the same insecticides may be used, but they should be dissolved in a refined petroleum oil, like mineral spirits.

As the *Lyctus* beetles lay their eggs in the open pores of wood, infestation can be prevented by covering the entire surface of each piece of wood with a suitable finish.

Powder-post beetles in the family Anobiidae, depending on the species, infest hardwoods and softwoods. Their life cycle takes two to several years and they require a wood moisture content near or above 15 percent for viable infestation. Therefore, in most modern buildings the wood moisture content is generally too low for anobiids. When ventilation is inadequate or in more humid regions of the United States, wood components of a building can reach the favorable moisture conditions for anobiids. This is especially a problem in air-conditioned buildings where water condenses on cooled exterior surfaces. Susceptibility to anobiid infestation can be alleviated by lowering the moisture content of wood through improved ventilation and the judicious use of insulation and vapor barriers. Insecticides registered for use against these beetles are restricted generally for exterior

Table 17–1—*Types of damage caused by wood-attacking insects*

Type of damage	Description	Causal agent	Damage	
			Begins	Ends
Pin holes	Holes 1/100 to 1/4 in diameter, usually circular			
	A. Tunnels open			
	1. Holes 1/50 to 1/8 in (0.5 to 3 mm) in diameter, usually centered in dark streak or ring in surrounding wood	Ambrosia beetles	In living trees and unseasoned logs and lumber	During seasoning
	2. Holes variable sizes; surrounding wood rarely dark stained; tunnels lined with wood-colored substance	Timber worms	do. (not in lumber)	Before seasoning
	B. Tunnels packed with usually fine sawdust			
	1. Exit holes 1/32 to 1/16 in (0.8 to 1.6 mm) in diameter; in sapwood of large-pored hardwoods; loose floury sawdust in tunnels	Lyctid powder-post beetles	During or after seasoning	Reinfestation continues until sapwood destroyed
	2. Exit holes 1/16 to 1/8 in (1.6 to 3 mm) in diameter; primarily in sapwood, rarely in heartwood; tunnels loosely packed with fine sawdust and elongate pellets	Anobiid powder-post beetles	Usually after wood in use (in buildings)	Reinfestation continues; progress of damage very slow
	3. Exit holes 3/32 to 9/32 in (2.5 to 7 mm) in diameter; primarily sapwood of hardwoods, minor in softwoods; sawdust in tunnels fine to coarse and tightly packed	Bostrichid powder-post beetles	Before seasoning or if wood rewetted	During seasoning or redrying
	4. Exit holes 1/16 to 1/12 in (1.6 to 2 mm) in diameter; in slightly damp or decayed wood; very fine sawdust or pellets tightly packed in tunnels	Wood-boring weevils	In slightly damp wood in use	Reinfestation continues while wood is damp
Grub holes	Exit holes 1/8 to 1/2 in (3 to 13 mm) in diameter, circular or oval			
	A. Exit holes 1/8 to 1/2 in (3 to 13 mm) in diameter; circular; mostly in sapwood; tunnels with coarse to fibrous sawdust or it may be absent	Roundheaded borers (beetles)	In living trees and unseasoned logs and lumber	When adults emerge from seasoned wood or when wood kiln dried
	B. Exit holes 1/8 to 1/2 in (3 to 13 mm) in width; mostly oval; in sapwood and heartwood: sawdust tightly packed in tunnels	Flatheaded borers (beetles)	do.	Do.

Table 17 – 1—Types of damage caused by wood-attacking insects—Continued

Type of damage		Description	Causal agent	Damage Begins	Damage Ends
Grub holes (con.)	C.	Exit holes ca. 1/4 in (6 mm) in diameter; circular; in sapwood of softwoods, primarily pine; tunnels packed with very fine sawdust	Old house borer (a roundheaded borer)	During or after seasoning	Reinfestation continues in seasoned wood in use
	D.	Exit holes perfectly circular, 1/6 to 1/4 in (4 to 6 mm) in diameter; primarily in softwoods; tunnels tightly packed with coarse sawdust, often in decay-softened wood	Wood-wasps	In dying trees or fresh logs	When adults emerge from seasoned wood, usually in use, or when kiln dried
	E.	Nest entry hole and tunnel perfectly circular ca. 1/2 in (13 mm) in diameter; in soft softwoods in structures	Carpenter bees	In structural timbers, siding, etc.	Nesting reoccurs annually in spring at same and nearby locations
Network of galleries		Systems of interconnected tunnels and chambers	Social insects with colonies		
	A.	Walls look polished; spaces completely clean of debris	Carpenter ants	Usually in damp, partly decayed, or soft-textured wood in use	Colony persists unless prolonged drying of wood occurs
	B.	Walls usually speckled with mud spots; some chambers may be filled with "clay"	Subterranean termites	In wood structures	Colony persists
	C.	Chambers contain pellets; areas may be walled-off by dark membrane	Drywood termites (occasionally dampwood termites)	do.	Do.
Pitch pocket			Various insects	In living trees	In tree
Black check			Grubs of various insects	do.	Do.
Pith fleck			Fly maggots or adult weevils	do.	Do.
Gum spot			Grubs of various insects	do.	Do.

17-11

Table 17–1—*Types of damage caused by wood-attacking insects—Continued*

Type of damage	Description	Causal agent	Damage	
			Begins	Ends
Ring distortion		Larvae of defoliating insects or flatheaded cambium borers	do.	Do.
Bluing	Stained area over 1 in long	Staining fungi introduced by insects in trees or recently felled logs	With insect wounds	With seasoning

applications to avoid potential safety hazards indoors.

Beetles in the family Bostrichidae and weevils in the family Curculionidae are associated with wood moisture contents favorable for wood-infesting fungi and they may benefit nutritionally from the fungi. Thus, protection against these insects uses the same procedures as for protection against wood-decay fungi.

A roundheaded powder-post beetle, commonly known as the "old house borer," causes damage to seasoned pine floor joists. The larvae reduce the sapwood to a powdery or granular consistency, and make a ticking sound while at work. When mature, the beetles make an oval hole about 1/4 inch in diameter in the surface of the wood and emerge. Anobiid powder-post beetles, which make holes 1/16 to 1/8 inch in diameter, also cause damage to pine joists. Infested wood should be drenched with a solution of one of the currently recommended insecticides in a highly penetrating solvent. Beetles working in wood behind plastered or paneled walls can be eliminated by having a licensed operator fumigate the building.

Termites

Termites superficially resemble ants in size, general appearance, and habit of living in colonies. About 56 species are known in the United States. From the standpoint of their methods of attack on wood, they can be grouped into two main classes: (1) The ground-inhabiting or subterranean termites; and (2) the wood-inhabiting or nonsubterranean termites.

Subterranean Termites

Subterranean termites are responsible for most of the termite damage done to wood structures in the United States. This damage can be prevented. Subterranean termites are more prevalent in the Southern States than in the Northern States, where low temperatures do not favor their development (fig. 17—8). The hazard of infestation is greatest (1) beneath basementless buildings erected on a concrete-slab foundation or over a crawl space that is poorly drained and ventilated and (2) in any substructure wood close to the ground or an earth fill (e.g., an earth-filled porch).

The subterranean termites develop their colonies and maintain their headquarters in the ground. They build their tunnels through earth and around obstructions to get at the wood they need for food. They also must have a constant source of moisture. The worker members of the colony cause destruction of wood. At certain seasons of the year male and female winged forms swarm from the colony, fly a short time, lose their wings, mate, and, if successful in locating a suitable home, start new colonies. The appearance of "flying ants," or their shed wings, is an indication that a termite colony may be near and causing serious damage. Not all "flying ants" are termites; therefore suspicious insects should be identified before money is spent for their eradication (fig. 17−9).

Subterranean termites normally do not establish themselves in buildings by being carried there in lumber, but enter from ground nests after the building has been constructed; however, an introduced species, the Formosan termite, is adept at initiating above-ground infestations where wood remains wet for prolonged periods, such as from roof leaks. Telltale signs of subterranean termite presence are the earthen tubes or runways built by these insects over the surfaces of foundation walls to reach the wood above. Another sign is the swarming of winged adults early in the spring or fall. In the wood itself, the termites make galleries that generally follow the grain, leaving a shell of sound wood to conceal their activities. As the galleries seldom show on the wood surfaces, probing with an ice pick or knife is advisable if the presence of termites is suspected.

The best protection where subterranean termites are prevalent is to prevent them from gaining hidden access to a building. The foundations should be of concrete or other solid material through which the termites cannot penetrate. With brick, stone, or concrete blocks, cement mortar should be used, for termites can work through some other kinds of mortar. Also, it is a good precaution to cap the foundation with about 4 inches of reinforced concrete. Posts supporting first-floor girders should, if they bear directly on the ground, be of concrete. If there is a basement, it should be floored with concrete. Untreated posts in such a basement should rest on concrete piers extending a few inches above the basement floor. However, pressure-treated posts can rest directly on the basement floor. With the crawl-space type of foundation, wood floor joists should be kept at least 18 inches and girders 12 inches from the earth and good ventilation provided beneath the floor.

Moisture condensation on the floor joists and subflooring, which may cause conditions favorable to decay and contribute to infestation by termites, can be avoided by covering the soil below with a moisture barrier.

All concrete forms, stakes, stumps, and waste wood should be removed from the building site, for they are possible sources of infestation. In the main, the precautions effective against subterranean termites are also helpful against decay.

The principal method of protecting buildings in high termite hazard areas is to thoroughly treat the soil adjacent to the foundation walls and piers beneath the building with a soil insecticide. When concrete-slab floors are laid directly on the ground, all soil under the slab should be treated with an approved insecticide before the concrete is poured. Furthermore, insulation containing cellulose that is used as a filler in expansion joints should be impregnated with an approved

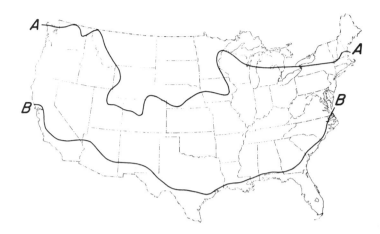

Figure 17−8—A, The northern limit of recorded damage done by subterranean termites in the United States; B, the northern limit of damage done by dry-wood termites.   (M134 686)

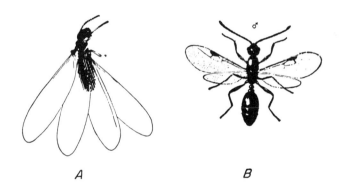

Figure 17−9—A, Winged termite B, winged ant (both greatly enlarged). The wasp waist of the ant and the long wings of the termite are distinguishing characteristics.   (M137 348)

chemical toxic to termites. Sealing the top 1/2 inch of the expansion joint with roofing-grade coal-tar pitch also provides effective protection from ground-nesting termites. New modifications in soil treatment and an insecticidal bait control method are currently under investigation and appear promising. Current references to termite control should be consulted to take advantage of the newer developments in termite control.

To control termites already in a building, break any contact between the termite colony in the soil and the woodwork. This can be done by blocking the runways from soil to wood, by treating the soil, or both. Guard against possible reinfestations by frequent inspections for telltale signs that were listed previously.

Nonsubterranean Termites

Nonsubterranean termites have been found only in a narrow strip of territory extending from central California around the southern edge of the continental United States to Virginia (fig. 17—9) and also in the West Indies and Hawaii. Their principal damage is confined to an area in southern California, to parts of southern Florida, notably Key West, and to the islands of Hawaii. They also are a localized problem in Arizona and New Mexico.

The nonsubterranean termites, especially the dry-wood type, do not multiply as rapidly as the subterranean termites, and have somewhat different colony life and habits. The total amount of destruction they cause in the United States is much less than that caused by the subterranean termites. The ability of dry-wood termites to live in dry wood without outside moisture or contact with the ground, however, makes them a definite menace in the regions where they occur. Their depredations are not rapid, but they can thoroughly riddle timbers with their tunnelings if allowed to work unmolested for many years.

In constructing a building in localities where the dry-wood type of nonsubterranean termite is prevalent, it is good practice to inspect the lumber carefully to see that it was not infested before arrival at the building site. If the building is constructed during the swarming season, the lumber should be watched during the course of construction, because infestation by colonizing pairs can easily take place. Because paint is a good protection against the entrance of dry-wood termites, exposed wood (except that which is preservative treated) should be kept covered with a paint film. Fine screen should be placed over any openings through which access might be gained to the interior unpainted parts of the buildings. As in the case of ground-nesting termites, dead trees, old stumps, posts, or wood debris of any kind that could serve as sources of infestation should be removed from the premises.

If a building is infested with dry-wood termites, badly damaged wood should be replaced. If the wood is only slightly damaged or difficult to replace, further termite activity can be arrested by injecting a small amount of an approved pesticidal dust or liquid formulation into each nest. Current recommendations for such formulations should be consulted.[2] Buildings heavily infested with nonsubterranean termites can be fumigated with success. This method is quicker and often cheaper than the use of poisonous liquids and dusts and does not require finding all of the colonies. It does not prevent the termites from returning because no poisonous residue is left in the tunnels. Fumigation is very dangerous and should be conducted only by licensed professional fumigators.

In localities where dry-wood termites do serious damage to

---

[2] See footnote 1.

posts and poles, the best protection for these and similar forms of outdoor timbers is full-length pressure treatment with a preservative.

Naturally Termite-Resistant Woods

Only a limited number of woods grown in the United States offer any marked degree of natural resistance to termite attack. The close-grained heartwood of California redwood has some resistance, especially when used above ground. Very resinous heartwood of southern pine is practically immune to attack, but it is not available in large quantities and is seldom used.

Carpenter Ants

Carpenter ants are black or brown. They occur usually in stumps, trees, or logs but sometimes damage poles, structural timbers, or buildings. One form is easily recognized by its giant size relative to other ants. Carpenter ants use wood for shelter rather than for food, usually preferring wood that is naturally soft or has been made soft by decay. They may enter a building directly, by crawling, or may be carried there in fuelwood. If left undisturbed, they can, in a few years, enlarge their tunnels to the point where replacement or extensive repairs are necessary. The parts of dwellings they frequent most often are porch columns, porch roofs, window sills, and sometimes the wood plates in foundation walls. The logs of rustic cabins are also attacked.

Precautions that prevent attack by decay and termites are usually effective against carpenter ants. Decaying or infested wood, such as logs or stumps, should be removed from the premises, and crevices present in the foundation or woodwork of the building should be sealed. Particularly, leaks in porch roofs should be repaired, because the decay that may result makes the wood more desirable to the ants.

When carpenter ants are found in a structure, any badly damaged timbers should be replaced. Because the ant needs high humidity in its immature stages, alterations in the construction may also be required to eliminate moisture from rain or condensation. In wood not sufficiently damaged to require replacement, the ants can be killed by injection of an approved insecticide into the nest galleries.

Carpenter Bees

Carpenter bees resemble large bumblebees, but the top of their abdomen is bare of hairs. The females make large (1/2-in diameter) tunnels into soft wood for nests. They partition the hole into cells with each cell provided with pollen and nectar for a single egg. Because they reuse nesting sites for many years, a nesting tunnel into a structural timber may be extended several feet and have multiple branches. In thin wood, such as siding, the holes may destroy the full thickness of the wood. They nest in stained wood and wood with thin paint films or light preservative salt treatments as well as in bare wood.

Control is aimed at discouraging the use of nesting sites in and near buildings. The tunnel may be injected with an insecticide labeled for bee control and after several days plugged. Treating the surface around the entry hole will discourage reuse of the tunnel during the spring nesting period. A good paint film or pressure preservative treatment protects exterior wood surfaces from nesting damage. Bare interior wood surfaces, such as in garages, can be protected by screens and tight-fitting doors.

Marine-Borer Damage and Control

Damage by marine-boring organisms to wood structures in salt or brackish waters is practically worldwide. Slight attack is sometimes found in rivers even above the region of brackishness. The rapidity of attack depends upon local conditions and the kinds of borers present. Along the Pacific, Gulf, and South Atlantic coasts of the United States attack is rapid, and untreated piling may be completely destroyed in a year or less. Along the coast of the New England States the rate of attack is slower but still sufficiently rapid, generally, to require protection of wood where long life is desired.

The principal marine borers from the standpoint of wood damage in the United States are described here. Control measures discussed in this section are those in use at the time this handbook was prepared. Regulations should be reviewed at the time control treatments are being considered so that approved practices will be followed.[3]

Shipworms

Shipworms are the most destructive of the marine borers. They are mollusks of various species that superficially are wormlike in form. The group includes several species of *Teredo* and several species of *Bankia*, which are especially damaging. These are readily distinguishable on close observation but are all very similar in several respects. In the early stages of their life they are minute, free-swimming organisms. Upon finding suitable lodgement on wood they quickly develop into a new form and bury themselves in the wood. A pair of boring shells on the head grows rapidly in size as the boring progresses, while the tail part or siphon remains at the original entrance. Thus, the animal grows in length and diameter within the wood but remains a prisoner in its burrow,

which it lines with a shell-like deposit. It lives on the wood borings and the organic matter extracted from the sea water that is continuously being pumped through its system. The entrance holes never grow large, and the interior of wood may be completely honeycombed and ruined while the surface shows only slight perforations. When present in great numbers, the borers grow only a few inches before the wood is so completely occupied that growth is stopped, but when not crowded, they can grow to lengths of 1 to 4 feet according to species.

Pholads

Another group of wood-boring mollusks is *pholads*, which clearly resemble clams and therefore are not included with the shipworms. These are entirely encased in their double shells. The *Martesia* are the best-known species, but a second group is the *Xylophaga*. Like the shipworms, the *Martesia* enter the wood when very small, leaving a small entrance hole, and grow larger as they burrow into the wood. They generally do not exceed 2-1/2 inches in length and 1 inch in diameter, but are capable of doing considerable damage. Their activities in the United States appear to be confined to the Gulf of Mexico.

Limnoria and *Sphaeroma*

Another distinct group of marine borers are crustaceans, which are related to lobsters and shrimp. The principal ones are species of *Limnoria* and *Sphaeroma*. Their attack differs from that of the shipworms and the *Martesia* in that it is quite shallow; the result is that the wood gradually is thinned through erosion by the combined action of the borers and water. Also, the *Limnoria* and *Sphaeroma* do not become imprisoned in the wood but may move freely from place to place.

Limnoria are small, about 1/8 to 1/6 inch long, and bore small burrows in the surface of wood. Although they can change their location, they usually continue to bore in one place. When great numbers are present, their burrows are separated by very thin walls of wood that are easily eroded by the motion of the water or damaged by objects floating upon it. This erosion causes the *Limnoria* to burrow continually deeper; otherwise the burrows would probably not become more than 2 inches long or more than 1/2 inch deep. As erosion is greatest between tide levels, piles heavily attacked by *Limnoria* characteristically wear within such levels to an hourglass shape. Untreated piling can be destroyed by *Limnoria* within a year in heavily infested harbors.

Sphaeroma are somewhat larger, sometimes reaching a length of 1/2 inch and a width of 1/4 inch. They resemble in general appearance and size the common sow bug or pill bug

---

[3] See footnote 1.

that inhabits damp places. *Sphaeroma* are widely distributed but not as plentiful as *Limnoria* and do much less damage. Nevertheless piles in some structures have been ruined by them. Occasionally they have been found working in fresh water. In types of damage, *Sphaeroma* action resembles that of *Limnoria*.

The average life of well-creosoted structures is many times the average life that could be obtained from untreated structures. However, even thorough creosote treatment will not always stop *Martesia*, *Sphaeroma*, and especially *Limnoria*.

Shallow or erratic creosote penetration affords but slight protection. The spots with poor protection are attacked, and from them the borers spread inward and destroy the untreated interior of the pile. Low retention fails to provide a reservoir of surplus preservative to compensate for depletion by evaporation and leaching.

When wood is to be used in salt water, avoidance of cutting or injuring the surface after treatment is even more important than when wood is to be used on land. No cutting or injury of any kind for any purpose should be permitted in the underwater part of the pile. Where piles are cut to grade above the waterline, the exposed surfaces should, of course, be protected from decay. This may be accomplished by in-place application of a wood preservative followed by a suitable capping compound.

Resistance to Marine Borers

No wood is immune to marine-borer attack, and no commercially important wood of the United States has sufficient marine-borer resistance to justify its use untreated in any important structure in areas where borers are active. The heartwood of several foreign species, such as greenheart, jarrah, azobe, and manbarklak, has shown resistance to marine-borer attack. Service records on these woods, however, do not always show uniform results and are affected by local conditions.

Protection of Permanent Structures

The best practical protection for piles in sea water where borer hazard is moderate is heavy treatment with coal-tar creosote or creosote coal tar solution. Where severe borer hazard exists, dual treatment (copper arsenate[4]-containing waterborne preservatives followed by coal-tar creosote) is recommended. The treatment must be thorough, the penetration as deep as possible, and the retention high to give satisfactory results in heavily infested waters. It is best to treat such piles by the full-cell process "to refusal"; that is, to force in all the preservative the piles can hold without using

treatments that cause serious damage to the wood. The retentions recommended in chapter 18 are minimum values; when maximum protection against marine borers is desired, as much more preservative as is practicable should be injected. For highest retentions it is necessary to air dry the piling before treatment. Details of treatments are discussed in chapter 18.

The life of treated piles is influenced by the thoroughness of the treatment, the care and intelligence used in avoiding damage to the treated shell during handling and installation, and the severity of borer attack. Differences in exposure conditions, such as water temperature, salinity, dissolved oxygen, water depth, and currents, tend to cause wide variations in the severity of borer attack even within limited areas. Service records show average-life figures of from 22 to 48 years on well-treated Douglas-fir piles in San Francisco Bay waters. In South Atlantic and Gulf of Mexico waters, creosoted piles are estimated to last 10 to 12 years, and frequently much longer. On the North Atlantic Coast, even longer life is to be expected.

Metal armor and concrete jacketing have been used with varying degrees of success for the protection of marine piles. The metal armor may be in the form of sheets, wire, or nails. Sheathing of piles with copper or muntz metal has been only partially successful, owing to difficulty in maintaining a continuous armor. Theft, damage in driving, damage by storm or driftwood, and corrosion have sooner or later let in the borers, and in only a few cases has long life been reported. Attempts during World War II to electroplate wood piles with copper were not successful. Concrete casings are now in greater use than metal armor and appear to provide better protection when high-quality materials are used and are carefully applied. Unfortunately, they are readily damaged by ship impact. For this reason, concrete casings are less practical for fender piles than for foundation piles that are protected from mechanical damage.

Jacketing piles by wrapping them with heavy polyvinyl plastic is one of the most recent forms of supplementary protection. If properly applied, it will kill any borers that may have already become established, by rendering stagnant the water in contact with the piles. Like other materials, the plastic jacket is subject to mechanical damage.

Protection of Boats

Wood barges and lighters have been constructed with planking or sheathing pressure treated with creosote to provide hull protection from marine borers, and the results have been favorable. Although coal-tar creosote is an effective preservative for protecting wood against marine borers in areas of

---

[4] See footnote 1.

moderate borer hazard, it has disadvantages in many types of boats. Creosote adds considerably to the weight of the boat hull, and its odor is objectionable to boat crews. In addition, antifouling paints are difficult to apply over creosoted wood.

Some copper bottom paints protect boat hulls against marine-borer attack, but the protection continues only while the coating remains unbroken. As it is difficult to maintain an unbroken coating of antifouling paint, the U.S. Navy has found it desirable to impregnate the hull planking of some wood boats with certain copper-containing preservatives.[5] Such preservatives, when applied with high retentions (1.5 to 2.0 pcf), have some effectiveness against marine borers and should help to protect the hull of a boat during intervals between renewals of the antifouling coating. These copper preservatives do not provide protection equivalent to that furnished by coal-tar creosote; their effectiveness in protecting boats is therefore best assured if the boats are dry docked at regular and frequent intervals and the antifouling coating maintained. However, the leach-resistant wood preservatives containing copper arsenates[5] have shown superior performance (at a retention of 2.5 pcf) to creosote in tests conducted in areas of severe borer hazard.

Plywood as well as plank hulls can be protected against marine borers by preservative treatment. The plywood hull presents a surface that can be covered successfully with a protective membrane of reinforced plastic laminate. Such coverings should not be attempted on wood that has been treated with a preservative carried in oil, because the bond will be unsatisfactory.

Selected References

Beal, R. H. Formosan invader. Pest Control. 35(2): 13−17; 1967.

Beal, R. H.; Maulderi, J. K.; Jones, S. C. Subterranean termites, their prevention and control in buildings. Home & Garden Bull. 64; Washington, DC: U. S. Department of Agriculture; rev. 1983.

Cassens, D. L.; Eslyn, W. E. Fungicides to prevent sapstain and mold on hardwood lumber. Madison, WI: Forest Products Journal. 31: 39−42; 1981.

Ebeling, W. Wood destroying insects and fungi. In: Urban entomology. Berkeley, CA: University of California, Division of Agricultural Science; 1975: 128−216.

Esenther, G. R.; Beal, R. H. Termite control: decayed wood bait. Sociobiology. 4(2): 215−222; 1979.

Eslyn, W. E.; Clark, J. W. Appraising deterioration in submerged piling. Materials und Organismen Supplement. 3: 43−52; 1976.

Eslyn, W. E.; Clark, J. W. Wood bridges—decay inspection and control. Agric. Handb. 557. Washington, DC: U.S. Department of Agriculture; 1979.

Graham, R. D.; Helsing, G. G. Wood pole maintenance manual: inspection and supplemental treatment of Douglas-fir and western redcedar poles. Res. Bull. 24. Corvallis, OR: Forest Research Lab, Oregon State University; 1979.

Greaves, H. Wood-inhabiting bacteria: General considerations. Commonwealth Scientific and Industrial Research Organization, Forest Products Newsletter. 359; 1969.

Hartley, C.; May, C. Decay of wood in boats. U. S. Dep. Agric. Forest Path. Spec. Release 8; U. S. Department of Agriculture, Forest Service; 1943.

Highley, T. L.; Eslyn, W. E. Using fumigants to control interior decay in waterfront timbers. Madison, WI: Forest Products Journal. 32: 32−34; 1982.

Highley, T. L.; Scheffer, T. C. Controlling decay in above-water parts of waterfront structures. Madison, WI: Forest Products Journal. 28: 40−43; 1978.

Hunt, G. M.; Garratt, G. A. Wood preservation, third edition. The American forestry series. New York: McGraw-Hill Book Co.; 1967.

Jones, E. B. G.; Eltringham, S. K., eds. Marine borers, fungi and fouling organisms of wood. Proceedings of Organisation for Economic Co-operation and Development; 1968 March 27−April 3; 1971.

Krishna, K.; Weesner, F. M., eds. Biology of termites. Vol. I. New York: Academic Press; 1969.

Krishna, K.; Weesner, F. M., eds. Biology of termites. Vol. II. New York: Academic Press; 1970.

Lee, K. E.; Wood, T. G. Termites and soils. New York: Academic Press; 1971.

Moore, Harry B. Wood-inhabiting insects in houses: their identification, biology, prevention, and control. Prepared as part of interagency agreement IAA−25−75 between the U. S. Department of Agriculture, Forest Service, and the Department of Housing and Urban Development; 1979.

National Pest Control Association. Carpenter bees. Tech. Release 3−63. Dunn Loring, VA: NPCA; 1963.

National Pest Control Association. The horntails. Tech. Release 14−64. Dunn Loring, VA: NPCA; 1964.

National Pest Control Association. Carpenter ants. Tech. Release ESPC 052101. Dunn Loring, VA: NPCA; 1976.

Rietz, R. C. Storage of lumber. Agric. Handb. 531. Washington, DC: U. S. Department of Agriculture; 1978.

Roff, J. W.; Cserjesi, A. J.; Swann, G. W. Prevention of sap stain and mold in packaged lumber. FORINTEK Canada Corp. Tech. Rep. 14R. Ottawa, ON: FORINTEK; 1980.

Scheffer, T. C.; Eslyn, W. E. Winter treatments protect birch roundwood during storage. Madison, WI: Forest Products Journal 26: 27−31; 1976.

Scheffer, T. C.; Verrall, A. F. Principles of protecting wood buildings from decay. Res. Pap. FPL 190. Madison, WI: U. S. Department of Agriculture, Forest Service, Forest Products Laboratory; 1973.

Sherwood, G. E.; TenWolde, A. Moisture movement and control in light frame structures. Madison, WI: Forest Products Journal. 32: 69−73; 1982.

Weesner, F. M. The termites of the United States, a handbook. Elizabeth, NJ: National Pest Control Association; 1965.

Wilcox, W. W. Review of literature on the effect of early stages of decay on wood strength. Wood and Fiber. 9: 252−257; 1978.

Williams, L. H. Anobiid beetles should be controlled. Pest Control. 41(6): 18,20,22,38,40,42,44; 1973.

---

[5] See footnote 1.

Chapter 18

Wood Preservation

Wood Preservation *

Wood can be protected from the attack of decay fungi, harmful insects, or marine borers by applying selected chemicals as wood preservatives. The degree of protection obtained depends on the kind of preservative used and on achieving proper penetration and retention of the chemicals. Some preservatives are more effective than others, and some are more adaptable to certain use requirements. The wood can be well protected only when the preservative substantially penetrates it, and some methods of treatment assure better penetration than others. There is also a difference in the treatability of various species of wood, particularly of their heartwood, which generally resists preservative treatment more than sapwood.

Good wood preservatives, applied at recommended retentions and with the wood satisfactorily penetrated, greatly increase the life of wood structures, often by 5 to 10 times. On this basis the annual cost of treated wood in service is greatly reduced below that of similar wood without treatment. In considering preservative treatment processes and wood species, the combination must provide the required protection for the conditions of exposure and life of the structure.

Wood Preservatives

Wood preservatives fall into two general classes: Oils, such as creosote and petroleum solutions of pentachlorophenol; and waterborne salts that are applied as water solutions.

Preservative Oils

Wood does not swell from preservative oils, but it may shrink if it loses moisture during the treating process. Creosote and solutions with heavier, less volatile petroleum oils often help protect wood from weathering, but may adversely influence its cleanliness, odor, color, paintability, and fire resistance in use. Preservative oils sometimes migrate from treated studs or subflooring along nails and discolor adjacent plaster or finish flooring. Volatile oils or solvents with oil borne preservatives, if removed after treatment, leave the wood cleaner than the heavier oils but may not provide as much protection. Wood treated with some preservative oils can be glued satisfactorily, although special processing or cleaning may be required to remove surplus oils from surfaces before spreading the adhesive.

Coal-Tar Creosote

Coal-tar creosote, a black or brownish oil made by distilling coal tar, is one of the more important and useful wood preservatives. Its advantages are: (1) High toxicity to wood-destroying organisms; (2) relative insolubility in water and low volatility, which impart to it a great degree of permanence under the most varied use conditions; (3) ease of

application; (4) ease with which its depth of penetration can be determined; (5) general availability and relative low cost (when purchased in wholesale quantities); and (6) long record of satisfactory use.

The character of the tar used, the method of distillation, and the temperature range in which the creosote fraction is collected all influence the composition of the creosote. The composition of the various coal-tar creosotes available, therefore, may vary to a considerable extent. Small differences in composition, however, do not prevent creosotes from giving good service; satisfactory results in preventing decay may generally be expected from any coal-tar creosote that complies with the requirements of standard specifications.

Although coal-tar creosote or creosote-coal-tar solutions are well suited for general outdoor service in structural timbers, they have properties that are disadvantageous for some purposes.

The color of creosote and the fact that creosote-treated wood usually cannot be painted satisfactorily make this preservative unsuitable where appearance and paintability are important.

The odor of creosoted wood is unpleasant to some persons. Also, creosote vapors are harmful to growing plants, and foodstuffs that are sensitive to odors should not be stored where creosote odors are present. Workmen sometimes object to creosoted wood because it soils their clothes and because it burns the skin of the face and hands of some individuals. With normal precautions to avoid direct skin contact with creosote, there appears to be no danger to the health of workmen handling or working near the treated wood. The Environmental Protection Agency should be contacted for more specific information on the subject.

Freshly creosoted timber can be ignited and will burn readily, producing a dense smoke. However, after the timber has seasoned some months, the more volatile parts of the oil disappear from near the surface, and the creosoted wood usually is little, if any, easier to ignite than untreated wood. Until this volatile oil has evaporated, ordinary precautions should be taken to prevent fires. On the other hand, timber that has been kept sound by creosote treatment is harder to ignite than untreated wood that has started to decay. A preservative other than creosote should be used where fire

[1] Mention of a chemical in this chapter does not constitute a recommendation; only those chemicals registered by the U.S. Environmental Protection Agency (EPA) may be recommended. Registration of preservatives is under constant review by EPA and the Department of Agriculture. Use only preservatives that bear an EPA Registration Number and carry directions for home and farm use. Preservatives such as creosote and pentachlorophenol should not be applied to the interior of dwellings which are occupied by humans. Because all preservatives are under constant review by EPA, a responsible State or Federal agency should be consulted as to the current status of any preservative.

---

* Revision by Lee R. Gjovik, Forest Products Technologist.

hazard is highly important, unless the treated wood is also protected from fire.

A number of specifications prepared by different organizations are available for creosote oils of different kinds. Although the oil obtained under most of these specifications will probably be effective in preventing decay, the requirements of some organizations are more exacting than others. Federal Specification TT−C−645 for coal-tar creosote, adopted for use by the U.S. Government, will generally prove satisfactory; under normal conditions, this specification can be met without difficulty by most creosote producers. The requirements of this specification are similar to those of the American Wood-Preservers' Association (AWPA) Standard P1 for creosote, which is equally acceptable to the user.

Federal Specification TT−C−645 provides for three classes of coal-tar creosote. Class I is for poles; class II is for ties, lumber, structural timbers, land or fresh-water piles, and posts; and class III is for piles, lumber, and structural timbers for use in coastal waters.

Coal-Tar Creosotes for Nonpressure Treatments

Special coal-tar creosotes are available for nonpressure treatments. They differ somewhat from regular commercial coal-tar creosote in (1) being crystal-free to flow freely at ordinary temperatures and (2) having low-boiling distillation fractions removed to reduce evaporation in thermal (hot-and-cold) treatments in open tanks. Federal Specification TT−C−655 covers coal-tar creosote for brush, spray, or open-tank treatments.

Other Creosotes

Creosotes distilled from tars other than coal tar are used to some extent for wood preservation, although they are not included in current Federal or AWPA specifications. These include wood-tar creosote, oil-tar creosote, and water-gas-tar creosote. These creosotes protect wood from decay and insect attack but are generally less effective than coal-tar creosote.

Tars

Coal tars are seldom used alone for preserving wood because good penetration is usually difficult to obtain and because they are less effective against wood-destroying fungi than the coal-tar creosotes. Service tests have demonstrated that surface coatings of tar are of little value. Coal tar has been used in the pressure treatment of crossties, but it has been difficult to get the highly viscous tar to penetrate wood satisfactorily. When good absorptions and deep penetrations are obtained, however, it is reasonable to expect a satisfactory degree of effectiveness from treatment with coal tar. The tar has been particularly effective in reducing checking in crossties in service.

Water-gas-tar is used less extensively than coal tar, but, in certain cases where the wood was thoroughly impregnated, the results were good.

Creosote Solution

For many years, either coal tar or petroleum oil has been mixed with coal-tar creosote, in various proportions, to lower preservative costs. These creosote solutions have a satisfactory record of performance, particularly for crossties where they have been most commonly used.

Federal Specification TT−C−650, ''Creosote-Coal-Tar Solution,'' covers five classes of creosote-coal-tar solutions. Class I contains not less than 80 percent coal-tar distillate (creosote) by volume; class II, 70 percent; class III, 60 percent; class IV, 50 percent; and class V not less than 60 nor more than 75 percent coal-tar distillate. Classes I and II are for land and fresh-water piles, posts, lumber, structural timber, and bridge ties. Classes III and IV are for crossties and switch ties. Class V is for piles, lumber, and structural timber used in coastal waters.

AWPA Standard P2 includes four creosote-coal-tar solutions that must contain not less than 80 percent by volume of coal-tar distillate for class A, 70 percent for class B, 60 percent for class C, and 50 percent for class D. In addition each solution must meet requirements as to physical and chemical properties. All classes are permitted for lumber, timber, and ties used in above-ground, soil, or fresh water exposure. Classes A and B are also used for the pressure treatment of poles.

AWPA Standard P12 covers a creosote-coal-tar solution for the treatment of marine (coastal waters) piles and timbers. Federal Specification TT−W−568 and AWPA Standard P3 stipulate that creosote petroleum oil solutions shall contain not less than 50 percent (by volume) of coal-tar creosote and the petroleum oil shall meet the requirements of AWPA's Standard P4.

Creosote-coal-tar solutions, compared to straight creosote, tend to reduce weathering and checking of the treated wood. The solutions may have a greater tendency to accumulate on the surface of the treated wood (bleed) and may penetrate the wood with greater difficulty, because they generally are more viscous than straight creosote. Higher temperatures and pressures during treatment, when they can safely be used, will often improve penetration of high viscosity solutions.

Even though petroleum oil and coal tar are less toxic to wood destroying organisms than straight creosote, and their mixtures with creosote are also less toxic in laboratory tests, a reduction in toxicity does not imply less preservative protection. Creosote petroleum solutions and creosote-coal-tar solutions help to reduce checking and weathering of the treated wood. Frequently posts and ties treated with standard formulations of these solutions have shown better service than those similarly treated with straight coal-tar creosote.

Pentachlorophenol Solutions

Water-repellent solutions containing chlorinated phenols,

principally pentachlorophenol, in solvents of the mineral spirits type, were first used in commercial treatments of wood by the millwork industry about 1931. Commercial pressure treatment with pentachlorophenol in heavy petroleum oils started on poles about 1941, and considerable quantities of various products were soon pressure treated. AWPA Standard P8 and Federal Specification TT−W−570 define the properties of pentachlorophenol and AWPA Standard P9 covers solvents for oil-borne preservatives. A commercial process using pentachlorophenol dissolved in liquid petroleum gas was introduced in 1961.

Pentachlorophenol solutions for wood preservation generally contain 7.5 percent (by weight) of this chemical although solutions with volatile solvents may contain lower or higher concentrations. The performance of pentachlorophenol and the properties of the treated wood are influenced by the properties of the solvent used. The heavy petroleum solvent included in AWPA Standard P9 type A is preferable for maximum protection, particularly where the wood treated with pentachlorophenol is used in contact with the ground. Studies are underway to determine the properties and efficacy of wood treated with a water-dispersible and water-soluble pentachlorophenol.

The heavy oils remain in the wood for a long time and do not usually provide a clean or paintable surface. The volatile solvents, such as liquefied petroleum gas and methylene chloride, are used with pentachlorophenol when the natural appearance of the wood must be retained and the treated wood requires a paint coating or other finish. Because of the toxicity of pentachlorophenol, care is necessary to avoid excessive personal contact with the solution or vapor in handling and using it.

A "bloom" preventive, such as ester gum or oil-soluble glycol, is generally required with volatile solvents to prevent crystals of pentachlorophenol from forming on the surface of the treated wood. Brushing or washing the surface with hot water or an alkaline solution has been used to remove the crystalline deposits.

The results of pole service and field tests on wood treated with 5 percent pentachlorophenol in a heavy petroleum oil are similar to those with coal-tar creosote. This similarity has been recognized in the preservative retention requirements of treatment specifications. Pentachlorophenol is ineffective against marine borers and is not recommended for the treatment of marine piles or timbers used in coastal waters.

Water-Repellent Preservatives

Preservative systems containing water-repellent components are sold under various trade names, principally for the dip or equivalent treatment of window sash and other millwork. Federal Specification TT−W−572 stipulates that such preservatives be dissolved in volatile solvents, such as mineral spirits, that do not cause appreciable swelling of the wood, and that the treated wood be paintable and meet a performance test on water repellency. In pressure treatment with water-repellent preservative, however, considerable difficulty has been experienced in removing residual solvents and obtaining acceptable paintability.

The preservative chemicals in Federal Specification TT−W−572 may be one of the following: (1) not less than 5 percent of pentachlorophenol, (2) not less than either 1 or 2 percent (for tropical conditions) of copper in the form of copper naphthenate, or (3) not less than 0.045 percent copper in the form of copper-8-quinolinolate (for uses where foodstuffs will be in contact with the treated wood). The National Wood Window & Door Association (NWWDA) standard for water repellent preservative treatment for millwork, IS 4–81, permits other preservatives provided their toxicity properties are as high as those of 5 percent (by weight) pentachlorophenol solution. Mixtures of other chlorinated phenols with pentachlorophenol meet this requirement according to tests by NWWDA.

Water-repellent preservative containing copper-8-quinolinolate has been used in nonpressure treatment of wood containers, pallets, and other products for use in contact with foods. That preservative is also included in AWPA Standard P8. Here it is intended for use in volatile solvents to pressure-treat lumber for decking of trucks and cars or for related uses involving harvesting, storage, and transportation of foods.

Effective water-repellent preservatives will retard the ingress of water when wood is exposed above ground. They therefore help reduce dimensional changes in the wood due to moisture changes when the wood is exposed to rainwater or dampness for short periods. As with any wood preservative, their effectiveness in protecting wood against decay and insects depends upon the retention and penetration obtained in application.

Waterborne Preservatives

Standard wood preservatives used in water solution include acid copper chromate, ammoniacal copper arsenate, chromated copper arsenate (types I, II, and III), chromated zinc chloride, and fluor chrome arsenate phenol. These preservatives are often employed when cleanliness and paintability of the treated wood are required. The chromated zinc chloride and fluor chrome arsenate phenol formulations are not as leach resistant as the other waterborne preservatives or oils and, therefore, are recommended for above-ground light-duty uses only. Several formulations involving combinations of copper, chromium, and arsenic have shown high resistance to leaching and very good performance in service. The ammoniacal copper arsenate and chromated copper arsenate are now included in specifications for such items as building foundations, building poles, utility poles, marine piles, and piles for land and fresh water use.

Test results based on sea-water exposure have shown that dual treatment (waterborne copper-containing salt preservatives followed by coal-tar-creosote) is possibly the most effective method of protecting wood against all types of marine borers. The AWPA standards have recognized this process as well as the treatment of marine piles with high retentions of ammoniacal copper arsenate or chromated copper arsenate. The recommended treatment and retention in pounds per cubic foot for round timber piles exposed to severe marine borer hazard are shown in table 18−1.

Waterborne preservatives leave the wood surface comparatively clean, paintable, and free from objectionable odor. With several exceptions, they must be used at low treating temperatures (100 to 150 °F) because they are unstable at the higher temperatures. This restriction may involve some difficulty when higher temperatures are needed to obtain good treating results in such woods as Douglas-fir. Because water is added in the treatment process, the wood must be dried afterward to the moisture content required for the end use intended.

Waterborne preservatives, in the retentions normally specified for wood preservation, decrease the danger of ignition and rapid spread of flame. Formulations with copper and chromium tend to prolong glowing combustion in carbonized wood.

Acid Copper Chromate

Acid copper chromate (Celcure) contains, according to Federal Specification TT−W−546 and AWPA Standard P5, 31.8 percent copper oxide and 68.2 percent chromic acid. Equivalent amounts of copper sulfate, potassium dichromate, or sodium dichromate may be used in place of copper oxide. Tests on stakes and posts exposed to decay and termite attack indicate that wood well impregnated with Celcure gives good service. Tests by the Forest Products Laboratory and the U.S. Navy showed that wood thoroughly impregnated with at least 0.5 pcf of Celcure has some resistance to marine borer attack. The protection against marine borers, however, is much less than that provided by a standard treatment with creosote.

Ammoniacal Copper Arsenate

According to Federal Specification TT−W−549 and AWPA Standard P5, ammoniacal copper arsenate (Chemonite) should contain approximately 49.8 percent copper oxide or an equivalent amount of copper hydroxide, 50.2 percent of arsenic pentoxide. In order to improve the solubility of this system, 1.7 percent of acetic acid may be added. The net retention of preservative is calculated as pounds of copper oxide plus arsenic pentoxide per cubic foot of wood treated within the proportions in the specification.

Service records on structures treated with ammoniacal copper arsenate show that this preservative provides very good protection against decay and termites. High retentions of preservative will provide extended service life to wood exposed to the marine environment, provided pholad-type borers are not present.

Chromated Copper Arsenate

Types I, II, and III of chromated copper arsenate (CCA) are covered in Federal Specification TT−W−550 and AWPA Standard P5. The compositions of the three types according to that Federal specification are given in table 18−2. The

Table 18−1—*Preservative treatment and retention necessary to protect round timber piles from severe marine borer attack*

Treatment	Southern pine, red pine	Coastal Douglas-fir	AWPA standard
	Pounds per cubic foot		
Limnoria tripunctata only:			
Ammoniacal copper arsenate	2.50	2.50	C3 C18
Chromated copper arsenate	2.50	2.50	C3 C18
Limnoria tripunctata and			
Pholads (dual treatment):			
First treatment:			
Ammoniacal copper arsenate	1.00	1.00	C3 C18
Chromated copper arsenate	1.00	1.00	C3 C18
Second treatment:			
Creosote	20.0	20.0	C3 C18
Creosote-coal-tar	20.0	Not recommended	C3 C18

above specification permits substitution of potassium or sodium dichromate for chromium trioxide; copper sulfate, basic copper carbonate, or copper hydroxide for copper oxide; and arsenic acid or sodium arsenate for arsenic pentoxide.

CCA Type I—Service data on treated poles, posts, and stakes installed in the United States since 1938 have shown excellent protection by CCA Type I against decay fungi and termites.

CCA Type II—CCA Type II has been used commercially in Sweden since 1950 and now throughout the world. It was included in stake tests in the United States in 1949 and is giving excellent protection. Commercial use of this preservative in the United States started in 1964.

CCA Type III—Composition of CCA Type III was arrived at by AWPA technical committees in encouraging a single standard for chromated copper arsenate preservatives. Commercial preservatives of similar composition have been tested and used in England since 1954 and more recently in Australia, New Zealand, Malaysia, and in various countries of Africa and Central Europe and are performing very well.

High retentions of the three types of chromated copper arsenate preservatives will provide good resistance to *Limnoria* and *Teredo* marine borer attack.

Chromated Zinc Chloride

Chromated zinc chloride is covered in Federal Specification TT−W−551 and in AWPA Standard P5. Chromated zinc chloride (FR)[2] is included as a fire-retarding chemical.

Chromated zinc chloride was developed about 1934. The specifications require that it contain 80 percent of zinc oxide and 20 percent of chromium trioxide. Sodium chromate may be substituted for chromium trioxide and zinc chloride for zinc oxide. The preservative is only moderately effective in contact with the ground or in wet installations but has performed well under somewhat drier conditions. Its principal advantages are its low cost and ease of handling at treating

Table 18−2—*Compositions of the three types of chromated copper arsenate [1]*

Component	Type I	Type II	Type III
	Parts by weight		
Chromium trioxide	61	35.3	47
Copper oxide	17	19.6	19
Arsenic pentoxide	22	45.1	34

[1] As covered in Federal Specification TT−W−550 and AWPA Standard P5.

plants. Most standards do not permit this preservative to be used in ground contact.

Chromated zinc chloride (FR) contains 80 percent of chromated zinc chloride, 10 percent of boric acid, and 10 percent of ammonium sulfate. Retentions of from 1-1/2 to 3 pcf of wood provide combined protection from fire, decay, and insect attack.

Fluor Chrome Arsenate Phenol

The composition of fluor chrome arsenate phenol (FCAP) is included in Federal Specification TT−W−535 and the AWPA Standard P5. The active ingredients of this preservative are:

Ingredient	Percent
Fluoride	22
Chromium trioxide	37
Arsenic pentoxide	25
Dinitrophenol	16

To avoid objectionable staining of building materials, sodium pentachlorophenate is sometimes substituted in equal amounts for the dinitrophenol.

Sodium or potassium fluoride may be used as a source of fluoride. Sodium chromate or dichromate may be used in place of chromium trioxide. Sodium arsenate may be used in place of arsenic pentoxide.

FCAP type I (Wolman salts) and FCAP type II (Osmosalts) have performed well in above-ground wood structures but generally are not recommended for ground-contact use.

Preservative Effectiveness

Preservative effectiveness is influenced not only by the protective value of the preservative chemical itself, but also by the method of application and extent of penetration and retention of the preservative in the treated wood. Even with an effective preservative, good protection cannot be expected with poor penetration and substandard retentions. The species of wood, proportion of heartwood and sapwood, heartwood penetrability, and moisture content are among the important variables influencing the results of treatment. For various wood products, the preservatives and retentions listed in Federal Specification TT−W−571 are given in table 18−3.

Results of service tests on various treated products that show the effectiveness of different wood preservatives are published periodically in Forest Products Laboratory Research Note FPL−068 and elsewhere. Few service tests, however, include a variety of preservatives under comparable condi-

---

[2] Designation for fire retardant.

Table 18–3—Preservatives and minimum retentions for various wood products[1]

- - - - - - - - - Pounds per cubic foot - - - - - - - - -

Form of product and service condition	Coal-tar creosote	Creosote-coal-tar solution	Creosote-petroleum solution	Pentachlorophenol In heavy petroleum	Pentachlorophenol In light petroleum	Pentachlorophenol In volatile solvents	Acid copper chromate[2]	Ammoniacal copper arsenate[2]	Chromated copper arsenate[2] Types I, II, or III	AWPA Standard
A. Ties (crossties and switch ties)	7–10	7–10	7–10	0.35–.50	—	—	—	—	—	C2 & C6
B. Lumber, plywood, and structural timbers (including glued laminated)	—	—	—	—	—	—	—	—	—	C2,C9, C14, C18, & C20
(1) For use in coastal waters:[3] Lumber (under 5 in thick) Timbers (5 in or thicker):										
Southern pine	20–25	—	—	—	—	—	—	2.50	2.50	—
Coast Douglas-fir and western hemlock	20–25	—	—	—	—	—	—	2.50	2.50	—
Plywood	25	—	—	—	—	—	—	2.50	2.50	—
(2) For use in fresh water, in contact with ground, or for important structural members not in contact with ground or water	7–12	7–12	7–12	.50–.60	.62–.75	.62–.75	—	.60	.60	—
Glued-laminated timbers or laminates	6–12	6–12	6–12	.60	.75	.75	—	.60	.60	—
(3) For other uses not in contact with ground or water[4]	6–12	6–12	6–12	.30–.40	.30–.40	.30–.40	0.25–0.50	.250	.250	—
C. Piles	—	—	—	—	—	—	—	—	—	C3,C14, & C18
(1) For use in coastal waters:[3] Southern pine	25	25	—	—	—	—	—	2.50	2.50	—
Coast Douglas-fir	22	—	—	—	—	—	—	2.50	2.50	—
(2) For land or fresh-water use: Southern; and other pines	12	12	12	.60	—	—	—	.80	.80	C3 & C14
Douglas-fir and western larch	17	17	17	.85	—	—	—	1.00	1.00	—

18-7

Table 18–3—*Preservatives and minimum retentions for various wood products[1]*—Continued

Form of product and service condition	Coal-tar creosote	Creosote-coal-tar solution	Creosote-petroleum solution	Pentachlorophenol — In heavy petroleum	Pentachlorophenol — In light petroleum	Pentachlorophenol — In volatile solvents	Acid copper chromate[2]	Ammoniacal copper arsenate[2]	Chromated copper arsenate[2] Types I, II, or III	AWPA Standard
	---------------------- Pounds per cubic foot ----------------------									
D. Poles (utility)										
Southern and ponderosa pine	9.0	—	—	.45	.56	.56	—	—	—	C4
Red pine	13.5	—	—	.68	.85	.85	—	.60	.60	
Jack and lodgepole pine	16.0	—	—	.80	1.00	1.00	—	.60	.60	—
Coast Douglas-fir	12.0	—	—	.60	.75	.75	—	.60	.60	—
Interior Douglas-fir and western larch	16	—	—	.80	1.00	1.00	—	.60	.60	—
Western redcedar	16	—	—	.80	1.00	1.00	—	.60	—	—
Western redcedar, northern white-cedar, Alaska-cedar, lodgepole pine ("thermal" or hot-and-cold process)	20	—	—	1.0	—	—	—	—	—	C7,C8,& C10
E. Poles (building, round)	12–13.5	—	—	.60	—	—	—	.60	.60	C23
F. Posts (round)	—	—	—	—	—	—	—	—	—	C5 & C16
Fence	6	6	7	30	.38	.38	.50	.40	.40	—
Building	12	—	—	.60	—	—	—	.60	.60	C23

[1] Retentions for lumber, timber, plywood, piles, poles, and fenceposts are determined by assay of borings of a number and location as specified in Federal Specification TT–W–571 or in the Standards of the American Wood-Preservers' Association referenced in last column.

[2] All waterborne preservative retentions are specified on an oxide basis.

[3] Dual treatments are recommended when marine borer activity is known to be high (see AWPA Standards C2, C3, C14, and C18 for details).

[4] Additional preservatives are recommended for this use, and their retention levels, include plain chromated zinc chloride, 0.46 pcf; and fluor chrome arsenic phenol 0.22 pcf.

NOTE: Minimum retentions are those included in Federal Specification TT–W–571 and Standards of the American Wood-Preservers' Association. The current issues of these specifications should be referred to for up-to-date recommendations and other details. In many cases the retention is different depending on species and assay zone.

tions of exposure. Furthermore, service tests may not show a good comparison between different preservatives due to the difficulty in controlling the above-mentioned variables. Such comparative data under similar exposure conditions, with various preservatives and retentions, are included in Forest Products Laboratory stake tests on southern pine sapwood. A summary of these FPL results is included in table 18−4.

Penetrability of Different Species

The effectiveness of preservative treatment is influenced by the penetration and distribution of the preservative in the wood. For maximum protection it is desirable to select species for which good penetration is best assured. The heartwood is commonly difficult to treat. With round members such as poles, posts, and piles, the penetrability of the sapwood is important in achieving a protective outer zone around the heartwood.

Examples of species with sapwood that is easily penetrated when it is well dried and pressure treated are the pines, coast Douglas-fir, western larch, Sitka spruce, western hemlock, western redcedar, northern white-cedar, and white fir (*A. concolor*). Examples of species with sapwood and heartwood somewhat resistant to penetration are red and white spruces and Rocky Mountain Douglas-fir. Cedar poles are commonly incised to obtain satisfactory preservative penetration.

The sapwood and heartwood of several hardwood species, such as blackjack oak, some of the lowland red oaks, and aspen often present a problem in getting uniform preservative penetration.

The heartwood of most species resists penetration of preservatives although well-dried white fir, western hemlock, northern red oak, the ashes, and tupelo are examples of species with heartwood reasonably easy to penetrate. The southern pines, ponderosa pine, redwood, Sitka spruce, coast Douglas-fir, beech, maples, and birches are examples of species with heartwood moderately resistant to penetration.

Preparing Timber for Treatment

For satisfactory treatment and good performance thereafter, the timber must be sound and suitably prepared. Except in specialized treating methods involving unpeeled or green material, the wood should be well peeled and either seasoned or conditioned in the cylinder before treatment. It is also highly desirable that all machining be completed before treatment. Machining may include incising to improve the preservative penetration in woods that are resistant to treatment, as well as the operations of cutting, framing, or boring of holes.

Peeling

Peeling round or slabbed products is necessary to enable the wood to dry quickly enough to avoid decay and insect damage and to permit the preservative to penetrate satisfactorily. Even strips of the thin inner bark may prevent penetration. Patches of bark left on during treatment usually fall off in time and expose untreated wood, thus permitting decay to reach the interior of the member.

Careful peeling is especially important for wood that is to be treated by a superficial method. In the more thorough processes some penetration may take place both lengthwise and tangentially in the wood, and consequently small strips of bark are tolerated in some specifications. Processes in which a preservative is forced or permitted to diffuse through green wood lengthwise do not require peeling of the timber.

Machines of various types have been developed for peeling round timbers such as poles, piles, and posts (fig. 18−1).

Drying

For treatment with waterborne preservatives by certain diffusion methods, high moisture content may be permitted. For treatment by other methods, however, drying before treatment is essential. Drying before treatment opens up the checks before the preservative is applied, thus increasing penetration, and reduces the risk of checks opening after treatment and exposing unpenetrated wood. Good penetration of preservative is possible with wood at a moisture content as high as 40 to 60 percent, but serious checking after treatment can result when wood at that moisture level dries.

For large timbers and railroad ties, air drying is a widely used method of conditioning. Despite the greater time, labor, and storage space required, air drying is generally the cheapest and most effective method, even for pressure treatment. However, wet, warm climatic conditions make it difficult to air dry wood adequately without objectionable infection by stain, mold, and decay fungi. Such infected wood is often highly permeable; in rainy weather it can absorb a large quantity of water, which in turn prevents satisfactory treatment.

How long the timber must be air dried before treatment depends on the climate, location, and condition of the seasoning yard, methods of piling, season of the year, size, and species of the timbers. The most satisfactory seasoning practice for any specific case will depend on the individual drying conditions and the preservative treatment to be used. Treating specifications therefore are not always specific as to moisture content requirements.

To prevent decay and other forms of fungus infection during air drying, the wood should be cut and dried when conditions are less favorable for fungus development (see ch. 17).

Table 18–4—Results of Forest Products Laboratory studies on stakes pressure-treated with commonly used wood preservatives— stakes 2 by 4 by 18 inches of southern pine sapwood, installed at Harrison Experimental Forest, MS

Preservative	Average retention [1]	Average life or condition at last inspection
	Pcf	
Untreated stakes	--	1.8 to 3.6 yr
Acid copper chromate	0.13	11.6 yr
	.14	6.1 yr
	.25	60% failed after 15 yr
	.26	20% failed after 36 yr
	.29	4.6 yr
	.37	40% failed after 36 yr
	.50	40% failed after 15 yr
	.76	20% failed after 15 yr
Ammoniacal copper arsenate	.24	40% failed after 37 yr
	.25	10% failed after 15 yr
	.51	No failures after 37 yr
	.97	No failures after 37 yr
	1.25	No failures after 37 yr
Chromated copper arsenate		
Type I	.15	70% failed after 36 yr
	.22	20% failed after 15 yr
	.29	No failures after 36 yr
	.44	No failures after 36 yr
Type II	.23	20% failed after 15 yr
	.26	No failures after 32 yr
	.37	No failures after 32 yr
	.52	No failures after 32 yr
	.79	No failures after 32 yr
	1.04	No failures after 32 yr
Chromated zinc chloride	.30	14.2 yr
	.47	20.2 yr
	.46	60% failed after 15 yr
	.62	20.1 yr
	.62	40% failed after 15 yr
	.92	80% failed after 30 yr
	.96	40% failed after 15 yr
	1.78	40% failed after 30 yr
	3.67	No failures after 30 yr
Copper naphthenate:		
0.11 percent copper in No. 2 fuel oil	10.3	15.9 yr
.29 percent copper in No. 2 fuel oil	10.2	21.8 yr
.57 percent copper in No. 2 fuel oil	10.6	27.2 yr
.86 percent copper in No. 2 fuel oil	9.6	80% failed after 40 yr

Table 18−4—Results of Forest Products Laboratory studies on stakes pressure-treated with commonly used wood preservatives— stakes 2 by 4 by 18 inches of southern pine sapwood, installed at Harrison Experimental Forest, MS—Continued

Preservative	Average retention [1]	Average life or condition at last inspection
	Pcf	
Copper-8-quinolinolate:		
0.1 percent in Stoddard solvent	.01	5.3 yr
.2 percent in Stoddard solvent	.02	4.2 yr
.6 percent in Stoddard solvent	.06	5.6 yr
1.2 percent in Stoddard solvent	.12	7.8 yr
.15 percent in AWPA P9 heavy oil	.01	10% failed after 18 yr
.3 percent in AWPA P9 heavy oil	.03	No failures after 18 yr
.6 poroont in AWPA P9 hoavy oil	.06	No failuroc aftor 18 yr
1.2 percent in AWPA P9 heavy oil	.12	No failures after 18 yr
	3.3	24.9 yr
	4.1	14.2 yr
	4.2	17.8 yr
	4.6	21.3 yr
	7.8	60% failed after 40-½ yr
	8.0	60% failed after 41-½ yr
Creosote, coal-tar (regular type)	8.3	20% failed after 32 yr
	10.0	70% failed after 41 yr
	11.8	20% failed after 41-½ yr
	13.2	20% failed after 40-½ yr
	14.5	No failures after 41 yr
	16.5	No failures after 41-½ yr
Cresote, coal-tar (special types):		
Low residue, straight run	8.0	17.8 yr
Medium residue, straight run	8.0	18.8 yr
High residue, straight run	7.8	20.3 yr
Medium residue:		
Low in tar acids	8.1	19.4 yr
Low in naphthalene	8.2	21.3 yr
Low in tar acids and naphthalene	8.0	18.9 yr
Low residue, low in tar acids and naphthalene	8.0	19.2 yr
High residue, low in tar acids and naphthalene	8.2	20.0 yr
	5.3	80% failed after 34 yr
	8.0	18.9 yr
English vertical retort	10.1	50% failed after 34 yr
	15.0	No failures after 34 yr

Table 18—4—Results of Forest Products Laboratory studies on stakes pressure-treated with commonly used wood preservatives— stakes 2 by 4 by 18 inches of southern pine sapwood, installed at Harrison Experimental Forest, MS—Continued

Preservative	Average retention [1]	Average life or condition at last inspection
	Pcf	
Creosote, coal-tar (special types)—con.		
English coke oven	4.7	16.3 yr
	7.9	13.6 yr
	10.1	70% failed after 34 yr
	14.8	70% failed after 34 yr
Fluor chrome arsenate phenol	.12	10.2 yr
	.19	18.0 yr
	.22	18.3 yr
	.31	80% failed after 22 yr
	.38	24.1 yr
	.47	10% failed after 22 yr
Pentachlorophenol (various solvents):		
Liquefied petroleum gas	.14	80% failed after 20-½ yr
	.19	80% failed after 20-½ yr
	.34	No failures after 10-½ yr
	.34	40% failed after 18 yr
	.49	No failures after 18 yr
	.58	No failures after 20-½ yr
	.65	No failures after 18 yr
AWPA P9 (heavy petroleum)	.11	20% failed after 20-½ yr
	.19	No failures after 20-½ yr
	.29	No failures after 20-½ yr
	.53	No failures after 18 yr
	.67	No failures after 20-½ yr
Stoddard solvent (mineral spirits)	.14	13.8 yr
	.18	16.1 yr
	.20	9.5 yr
	.20	13.7 yr
	.38	No failures after 20-½ yr
	.40	15.5 yr
	.67	No failures after 20-½ yr
Heavy gas oil (mid-United States)	.20	33% failed after 34-½ yr
	.40	20% failed after 34-½ yr
	.60	10% failed after 34-½ yr
No. 4 aromatic oil (West Coast)	.21	90% failed after 32 yr
	.41	30% failed after 32 yr

Table 18 – 4—*Results of Forest Products Laboratory studies on stakes pressure-treated with commonly used wood preservatives— stakes 2 by 4 by 18 inches of southern pine sapwood, installed at Harrison Experimental Forest, MS***—Continued**

Preservative	Average retention [1]	Average life or condition at last inspection
	Pcf	
Tributyltin oxide:		
	.015	6.3 yr
	.025	4.5 yr
Stoddard solvent	.030	7.2 yr
	.045	7.4 yr
	.047	7.0 yr
AWPA P9 (heavy petroleum)	.024	70% failed after 22 yr
	.048	50% failed after 22 yr
	4.0	7.6 yr
	4.1	4.4 yr
	4.7	12.9 yr
	7.7	90% failed after 30 yr
	7.9	70% failed after 34-½ yr
	8.0	60% failed after 18 yr
	8.0	40% failed after 22 yr
Petroleum solvent controls	8.0	14.6 yr
	8.1	3.4 yr
	8.5	40% failed after 18 yr
	9.8	6.3 yr
	12.0	17.1 yr
	12.1	10% failed after 34-½ yr
	19.4	9.1 yr

[1] All waterborne salt preservative retentions are based on oxides.

If this is impossible, chances for infection can be minimized by prompt conditioning of the green material, careful piling and roofing during air drying, and pretreating the green wood with preservatives to protect it during air drying.

Lumber of all species, as well as southern pine poles, is often kiln dried before treatment, particularly in the southern United States where proper air seasoning is difficult. Kiln drying has the important added advantage of quickly reducing moisture content and thereby reducing transportation charges on poles.

Conditioning Green Products for Pressure Treatment

Plants that treat wood by pressure processes can condition green material by means other than air drying and kiln drying. Thus, they avoid a long delay and possible deterioration of the timber before treatment.

When green wood is to be treated under pressure, one of several methods for conditioning may be selected. The steaming-and-vacuum process is employed mainly for southern pine, while the Boulton or boiling-under-vacuum process is used for Douglas-fir and sometimes for hardwoods.

In the steaming process the green wood is steamed in the treating cylinder for several hours, usually at a maximum temperature of 245 °F. When the steaming is completed, a vacuum is immediately applied. During the steaming period the outer part of the wood is heated to a temperature approaching that of the steam; the subsequent vacuum lowers the boiling point so part of the water is evaporated or is forced out of the wood by the steam produced when the vacuum is applied. The steaming and vacuum periods employed depend upon the wood size, species, and moisture content. Steaming and vacuum usually reduce the moisture content of green wood slightly, and the heating assists greatly in getting the preservative to penetrate. A sufficiently long steaming period will also sterilize the wood.

In the Boulton or boiling-under-vacuum method of partial seasoning, the wood is heated in the oil preservative under vacuum, usually at temperatures of about 180 to 220 °F. This temperature range, lower than that of the steaming process, is a considerable advantage in treating woods that are especially susceptible to injury from high temperatures. The Boulton method removes much less moisture from heartwood than from the sapwood.

A third method of conditioning known as "vapor drying" has been patented and is used for seasoning railroad ties and other products. In the treating cylinder the green wood is exposed to hot vapors produced by boiling an organic solvent, such as xylene; the vapors are then condensed. As condensation takes place, the latent heat of the solvent is given up and the moisture vaporizes. The resulting mixed vapors of water

and the solvent are then passed through a condenser so the water can be separated and drained away and the solvent recovered and reused. The best results are obtained with solvents that have a boiling range of 280 to 320 °F. A small quantity of chemical remains in the wood. The wood is treated by standard pressure methods after the conditioning is completed.

Incising

Wood that is resistant to penetration by preservatives is often incised before treatment to permit deeper and more uniform penetration. In incising, lumber and timbers are passed through rollers equipped with teeth that sink into the wood to a predetermined depth, usually 1/2 to 3/4 inch. The teeth are spaced to give the desired distribution of preservative with the minimum number of incisions. A machine of different design is required for deeply incising the butts of poles, usually 2.5 inches (fig. 18−2).

The effectiveness of incising depends on the fact that preservatives usually penetrate into wood much farther in a longitudinal direction than in a direction perpendicular to the faces of the timber. The incisions expose end-grain surfaces and thus permit longitudinal penetration. It is especially effective in improving penetration in the heartwood areas of sawed surfaces.

Incising is practiced chiefly on Douglas-fir, western hemlock, and western larch ties and timbers for pressure treatment and on poles of cedar and Douglas-fir.

Cutting and Framing

All cutting, framing, and boring of holes should be done before treatment. Cutting into the wood in any way after treatment will frequently expose the untreated interior of the timber and permit ready access to decay fungi or insects.

It is much more practical than is commonly supposed to design wood structures so all cutting and framing may be done before treatment. Railroads have followed the practice extensively and find it not only practical but economical. Many wood-preserving plants are equipped to carry on such operations as the adzing and boring of crossties; gaining, roofing, and boring of poles; and the framing of material for bridges and for specialized structures such as water tanks and barges.

Treatment of the wood with preservative oils involves little or no dimensional change. With waterborne preservatives, however, some change in the size and shape may occur even though wood is redried to the moisture content it had before treatment. If precision fitting is necessary, the wood is cut and framed before treatment to its approximate final dimen-

Table 18–4—_Results of Forest Products Laboratory studies on stakes pressure-treated with commonly used wood preservatives— stakes 2 by 4 by 18 inches of southern pine sapwood, installed at Harrison Experimental Forest, MS—_Continued

Preservative	Average retention [1]	Average life or condition at last inspection
	Pcf	
Creosote, coal-tar (special types)—con.		
English coke oven	4.7	16.3 yr
	7.9	13.6 yr
	10.1	70% failed after 34 yr
	14.8	70% failed after 34 yr
Fluor chrome arsenate phenol	.12	10.2 yr
	.19	18.0 yr
	.22	18.3 yr
	.31	80% failed after 22 yr
	.38	24.1 yr
	.47	10% failed after 22 yr
Pentachlorophenol (various solvents):		
Liquefied petroleum gas	.14	80% failed after 20-½ yr
	.19	80% failed after 20-½ yr
	.34	No failures after 10-½ yr
	.34	40% failed after 18 yr
	.49	No failures after 18 yr
	.58	No failures after 20-½ yr
	.65	No failures after 18 yr
AWPA P9 (heavy petroleum)	.11	20% failed after 20-½ yr
	.19	No failures after 20-½ yr
	.29	No failures after 20-½ yr
	.53	No failures after 18 yr
	.67	No failures after 20-½ yr
Stoddard solvent (mineral spirits)	.14	13.8 yr
	.18	16.1 yr
	.20	9.5 yr
	.20	13.7 yr
	.38	No failures after 20-½ yr
	.40	15.5 yr
	.67	No failures after 20-½ yr
Heavy gas oil (mid-United States)	.20	33% failed after 34-½ yr
	.40	20% failed after 34-½ yr
	.60	10% failed after 34-½ yr
No. 4 aromatic oil (West Coast)	.21	90% failed after 32 yr
	.41	30% failed after 32 yr

Table 18—4—*Results of Forest Products Laboratory studies on stakes pressure-treated with commonly used wood preservatives— stakes 2 by 4 by 18 inches of southern pine sapwood, installed at Harrison Experimental Forest, MS***—Continued**

Preservative	Average retention [1]	Average life or condition at last inspection
	Pcf	
Copper-8-quinolinolate:		
0.1 percent in Stoddard solvent	.01	5.3 yr
.2 percent in Stoddard solvent	.02	4.2 yr
.6 percent in Stoddard solvent	.06	5.6 yr
1.2 percent in Stoddard solvent	.12	7.8 yr
.15 percent in AWPA P9 heavy oil	.01	10% failed after 18 yr
.3 percent in AWPA P9 heavy oil	.03	No failures after 18 yr
.6 percent in AWPA P9 heavy oil	.06	No failures after 18 yr
1.2 percent in AWPA P9 heavy oil	.12	No failures after 18 yr
	3.3	24.9 yr
	4.1	14.2 yr
	4.2	17.8 yr
	4.6	21.3 yr
	7.8	60% failed after 40-½ yr
	8.0	60% failed after 41-½ yr
Creosote, coal-tar (regular type)	8.3	20% failed after 32 yr
	10.0	70% failed after 41 yr
	11.8	20% failed after 41-½ yr
	13.2	20% failed after 40-½ yr
	14.5	No failures after 41 yr
	16.5	No failures after 41-½ yr
Cresote, coal-tar (special types):		
Low residue, straight run	8.0	17.8 yr
Medium residue, straight run	8.0	18.8 yr
High residue, straight run	7.8	20.3 yr
Medium residue:		
Low in tar acids	8.1	19.4 yr
Low in naphthalene	8.2	21.3 yr
Low in tar acids and naphthalene	8.0	18.9 yr
Low residue, low in tar acids and naphthalene	8.0	19.2 yr
High residue, low in tar acids and naphthalene	8.2	20.0 yr
	5.3	80% failed after 34 yr
English vertical retort	8.0	18.9 yr
	10.1	50% failed after 34 yr
	15.0	No failures after 34 yr

Figure 18 – 1—Machine peeling of poles. Here the outer bark had been removed by hand and the inner bark is being peeled by machine. Frequently all the bark is removed by machine.

(M130 271)

(M141 288)

Figure 18 – 2—Deep incising permits better penetration.

sions to allow for slight surfacing, trimming, and reaming of bolt holes. Grooves and bolt holes for timber connectors are cut before treatment and can be reamed out if necessary after treatment.

Applying Preservatives

Wood-preserving methods are of two general types: (1) Pressure processes, in which the wood is impregnated in closed vessels under pressures considerably above atmospheric, and (2) nonpressure processes, which vary widely as to procedures and equipment used. Pressure processes generally provide a closer control over preservative retentions and penetrations, and usually provide greater protection than nonpressure processes. Some nonpressure methods, however, are better than others and are occasionally as effective as pressure processes in providing good preservative retentions and penetrations.

Pressure Processes

In commercial practice, wood is most often treated by immersing it in preservative in high-pressure apparatus and applying pressure to drive the preservative into the wood. Pressure processes differ in details, but the general principle is the same. The wood, on cars or trams, is run into a long steel cylinder (fig. 18–3), which is then closed and filled with preservative. Pressure forces preservative into the wood until the desired amount has been absorbed. Considerable preservative is absorbed, with relatively deep penetration. Three processes, the full-cell, modified full-cell, and empty-cell, are in common use.

Full-Cell

The full-cell (Bethel) process is used when the retention of a maximum quantity of preservative is desired. It is a standard procedure for timbers to be treated full-cell with creosote when protection against marine borers is required. Waterborne preservatives are generally applied by the full-cell process, and control over preservative retention is obtained by regulating the concentration of the treating solution.

Steps in the full-cell process are essentially:

1. The charge of wood is sealed in the treating cylinder, and a preliminary vacuum is applied for 1/2 hour or more to remove the air from the cylinder and as much as possible from the wood.

2. The preservative, at ambient or elevated temperature depending on the system, is admitted to the cylinder without breaking the vacuum.

3. After the cylinder is filled, pressure is applied until the wood will take no more preservative or until the required retention of preservative is obtained.

4. When the pressure period is completed, the preservative is withdrawn from the cylinder.

5. A short final vacuum may be applied to free the charge from dripping preservative.

When the wood is steamed before treatment, the preservative is admitted at the end of the vacuum period that follows steaming. When the timber has received preliminary conditioning by the Boulton or boiling-under-vacuum process, the cylinder can be filled and the pressure applied as soon as the conditioning period is completed.

A pressure treatment referred to commercially as the "Cellon" process usually employs the full-cell process. It uses a preservative such as pentachlorophenol in highly volatile liquefied petroleum gas, such as butane or propane, which are gases at atmospheric pressure and ordinary temperatures. A cosolvent is employed to obtain the required concentration of preservative in the treating liquid.

For closer control over preservative retention during the Cellon process, the empty-cell process may be used. If so, a noncombustible gas, such as nitrogen, is substituted for air during the initial air pressure in the conventional Rueping process.

Modified Full-Cell

The modified full-cell is basically the same as the full-cell process except for the amount of initial vacuum. The modified full-cell process uses lower levels of vacuum, the actual amount is determined by the species and the final retention desired. This process is used only on material 2 inches or less in thickness.

Empty-Cell

The objective of empty-cell treatment is to obtain deep penetration with a relatively low net retention of preservative. For treatment with oil preservatives, the empty-cell process should always be used if it will provide the desired retention. Two empty-cell processes, the Rueping and the Lowry, are commonly employed; both use the expansive force of compressed air to drive out part of the preservative absorbed during the pressure period.

The Rueping empty-cell process has been widely used for many years in both Europe and the United States. The following general procedure is employed:

1. Air under pressure is forced into the treating cylinder, which contains the charge of wood. The air penetrates some species easily, requiring but a few minutes' application of pressure. In the treatment of the more resistant species, common practice is to maintain air pressure from 1/2 to 1 hour before admitting the preservative, but the necessity for long air-pressure periods does not seem fully established. The air pressures employed generally range between 25 and 100 psi; depending on the net retention of preservative desired and the resistance of the wood.

Figure 18–3—Interior view of treating cylinder at wood-preserving plant, with a load about to come in.

(M113 859)

2. After the period of preliminary air pressure, preservative is forced into the cylinder. As the preservative is pumped in, the air escapes from the treating cylinder into an equalizing or Rueping tank, at a rate that keeps the pressure constant within the cylinder. When the treating cylinder is filled with preservative, the treating pressure is raised above that of the initial air and is maintained until the wood will take no more preservative, or until enough has been absorbed to leave the required retention of preservative in the wood after the treatment.

3. At the end of the pressure period the preservative is drained from the cylinder, and surplus preservative removed from the wood with a final vacuum. The amount recovered may be from 20 to 60 percent of the gross amount injected.

18-17

The Lowry is often called the empty-cell process without initial air pressure. Preservative is admitted to the cylinder without either an initial air pressure or a vacuum, and the air originally in the wood at atmospheric pressure is imprisoned during the filling period. After the cylinder is filled with the preservative, pressure is applied, and the remainder of the treatment is the same as described for the Rueping treatment.

The Lowry process has the advantage that equipment for the full-cell process can be used without other accessories; the Rueping process usually requires additional equipment, such as an air compressor and an extra cylinder or Rueping tank for the preservative, or a suitable pump to force the preservative into the cylinder against the air pressure. Both processes, however, have advantages, and both are widely and successfully used.

Another treatment system uses methylene chloride as a carrier for pentachlorophenol with the Lowry process. Approximately 90 percent of the methylene chloride is recovered in the treating process, which leaves essentially dry pentachlorophenol in the wood. The removal of methylene chloride is accomplished by steam heating in a fairly long recovery cycle, usually over 12 hours.

With poles and other products where bleeding of preservative oil is objectionable, the empty-cell process is followed by either heating in the preservative (expansion bath) at a maximum temperature of 220 °F or a final steaming, for a specified time limit at a maximum temperature of 240 °F, prior to the final vacuum.

Treating Pressures and Preservative Temperatures

The pressures used in treatments vary from about 50 to 250 psi, depending on the species and the ease with which the wood takes the treatment; most commonly they range from about 125 to 175 psi. Many woods are sensitive to high treating pressures, especially when hot. AWPA standards, for example, permit a maximum pressure of 150 psi in the treatment of Douglas-fir, 125 psi for redwood, and 100 psi for western redcedar poles. In commercial practice even lower pressures are frequently used on such woods.

AWPA specifications commonly require that the temperature of creosote and creosote solutions during the pressure period shall be not more than 210 °F. Pentachlorophenol solutions may be applied at somewhat lower temperatures. Since high temperatures are much more effective than low temperatures for treating resistant wood, it is common practice to use average temperatures between 190 and 200 °F with creosote and creosote solutions. With a number of water-borne preservatives, however, especially those containing chromium salts, maximum temperatures are limited to 120 to 140 °F to avoid premature precipitation of the preservative.

Preservative Penetration and Retention

Penetration and retention requirements are equally impor-

tant in determining the quality of preservative treatment.

Penetrations vary widely, even in pressure-treated material. In most species, heartwood is more difficult to penetrate than sapwood. In addition, species differ greatly in the degree to which their heartwood may be penetrated. Incising tends to improve penetration of preservative in many refractory species, but those highly resistant to penetration will not have deep or uniform penetration even when incised. Penetrations in unincised heart faces of these species may occasionally be as deep as 1/4 inch, but often are not more than 1/16 inch.

Long experience has shown that even slight penetrations have some value, although deeper penetrations are highly desirable to avoid exposing untreated wood when checks occur, particularly for important members of high replacement cost. The heartwood of coast-type Douglas-fir, southern pine, and various hardwoods, while resistant, will frequently show transverse penetrations of 1/4 to 1/2 inch and sometimes considerably more.

Complete penetration of the sapwood should be the ideal in all pressure treatments. It can often be accomplished in small-size timbers of various commercial woods, and with skillful treatment it may often be obtained in piles, ties, and structural timbers. Practically, however, the operator cannot always ensure complete penetration of sapwood in every piece when treating large pieces of round material with thick sapwood, for example poles and piles. Specifications therefore permit some tolerance; for instance, AWPA Standard C4 on southern pine poles requires that 2.5 inches, or 85 percent of the sapwood thickness, be penetrated in not less than 18 out of 20 poles sampled in a charge. This applies only to the smaller class of poles. The requirements vary somewhat depending on the species, size, class, and specified retentions.

At one time all preservative retentions were specified in terms of the weight of preservative per cubic foot of wood treated, based on total weight of preservative retained and the total volume of wood treated in a charge. This is commonly called gauge retention. Federal specifications for most products, however, stipulate a minimum retention of preservative as determined from chemical analysis of borings from specified zones of the treated wood, known as results-type specification.

The preservatives and minimum retentions listed in Federal Specification TT−W−571 are shown in table 18−3. Because the figures given in this table are minimums, it may often be desirable to use higher retentions. Higher preservative retentions are justified in products to be installed under severe climatic or exposure conditions. Heavy-duty transmission poles and items such as structural timbers and house foundations, with a high replacement cost, are required to be treated to higher retentions. Correspondingly deeper penetration is also necessary for the same reasons.

It may be necessary to increase retentions to assure satisfactory penetration, particularly when the sapwood is either unusually thick or is somewhat resistant to treatment. To reduce bleeding of the preservative, however, it may be desirable to use preservative-oil retentions lower than the stipulated minimum. Treatment to refusal is usually specified for woods that are resistant to treatment and will not absorb sufficient preservative to meet the minimum retention requirements. However, such a requirement does not assure adequate penetration of preservative and cannot be considered as a substitute for more thorough treatment.

Nonpressure Processes

The numerous nonpressure processes differ widely in the penetrations and retentions of preservative attained and consequently in the degree of protection they provide to the treated wood. When similar retentions and penetrations are achieved, wood treated by a nonpressure method should have a service life comparable to that of wood treated by pressure. Nevertheless, results of nonpressure treatments, particularly those involving superficial applications, are not generally as satisfactory as pressure treatment. The superficial processes do serve a useful purpose when more thorough treatments are impractical or exposure conditions are such that little preservative protection is required.

Nonpressure methods, in general, consist of: (1) Superficial applications of preservative oils by brushing, or brief dipping; (2) soaking in preservative oils or steeping in solutions of waterborne preservatives; (3) diffusion processes with waterborne preservatives; (4) various adaptations of the thermal or hot-and-cold bath process; (5) vacuum treatment; and (6) a variety of miscellaneous processes.

Superficial Applications

The simplest treatment is to apply the preservative—creosote or other oils—to the wood with a brush or by dip. Oils that are thoroughly liquid when cold should be selected, unless it is possible to heat the preservative. The oil should be flooded over the wood, rather than merely painted upon it. Every check and depression in the wood should be thoroughly filled with the preservative, because any untreated wood left exposed provides ready access for fungi. Rough lumber may require as much as 10 gallons of oil per 1,000 square feet of surface, but surfaced lumber requires considerably less. The transverse penetrations obtained will usually be less than 1/10 inch although, in easily penetrated species, end grain (longitudinal) penetration is considerably greater.

Brush and dip treatments should be used only when more effective treatments cannot be employed. The additional life obtained by such treatments over that of untreated wood will be affected greatly by the conditions of service; for wood in contact with the ground, it may be from 1 to 5 years.

Dipping for a few seconds to several minutes in a preservative oil gives greater assurance (than brushing) that all surfaces and checks are thoroughly coated with the oil; usually it results in slightly greater penetrations. It is a common practice to treat window sash, frames, and other millwork, either before or after assembly, by dipping in a water-repellent preservative. Such treatment is covered by NWWDA IS 4–81 which also provides for equivalent treatment by the vacuum process. The amount of preservative used may vary from about 6 to 17 gallons per thousand board feet (0.5 to 1.5 pcf) of millwork treated.

The penetration of preservative into end surfaces of ponderosa pine sapwood is, in some cases, as much as 1 to 3 inches. End penetration in such woods as the heartwood of southern pine and Douglas-fir, however, is much less. Transverse penetration of the preservative applied by brief dripping is very shallow, usually only a few hundredths of an inch. Since the exposed end surfaces at joints are the most vulnerable to decay in millwork products, good end penetration is especially advantageous. Dip applications provide very limited protection to wood used in contact with the ground or under very moist conditions, and they provide very limited protection against attack by termites. They do have value, however, for exterior woodwork and millwork that is painted, that is not in contact with the ground, and that is exposed to moisture only for brief periods at a time.

Cold Soaking and Steeping

Cold soaking well-seasoned wood for several hours or days in low-viscosity preservative oils or steeping green or seasoned wood for several days in waterborne preservatives have provided varying success on fenceposts, lumber, and timbers.

Pine posts treated by cold soaking for 24 to 48 hours or longer, in a solution containing 5 percent of pentachlorophenol in No. 2 fuel oil, have shown an average life of 16 to 20 years or longer. The sapwood in these posts was well penetrated and preservative solution retentions ranged from 2 to 6 pcf. Most species do not treat as satisfactorily as the pines by cold soaking, and test posts of such woods as birch, aspen, and sweetgum treated by this method have failed in much shorter times.

Preservative penetrations and retentions obtained by cold soaking lumber for several hours are considerably better than those obtained by brief dipping of similar species. Preservative retentions, however, seldom equal those obtained in pressure treatment except in cases such as sapwood of pines that has become highly absorptive through mold and stain infection.

Steeping with waterborne preservatives has very limited use in the United States but has been employed for many years in Europe. In treating seasoned wood both the water and the preservative salt in the solution soak into the wood.

With green wood, the preservative enters the water-saturated wood by diffusion. Preservative retentions and penetrations vary over a wide range, and the process is not generally recommended when more reliable treatments are practical.

Diffusion Processes

In addition to the steeping process, diffusion processes are used with green or wet wood. These processes employ water-borne preservatives that will diffuse out of the water of the treating solution or paste into the water of the wood.

The double-diffusion process developed by the Forest Products Laboratory has shown very good results in fence post tests and 2 by 4 stake tests particularly on full-length immersion treatments. It consists of steeping green or partially seasoned wood first in one chemical solution and then in another (fig. 18–4). The two chemicals diffuse into the wood and then react to precipitate an effective preservative with high resistance to leaching. The process has had commercial application in cooling towers and for fence posts where preservative protection is needed to avoid early replacement.

Other diffusion processes involve applying preservatives to the butts or around the groundline of posts or poles. In standing-pole treatments the preservative may be injected into the pole at groundline with a special tool, applied on the pole surface as a paste or bandage (fig. 18–5), or poured into holes bored in the pole at the groundline. These treatments have recognized value for application to untreated standing poles and to treated poles where preservative retentions are determined to be inadequate.

Adaptations of Thermal Process

Hot-and-cold bath (referred to commercially as "thermal") treatment with coal-tar creosote or pentachlorophenol in heavy petroleum oil is one of the more effective of the nonpressure processes. The thoroughness of treatment in some cases approaches that of the pressure processes. The wood is heated in the preservative in an open tank for several hours, then quickly submerged in cold preservative and allowed to remain for several hours.

During the hot bath, the air in the wood expands and some is forced out. Heating the wood also improves the penetration of the preservative. In the cooling bath, the air in the wood contracts and a partial vacuum is created, so liquid is forced into the wood by atmospheric pressure. Some preservative is absorbed by the wood during the hot bath, but more is taken up during the cooling bath.

The chief use of the hot-and-cold process is for treating poles of some thin sapwood species, such as incised western redcedar and lodgepole pine, for utility poles (fig. 18–6). The process is also useful for fenceposts and for lumber or timbers for other purposes when circumstances do not permit the more effective pressure treatments. Coal-tar creosote and pentachlorophenol solutions are the preservatives ordinarily chosen for posts and poles. For the preservatives that cannot safely be heated, the process must be modified.

With coal-tar creosote, hot-bath temperatures up to 235 °F may be employed, but usually a temperature of 210 to 220 °F is sufficient. In the commercial treatment of cedar poles, temperatures of from 190 to 235 °F, for not less than 6 hours, are specified with creosote and pentachlorophenol solutions. In the cold bath or cooling bath the specified temperature is not less than 90 °F nor more than 150 °F for not less than 2 hours.

The immersion time in both baths must be governed by the ease with which the timber takes treatment. With well-seasoned timber that is moderately easy to treat, a hot bath of 2 or 3 hours and a cold bath of like duration is probably sufficient. Much longer periods are required with resistant woods. With preservative oils, the objective is to obtain as deep penetration as possible, but with a minimum amount of oil.

Preservative retentions are often very high in the hot-and-cold bath treatments of posts of woods such as southern pine, particularly if those posts contain molds, blue stain, and incipient decay. One method of reducing preservative retentions is to employ a final heating or "expansion" bath with the creosote at 200 to 220 °F for an hour or two, and to remove the wood while the oil is hot. This second heating expands the oil and air in the wood, and some of the oil is thus recovered. The expansion bath also leaves the wood cleaner than when it is removed directly from cold oil.

Vacuum Process

The vacuum process has been used to treat millwork with water-repellent preservatives and construction lumber with waterborne and water-repellent preservatives.

In treating millwork, the objective is to use a limited quantity of water-repellent preservative and obtain retentions and penetrations similar to those obtained by dipping for 3 minutes. The treatment is included in NWWDA, IS 4–81 for "Water-Repellent Preservative Nonpressure Treatment of Millwork." Here a quick, low initial vacuum is followed by brief immersion in the preservative, and then a high final or recovery vacuum. The treatment is advantageous over the 3-minute dip treatment because the surface of wood is quickly dried— thus expediting the glazing, priming, and painting operations. The vacuum treatment is also reported to be less likely than dip treatment to leave objectionably high retentions in bacteria-infected wood referred to as "sinker stock."

For buildings, lumber has been treated by the vacuum process, either with a waterborne preservative or a water-repellent pentachlorophenol solution, with preservative retentions usually lower than those required for pressure treatment. The process differs from that used in treating millwork in employing a higher initial vacuum and a long immersion or soaking period.

A study of the process by the Forest Products Laboratory employed an initial vacuum of 27.5 inches of mercury for 30 minutes, a soaking period of 8 hours, and a final or recovery vacuum of 27.5 inches for 2 hours. The study showed good penetration of preservative in the sapwood of dry lumber of easily penetrated species such as the pines; however, in heartwood and unseasoned sapwood of pine and heartwood of seasoned and unseasoned coast Dougas-fir, penetration was much less than that obtained in pressure treatment. Preservative retention was less controllable in vacuum than in empty-cell pressure treatment. Good control over retentions is possible, in vacuum treatment with a waterborne preservative, by adjusting concentration of the treating solution.

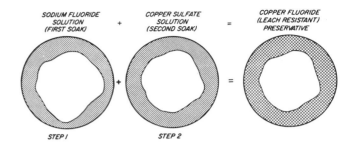

Figure 18-4—Double diffusion process steps.　　(M84 5705)

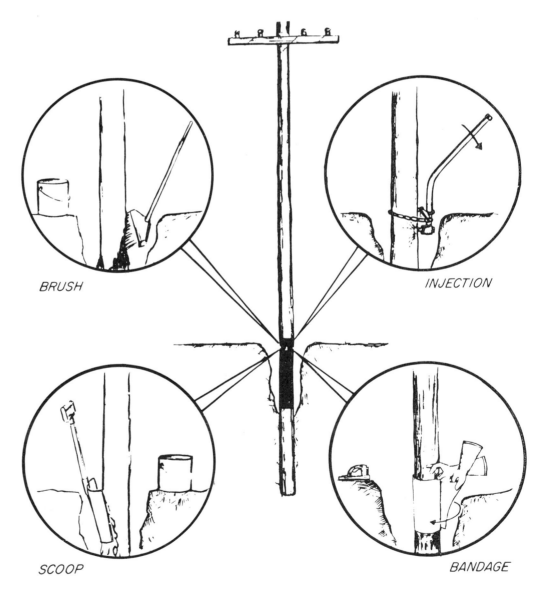

Figure 18-5—Methods of applying groundline treatments.

(ML84 5822)

Figure 18–6—A commercial plant for the hot- and
cold-bath (thermal) treatment of utility poles. (M130 272)

Miscellaneous Nonpressure Processes

A number of other nonpressure methods of various types
have been used to a limited extent. Several of these involve
the application of waterborne preservatives to living trees.
The Boucherie process for the treatment of green, unpeeled
poles has been used for many years in Europe. The process
involves attaching liquid-tight caps to the butt ends of the
poles. Then, through a pipeline or hose leading to the cap, a
waterborne preservative is forced into the pole under hydrostatic pressure.

A tire-tube process is a simple adaptation of the Boucherie
process used for treating green, unpeeled fenceposts. In this
treatment a section of used inner tube is fastened tightly
around the butt end of the post to make a bag that holds a
solution of waterborne preservative.

Effect of Treatment on Strength

Coal-tar creosote, creosote-coal-tar mixtures, creosote-
petroleum oil mixtures, and pentachlorophenol dissolved in
petroleum oils are practically inert to wood and have no
chemical influence that would affect its strength. Chemicals
commonly used in waterborne salt preservatives, including
chromium, copper, arsenic, and ammonia are reactive with
wood. Thus, they are potentially damaging to mechanical
properties and can cause corrosion of mechanical fasteners.
At retention levels required for protection in ground contact,
mechanical properties are essentially unchanged, except that
work to maximum load in bending, impact bending, and
toughness are reduced somewhat. Heavy salt loadings required
for marine application may reduce bending strength by 10
percent or more and work properties up to 50 percent.

Further reductions in mechanical properties may be observed
if the treating and subsequent drying processes are not controlled within acceptable limits. Factors that influence the
effect of the treating process on strength include (1) species
of wood, (2) size and moisture content of the timbers treated,
(3) heating medium used and its temperature, (4) length of
the heating period in conditioning the wood for treatment and
time the wood is in the hot preservative, and (5) amount of
pressure used. Most important of those factors are the severity and duration of the heating conditions used. The effect of
temperature on the strength of wood is covered in chapter 4.

Handling and Seasoning Timber After Treatment

Treated timber should be handled with sufficient care to
avoid breaking through the treated areas. The use of pikes,
cant hooks, picks, tongs, or other pointed tools that dig deeply
into the wood should be prohibited. Handling heavy loads of
lumber or sawed timber in rope or cable slings may crush the
corners or edges of the outside pieces. Breakage or deep
abrasions may also result from throwing the lumber or drop-
ping it. If damage results, the exposed places should be
retreated as thoroughly as conditions permit. Long storage of
treated wood before installation should be avoided because
such storage encourages deep and detrimental checking and
may also result in significant loss of some preservatives.
Treated wood that must be stored before use should be covered for protection from the sun and weather.

Although cutting wood after treatment is highly undesirable,
it cannot always be avoided. When cutting is necessary, the
damage may be partly overcome in timber for land or fresh-
water use by a thorough application to the cut surface of a
grease containing 10 percent pentachlorophenol. This provides a protective reservoir of preservative on the surface,
some of which may slowly migrate into the end grain of the
wood. Thoroughly brushing the cut surfaces with two coats
of hot creosote is also helpful, although brush coating cut
surfaces gives little protection against marine borers. A special device is available for pressure treating bolt holes bored
after treatment. For wood treated with waterborne preservatives, where the use of creosote or pentachlorophenol solution on the cut surfaces is not practicable, a 5 percent solution
of the waterborne preservative in use may be substituted.

For treating the end surfaces of piles where they are cut off
after driving, at least two generous coats of creosote should
be applied. A coat of asphalt or similar material may be well

applied over the creosote, followed by some protective sheet material, such as metal, roofing felt, or saturated fabric, fitted over the pile head and brought down the sides far enough to protect against damage to the top treatment and against the entrance of storm water. AWPA Standard M4 contains instructions for the care of pressure-treated wood after treatment.

Wood treated with preservative oils should generally be installed as soon as practicable after treatment, but sometimes cleanliness of the surface can be improved by exposure to the weather for a limited time before use. Waterborne preservatives or pentachlorophenol in a volatile solvent, however, are best suited to uses where cleanliness or paintability are of great importance.

With waterborne preservatives, seasoning after treatment is important for wood to be used in buildings or other places where shrinkage after placement in the structure would be undesirable. Injecting waterborne preservatives puts large amounts of water into the wood, and considerable shrinkage is to be expected as subsequent seasoning takes place. For best results, the wood should be dried to approximately the moisture content it will ultimately reach in service. During drying, the wood should be carefully piled and, whenever possible, restrained by sufficient weight on the top of the pile to avoid warping.

With some waterborne preservatives, seasoning after treatment is recommended for all treated wood. During this seasoning period, volatile chemicals escape and the chemical reactions are completed within the wood; thus, the resistance of the preservative to leaching by water is increased.

Quality Assurance for Treated Wood

Treating Conditions and Specifications

Specifications on the treatment of various wood products by pressure processes and on the hot-and-cold bath (thermal) treatment of cedar poles have been developed by AWPA. These specifications limit pressures, temperatures, and time during conditioning and treatment to avoid conditions that will cause serious injury to the wood. They also contain minimum requirements as to preservative penetrations and retentions and recommendations for handling wood after treatment, to provide a quality product.

The specifications are rather broad in some respects, allowing the purchaser some latitude in specifying the details of his individual requirements. The purchaser should exercise great care, however, not to limit the operator of the treating plant so he cannot do a good treating job, and not to require treating conditions so severe that they will damage the wood. Federal Specification TT−W−571 lists treatment practices for use on U.S. Government orders for treated wood products; other purchasers have specifications similar to those of AWPA.

Inspection

Inspection of timber for quality and grade before treatment is desirable. Grademarked lumber, plywood, and timber can be obtained in many instances. When inspection prior to treatment is impractical, the purchaser can usually inspect for quality and grade after treatment; if this is to be done, however, it should be made clear in the purchase order.

Currently, the inspection of treatment of complete charges is generally specified at the time of treatment at the treating plant; however, the option is generally available whereby the purchaser could determine the quality of treatment from selected samples of the treated product at destination or within a specified time after treatment. The purchaser should recognize, however, that a sample selected from a charge or lot at the treating plant is likely to be different from a sample taken at destination from a few items from a much larger lot of treated material. Furthermore, the nature and quantity of the preservative in the wood change as the period of service increases, so samples of treated wood taken at treatment may not be the same as those taken later. Destination inspection requires consideration of these questions and the details are spelled out in Section 4 of Federal Specification TT−W−571.

The treating industry, with the assistance of the Federal Housing Administration and the Forest Products Laboratory, has developed a quality-control and grademarking program for treated products, such as lumber, timbers, plywood, and marine piles. This quality-control program, administered through the American Wood Preservers' Bureau, promises to assist the user in securing well-treated material; otherwise, the purchaser must either accept the statements or certificate of the treating-plant operator or have an inspector at the treating plant to inspect the treated products and ensure compliance with the specifications. Railroad companies and other corporations that purchase large quantities of treated timber usually maintain their own inspection services. Commercial inspection and consulting service is available for purchasers willing to pay an inspection fee but not using enough treated timber to justify employing inspectors of their own. Experienced, competent, and reliable inspectors can assure compliance with material and treating standards and thus reduce risk of premature failure of the material.

Penetration measurements should be made at the treating plant if inspection service is provided, but can be made by the purchaser at a designated time after the timber has been treated. They give about the best single measure of the thoroughness of the treatment. It is also important that all treated material be branded, dye stamped, or labeled in accordance with AWPA Standard C1.

The depth of penetration of creosote and other dark-colored preservatives can be determined directly by observing a core removed by an increment borer. The core should usually be taken at about midlength of the piece, or at least several feet from the end of the piece, to avoid the unrepresentative end portion that is sometimes completely treated by end penetration. Since preservative oils tend to creep over cut surfaces, the observation should be made promptly after the borer core is taken. Holes made for penetration measurements should be tightly filled with thoroughly treated wood plugs.

The penetration of preservatives that are practically colorless must be determined by chemical dips or sprays that show the penetration by color reactions.

How to Purchase Treated Wood

To obtain a treated wood product of high quality the purchaser should use the appropriate specifications. Specifications and standards of importance here are: Federal Specification TT−W−571, "Wood Preservation—Treating Practices"; F.S. TT−W−572, "Wood Preservation—Water Repellent"; Official Quality Control Standards of the American Wood Preservers' Bureau; and the American Wood-Preservers' Association Book of Standards. The inspection of material for conformity to the minimum requirements listed in the above specifications should be in accordance with the American Wood Preservers' Standard M2, "Standard for Inspection of Treated Timber Products."

Selected References

American Wood-Preservers' Association. Annual proceedings. (Reports of Preservations and Treatment Committees contain information on new wood preservatives considered in the development of standards.) Stevensville, MD: AWPA; (see current edition).

American Wood-Preservers' Association. Book of Standards. (Includes standards on preservatives, treatments, methods of analysis, and inspection.) Stevensville, MD: AWPA; (see current edition).

American Wood-Preservers' Bureau. Quality Control Standards. (Includes commodities such as building poles, piling, lumber, timber and plywood.) Arlington, VA: AWPB; (see current edition).

Baechler, R. H.; Blew, J. O.; Roth, H. G. Studies on the assay of pressure-treated lumber. Proceedings of AWPA. 58: 21−34; 1962.

Baechler, R. H.; Gjovik, L. R.; Roth, H. G. Assay zones for specifying preservative-treated Douglas-fir and southern pine timbers. Proceedings of AWPA. 65: 114−123; 1969.

Baechler, R. H.; Gjovik, L. R.; Roth, H. G. Marine tests on combination-treated round and sawed specimens. Proceedings of AWPA. 66: 249−257; 1970.

Baechler, R. H.; Roth, H. G. The double-diffusion method of treating wood: A review of studies. Madison, WI: Forest Products Journal. 14(4): 171−178; 1964.

Blew, J. O.; Davidson, H. L. Preservative retentions and penetration in the treatment of white fir. Proceedings of AWPA. 67: 204−221; 1971.

Boone, R. S.; Gjovik, L. R.; Davidson, H. L. Treatment of sawn hardwood stock with double-diffusion and modified double-diffusion methods. Res. Pap. FPL 265. Madison, WI: U.S. Department of Agriculture, Forest Service, Forest Products Laboratory; 1976.

Gaby, L. I.; Gjovik, L. R. Treating and drying composite lumber with waterborne preservatives: Part I. Short specimen testing. Madison, WI: Forest Products Journal. 34(2) 23−26; 1984.

Gjovik, L. R. Pretreatment molding of southern pine: Its effect on permanence and performance of preservatives exposed in sea water. Proceedings of AWPA. 73: 142−153; 1977.

Gjovik, L. R. Treatability of southern pine, Douglas-fir and Engelmann spruce heartwood with ammoniacal copper arsenate and chromated copper arsenate. Proceedings of AWPA. 79: 18−30; 1983.

Gjovik, L. R. Wood preservatives and the environment: Treated wood. Proceedings of AWPA. 70: 114−115; 1974.

Gjovik, L. R.; Baechler, R. H. Selection, production, procurement and use of preservative treated wood. Gen. Tech. Rep. FPL−15. Supplementing Federal Specification TT−W−571. Madison, WI: U.S. Department of Agriculture, Forest Service, Forest Products Laboratory; 1977.

Gjovik, L. R.; Baechler, R. H. Treated wood foundations for buildings. Madison, WI: Forest Products Journal. 20(5): 45−48; 1970.

Gjovik, L. R.; Davidson, H. L. Comparison of wood preservatives in Mississippi post study. Res. Note FPL−01. Madison, WI: U.S. Department of Agriculture, Forest Service, Forest Products Laboratory; (see current edition).

Gjovik, L. R.; Davidson, H. L. Service records on treated and untreated posts. Res. Note FPL−068. Madison, WI: U.S. Department of Agriculture, Forest Service, Forest Products Laboratory; 1975.

Gjovik, L. R.; Gutzmer, D. I. Comparison of wood preservatives in stake tests. Res. Note FPL−02. Madison, WI: U.S. Department of Agriculture, Forest Service, Forest Products Laboratory; (see current edition).

Gjovik, L. R.; Micklewright, J. T. Wood preservation: How important is the wood preserving industry. Southern Lumberman. 243 (3028); 1982.

Gjovik, L. R.; Roth, H. G.; Davidson, H. L. Treatment of Alaskan species by double-diffusion and modified double-diffusion methods. Res. Pap. FPL 182. Madison, WI: U.S. Department of Agriculture, Forest Service, Forest Products Laboratory; 1972.

Gjovik, L. R. et al. Biologic and Economic Assessment of Pentachlorophenol, Inorganic Arsenicals, and Creosote. Volume I: Wood preservatives. Tech. Bull 1658−1. Washington, DC: U.S. Department of Agriculture, in cooperation with State Agricultural Experiment Stations, Cooperative Extension Service, other state agencies and the Environmental Protection Agency. 1980.

Hunt, George M.; Garratt George A. Wood preservation. 3rd ed. The American Forestry Series. New York: McGraw-Hill; 1967.

Johnson, B. R.; Gutzmer, D. I. Marine exposure of preservative-treated small wood panels. Res. Pap. FPL 399. Madison, WI: U.S. Department of Agriculture, Forest Service, Forest Products Laboratory, 1981.

Micklewright, J. T.; Gjovik, L. R. Wood preserving statistics: Update. Proceedings of AWPA. 77: 143−147; 1981.

National Forest Products Association. The all-weather wood foundation. NFPA Tech. Rep. 7; Washington, DC; (see current edition).

National Forest Products Association. All-weather wood foundation system, design fabrication installation manual. NFPA report; Washington DC; (see current edition).

National Woodwork Manufacturers Association. Industry standard for water-repellent preservative treatment for millwork. IS 4−81. Chicago, IL: NWMA; 1981.

U.S. Federal Supply Service. Wood preservation treating practices. Federal Specification TT−W−571; USFSS, Washington, DC; (see current edition).

U.S. Federal Supply Service. Wood preservatives: Water-repellent. Federal Specification TT−W−572; USFSS, Washington, DC; (see current edition).

Chapter 19

Round Timbers and Ties

Round Timbers and Ties*

Round timbers and ties represent some of the most efficient uses of our forest resources. They require a minimum of processing from harvesting the tree to the marketing of a structural commodity. Poles and piles are debarked or peeled, seasoned, and often treated with preservative prior to use as structural members. Construction logs are usually shaped to facilitate construction. Ties, used for railroads, landscaping, and mining, are slab-cut to provide flat surfaces. The relative economy of these products, as compared to potential alternatives, results in their popular use nationwide.

Standards and Specifications

Material standards and specifications listed in table 19—1 have been created by the joint efforts of producers and users to ensure compatibility between product quality and end use. These guidelines include recommendations for production, treatment, and engineering design. They are updated periodically to conform to changes in material and design technology.

Material Requirements

Round timber and tie material requirements vary with intended use. The ability to meet these requirements varies with species; thus species selection is important.

Five main factors that should be considered include availability, form, weight, durability, and strength. Availability reflects the economic feasibility of procuring members of the required size and grade. Form or physical appearance is often the basis for judging acceptability. Weight affects shipping and handling costs and is a function of volume, moisture content, and wood density. Durability is directly related to expected service life and is a function of treatability as well as natural decay resistance. Finally, regardless of the application, any structural member must be strong enough to resist imposed loads with a reasonable factor of safety. Material specifications available for most applications of round timbers and ties contain guidelines for evaluating these factors.

Availability

Material evaluation begins with an assessment of availability. For some applications, local species of timber may be readily available. Many applications of round timbers and ties, however, involve quality and quantity requirements that cannot easily be met with local species. These requirements vary with the application.

Poles

Most structural applications of poles require timbers that are relatively straight and free of large knots. Poles used to support electric utility distribution and transmission lines (fig. 19—1) range in length from 30 to 125 feet and must be readily available in large quantities. Those used for pole buildings rarely exceed 30 feet in length.

Hardwood species can be used for poles when the trees are of suitable size and form; their use is limited, however, by their weight, excessive checking, and by lack of experience in preservative treatment of hardwoods. Thus, most poles are softwoods.

The southern pines (principally loblolly, longleaf, shortleaf, and slash) account for the highest percentage of poles treated in the United States. The thick and easily treated sapwood of these species, their favorable strength properties and form, and their availability in popular pole sizes over a wide area account for their extensive use. In longer lengths, southern pine poles are in limited supply so Douglas-fir, and to some extent western redcedar, are used to meet requirements for 50-foot and longer transmission poles.

Douglas-fir is used throughout the United States for transmission poles and is used in the Pacific Coast region for distribution and building poles. Because the heartwood of Douglas-fir is resistant to preservative penetration and has limited decay and termite resistance, serviceable poles need a well-treated shell of sapwood that is free of checking. To minimize checking after treatment, poles should be adequately seasoned or conditioned before treatment. With these precautions the poles should compare favorably with treated southern pine poles in serviceability.

A small percentage of the poles treated in the United States are of western redcedar, produced mostly in British Columbia. The number of poles of this species used without treatment is not known but is considered to be small. Used primarily for utility lines in the northern and western United States, well-treated redcedar poles give a service life that compares favorably with other pole species and could be used effectively in pole type buildings.

Lodgepole pine is also used in small quantities for treated poles. This species is used both for utility lines and for pole-type buildings. It has a good service record when well-treated. Special attention is necessary, however, to obtain poles with sufficient sapwood thickness to ensure adequate penetration of preservative, since the heartwood is not usually penetrated and is not decay resistant. The poles must also be well seasoned prior to treatment to avoid checking and exposure of unpenetrated heartwood to attack by decay fungi.

Western larch poles produced in Montana and Idaho came into use after World War II because of their favorable size,

---

* Revision by Ronald Wolfe, General Engineer.

shape, and strength properties. Western larch requires preservative treatment full length for use in most areas and, as in the case of lodgepole pine poles, must be selected for adequate sapwood thickness and must be well seasoned prior to treatment.

Other species occasionally used for poles are listed in the ANSI 0 5.1 standard (American National Standards Institute). These minor species make up a very small portion of pole production and are used locally.

Glued laminated or glulam poles are also available for use where special sizes or shapes are required. Standard ANSI 0 5.2 provides guidelines for specifying these poles.

Piles

Material available for timber piles is more restricted than that for poles. Most timber piles used in the eastern half of the United States are southern pine, while those used in the western United States are coast Douglas-fir. Oak, red pine, and cedar piles are also referenced in timber pile literature but are not as widely used as southern pine and Douglas-fir.

Construction Logs

Availability is a minor problem for construction logs. Any species that produces straight log sections longer than 12 feet with diameters exceeding 6 inches can be used for log buildings. Strength, length, and treatability are not as important for construction logs as they are for poles and piles. Availability concerns of log home manufacturers primarily involve stumpage and handling costs. Volume builders use species that can be obtained in large quantities. Small build-

Figure 19–1—Round timber poles form the major structural element in these transmission structures. (Photo courtesy of Koppers Co.) (MC84 9064)

Table 19–1—*Standards and specifications for round timbers and ties*

	Material requirements	Preservative treatment	Engineering design stresses	
			Procedures	Design values
Utility poles	ANSI O5.1	TT–W–571 AWPA C1,C4,C8,C10	—	ANSI O5.1
Construction poles	ANSI O5.1	TT–W–571 AWPA C23	ASTM D 3200	ASAE EP 388
Piles	ASTM D 25	TT–W–571 AWPA C1,C3	ASTM D 2899	NDS
Construction logs	(See material supplier)	—	ASTM D 3957	(See material supplier)
Ties	AREA	TT–W–571 AWPA C2,C6 AREA	—	AREA

Figure 19–2—Log homes are made in a variety of shapes and sizes. For most prefabricated log homes, the logs are machined to give uniform section along their length, resulting in easier construction, improved resistance to air infiltration, and better structural interaction.

(M84 0375-12)

ers and do-it-yourselfers, however, often use locally available species (fig. 19–2).

Ties

The most important availability consideration for railroad crossties is quantity. Ties are produced from most native species of timber that yield log lengths greater than 8 feet with diameters greater than 7 inches. The American Railway Engineering Association (AREA) lists 26 U.S. species that may be used for ties. Thus, the tie market provides a use for many low-grade hardwood and softwood logs.

Form

Natural growth properties of trees play an important role in their use as structural round timbers. Three important form considerations are cross-section dimensions, straightness, and the presence of surface characteristics such as knots.

Poles and Piles

Standards for poles and piles have been written with the assumption that trees have a round cross section with a circumference that decreases linearly with height. Thus, the shape of a pole or pile is often assumed as the frustum of a cone. Actual measurements of tree shape indicate that taper is rarely linear and often varies with location along the height of the tree. Average taper values from the ANSI 0 5.1 standard are shown in table 19–2 for the more popular pole species. Guidelines to account for the effect of taper on the location of the critical section above groundline are given in ANSI 0 5.1. The standard also tabulates pole dimensions for up to 15 size classes of 11 major pole species.

Taper also affects construction detailing of pole buildings. Where siding or other exterior covering is applied, poles are generally set with the taper to the interior side of the structures to provide a vertical exterior surface (fig. 19–3). Another common practice is to modify the round poles by slabbing to provide a continuous flat face. The slabbed face permits more secure attachment of sheathing and framing members and facilitates the alignment and setting of intermediate wall and corner poles. The slabbing consists of a minimum cut to provide a single continuous flat face from groundline to top of intermediate wall poles and two continuous flat faces at right angles to one another from groundline to top of corner poles. It should be recognized that preservative penetration is generally limited to the sapwood of most species; therefore slabbing, particularly in the groundline area of poles with thin sapwood, may result in somewhat less protection than that of an unslabbed pole. All cutting and sawing should be confined to that portion of the pole above groundline and should be performed before treatment.

ASTM D 25 provides tables of pile sizes for either friction piles or end-bearing piles. Friction piles rely on skin friction rather than tip area for support, whereas end-bearing piles resist compressive force at the tip. For this reason, a friction pile is specified by butt circumference and may have a smaller tip than an end-bearing pile. Conversely, end-bearing piles are specified by tip area and butt circumference is minimized.

Straightness of poles or piles is determined by two form properties—sweep and crook. Sweep is a measure of bow or gradual deviation from a straight line joining the ends of the pole or pile. Crook is an abrupt change in direction of the centroidal axis. Limits on these two properties are specified in both ANSI 05.1 and ASTM D 25.

Construction Logs

ASTM D 3957 is a guide for establishing stress grades for construction logs used in log buildings. These structural mem-

Figure 19–3—Poles are economic foundation and wall systems for agricultural and storage buildings.   (M84 0327)

Figure 19–4—Construction logs can be formed in a variety of shapes for log homes. Vertical surfaces may be varied for esthetic purposes, while the horizontal surfaces generally reflect structural and thermal considerations.   (ML84 5809)

bers may have a variety of cross-sectional shapes (fig. 19–4) and they are usually milled so that their shape is uniform along their length. Due to the variety of shapes, the standard recommends stress grading on the basis of the largest rectangular section that can be inscribed totally within the log section. The standard also provides commentary on the effects of knots and slope of grain.

Logs are also used as stringers for bridges in remote logging areas. Although no consensus standard is available for specifying and designing log stringers, a design guide has been prepared by the USDA Forest Service.

Table 19–2—*Circumference taper (inch change in circumference per foot)* [1]

Western redcedar	0.38
Ponderosa pine	.29
Jack, lodgepole, and red pine	.30
Southern pine	.25
Douglas-fir, larch	.21
Western hemlock	.20

[1] Taken from ANSI 05.1.

Ties

As with construction logs, railroad ties are commonly shaped to a fairly uniform section along their length. The AREA publishes specifications for the sizes, which include seven size classes ranging from 5 by 5 inches to 7 by 10 inches. These tie classes may be ordered in any of three standard lengths—8 feet, 8-1/2 feet, or 9 feet.

Weight and Volume

The weight of round timber depends on the species, size, moisture content, and preservative treatment. Weights per cubic foot of the various species of wood may be estimated from information given in chapter 3.

Many models have been developed for estimating the volume of round timbers. Two methods given in American Wood-Preservers' Association (AWPA) Standard F3 include:

$$V = 3L\left(\frac{C_m}{\pi}\right)^2 0.001818 \qquad (19-1)$$

where V is volume in cubic feet, L is length in feet, and C_m is midlength circumference in inches, and

$$V = 0.001818L(D^2 + d^2 + Dd) \qquad (19-2)$$

where D is the top diameter and d is the butt diameter. V and L definitions are the same as for formula (19−1). If equation (19−2) is used, the volume (V) obtained for certain species must be multiplied by a correction factor, as indicated below:

Oak piles	0.82
Southern pine piles	.93
Southern pine and red pine poles	.95

Formula (19−1) is the AWPA official method, except for Douglas-fir for which either method can be used. Volume tables for both methods are given in AWPA Standard F3.

The volume of a round timber shows little difference whether green or dry. Drying of round timbers causes checks to open, but there is little reduction of the gross diameter of the pole.

The volume of shaped timbers such as construction logs and crossties is simply the cross-sectional area times the length.

Wood weight must be adjusted to account for the effects of preservative treatment and moisture content. Recommended preservative retentions are listed in table 18−3 in pounds per cubic foot. Thus, knowing the volume, the preservative weight can be approximated by multiplying volume by the recom-

mended preservative retention. The weight of moist wood is given in chapter 3. This weight can also be determined by multiplying the approximate dry weight by one plus the moisture content (decimal equivalent).

Durability

For most applications of round timber, durability is primarily a question of decay resistance. Some species are noted for their natural decay resistance; however, even these may require preservative treatment, depending upon the environmental conditions under which the material is used and the required service life. For some applications, natural decay resistance is sufficient. This is the case for temporary piles, marine piles in fresh water entirely below the permanent water level, and construction logs used in building construction. Any construction logs in ground contact should be pressure-treated, and logs within two or three levels above a concrete foundation should be brush-treated with a waterborne salt solution.

Preservative Treatment

Federal Specification TT−W−571 covers the inspection and treatment requirements for various wood products including poles, piles, and ties. This specification refers to the AWPA Standards C1 and C3 for pressure treatment, C2 and C6 for treatment of ties, C8 for full-length thermal (hot and cold) treatment of western redcedar poles, C10 for full-length thermal (hot and cold) treatment of lodgepole pine poles, and C23 for pressure treatment of construction poles. The AREA specifications for crossties and switch ties also cover preservative treatment. Retention and types of various preservatives recommended for various applications are given in table 18−3.

Inspection and treatment of poles in service has been effective in prolonging the useful life of untreated poles and those with inadequate preservative penetration or retention. The Department of Forest Products at Oregon State University has published guidelines for developing an inservice pole maintenance program.

Service Life

Service conditions for round timbers and ties vary from mild for construction logs to severe for crossties. Construction logs used in log homes may last indefinitely if kept dry and properly protected from insects. Most railroad ties, on the other hand, are continually in ground contact and are subject to mechanical damage.

Poles

The life of poles can vary within wide limits, depending upon properties of the pole, preservative treatments, service conditions, and maintenance practices. In distribution or transmission line supports, however, service life is often limited by obsolescence of the line rather than the physical life of the pole.

It is common to report the "average" life of untreated or treated poles based on observations over a period of years. These "average life" values are useful as a rough guide to what physical life may be expected from a group of poles, but it should be kept in mind that, within a given group, 60 percent of the poles will have failed before reaching an age equal to the "average life."

Early or premature failure of treated poles can generally be attributed to one or more of three factors: (1) Poor penetration and distribution of preservative; (2) an inadequate retention of preservative; or (3) use of a substandard preservative.

Western redcedar is one species recognized as having a naturally decay-resistant heartwood. If used without treatment, however, the average life is somewhat less than 20 years. Properly treated southern pine may last 35 years or longer.

Piles

The expected life of a pile is also determined by treatment and use. Wood that remains completely submerged in water does not decay although bacteria may cause some degradation; therefore, decay resistance is not necessary in piles so used, but it is necessary in any part of the piles that may extend above the permanent water level. When piles that support the foundations of bridges or buildings are to be cut off above the permanent water level, they should be treated to conform to recognized specifications such as Federal Specification TT−W−571 and AWPA Standards C1 and C3. The untreated surfaces exposed at the cutoffs should also be given protection by thoroughly brushing the cut surface with coal-tar creosote. A coat of pitch, asphalt, or similar material may then be applied over the creosote and a protective sheet material, such as metal, roofing felt, or saturated fabric, fitted over the pile cut-off in accordance with AWPA Standard M4. Correct application and maintenance of these materials are critical in maintaining the integrity of piles.

Piles driven into earth that is not constantly wet are subject to about the same service conditions as apply to poles, but are generally required to last longer. Preservative retention requirements for piles are therefore higher than for poles (see table 18−3).

Piles used in salt water are subject to destruction by marine borers even though they do not decay below the waterline. The best practical protection against marine borers has been a treatment first with a waterborne preservative, followed after seasoning with a creosote treatment. Other preservative treatments of marine piles are covered in Federal Specification TT−W−571 and AWPA Standard C3 (table 18−3).

Ties

The life of ties in service depends on their ability to resist decay and mechanical destruction. Under sufficiently light traffic, heartwood ties of naturally durable wood, even if of low strength, may give 10 or 15 years' average service without preservative treatment; under heavy traffic without adequate mechanical protection the same ties might fail in 2 or 3 years. Advances in preservatives and treatment processes, coupled with increasing loads, are shifting the primary cause of tie failure from decay to mechanical damage. Well-treated ties, properly designed to carry intended loads, should last from 25 to 40 years on the average. Records on life of treated and untreated ties are occasionally published in the annual proceedings of AREA and AWPA.

Strength Properties

Allowable strength properties of round timbers have been developed and published in several standards. In most cases, published values are based on strength of small clear test samples. Allowable stresses are derived by adjusting small clear values for effects of growth characteristics, conditioning, shape, and load conditions as discussed in applicable standards. In addition, published values for some species of poles and piles reflect results of full-size tests.

Poles

Most poles are used as structural members in support structures for distribution and transmission lines. For this application, poles may be designed as single-member or guyed cantilevers or as structural members of a more complex structure. Specifications for wood poles used in single pole structures have been published by ANSI in Standard 05.1, "Specifications and Dimensions of Wood Poles." Guidelines for the design of pole structures are given in the ANSI C2 National Electric Safety Code (NESC).

The ANSI 05.1 standard gives values for fiber stress in bending for species commonly used as transmission or distribution poles. These values represent the near-ultimate fiber stress for poles used as cantilever beams. For most species, these values are based partly on full-sized pole tests and include adjustments for moisture content and pretreatment conditioning. The values in ANSI 05.1 are compatible with the ultimate strength design philosophy of the NESC, but they are not compatible with the working stress design philosophy of the National Design Specification (NDS).

Reliability-based techniques are being developed for the design of distribution-transmission line systems. This approach requires a strong data base on the performance of pole structures; however, it should result in more efficient use of poles.

A second use of poles is in the construction of buildings. These buildings are primarily farm structures and industrial buildings where the poles serve as primary skeletal members to form the walls and support the roof system. Specifications for wood poles for farm buildings are given by the American Society of Agricultural Engineers (ASAE) in their Standard EP 388, "Design Properties of Round, Sawn, and Laminated Preservatively Treated Construction Poles."

The ASAE standard lists design stress values that have been adjusted to account for variability, seasoning, and environmental effects as well as factor of safety.

Piles

Bearing loads on piles are sustained by earth friction along the sides of the pile, by bearing of the tip on a solid stratum, or by a combination of the two. Wood piles, because of their tapered form, are particularly efficient in supporting loads by side friction. Bearing values that depend upon side friction are related to the stability of the soil and generally do not approach the ultimate strength of the pile. Where wood piles sustain foundation loads by bearing of the tip on a solid stratum, loads may be limited by the compressive strength of the wood parallel to the grain. If a large proportion of the length of a pile extends above ground, its bearing value may be limited by its strength as a long column. Side loads may also be applied to piles extending above ground. In such instances, however, bracing is often used to reduce the unsupported column length or to resist the side loads.

The most critical loads on piles often occur during driving. Under hard driving conditions, piles that are too dry (less than 18 percent moisture content at 2-in depths) have been observed to literally explode under the force of the driving hammers. Steel banding is recommended as a means of increasing resistance to splitting, and driving the piles into predrilled holes reduces driving stresses.

The reduction in strength of a wood column resulting from crooks, eccentric loading, or any other condition that will result in combined bending and compression is not so great as would be predicted using the NDS interaction equations. This does not imply that crooks and eccentricity should be without restriction, but it should relieve anxiety as to the influence of crooks, such as those found in piles.

Design procedures for eccentrically loaded columns are given in chapter 8.

There are several ways of determining bearing capacity of piles. Engineering formulas can be used for estimating bearing values from the penetration under blows of known energy from the driving hammer. Some engineers prefer to estimate bearing capacity from experience or observation of the behav-

ior of pile foundations under similar conditions or from the results of static-load tests.

Working stresses for piles are governed by building code requirements and by recommendations of ASTM D 2899. This standard gives recommendations for adjusting small clear strength values listed in ASTM D 2555 for use in the design of full-size piles. In addition to adjustments for properties inherent to the full-sized pile, the D 2899 standard also provides recommendations for adjusting allowable stresses for the effects of pretreatment conditioning.

Design stresses for timber piles are tabulated in the National Design Specification (NDS) for wood construction. The NDS values include adjustments for the effects of moisture content, load duration, and preservative treatment. Recommendations are also given to adjust for lateral support conditions and factor of safety.

Construction Logs

ASTM D3957−80 provides a method of establishing stress grades for structural members of any of the more common log configurations. Manufacturers can use this standard to develop grading specifications and derive engineering design stresses for their construction logs.

Ties

Railroad cross and switch ties have historically been overdesigned from the standpoint of rail loads. Tie life depended largely upon deterioration rather than mechanical damage. However, due to advances in decay-inhibiting treatment and increased axle loads, adequate structural design is becoming more important in reducing tie replacement costs.

Rail loads induce stresses in bending and shear as well as in compression perpendicular to the grain in railroad ties. The AREA manual gives recommended limits on ballast bearing pressure and allowable stresses for crossties. This information may be used by the designer to determine adequate tie size and spacing to avoid premature failure due to mechanical damage.

Specific gravity and compressive strength parallel to the grain are also important properties to consider in evaluating crosstie material. These properties indicate the wood's resistance to both pullout and lateral thrust of spikes.

Selected References

General

American Wood-Preservers' Association. Book of standards (includes standards on pressure and thermal treatment of poles, piles, and ties). (American Wood-Preserver's Bureau official quality control standards.) Bethesda, MD: AWPA; (see current edition).

Canadian Department of Forestry. The air seasoning of timbers, poles, and ties. Canadian Department of Forestry Publ. No. 1030. Vancouver, BC, Canada; 1963.

U. S. Federal Supply Service. Poles and piles, wood. Federal specification MM−P−371c−ties, railroad (cross and switch); Federal Specification MM−T−371d−wood preservation: treating practice; Federal Specification TT−W−571. Washington, DC: USFSS; (see current edition).

Poles

American National Standards Institute. Specifications and dimensions for wood poles. ANSI 0 5.1; New York: ANSI; (see current edition).

American National Standards Institute. National electrical safety code. ANSI C 2. New York: Institute of Electrical and Electronics Engineering, Inc.; 1981.

American National Standards Institute. Structural glued laminated timber for utility structures. ANSI 0 5.2; New York: ANSI; (see current edition).

American Society of Agricultural Engineers. Design properties of round, sawn, and laminated preservatively treated construction poles. ASAE EP 388. In: 1982 Agricultural engineers yearbook. St Joseph, MI: ASAE; 1982: 301−302.

American Society for Testing and Materials. Standard specification and methods for establishing recommended design stresses for round timber construction poles. ASTM D 3200. Philadelphia, PA: ASTM; (see current edition).

American Society for Testing and Materials. Standard methods of static tests of wood poles. ANSI/ASTM D 1036−58. Philadelphia, PA: ASTM; (reapproved 1978).

Electric Power Research Institute. Probability-based design of wood transmission structures.
a. Volume 1: Strength and stiffness of wood utility poles.
b. Volume 2: Analysis and probability-based design of wood utility structures.
c. Volume 3: Users manual. Poleda-80. Pole design and analysis. Prepared by Research Institute of Colorado for EPRI; EL−2040, vols. 1−3, Proj. 1352−1; Sept. 1981.

Graham, Robert D.; Helsing, G. G. Wood pole maintenance manual: Inspection and supplemental treatment of Douglas-fir and western redcedar poles. Res. Bull. 24. Corvallis, OR: Oregon State University, Forest Research Laboratory; 1979.

Patterson, Donald. Pole building design. Washington, DC: American Wood Preservers Institute; 1969.

Thompson, Warren S. Effect of steaming and kiln drying on properties of southern pine poles—Part I—mechanical properties. Madison, WI: Forest Products Journal. 19(1): 21−28; 1969.

Wood, L. W.; Erickson E. C. O.; Dohr, A. W. Strength and related properties of wood poles. Philadelphia PA: American Society for Testing and Materials; 1960.

Wood, L. W.; Markwardt, L. J. Derivation of fiber stresses from strength values of wood poles. Res. Pap. FPL 39. Madison, WI: U.S. Department of Agriculture, Forest Service, Forest Products Laboratory; 1965.

Piles

American Society for Testing and Materials. Standard specification for round timber piles. ASTM D 25. Philadelphia, PA: ASTM; (see latest edition).

American Society for Testing and Materials. Establishing design stresses for round timber piles. ASTM D 2899. Philadelphia, PA: ASTM; (see latest edition).

American Wood Preservers Institute. Pile foundation know-how. Washington, DC: AWPI; 1969.

Armstrong, R. M. Structural properties of timber piles—Behavior of deep foundations. ASTM STP 670. Philadelphia, PA: ASTM; 1979: 118−152.

National Forest Products Association. National design specification for wood construction. Washington, DC: NFPA (see current edition).

Thompson, Warren S. Factors affecting the variation in compressive strength of southern pine piling. Washington, DC: American Wood-Preservers' Association; 1969.

Construction Logs

American Society for Testing and Materials. Standard methods for establishing stress grades for structural members used in log buildings. ASTM D 3957−80. Philadelphia, PA: ASTM; 1980.

Muchmore, F. W. Design guide for native log stringer bridges. Juneau, AK: U.S. Department of Agriculture, Forest Service, Region 10.

Rowell, R. M.; Black, J. M.; Gjovik, L. R.; Feist, W. C. Protecting log cabins from decay. Gen. Tech. Rep. FPL−11. Madison, WI: U.S. Department of Agriculture, Forest Service, Forest Products Laboratory; 1977.

Ties

American Railway Engineering Association. Ties and wood preservation. 1981−82 Ties manual for railway engineers: Vol I, Ch. 3. Chicago IL: AREA (looseleaf manual, revised annually).

Raymond, Gerald P. Railroad wood tie design and behavior. Transportation Engineering Journal.; 1977.

Chapter 20

Moisture Movement and Thermal Insulation in Light-Frame Structures

Moisture Movement and Thermal Insulation in Light-Frame Structures*

Moisture patterns within buildings and building components have a major effect on the performance of light-frame structures. These moisture patterns are greatly affected by thermal insulation and other energy conservation measures. Because of this interaction of moisture and temperature, the two subjects must be treated together. The amount of water vapor that can be suspended in air of a given volume depends on the temperature of that air. If air temperature is lowered below the point where it becomes saturated (dewpoint temperature), the water vapor will condense as water or frost. Likewise if the air contacts any surface that is below dewpoint temperature, water will form on that surface. Because wood and wood products are hygroscopic, condensation on these materials is absorbed and may result in swelling and the consequent buckling, paint peeling, and leaching of extractives as well as creating a potential for the growth of mildew or decay. Energy conservation measures, such as adding insulation and reducing air leakage in the building, increase the potential for condensation problems by causing higher indoor humidities and greater temperature differences between indoor and outdoor surfaces.

Thermal insulation and other energy conservation measures are essential to the economical operation of buildings and have important effects on the comfort of occupants. Wood itself is a relatively good insulator, but commercial insulating materials are usually incorporated into components such as walls, ceilings, and floors for greater efficiency. Commercial insulations are manufactured in a number of types, and each has advantages for specific applications. The management of indoor moisture is also essential to the comfort of occupants and economical operation of buildings. The most critical aspect of moisture management is the prevention of indoor humidity high enough to create a potential for staining from surface condensation as well as for swelling of members or for the growth of decay in concealed parts of building components. In addition, the application of vapor retarders and inclusion of ventilation and venting are necessary for the prevention of serious moisture problems in light-frame structures.

Modes of Heat Transfer

Heat seeks to attain a balance with surrounding conditions, just as water will flow from a higher to a lower level. When occupied buildings are heated to maintain inside temperature in the comfort range, a difference in temperature exists between inside and outside. Heat will therefore be transferred through walls, floors, ceilings, windows, and doors at a rate related to the temperature difference and to the resistance to heat flow of intervening materials. The transfer of heat takes place by one or more of three methods--conduction, convection, and radiation (fig. 20−1).

Conduction is the mode of heat flow through solid materials; for example, the conduction of heat along a metal rod when one end is heated in a fire. Convection involves transfer of heat by air currents; for example, air moving across a hot radiator carries heat to other parts of the room or space. Heat also may be transmitted from a warm body to a cold body by radiation. Heat obtained directly from a heat source, such as a fire, is radiant heat.

Heat transfer through a structural assembly composed of a variety of materials may be by one or more of the three methods described. Consider a frame house with an exterior wall composed of gypsum board, 2- by 4-inch studs, batt insulation, sheathing, sheathing paper, and bevel siding. In such a house, heat is transferred from the room air to the gypsum board by radiation, conduction, and convection, and through the gypsum board by conduction. Heat transfer across the insulated stud space is mainly by conduction, but may include convection. By conduction, it moves from the back of the gypsum board through the insulation to the colder sheathing. If the insulation is not too dense, convection also occurs. The air warmed by the gypsum board moves upward on the warm side of the stud space, and that cooled by the sheathing moves downward on the cold side. Heat transfer through sheathing, sheathing paper, and siding is by conduction. Some small air spaces will be found back of the siding, and the heat transfer across these spaces is principally by radiation. Through the studs from gypsum board to sheathing, heat is transferred by conduction; and from the outer surface of the wall to the atmosphere, it is transferred mainly by convection and radiation.

The thermal conductivity of a material is inversely proportional to the insulating value of that material. Heat conductivity in a homogeneous material is customarily measured as the amount of heat (in British thermal units) that will flow in 1 hour through a layer of the material which is 1 foot square and 1 inch thick, where the faces of the layer have a temperature difference of 1 °F. Heat conductivity is usually expressed by the symbol k.

Where a material is not homogeneous in structure, such as one containing air spaces like hollow tile, the term *conductance* is used instead of conductivity. The conductance, usually designated by the symbol C, is the amount of heat (in Btu's) that will flow in 1 hour through 1 square foot of the material or combination of materials per 1 °F temperature difference between surfaces of the material. (A dead air space with or without a reflective surface may also be rated for conductance by the same method.)

Resistivity and resistance (direct measures of the insulating value) are the reciprocals of transmission (conductivity or

---

* Revision by Gerald Sherwood, General Engineer.

conductance) and are represented by the symbol R. Resistivity, which is unit resistance, is the reciprocal of k and is given the same symbol R as resistance in the technical literature, because resistances are added together to calculate the total for any construction. This is the same R commonly used to rate commercial insulation. The overall coefficient of heat transmission through a wall or similar unit air to air, including surface resistances, is represented by the symbol U. U defines the transmittance in Btu's per hour, per square foot, per 1 °F temperature differential. Thus, the total equivalent resistance of a construction section is the $R = \frac{1}{U}$.

R values for insulation are usually given by the manufacturer. For other materials of construction, R values are established by the American Society of Heating, Refrigerating, and Air Conditioning Engineers (ASHRAE) and are presented in their current handbook of fundamentals along with detailed instructions for calculating overall heat transfer rates.

Overview of Insulating Materials

The R value of a material depends on the material composition, its density, and factors such as the material's mean temperature and moisture content. Another important aspect of a material's thermal performance is capacitance, or its ability to store heat. This depends largely on the mass of the material. The overall rate of heat transfer through a combination of materials is further affected by surface conditions such as roughness, color, and reflectivity, as variations in these properties will lead to variations in the component of the total heat loss by convection, conduction, and radiation.

Materials used for construction and insulation of buildings can be divided into several types depending on characteristics which have a major influence on their thermal properties. Fibrous materials are in the form of loose fill, blankets, or batts specifically for use as thermal insulation. Foams are available in rigid sheets, loose beads, or for foaming in place. Materials of construction include wood, metal, glass, and various panel products. Masonry materials are considered in a separate category because of their weight and consequent high-heat capacitance.

Building Insulation

Fibrous Materials

Rock wool, glass fiber, or cellulose are the most common fibrous insulating materials. These materials are often further classified in two categories: (1) vegetable fiber, such as cellulose, is made from wood; (2) mineral fiber is derived from nonorganic materials, and includes rock wool and glass fiber.

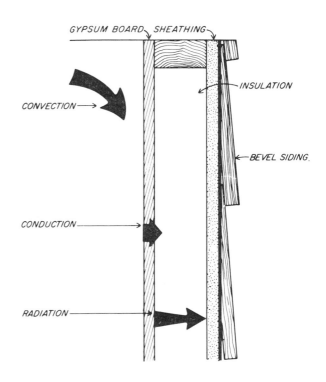

Figure 20—1—Modes of heat transfer in a light-frame wall. Convection involves the transference of heat by air currents from a warm to a colder zone. Conduction refers to the transmission of heat through a conductor, such as metal. In radiation the waves do not heat space in which they move but, when they come in contact with a colder surface, the waves are converted to heat. (ML84 5804)

Fibrous materials are available in the forms of blanket, batt, and loose fill. Blankets and batts are generally in a specific thickness and have a higher resistance to heat transfer than the same material as loose fill. Batts are semirigid, purchased flat, and may be installed in structural spaces with a friction fit which holds them in place. Blankets are quite flexible and are purchased in rolls. Loose fill is available in bags for pouring, as into an attic by the homeowner, or it may be blown into structural cavities by commercial crews. The quality of application affects the resulting density and thus the thermal resistance of the end product.

A maximum resistance value for fibrous insulations is obtained when the fibers are spaced uniformly and are perpendicular to the direction of heat flow. The fiber diameter also affects the insulation value; smaller fiber diameters at the same density have higher resistance values. The type and amount of binder, which influences the bond or contact of the fibers, may also affect the thermal conductivity.

Density affects the thermal resistance of all materials, but fibrous insulating materials are especially influenced by density because they can usually be compressed or expanded at the time of construction. The resistance to heat flow depends mainly on air trapped in and between the fibers. As the insulation is compressed, more fibrous materials are required to achieve the same thickness; the compression creates more air spaces and thus more resistance to heat flow. However, at some density part of the air spaces close and the fibrous material begins to lose insulating value. The optimum density varies with the fiber diameter, and it is achieved when the insulation in place can support its own weight without settling.

Although compressing a fibrous blanket or batt may increase the thermal resistance per unit thickness of the resulting material, it will decrease the overall resistance. For example: A 6-inch blanket compressed into a 3-1/2-inch wall cavity may have a slightly higher resistance than a standard 3-1/2-inch blanket; however, its resistance will not be as high as it would have been in a 6-inch space.

Foams

Commonly used foams in building construction are polystyrene, polyurethane, and isocyanurate. The insulating value of foams is derived from cells filled with air or other gases. Heat transfer is mainly by conduction across these cells. Optimum thermal resistance is achieved by a specific combination of cell size and density. The thermal resistance is also greatly affected by the kind of gas in the cells.

Polystyrene is commonly manufactured as rigid sheets and may be extruded or molded. Extruded polystyrene, with closed cells which trap gases in the material, has a higher resistance than molded polystyrene with open cells. Polystyrene beads are also available as a loose-fill insulation which has larger cells with lower resistance than the rigid forms.

Polyurethane and isocyanurate may be produced in rigid sheets, foamed in place into cavities, or applied to surfaces of building components as a fast-setting foam. These foams have 90 percent or more closed cells, and retain fluorocarbon gases for extended periods of time. However, as air permeates into cells and dilutes the fluorocarbon gas, the thermal resistance slowly decreases. Some factors affecting the rate of air permeation are environmental temperature, thickness, and surface protection. Covering both surfaces with gas-impermeable membranes greatly reduces the rate of air permeation and slows the deterioration of thermal resistance.

Materials of Construction

Density of construction materials is the major influence on their thermal conductivity. Very dense materials such as concrete and glass have high conductivities. Wood and fiber panel products have lower conductivities because of a porous structure that includes voids in and between fibers. In most construction systems, a high percentage of insulation is provided by nonstructural insulating materials, so conductivity of the construction materials is not critical. However, the high-conductivity materials must be used with some discretion to avoid thermal bridges—that is, dense materials extending all the way through the building component—between warm and cold faces.

Most masonry has a relatively high density, and consequently a high thermal conductivity. However, its capacitance, or ability to store heat, may offer some advantage for certain climates especially during the air-conditioning season. Where there is a large diurnal cycling of temperature, the material can slowly collect heat during the day and release heat to the inside during the night. If the masonry is sufficiently cooled during the night cycle, the building interior will remain cool during the day while the masonry is storing heat that would otherwise be transferred to the interior.

Thermal conductivity of concrete or manufactured masonry units, such as concrete block, can be reduced by using a lightweight aggregate, but this also reduces heat storage capacity.

Environmental Conditions

Temperature

Thermal conductivity is affected by the mean temperature of an insulating material during rating under test conditions. Most of the insulating values are obtained at 75 °F. However, for a meaningful comparison of thermal efficiency of materials, data must be from tests in the temperature range which the material will achieve in service. Each type of insulation varies differently with temperature. Some have a decreasing conductivity with decreasing temperature, while the reverse effect occurs in others (fig. 20−2).

Moisture

Most building materials are hygroscopic, which means that their moisture content depends on the relative humidity of the surrounding air. Of the available insulation materials only those made from wood fibers are considered hygroscopic. The moisture content in other insulation materials such as mineral fiber and plastic foams is quite low. However, local condensation or liquid absorption can raise the moisture content.

Water present in insulation increases thermal conductivity. The extent depends on the amount of moisture present.

Moisture and its Effects

Moisture has a major effect on human comfort as well as on the performance of most building materials. It is not always possible under various weather conditions to maintain the most desirable range of indoor humidity for both human comfort and performance of indoor materials. Also, ideal indoor humidity for human comfort can create serious problems of condensation on cold surfaces and within building components when outdoor temperatures are low. Therefore, compromise is often necessary during cold weather to avoid excessive maintenance of the building. When average outdoor temperatures drop below about 35 °F, indoor relative humidity should not exceed 40 percent to avoid a condensation hazard in building components. Most people are comfortable with humidities of 30 to 40 percent, which is high enough to avoid damage to indoor materials.

Conditions for Human Comfort

The human comfort range for humidity is directly related to temperature. Generally the optimum condition is the combination that allows moisture to evaporate from the body at a rate which maintains ideal body temperature. Physical activity generates heat and thus requires a faster evaporation rate to maintain comfortable body temperature. Clothing restricts the escape of heat and moisture, and so has a major effect on the comfort range. The human body can make limited adjustments with long-term exposure to certain conditions, but is less able to adjust quickly to sudden or frequent changes. Very low humidity during the heating seasons may result in dry skin and respiratory irritations. This effect is usually not perceptible at indoor temperatures of about 70 °F when relative humidity is 30 percent or higher.

Effect of Moisture on Interior Materials

The primary interior materials affected by moisture changes are wood and wood products. Dry conditions cause shrinkage with consequent loosening of joints or opening of cracks. High-humidity conditions cause buckling, particularly in thin panel products. If indoor air has a dewpoint temperature higher than wall or window surface temperature, moisture will condense, resulting in stains on the window sash as well as mildew growth on walls, ceilings, or other cold surfaces. Dimensional changes in either direction cause failure of some paints and finishes, and cycling between wet and dry may loosen nails.

To prevent these problems, moisture content of wood must be kept reasonably constant through seasonal changes. In much of the United States, interior wood is at 8 to 9 percent moisture content during summer. Indoor winter conditions of 70 °F and 30 to 40 percent relative humidity should result in an equilibrium moisture content of 7 to 8 percent, so the slight difference should not cause problems. Essentially the same moisture content is found in the Southeast in buildings that are air-conditioned. In the dry Southwest, moisture con-

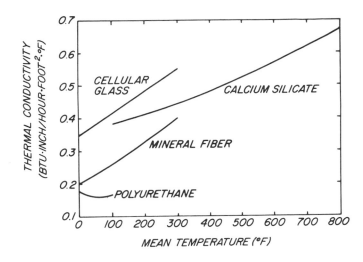

Figure 20–2—Typical variation of thermal conductivity of various insulating materials with mean temperature (Adapted from ASHRAE Handbook, 1977).

(ML84 5800)

tent may be as low as 6 percent in the summer, but the 7 to 8 percent target for the heating season is still practical. A good target is to limit variation to 3 percent moisture content in order to prevent damage to interior wood trim, wood furniture, and other wood products. Winter indoor conditions of 30 to 40 percent relative humidity will generally accomplish this.

Durability of Concealed and Exterior Materials

Cold weather condensation has contributed to decay in structural wood components of roofs and floors. Although decay in walls is rare, paint failures and stains on siding as well as distortion of siding do occur. Dimensional changes in thin panel products cause buckling which affects the durability of roofing or siding.

Specific conditions are necessary for growth of wood decay fungi. For the most part, decay is relatively slow at temperatures below 50 °F and much above 90 °F and wood must have reached fiber saturation point. This means the wood must be at 30 percent moisture content or higher. However, when readings of 20 percent moisture content are made, some parts of the member are often saturated; so 20 percent is frequently recognized as a danger point. Free water is necessary to develop saturation conditions in wood, so condensation that remains for a long period of time on the surface of wood is required for decay.

The design target for preventing decay is to limit the presence of condensation: (1) to periods of extreme cold when temperatures at the location of the water are below 40 °F, or (2) to daily cycling of condensation and evaporation in locations such as attics where there is a daily cycling of temperature.

To prevent problems of distortion in siding or sheathing, moisture changes in these materials should be limited to 5 percent moisture content. This limit should also prevent dimensional change in wood siding that might cause paint failures or splitting of the siding.

The amount of moisture that can be stored by various wall materials has been recognized as an important variable in moisture control. When moisture storage is available, the effects of condensation may be considerably alleviated. Design trends are oriented to more accurate moisture storage effects including long-term (annual or longer) moisture storage balance.

Moisture Control

To avoid moisture damage to building materials, certain preventative measures are required where major temperature differentials exist between indoors and outdoors. Consideration of moisture transfer mechanisms reveals that management of indoor relative humidity is the most effective control measure against condensation. Construction practices such as the use of vapor retarders and venting of structural spaces are important moisture control measures, but they will not prevent moisture damage where unreasonably high indoor humidities exist. A combination of indoor moisture management and good construction practices is required to prevent moisture damage in buildings.

Moisture Transfer Mechanisms

Moisture in buildings occurs as water vapor, water, or, in some cases in colder climates, as ice. Water can move from one location to another by gravity or capillary action. Water vapor transfer takes place by diffusion or convection or, as in most cases, by a combination of these mechanisms.

Diffusion always takes place in one direction: Vapor diffuses from a location with a high vapor density to one with a lower density. Fick's law states that the rate of vapor transfer is proportional to the difference in vapor density. However, when describing vapor transport through building components, Fick's law is usually expressed in terms of vapor pressures instead of densities.

Convection takes place when moisture is carried with an airflow. The direction of the flow does not depend on vapor density differentials and may be opposite to the direction of the diffusion flow. The amount of vapor carried depends on the rate of flow of air and the moisture content of the air. The air pressure differentials, which drive the air currents, may be caused by wind, fans, stoves and furnaces, and the stack effect.

Fans not only produce air currents within a building but also create pressure differentials across outside walls, doors, and windows, causing exfiltration of indoor air or infiltration of outside air, depending on the location in the building. Combustion air requirements for gas, oil, and wood stoves and furnaces lower indoor air pressure, causing infiltration of outside air.

The stack effect is caused by the difference between indoor and outdoor temperatures. The lower density of the warm inside air creates air infiltration into the lower part of the building and exfiltration from the upper part. In buildings with an attic, the stack effect forces indoor air into the attic and outdoor air infiltrates through walls, doors, and windows. A chimney also drastically changes air pressures and flows by lowering air pressures throughout the building.

Wind pressures are an additional source of air infiltration and exfiltration. Wind pressure differentials across walls and roof depend on windspeed, wind direction, terrain, and shape of the building. Outdoor air pressures are generally greater than indoor pressures at the windward side and lower than indoor pressures elsewhere. This is likely to create exfiltration at all other sides of the building.

The actual airflows depend on the combined pressure differentials caused by heating equipment, wind, and stack effect and are extremely difficult to predict. The actual combined effect certainly will be smaller than the sum of the separate effects. The air change rate can be measured with the tracer gas method, but such measurements do not show the magnitude or direction of air and moisture flow through individual building components. An alternative method, fan pressurization or depressurization, does not yield air exchange rates, but does indicate construction air tightness. Major areas of leakage can often be identified by this method.

Because of the many variables and uncertainties of moisture migration in buildings and building components, even the simplest analysis methods are relatively complicated and time consuming. Simulation models are being developed, but extensive long-term testing will be required to verify these. Currently the most commonly used design method is the moisture profile or dewpoint method, which is presented in detail in the ASHRAE handbook. This method is based entirely on vapor diffusion theory, thus ignoring any convection effects or liquid movement. Assuming steady-state conditions, the temperatures can be calculated at points within a wall or ceiling from indoor and outdoor temperatures and the thermal resistances of each layer of material. Each temperature corresponds with a saturation vapor pressure at that point. Similarly, actual vapor pressures at those points can be computed from indoor and outdoor vapor pressures and vapor flow resistances. Condensation occurs when the calculated vapor pressure exceeds the saturation vapor pressure. The rate of moisture accumulation may then be calculated from the difference between vapor flow to and from the condensing surface.

An alternative analysis method has been developed in Germany and is known as the Kieper method. It has some clear advantages over the traditional moisture profile method, yet it has not found widespread acceptance. It allows rapid graphic evaluation of different wall designs under any given environmental conditions. The method involves the use of a series of overlays that are not commercially available in the United States. The MOISTWALL program for a programmable calculator is another alternative, based on the same principles as the Kieper method. Both methods are entirely based on diffusion theory, ignoring any air convection effects or liquid transfer.

Indoor Humidity Control

Maintenance of reasonably low interior humidity is the most effective way to prevent condensation in the winter. Indoor humidity depends on the balance between moisture gains from sources within the house and the rate at which moisture can escape to the outdoors. The most common sources for moisture include human respiration, showers, laundry, and damp basements or crawl spaces. Most moisture is removed with the air escaping from the house. Vapor retarders generally have little effect on indoor humidity unless they contribute substantially to the airtightness of the house by acting as an air infiltration barrier.

Older homes have often required humidification because of high air exchange rates. In those homes setting the humidifier control no higher than 40 percent accomplishes a simple and effective humidity control.

Newer, more energy-efficient homes with air infiltration rates below 0.5 air change per hour often have reached a more delicate moisture balance. Most do not need any humidification and, in some, humidity levels are too high in the winter, resulting in mildew and condensation on windows or other cold surfaces. In these houses increasing ventilation is generally the most effective strategy, unless a single major moisture source can be identified and easily eliminated. Rather than increasing natural ventilation by removing air/vapor retarders or weather stripping, it is more energy efficient to provide for mechanical ventilation with fans or air-to-air heat exchangers. This offers the maximum humidity control, while minimizing energy losses.

Vapor Retarders

Early attempts at insulating buildings in the 1930's revealed the potential for condensation on cold side surfaces. Reduction in heat loss resulted in lowering the temperature of exterior materials below the dewpoint temperature of the inside air. The solution appeared to be to provide a barrier that would prevent indoor moisture from entering walls or ceiling, and so the term ''vapor barrier'' was created. The term ''vapor retarder'' was adopted to prevent misconceptions because vapor barriers are often thought to stop all moisture movement even though they only reduce the rate of movement.

Vapor retarders are rated for permeance. An accepted unit of permeance is a perm, or 1 grain per square foot per hour per inch of mercury difference in vapor pressure between the two sides. An early definition of ''vapor barrier'' was any material with a perm rating of less than 1. Although this definition still persists, current building materials and methods often require much lower perm ratings for vapor retarders. Four-mil polyethylene film, which is commonly used, has a perm rating of 0.08.

The integrity of vapor retarders is critical to their performance. Punctures, tears, and other discontinuities, negate their effectiveness in preventing diffusion of water vapor. Vapor retarders are also effective draft stops and may prevent movement of water vapor carried by air as well as movement by diffusion. In buildings much of the moisture transfer into and through structural spaces is by air movement. Air leaks at electrical outlets and ceiling fixtures, and around windows, doors, flues, and plumbing stacks, allow moisture to completely bypass the vapor retarder.

Vapor retarders may be in the form of structural materials, flexible sheets, or coatings. The most common application in new construction is flexible sheets. Coatings, such as vapor retarder paint, are often more convenient for retrofit because they can be applied to exposed surfaces. Materials sold as vapor retarders are often rated for permeance by the manufacturer. Perm ratings for these and other materials of construction are established by ASHRAE and are presented in their current handbook of fundamentals.

Average winter temperature and its duration is a major factor in cold-weather condensation, so this has been used as a basis for establishing condensation zones as shown in figure 20-3. The dash lines are -20 °F, 0 °F, and 20 °F isotherms of winter design temperature. Zone I roughly includes areas with design temperatures of -20 °F or colder; Zone II, design temperatures of 0 to -20 °F; and Zone III, those at 0 °F and warmer. Within each zone, similar degrees of condensation trouble are expected, and similar corrective measures apply. Vapor retarders are recommended as near as possible to the indoor (or heated space) face of all walls in Zones I, II, and

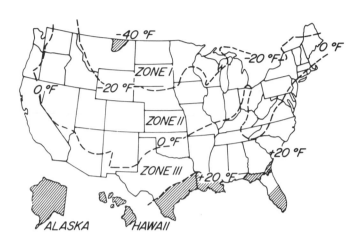

Figure 20-3—Condensation zones in the United States. (ML84 5853)

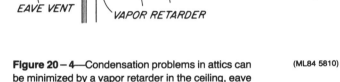

Figure 20-4—Condensation problems in attics can be minimized by a vapor retarder in the ceiling, eave vents, and ridge vents. (ML84 5810)

III. Vapor retarders are also recommended for ceilings in Zones I and II. In the warm, humid areas of the south (cross-hatched areas) that have constant air-conditioning requirements for long periods of time, vapor retarder requirements have not been firmly established. In the extremely cold climate of Alaska, vapor retarders are required, and special care is required in their installation. Local building authorities or other recognized sources should be contacted for good practice in those areas.

Soil covers placed under buildings are often referred to as vapor retarders. Even when they are in contact with wet soil, they are actually preventing water in the soil from evaporating into the crawl space. The use of soil covers in crawl spaces as well as under concrete slabs on grade is good practice for keeping soil moisture out of the building. Soil covers are usually roll roofing or heavy polyethylene. It is critical that the material be tough enough to prevent tearing. Joints in the material should lap at least 2 inches, and edges should be held in place with sand, bricks, or stones.

Venting of Structural Spaces

Even before the extensive use of insulation, the necessity of venting of structural spaces was recognized. Crawl spaces under floors have always required venting to carry away moisture from the soil. Ceiling insulation has resulted in low attic temperatures with the consequent potential for condensation if moisture is not vented to the outdoors. Vapor retarders have a major influence on the amount of venting required for structural spaces. Crawl space venting can be reduced to one-tenth when a soil cover vapor retarder is used, and some

standards permit attic venting to be reduced by one-half when a ceiling vapor retarder is used.

Venting is best accomplished in attics by placing outlet vents near the peak and inlet vents at eaves, so the stack effect keeps air moving continuously (fig. 20-4). Wherever venting is used, good distribution of air movement over the entire area is important.

Recommended venting for various types of roofs are shown in table 20-1. These venting rates are based on insulated ceilings. The net area refers to the total of all openings free from obstructions. The use of louvers and eight-mesh screen requires a gross area two and a quarter times that listed. Vapor retarders are indicated as requirements under certain conditions. It is also necessary that stray openings from walls into the attic or around a loose-fitting attic door be avoided. The stack effect allows a large inflow of warm air from the indoors, which carries water vapor into the attic.

Crawl spaces under dwellings where the earth is damp and uncovered require a high rate of venting. At least four openings, one at each corner placed as high as possible, should be provided. Their total net area may be calculated:

$$a = (2L/100) + (A/300)$$

where

L = perimeter of crawl space, feet
A = area of crawl space, square feet
a = total net area of all vents (or gross area if a 4-mesh screen is used), square feet

This rate of venting is usually sufficient, but cools the first floor so much that floor insulation is needed. A better control

20-8

measure is a cover on the damp ground. This cover may be a concrete slab, heavy roll roofing, or 0.004— to 0.006-inch-thick polyethylene plastic film laid on a graded surface, with its edge lapped 2 inches but not necessarily cemented. With this retarder, the vent area may be reduced to 10 percent of that calculated by the above equation.

Field Performance

The critical test of any design method is field performance. Although design for moisture control has been more empirical than analytical, the concepts and theories discussed above have had a major influence. General observations and field studies have revealed some critical variables.

Critical Variables

One of the most critical variables for prevention of moisture problems is indoor relative humidity. Where major cold weather condensation has been observed, indoor relative humidity was frequently high. Problems of mildew indoors and paint failures outdoors often occur only on outside walls of bathrooms or kitchens where indoor humidities are highest.

The amount of concrete used in construction also has a major effect on indoor moisture the first year after a building is constructed. Large quantities of water are used in concrete foundation walls and slabs. Excess water is released to the air as the concrete cures during the first few months. If the building is completed in the fall, humidity control may be a particular problem the first winter. The building should be frequently opened for ventilation until humidity levels are reduced. The problem is often eliminated during subsequent seasons.

Air leakage into structural spaces is the major mode of moisture movement into concealed spaces. It is critical to limit air leaks at all joints, around stacks and floors, and through openings such as electrical outlets.

Penetrations in the vapor retarder are also critical variables for moisture control. Punctures, tears, and other discontinuities allow moisture passage by diffusion and air leakage. Vapor retarders should be as complete as possible.

Outdoor temperature is a variable that cannot be controlled, but it has a major influence on field performance of buildings. Condensation occurs only on materials with temperatures below the dewpoint temperature of adjacent air. The potential for condensation problems is greatest at lowest temperatures.

Table 20 – 1—*Recommended good practice for loft and attic ventilation*[1] *(total net area of ventilation should be 1/300th of area enclosed within building lines at eave level)*

Roof style and slope	Ventilation achieved for condensation Zone I and II[2]	Zone III
Flat (slope 1 in 4 or flatter)	By uniform distribution at eaves, plus vapor retarder in top-story ceiling. Free circulation necessary through all spaces.	Same as I.
Gable (slope steeper than 1 in 4)	By at least 2 louvers on opposite sides, located near the ridge, plus vapor retarder in top-story ceiling.	Same as I but no vapor retarder.
Hip	1/600th distributed uniformly at eaves and 1/600th located at ridge with all spaces interconnected, plus vapor retarder in top-story ceiling.	Same as I but no vapor retarder.
Gable or hip with area to be occupied	1/600th distributed uniformly at eaves and 1/600th located at ridge with all spaces interconnected. Vapor retarder used on warm side of top full-story ceiling, dwarf walls, sloping part of roof, and attic story ceiling.	Same as I but no vapor barrier if insulation is omitted.

[1] For summer comfort in many areas, increased ventilation may be desirable. For winter comfort insulation is recommended between a living space and a loft or attic ventilated at these rates.

[2] Zone numbers shown in figure 20 – 3.

Retrofit

Moisture control is more difficult to add than to build in at the time of construction, but some preventative or remedial resources can be added. In most retrofits it is not feasible to reduce air leakage to a level that would result in excessively high relative humidities. However, the added insulation makes outside surfaces colder and thus increases the condensation potential. Attic ventilation can often be added or increased to carry off moisture entering the attic by air leakage and by diffusion through the ceiling. Walls are more dependent upon vapor retarders for moisture control. Most older homes have several coats of oil base paint on the walls and this gives some resistance to water vapor transfer. Vapor retarder paints can be applied for added resistance, especially on walls exposed to high humidities such as in bathrooms. Even with this vapor protection, air leakage at baseboards, electrical outlets, and around windows carries moist air into the wall cavity. Eliminating this air leakage is more critical to water vapor control than adding vapor resistance to the exposed wall surface alone.

Preventative or remedial measures dictated by retrofit technology for moisture control are:

1. Eliminate, as much as possible, air leakage from indoors into walls and attics.

2. Add vapor retarder paint to walls of bathrooms or other high-humidity areas.

3. Prevent indoor humidity from exceeding 40 percent during winter and keep even lower when outdoor temperatures are 10 °F or lower.

Selected References

American Society of Heating, Refrigerating, and Air-Conditioning Engineers. 1981 handbook of fundamentals. Atlanta GA: ASHRAE; 1981.

Anderson, L. O.; Sherwood, G. E. Condensation problems in your house: prevention and solution. Agric. Inf. Bull. 373. Madison, WI: U.S. Department of Agriculture; 1974.

Duff, J. E. The effect of air conditioning on the moisture conditions in wood walls. Res. Pap. SE−78. Asheville, NC: U.S. Department of Agriculture, Forest Service, Southeast Forest Experiment Station; 1971.

Hunt, C. M. Air infiltration: A review of some existing measurement techniques and data. In: Building air change rate and infiltration measurements. ASTM−STP−719. Philadelphia, PA: American Society for Testing and Materials; 1980.

Sherwood, G. E.; Hans, G. E. Energy efficiency in light-frame wood construction. Res. Pap. FPL 317. Madison, WI: U.S. Department of Agriculture, Forest Service, Forest Products Laboratory; 1979.

Sherwood, G. E.; Peters, C. C. Moisture conditions in walls and ceilings of a simulated older home during winter. Res. Pap. FPL 290. Madison, WI: U.S. Department of Agriculture, Forest Service, Forest Products Laboratory; 1977.

Sherwood, G. E.; TenWolde, A. Movement and management of moisture in light-frame structures. In: Proceedings, Wall and Floor Systems: Design and Performance of Light-Frame Structures. Madison, WI: Forest Products Research Society; 1982.

TenWolde, A. The Kieper and MOISTWALL moisture analysis methods for walls. Proceedings of the ASHRAE/DOE Conference. ASHRAE SP 38. Atlanta, GA: American Society of Heating, Refrigerating, and Air-Conditioning Engineers; 1983.

Verrall, A. F. Condensation in air-cooled buildings. Madison, WI: Forest Products J. 12: 531−536; 1962.

Chapter 21

Insulation Board, Hardboard, Medium-Density Fiberboard, and Laminated Paperboards

Insulation Board, Hardboard, Medium-Density Fiberboard, and Laminated Paperboards[*]

This group of panel materials are all reconstituted wood (or some other lignocellulose like bagasse) in that the wood is first reduced to fibers or fiber bundles and then put back together by special forms of manufacture into panels of relatively large size and moderate thickness. These board or panel materials in final form retain some of the properties of the original wood but, because of the manufacturing methods, gain new and different properties from those of the wood. Because they are manufactured, they can be and are "tailored" to satisfy a use-need, or a group of needs.

Another group of panel materials based on particles rather than fibers is described in chapter 22.

Fiber-based panel products are made essentially by breaking wood down through thermal-mechanical processes to its fibers. These fibers are interfelted in the reconstitution process and are characterized by a bond produced by that interfelting. They are frequently classified as fibrous-felted board products. At certain densities under controlled conditions of hot-pressing, rebonding of the lignin effects a further bond in the panel product produced. Binding agents and other materials may be added during manufacture to increase strength, resistance to fire, moisture, or decay, or to improve some other property. Among the materials added are rosin, alum, asphalt, paraffin, synthetic and natural resins, preservative and fire-resistant chemicals, and drying oils. Wax sizing is commonly added to improve water resistance.

Since fiber-based panel products are produced from small components of wood, the raw material need not be in log form. Many processes for manufacture of board materials start with wood in the form of pulp chips. Coarse residues from other primary forest products manufacture therefore are an important source of raw materials for fiber-based panel products. Bagasse, the fiber residue from sugarcane, and wastepaper are used also as raw material for board products.

Fiber-based panel materials are broadly divided into four groups—insulation board (the lower density products), hardboard, medium-density fiberboard, and laminated paperboard. The dividing point between an insulation board and a hardboard and medium-density fiberboard is a specific gravity of 0.5 (about 31 pcf). Laminated paperboards require a special classification because the density of these products is slightly greater than insulation board, but at the low end of hardboards. Because laminated paperboards are made by laminating together plies of paper about 1/16 inch thick, they have different properties along the direction of plies than across the machine direction. Other fiber-based panel products have nearly equal properties along and across the panel. Practically, because of the range of uses and specially developed products within the broad classification, further breakdowns are necessary to classify the various products adequately. The following breakdown by density places the fiber-based panel products in their various groups:

	Specific Gravity	Density (Pcf)
Insulation board	0.16 to 0.5	10 to 31
Hardboard	.5 to 1.45	31 to 90
Medium-density hardboard	.5 to .8	31 to 50
High-density hardboard	.8 to 1.28	50 to 80
Special densified hardboard	1.35 to 1.45	84 to 90
Medium-density fiberboard	.5 to .88	31 to 55
Laminated paperboard	.5 to .59	31 to 37

Properties of the various fiber-based panel products are determined according to ASTM standards, and to a considerable extent these properties either suggest or limit the uses. In the following sections the fiber-based panel materials are divided into the various categories suggested by kind of manufacture, properties, and use.

Manufacture, Properties, and Uses of Insulating Boards

Insulating board is a generic term for a homogeneous panel made from lignocellulose fibers (usually wood or bagasse) that have been interfelted and consolidated into homogeneous panels having a density of less than 31 pcf and more than 10 pcf. Other ingredients may be added during manufacture to provide specific physical properties. Insulation board is dried in an oven but not consolidated under pressure during manufacture.

There are many different types of insulating board, with different names and intended uses (table 21–1). Nominal dimensions of the different types of insulating boards are presented in table 21–2, and their minimum physical properties in table 21–3.

Sheathing is regularly manufactured in three grades: Regular density, intermediate, and nail base. Regular-density sheathing is usually about 18 pcf in density, and when the 2-by 8-foot material is used as sheathing, it is applied with long edges horizontal. The 4-foot-wide material is recommended for application with long edges vertical. When 25/32-inch-thick regular-density sheathing is applied with the long edges vertical and adequate fastening (either nails or staples) around the perimeter and along intermediate framing, requirements for racking resistance of the wall construction are usually satisfied. Horizontal applications with the 25/32-inch material require additional bracing in the wall system to meet code requirements for rigidity, as do some applications of the 1/2-

---

[*] Revision by Gary C. Myers, Forest Products Technologist.

Table 21 – 1—*Types, classes, and intended uses of insulating board*

Type	Class	Name	Intended use
I	—	Sound deadening board	In wall assemblies to control sound transmission
II	—	Building board	As a base for interior finishes
III	—	Insulating formboard	As a permanent form for poured-in-place reinforced gypsum or light-weight concrete aggregate roof construction
IV		Sheathing:	
	1	Regular-density	As wall sheathing in frame construction where method of application or thickness determines adequacy of racking resistance
	2	Intermediate-density	As wall sheathing where usual method of application provides adequate racking resistance
	3	Nail-base	As wall sheathing where usual method of application provides adequate racking resistance, and where exterior siding materials, such as wood or asbestos shingles, can be directly applied with special nails
V	—	Shingle backer	As an undercoursing for wood or asbestos cement shingles
VI	—	Roof insulating board	As above-deck insulation under built-up roofing
VII		Ceiling tiles and panels:	
	1	Nonacoustical	As decorative wall and ceiling coverings
	2	Acoustical	As decorative, sound-absorbing wall and ceiling coverings
VIII	—	Insulating roof deck	As roof decking for flat, pitched, or shed-type open-beamed, ceiling-roof construction
IX	—	Insulating wallboard	As a general-purpose product used for decorative wall and ceiling covering

Source: PS 57 – 73.

inch-thick regular-density sheathing applied with long edges vertical. Intermediate sheathing is usually about 22 pcf in density; nail base is about 25. Nail-base sheathing has adequate nail-holding strength so that asbestos and wood shingles for weather course (siding) can be attached directly to the nail-base sheathing with special annular grooved nails. With the other grades of sheathing, siding materials must be nailed directly to framing members or to nailing strips attached through the sheathing to framing. Because the method and amount of fastening is critical to racking resistance, local building codes should be consulted for requirements in different areas.

Ceiling tile and lay-in panels are an important use for structural insulating board. Such board has a paint finish applied in the factory for decoration and to provide resistance to flame spread. Interior-finish insulating board, when perforated or provided with special fissures or other sound traps, will also provide a substantial reduction in noise reflectance. The fissures and special sound traps are designed to provide improved appearance over that of the conventional perforations while satisfying the requirements for sound absorption. The manufacturers of insulation board long have recognized the appeal of esthetically pleasing ceiling finishes. Each of them offers finishes in designs that blend with either traditional or contemporary architecture and furnishings.

Generally ceiling tiles are 12 by 12 or 12 by 24 inches in size, 1/2 inch thick, and have tongue and groove or butt and chamfered edges. They are applied to nailing strips with nails, staples, or special mechanical fastenings, or directly to a surface with adhesives.

A panel product similar to tile, but nominally 24 by 24 or 24 by 48 inches, is gaining popularity. These panels, commonly called "lay-in ceiling panels," are installed in metal tees and angles in suspended ceiling systems. These lay-in

Table 21 – 2—*Nominal dimensions of insulating board*

Type of insulating board	Nominal dimensions		
	Width	Length	Thickness
	- Inches -		
Type I	48	96 or 108	1/2
Type II	48	96, 120, or 144	1/2
Type III	24, 32, 48	48 to 144	1
Type IV:			
Class 1	24	96	1/2 or 25/32
	48	96 or 108	1/2 or 25/32
Classes 2 and 3	48	96 or 108	1/2
Type V	11-3/4, 13-1/2, or 15	48	5/16 or 3/8
	23	47	1/2, 1, 1-1/2, 2, 2-1/2, or 3
Type VI	24	48	1/2, 1, 1-1/2, 2, 2-1/2, or 3
Type VII:			
Class 1	12	12 or 24	1/2, 9/16, or 5/8
	12	96 or 120	1/2
	16	16 or 32	1/2, 9/16, or 5/8
	24	24 or 48	1/2, 9/16, or 5/8
	48	96, 120, or 144	1/2
Class 2	12	12 or 24	1/2, 9/16, or 5/8
	24	24 or 48	1/2, 9/16, or 5/8
Type VIII	24	96	1-1/2, 2, or 3
Type IX	48	96 or 120	3/8

Source: PS 57–73.

21-4

Table 21 – 3—*Physical properties of insulating board*

Property	Type I	Type II	Type III	Type IV Class 1 1/2 inch thick	Type IV Class 1 25/32 inch thick	Type IV Class 2	Type IV Class 3	Type V 3/8 inch thick	Type V 5/16 inch thick	Type VI— 1/2 inch thick	Type VII¹	Type VIII	Type IX
Thermal conductivity, "k," average maximum (Btu · in/h · ft² · °F at 75 ± 5 °F)	0.38	0.38	0.40	0.40	0.40	0.44	0.48	0.40	0.40	—	0.38	0.40	0.40
Transverse strength either direction, average minimum (lb)													
Dry	12	12	37	14	25	17	25	6	6	7	10	NR	6
Wet	NR	NR	18	NR	NR	NR	NR	NR	NR	NR	NR	NR	NR
Modulus of rupture, average minimum (psi)													
Dry	240	240	190	275	200	340	500	200	240	140	200	225	200
Wet	NR	NR	95	NR	NR	NR	NR	NR	NR	NR	NR	NR	NR
After accelerated aging	NR	NR	NR	NR	NR	NR	NR	NR	NR	NR	NR	²50 percent of dry value	NR
Modulus of elasticity, average minimum (psi x 10³)	NR	NR	NR	NR	NR	NR	NR	NR	NR	NR	NR	40	NR
Deflection span ratio, average maximum	NR	NR	NR	NR	NR	NR	NR	NR	NR	NR	NR	1:240	NR
Deflection at specified minimum load, average maximum (in)													
Dry	0.85	0.85	0.16	0.75	0.56	0.75	0.65	1.18	1.18	1.25	NR	NR	NR
Wet	NR	NR	0.11	NR	NR	NR	NR	NR	NR	NR	NR	NR	NR
Tensile strength parallel to surface, average minimum (psi)	150	150	150	150	150	200	300	150	150	50	150	150	150

Table 21–3—*Physical properties of insulating board—Continued*

Property	Type I	Type II	Type III	Type IV Class 1 1/2 inch thick	Type IV Class 1 25/32 inch thick	Type IV Class 2	Type IV Class 3	Type V 3/8 inch thick	Type V 5/16 inch thick	Type VI—1/2 inch thick	Type VII[1]	Type VIII	Type IX
Tensile strength perpendicular to surface, average minimum (psf)	600	600	600	600	600	800	1,000	600	600	500	600	600	600
Water absorption by volume, average maximum (percent)	7	7	10	7	7	15	12	7	7	10	NR	10	15
Linear expansion, 50–90 percent relative humidity, average maximum (percent)	0.5	0.5	0.5	0.5	0.5	0.6	0.6	0.5	0.5	0.5	0.5	0.5	0.5
Vapor permeance, average minimum (grains/h ft² in Hg pressure differential)	NR	NR	5	5	5	5	5	5	5	NR	NR	[3]0.5	NR
Direct nail withdrawal resistance, average minimum (lb/nail) Dry	NR	NR	NR	NR	NR	NR	40	NR	NR	NR	NR	NR	NR
Soaked	NR	NR	NR	NR	NR	NR	25	NR	NR	NR	NR	NR	NR
Racking load, average minimum (lb) Dry	NR	NR	NR	NR	5,200	5,200	5,200	NR	NR	NR	NR	NR	NR
Wet	NR	NR	NR	NR	4,000	4,000	4,000	NR	NR	NR	NR	NR	NR
Flame-spread index, finish surface, maximum	NR	200	NR	NR	NR	NR	NR	NR	NR	NR	200	200	200

[1] The physical properties listed for acoustical material, except for flame spread, apply to the base material before punching, drilling, perforating, or embossing. NR = Not required for this product.

[2] For example, if the dry modulus of rupture is found to be 300 psi, then the modulus of rupture after accelerated aging must be not less than 150 psi.

[3] Average maximum. For products without a vapor barrier, there is no requirement for vapor permeance.

Source: PS 57–73.

panels are usually 1/2 inch thick and are supported in place along all four edges. They are frequently used in combination with translucent plastic panels that conceal light fixtures (fig. 21–1).

Finishes and perforation treatments for sound absorption are the same as for regular ceiling tile. Producers of insulating board are extending their manufacture to specially embossed ceiling panels that can be applied with butt-joint edges and ends that present an essentially unbroken surface, and factory-finished panels that look like real wood planks. Plastic films are being used increasingly for surfacing ceiling tile for applications in kitchens and bathrooms where repeated washability and resistance to moisture is desired. These products are especially adaptable for remodeling.

Manufacture, Properties, and Uses of Hardboard

Hardboard is a generic term for a panel manufactured primarily from interfelted lignocellulose fibers which are consolidated under heat and pressure in a hot press to a density of 31 pcf or greater. Other materials may be added to improve certain properties, such as stiffness, hardness, finishing properties, resistance to abrasion, and moisture, as well as to increase strength, durability, and utility. Hardboards are further subdivided into medium-density and high-density materials. Both are manufactured as previously defined, but a medium-density hardboard has a density between 31 and 50 pcf, and the high-density hardboard has a density greater than 50 pcf.

High-density and medium-density conventional hardboards are manufactured in several ways, and the result is reflected in the appearance of the final product. Hardboard is described as being S−1−S (screen-backed) or S−2−S (smooth two sides). When the mat from which the board is made is formed from a water slurry (wet-felted) and the wet mat is hot-pressed, a screen is required to permit steam to escape. In the final board the reverse impression of the screen is apparent on the back of the board, hence the screen-back designation. A screen is similarly required with mats formed from an air suspension (air-felted) when moisture contents are sufficiently high going into the hot press so that venting is required.

In some variations of hardboard manufacture, a wet-felted mat is dried before being hot-pressed. With this variation it is possible to hot-press without using the screen, and an S−2−S board is produced. In air-felting hardboard manufacture, it is possible also to press without the screen, if moisture content of mats entering the hot press is low. In a new adaptation of pressing hardboard mats, a caul with slots or small circular holes is used to vent steam; the board produced has a series of

Figure 21 – 1—A lay-in ceiling panel being installed in a suspended ceiling system. (M84 0280-11)

small ridges or circular nubbins which, when planed or sanded off, yield an S−2−S board.

Prefinished paneling and siding products account for about 65 percent of the current product mix. An additional 25 percent is for industrial uses, including cut-to-size and molded products.

Medium-Density Hardboard

Medium-density hardboard is manufactured by the conventional methods used for other hardboard and is tailored for use as house siding. Medium-density hardboard for house siding use is mostly 7/16 inch thick and is fabricated for application as either panel or lap siding.

Panel siding is 4 feet wide and commonly furnished in 8-, 9-, or 10-foot lengths. Surfaces may be grooved 2 inches or more on center parallel to the long dimension to simulate reversed board and batten or may be pressed with ridges simulating a raised batten.

Lap siding is frequently 12 inches wide with lengths to 16 feet and is applied in the same way as conventional wood lap siding. Some manufacturers offer their lap siding products with special attachment systems that provide either concealed fastening or a wider shadow line at the bottom of the lap.

Most siding is furnished with some kind of a factory-applied finish. At least the surface and edges are given a prime coat of paint. Finishing is completed later by application of at least one coat of paint. Two coats of additional paint, one of a second primer and one of topcoat, provide for a longer inter-

21-7

Figure 21–2—Examples of medium-density hard- (M84 0280-2)
board commonly used for exterior siding. Top two
examples have textured surfaces to simulate wood
grain, and bottom example is smooth surfaced.

val before repainting. There is a trend for complete prefinishing of medium-density hardboard siding. The complete prefinishing ranges from several coats of liquid finishes to cementing various films to the surfaces and edges of boards. Surfaces of medium-density hardboard for house siding range from very smooth to textured, many simulating weathered wood with the latewood grain raised as though earlywood has been eroded away. Three samples of hardboard siding are presented in figure 21–2, showing the different surfaces.

Small amounts of medium-density hardboard are prefinished for interior paneling along with the high-density hardboards. Siding is the most important use and others will not become extensive until that market is fully developed and exploited. The experience with medium-density hardboard has been good. Dimensional movement with moisture change has not produced major problems in service. When hardboard siding is stored, applied, and finished with high-quality paints in accordance with the manufacturers or American Hardboard Association recommendations, it has required little paint maintenance. Proper finishing, maintenance, and refinishing procedures for hardboard siding are covered in chapter 16 and in American Hardboard Association literature. Code authorities and others have recognized the evidence submitted by manufacturers on the performance of medium-density hardboard siding. A summary of the properties specified for this material in product standard PS 60–73 is presented in table 21–4.

High-Density Hardboard

Manufacture of high-density hardboard has grown rapidly since World War II. Numerous older uses are well established, and new ones are being developed continually. Property requirements are presented in table 21–5, which classifies hardboard by surface finish, thickness, and minimum physical properties.

Originally there were two basic qualities of high-density hardboard; standard and tempered. These are still the two qualities used in greatest quantity. Standard hardboard is a panel product with a density of about 60 to 65 pcf, usually unaltered except for humidification and trimming to size after hot-pressing. Tempered hardboard is a standard-quality hardboard that is treated with a blend of siccative resins (drying oil blends or synthetics) after hot-pressing. The resins are stabilized by baking after the board has been impregnated. Usually about 5 percent resin solids are required to produce a hardboard of tempered quality. Tempering improves water resistance, hardness, and strength appreciably, but embrittles the board and makes it less shock resistant.

A third hardboard, service quality, has become important. This is a product of lower density than standard, usually 50 to 55 pcf, made to satisfy needs where the higher strength of standard quality is not required. Because of its lower density, service-quality hardboard has better dimensional stability than the denser products.

When service hardboard is given the tempering treatment, it is classed as tempered service, and property limits have been set for specifications. It is used where water resistance is required but the higher strength of regular treatment is not. Underlayment is service-quality hardboard, nominally 1/4 inch thick, that is sanded or planed on the back surface to provide a thickness of not less than 0.200 inch.

These are the regular qualities of high-density hardboard; because a substantial amount of this hardboard is manufactured for industrial use, special qualities are made with different properties dictated by the specific use. For example, hardboard manufactured for concrete forms is frequently given a double tempering treatment. For some uses where high impact resistance is required, like backs of television cabinets, boards are formulated from specially prepared fiber and additives. Where special machining properties like die punching or post-forming requirements must be satisfied, the methods of manufacturing and additives used are modified to produce the desired properties.

Commercial thicknesses of high-density hardboard generally range from 1/8 to 1/2 inch. Not all thicknesses are produced in all grades. The thicknesses of 1/10 and 1/12 inch are regularly produced only in the standard grade. Tempered hardboards are produced regularly in thicknesses between 1/8 and 5/16 inch. Service and tempered service are regularly produced in fewer thicknesses, none less than 1/8 inch and not by all manufacturers or in screen-back and S–2–S types. The appropriate standard specification or source of material should be consulted for specific thicknesses of each kind.

High-density hardboards are produced in 4- and 5-foot

Table 21 – 4—*Physical properties of hardboard siding*

Property	Requirement
Percent water absorption based on weight (maximum average per panel)	
Primed	15
Unprimed	20
Percent thickness swelling (maximum average per panel)	
Primed	10
Unprimed	15
Percent linear expansion (maximum average per panel)	
Lap siding	0.38, for 0.325- to 0.375-in thickness
	0.40, for over 0.376-in thickness
Panel siding	0.36, for 0.220- to 0.265-in thickness
	0.38, for 0.325- to 0.375-in thickness
	0.40, for over 0.376-in thickness
Weatherability of substrate (maximum swell after five cycles) (in)	0.010 and no objectionable fiber raising
Sealing quality of primer coat	No visible flattening
Weatherability of primer coat	No checking, erosion, or flaking
Nailhead pullthrough (minimum average per panel) (lb)	150
Lateral nail resistance (minimum average per panel) (lb)	150
Modulus of rupture (minimum average per panel) (psi)	1,800 for 3/8- and 7/16-in-thick siding
	3,000 for 1/4-in-thick siding
Hardness (minimum average per panel) (lb)	450
Impact (minimum average per panel) (in)	9
Moisture content (pct)[1]	2–9 inclusive, and not more than 3 pct variance between any two boards in any shipment or order

[1] Because hardboard is a wood-base material, its moisture content will vary with environmental humidity conditions. When the environmental humidity conditions in the area of intended use are a critical factor, the purchaser should specify a moisture content range more restrictive than 2 to 9 percent, so that fluctuation in the moisture content of the siding will be kept to a minimum.

Source: PS 60–73.

widths with the more common width being 4 feet. Standard commercial lengths are 4, 6, 8, 12, and 16 feet with an 18-foot length being available in the 4-foot width. Most manufacturers maintain cut-to-size departments for special orders. Retail lumberyards and warehouses commonly stock 8-foot lengths, except for underlayment, which is usually 4 feet square.

About 10 percent of the hardboard used in the United States is imported. Foreign-made board may or may not be manufactured to the same standards as domestically produced products. Before substituting a foreign-made product in a use where specific properties are required, it should be determined that the foreign-made item has properties required for the use. Canadian products are usually produced to the same standards as United States products.

In addition to the standard smooth-surface hardboards, special products are made using patterned cauls so the surface is striated or produced with a relief to simulate ceramic tile, leather, basket weave, etched wood, or other texture. Hardboards are punched to provide holes for anchoring fittings for shelves and fixtures (perforated board) or with holes comprising 15 percent or more of the area for installation in ceilings with sound-absorbent material behind it for acoustical treatments or as air diffusers above plenums.

Table 21 − 5—*Classification of high-density hardboard by surface finish, thickness, and physical properties*

Class	Nominal thickness	Water resistance (maximum average per panel)		Modulus of rupture (minimum average per panel)	Tensile strength (minimum average per panel)	
		Water absorption based on weight	Thickness swelling		Parallel to surface	Perpendicular to surface
	Inch	*Percent*		*Pounds per square inch*		
1	1/12	30	25	6,000	3,000	130
Tempered	1/10	25	20			
	1/8	25	20			
	3/16	25	20			
	1/4	20	15			
	5/16	15	10			
	3/8	10	9			
2	1/12	40	30	4,500	2,200	90
Standard	1/10	35	25			
	1/8	35	25			
	3/16	35	25			
	1/4	25	20			
	5/16	20	15			
	3/8	15	10			
3	1/8	35	30	4,500	2,000	75
Service-tempered	3/16	30	30			
	1/4	30	25			
	3/8	20	15			

More and more effort is being put forth by industry to modify and finish hardboard so it can be used in more ways with less "on-the-job" cost of installation and finishing and to permit industrial users a saving in final product. Most important is prefinishing, particularly wood graining, where the surface of the board is finished with lithographic patterns of popular cabinet woods printed in two or more colors.

The uses for hardboard are diverse. It has been claimed that "hardboard is the grainless wood of 1,000 uses, and can be used wherever a dense, hard panel material in the thicknesses as manufactured will satisfy a need better, or more economically than any other material." Because of its den-sity it is harder than most natural wood, and because of its grainless character it has nearly equal properties in all directions in the plane of the board. It is not so stiff nor as strong as natural wood along the grain, but is substantially stronger and stiffer than wood across the grain. Minimum specific properties presented in table 21–5 can be compared with similar properties for wood, wood-base panels, and other materials. Hardboard retains some of the properties of wood; it is hygroscopic and shrinks and swells with changes in moisture content.

Changes in moisture content due to service exposures may be a limiting factor in satisfactory performance. Correct

Table 21 – 5—*Classification of high-density hardboard by surface finish, thickness, and physical properties*—Continued

Class	Nominal thickness	Water resistance (maximum average per panel)		Modulus of rupture (minimum average per panel)	Tensile strength (minimum average per panel)	
		Water absorption based on weight	Thickness swelling		Parallel to surface	Perpendicular to surface
	Inch	Percent		Pounds per square inch		
4 Service	1/8	45	35	3,000	1,500	50
	3/16	40	35			
	1/4	40	30			
	3/8	35	25			
	7/16	35	25			
	1/2	30	20			
	5/8	25	20			
	11/16	25	20			
	3/4	20	15			
	13/16	20	15			
	7/8	20	15			
	1	20	15			
	1-1/8	20	15			
5 Industrialite	1/4	50	30	2,000	1,000	25
	3/8	40	25			
	7/16	40	25			
	1/2	35	25			
	5/8	30	20			
	11/16	30	20			
	3/4	25	20			
	13/16	25	20			
	7/8	25	20			
	1	25	20			
	1-1/8	25	20			

Source: ANSI/AHA A 135.4–1982.

application and attachment as well as prior conditioning to a proper moisture content will give satisfactory service, but improper application or conditioning precludes it. Proper moisture conditioning prior to assembly is of particular importance in glued assemblies.

Product development in hardboard has held generally to the line of class and type of board product, in contrast with structural insulating board which deals with specific items for particular uses. During the past few years, much of the success of hardboard resulted because the industry developed certain products for a specific use and had treatments, fabrication, and finishes required by the use. Typical are prefinished paneling, house siding, underlayment, and concrete form hardboard.

Many uses for hardboard have been listed, but generally they can be subdivided according to uses developed for construction, furniture and furnishings, cabinet and store fixture work, appliances, and automotive and rolling stock. Several examples of hardboard products are presented in figure 21—3.

In construction, hardboard is used as floor underlayment to provide a smooth undercourse under plastic or linoleum flooring, as a facing for concrete forms for architectural concrete, as facings for flush doors, as molded facings for interior doors, as insert panels and facings for garage doors, and as material punched with holes for wall linings in storage walls and in built-ins where ventilation is desired. High-density hardboard is being used as a shear-web material for box and I—beams, to be used as load-carrying members in building construction. Prefinished hardboard, either with baked finishes or the regular ones like those used generally in wood-grain printing, is used for wall lining in kitchens, bathrooms, family rooms, and recreation rooms.

In furniture, furnishings, and cabinetwork, conventional hardboard is used extensively for drawer bottoms, dust dividers, case goods and mirror backs, insert panels, television, radio and stereo cabinet sides, backs (die-cut openings for ventilation), and as crossbands and balancing sheets in laminated or overlay panels. Hardboard also has use as a core material for relatively thin panels overlaid with films and thin veneers, and as backup material for metal panels. In appliances other than television, radio, and stereo cabinets, it is used wherever the properties of the dense, hard sheet satisfy a need economically. Because it can be postformed to single curvature (and in some instances to mild double curvature) by the application of heat and moisture, it is used in components of appliances requiring that kind of forming.

In automobiles, trucks, buses, and railway cars, hardboard is commonly used in interior linings. Door and interior sidewall panels of automobiles are frequently hardboard, postformed, and covered with cloth or plastic. The base for sun visors is often hardboard, as are the platforms between seats and rear windows. Molded hardboard also has been used for three-dimensional-shaped components like door panels and armrests. Ceilings of station wagons and truck cabs are often enameled or vinyl-covered thin hardboard.

Special Densified Hardboard

This special building fiberboard product is manufactured mainly as diestock and electrical panel material. It has a density of 84 to 90 pcf and is produced in thicknesses between 1/8 and 2 inches in panel sizes of 3 by 4, 4 by 6, and 4 by 12 feet.

Special densified hardboard is machined easily with machine tools and its low weight as compared with metals (aluminum alloys about 170 pcf) makes it a useful material for templates and jigs for manufacturing. It is relatively stable dimensionally from moisture change because of low rates of moisture absorption. It is more stable for changes in temperature than the metals generally used for those purposes. The 1/8-inch-thick board is specially manufactured for use as lofting board, which is a surface on which small-scale plans of a boat design are projected to actual boat size.

As diestock, it finds use for stretch- and press-forming and spinning of metal parts, particularly when few of the manufactured items are required and where the cost of making the die itself is important in the choice of material.

The electrical properties of the special densified hardboard meet many of the requirements set forth by the National Electrical Manufacturers Association for insulation resistance and dielectric capacity in electrical components so it is used extensively in electronic and communication equipment.

Other uses where its combination of hardness, abrasion resistance, machinability, stability, and other properties are important include cams, gears, wear plates, laboratory work surfaces, and welding fixtures.

Manufacture, Properties, and Uses of Medium-Density Fiberboard

Medium-density fiberboard (MDF) is a panel product manufactured from lignocellulosic fibers combined with a synthetic resin or other suitable binder. The panels are manufactured to a density range of 31 to 55 pcf (0.50 to 0.88 specific gravity) by the application of heat and pressure by a process in which the interfiber bond is substantially created by the added binder. Other materials may have been added during manufacturing to improve certain properties.

The technology utilized to manufacture MDF is a combination of that used in the particleboard industry and that used in the hardboard industry. Consequently, there was much debate over the definition of the product. This was settled by the development of an American National Standard for Medium Density Fiberboard for Interior Use (ANSI A 208.2−1980), cosponsored by the American Hardboard Association and the National Particleboard Association. Minimum property requirements for MDF are presented in table 21−6. MDF fiberboard is available in thicknesses from 3/16 inch up to 1-1/2 inches, but most board manufactured is 3/4 inch thick and in the 44 to 50 pcf density range for applications in the industrial markets.

The furniture industry is by far the dominant MDF market. MDF is frequently taking the place of solid wood, plywood, and particleboard for many furniture applications. Compared to particleboard, it has a very smooth surface which facilitates wood-grain printing, overlaying with sheet materials, and veneering. MDF has tight edges which need not be edge-banded and can be routed and molded like solid wood, as illustrated in figure 21−4. Grain-printed and embossed, MDF is used in many furniture lines. The potential for MDF in other interior and exterior markets such as doors, moldings, exterior trim, and pallet decking is currently being explored by the industry. Many industry people expect MDF markets will expand significantly during the next decade.

Manufacture, Properties, and Uses of Laminated Paperboards

Laminated paperboards are made in two general qualities, an interior and a weather-resistant quality. The main differences between the two qualities are in the kind of bond used to laminate the layers together and in the amount of sizing used in the pulp stock from which the individual layers are made. For interior-quality boards, the laminating adhesives are commonly of starch origin while, for the weather-resistant board, synthetic resin adhesives are used. Laminated paperboard is regularly manufactured in thicknesses of 3/16, 1/4,

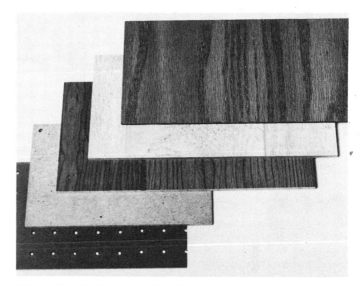

Figure 21−3—Examples of hardboard products, from top to bottom, are facing for flush doors, two types of wood grain printed paneling, standard hardboard, and pegboard. (M84 0280-3)

Figure 21−4—Example of medium-density fiberboard that has been embossed, printed with a wood grain, and finished for use as a cabinet drawer front. (M84 0280-6)

Table 21−6—*Property requirements of medium-density fiberboard*

Nominal thickness	Modulus of rupture	Modulus of elasticity	Internal bond (tensile strength perpendicular to surface)	Linear expansion	Screwholding	
					Face	Edge
In	*Psi*	*Psi*	*Psi*	*Pct*	*Lb*	*Lb*
13/16 and below	3,000	300,000	90	[1]0.30	325	275
7/8 and above	2,800	250,000	80	.30	300	225

[1] For boards having nominal thickness of 3/8 in or less, the linear expansion value shall be 0.35 percent.

Source: ANSI A 208.2−1980.

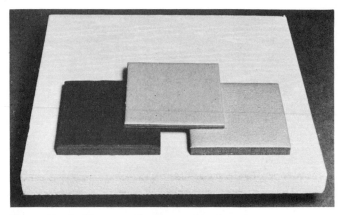

Figure 21—5—Three smaller pieces of laminated paperboard are 3/16- and 3/8-inch thickness and illustrate different surfaces. Bottom paperboard product is a fascia for exterior use.

(M84 0280-10)

and 3/8 inch for construction uses although for such industrial uses as dust dividers in case goods, furniture, and automotive liners, 1/8-inch thickness is common. Important properties are presented in table 21−7. Several examples of laminated paperboard products are presented in figure 21−5. A few other uses for laminated paperboards include mirror backing, toys and games, packaging, museum exhibits, photo murals, outdoor signs, and displays.

For building use, considerable amounts go into the prefabricated housing and mobile home construction industry as interior wall and ceiling finish. In the more conventional building construction market, interior-quality boards are also used for wall and ceiling finish, often in remodeling to cover cracked plaster. Some of the full-wall laminated paper panels meet and exceed racking requirements, when applied according to the manufacturer's instructions.

Table 21−7—*Strength and mechanical properties of laminated paperboard*[1]

Property	Value
Density (pcf)	32−33
Specific gravity	0.52−0.53
Modulus of elasticity (compression) (1,000 psi):	
Along the length of the panel[2]	300−390
Across the length of the panel[2]	100−140
Modulus of rupture (psi):	
Span parallel to length of panel[2]	1,400−1,900
Span perpendicular to length of panel[2]	900−1,100
Tensile strength parallel to surface:	
Along the length of the panel[2]	1,700−2,100
Across the length of the panel[2]	600−800
Compressive strength parallel to surface (psi):	
Along the length of the panel[2]	700−900
Across the length of the panel[2]	500−800
24-hour water absorption (pct by weight)	10−170
Linear expansion from 50 to 90 percent relative humidity (pct):[3]	
Along the length of the panel[2]	0.2−0.3
Across the length of the panel[2]	1.1−1.3
Thermal conductivity at mean temperature of 75 °F (Btu · in/h · ft^2 · °F)	0.51

[1] The data presented are general round-figure values, accumulated from numerous sources; for more exact figures on a specific product, individual manufacturers should be consulted or actual tests made. Values are for general laboratory conditions of temperature and humidity.

[2] Because of directional properties, values are presented for two principal directions, along the usual length of the panel (machine direction) and across it.

[3] Measurements made on material at equilibrium at each condition at room temperature.

Water-resistant grades are manufactured for use as sheathing, soffit linings, and other "exterior protected" applications like porch and carport ceilings. Soffit linings and lap siding are specially fabricated in widths commonly used and are prime coated with paint at the factory.

The common width of laminated paperboard is 4 feet, although 8-foot widths are available in 12-, 14-, 16-foot, and longer lengths for such building applications as sheathing entire walls. Laminated paperboards, for use where a surface is exposed, have the surface ply coated with a high-quality pulp to improve surface appearance and performance. Surface finish may be smooth or textured.

Selected References

Acoustical and Board Products Association. Product specification for sound deadening board in wall assemblies. ABPA—IB Spec. No. 4. Park Ridge, IL: ABPA; 1975.

American Hardboard Association. Application instructions for basic hardboard products. Palatine, IL: AHA; n.d.

American Hardboard Association. Finishing recommendations for new construction. Utilizing unprimed and primed hardboard siding. Palatine, IL: AHA; n.d.

American Hardboard Association. Hardboard paneling, 1/4 inch prefinished. Palatine, IL: AHA; n.d.

American Hardboard Association. Hardboard siding. Application and storage instructions. Palatine, IL: AHA; n.d.

American Hardboard Association. Maintenance tips for home owners with hardboard siding exteriors. Palatine, IL: AHA; n.d.

American Hardboard Association. Today's hardboard—and how to handle it. An AHA guide to selling hardboard siding and paneling. Palatine, IL: AHA; n.d.

American Hardboard Association. Today's hardboard—building better than ever. Palatine, IL: AHA; n.d.

American Hardboard Association. Today's hardboard for creative people who build almost anything. An AHA guide for industrial arts students and instructors. Palatine, IL: AHA; n.d.

American Hardboard Association. American National Standard ANSI/AHA A 135.4—1982. Basic hardboard. Palatine, IL: AHA; 1982.

American Hardboard Association. Voluntary product standard PS 57—73. Cellulosic fiber insulating board. Palatine, IL: AHA; 1973.

American Hardboard Association. Voluntary product standard PS 59—73. Prefinished hardboard paneling. Palatine, IL: AHA; 1973.

American Hardboard Association. Voluntary product standard PS 60—73. Hardboard siding. Palatine, IL: AHA; 1973.

American Hardboard Association. Hardboard siding. Recommendations for the use of hardboard siding in manufactured housing. Palatine, IL: AHA; 1978.

American Hardboard Association. Tileboard wall paneling. Palatine, IL: AHA; 1978.

American National Standards Institute. Medium density fiberboard for interior use. ANSI A 208.2—1980. Silver Spring, MD: National Particleboard Association; 1980.

American Society for Testing and Materials. Standard methods of testing insulating board (cellulosic fiber), structural and decorative. ASTM C 209—72. Philadelphia, PA: ASTM; 1972.

American Society for Testing and Materials. Standard specification for insulating board (cellulosic fiber), structural and decorative. ASTM C 208—72. Philadelphia, PA: ASTM; 1972.

American Society for Testing and Materials. Standard methods of test for structural insulating roof deck. ASTM D 2164—65 (reapproved 1977). Philadelphia, PA: ASTM; 1977.

American Society for Testing and Materials. Standard methods of evaluating the properties of wood-base fiber and particle panel materials. ASTM D 1037—78. Philadelphia, PA: ASTM; 1978.

American Society for Testing and Materials. Standard specification for structural insulating formboard (cellulosic fiber). ASTM C 532—66 (reapproved 1979). Philadelphia, PA: ASTM; 1979.

American Society for Testing and Materials. Standard specification for fiberboard nail-base sheathing. ASTM D 2277—75 (reapproved 1980). Philadelphia, PA: ASTM; 1980.

McNatt, J. D. Design stresses for hardboard—effect of rate, duration, and repeated loading. Madison, WI: Forest Products Journal. 20(1): 53—60; 1970.

McNatt, J. D. Hardboard-webbed beams: research and application. Madison, WI: Forest Products Journal. 30(10): 57—64; 1980.

Stern, R. K. Development of an improved hardboard-lumber pallet design. Res. Pap. FPL 387. Madison, WI: U.S. Department of Agriculture, Forest Service, Forest Products Laboratory; 1980.

Tuomi, R. L.; Gromala, D. S. Racking strength of walls: let-in corner bracing, sheet materials, and effect of loading rate. Res. Pap. FPL 301. Madison, WI: U.S. Department of Agriculture, Forest Service, Forest Products Laboratory; 1977.

Chapter 22

Wood-Base Particle Panel Materials

Wood-Base Particle Panel Materials*

The class of "wood-base particle panel materials" includes many subgroups known throughout the United States as particleboard, flakeboard, waferboard, or oriented strand board (OSB). These panel materials are similar because the wood raw material is first reduced to small fractions and then bonded back together, through specialized manufacturing methods, into panels of relatively large size (4 to 8 ft wide by 8 to 60 ft long) and moderate thicknesses (1/8 to 1-1/4 in). These board or panel materials in final form retain a few of the properties of the original wood, but because of the manufacturing methods, gain many new and different properties. As these are manufactured wood products, unlike solid wood, they can be and are tailored to satisfy the property requirements of a specific use or a broad group of end uses.

Particle panel products are defined as any wood-base panel product which is made from pieces of wood smaller than veneer sheets but larger than wood fiber. Information on veneered products may be found in chapters 10 and 11 and fiber panel products are discussed in chapter 21. In general, the name particleboard is used as a generic term for all particle panel products. The raw material for these products comes from a variety of sources including planer shavings, plywood mill waste, roundwood, sawdust, and pulp-type chips. The residues of planing operations constitute the source material for a large segment of the particleboard market for uses such as floor underlayment, furniture corestock, and molded items. Recently several types of particleboards, termed flakeboards, have been widely used as sheathing or single-layer floor panels. Depending upon the particle or flake size and the panel construction, flakeboards may be further classified as waferboard or OSB. Shown in figure 22–1 are three panel materials which can be broadly called particle panel products, and can be further classified as particleboard, waferboard, and OSB.

Particle Panel Processing Overview

Particle panel products are manufactured from residues of milling operations such as planer shavings, sawdust, and plywood trimmings, or alternatively the wood may be obtained from round logs or woods residue (such as branches, broken logs, and tops). Milling residues may be further reduced to desired size in a hammermill operation. Roundwood, the usual source of raw material for flakeboard, waferboard, or OSB, is usually heat conditioned in water prior to reduction to flakes in disk, drum, ring, or other flaking equipment (fig. 22–2). After this initial raw material production phase, the wood particles are called furnish for the process line (fig. 22–3).

The furnish is then dried to a uniform moisture content, usually ranging from 3 to 6 percent, and then is screened to segregate particle sizes. Fine particles may be used in the faces of traditional particleboard to produce a smooth surface, or they may be used as dryer fuel in flakeboard plants because fines are detrimental to the performance of structural panels. Adhesive binders at 3 to 7 percent (percent weight of dry wood) and wax at about 1 percent are either sprayed as a liquid or metered as a powdered blend while the wood particles are tumbled in a blender. Thermosetting urea-formaldehyde and phenol-formaldehyde adhesives are the major types of binders used, thus requiring the application of heat to cure them.

Two processes are commonly used for consolidation of the furnish into the final panel configuration, with the process depending on panel thickness.

For panels over 1/4 inch in final thickness, a flat press processing method is generally used. The blended furnish is formed into a mat of uniform height and moved into the platen press, where the mat is consolidated under controlled heat and pressure to a given average density. Pressing times depend upon many factors, but a typical process requires 3 to 6 minutes of press time to produce 1/2-inch-thick panels. Complete control of the mat forming process places particles according to their size, so that smooth-surface panels may be produced. In the processing of flakeboards, flakes may be aligned in layers through the mat thickness as a means of improving mechanical properties.

For panels 1/4 inch or less in thickness, the blended furnish may be deposited directly onto a large heated rotary drum, where the final mat-pressing or consolidation operation occurs.

Post-pressing operations for any of these panels may include cooling, trimming, sanding, and cutting to size.

Process Variable Effect on Particle Panel Properties

Particle Geometry

Board characteristics influenced by particle geometry involve most of the mechanical properties, nailholding and screwholding strength, surface smoothness, dimensional and weathering properties, and machining characteristics. In addition, the particle geometry influences most other process variables. The intricacy of the interactions is compounded by the fact that the furnish does not consist of one particle geometry but is a mix of many varying sizes.

Assessment of properties of homogeneous unaligned boards

---

* Revision by Theodore L. Laufenberg, General Engineer.

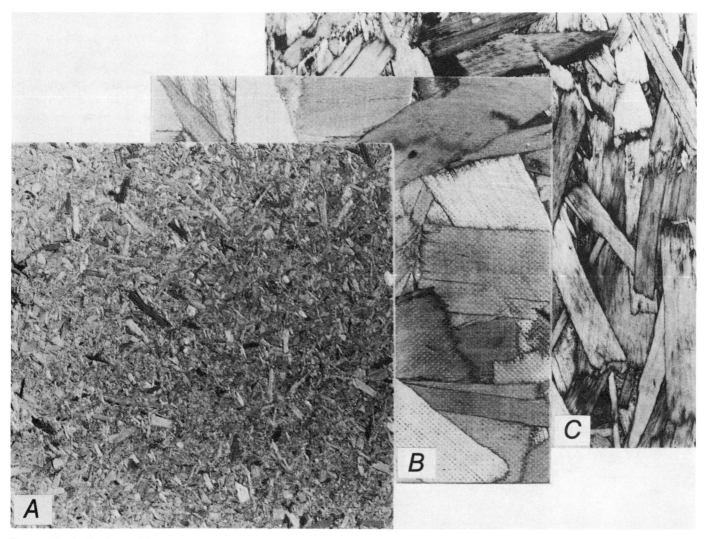

Figure 22–1—Basic particle panel products: *A* particleboard, *B* waferboard, and *C* oriented strand board.

made from particles having various geometries has shown that a most important parameter is the ratio of particle length along the fiber to particle thickness. A rule of thumb for obtaining panels with high bending, tensile, and compressive strength and stiffness is to have this ratio higher than 150. Tensile strength perpendicular to the plane of the panel (internal bond strength) is enhanced by thicker particles due to a relative increase in the amount of adhesive per unit surface area of the particles.

Dimensional stability of particle panel products is related to particle geometry in that this property is controlled by the orientations of particles relative to the board surface. Dimensional changes in board thickness due to changes in moisture content are larger for boards made with thick flakes. Dimensional changes in the plane of the board are reduced with longer particles.

Resin Content

If other variables are held constant, increasing resin content produces only a moderate improvement of bending, tensile, and compressive strengths in the plane of the board. The rate of improvement, however, is not the same for all properties and is affected by particle geometry. Internal bond strength and bond durability improve continuously with increased resin levels.

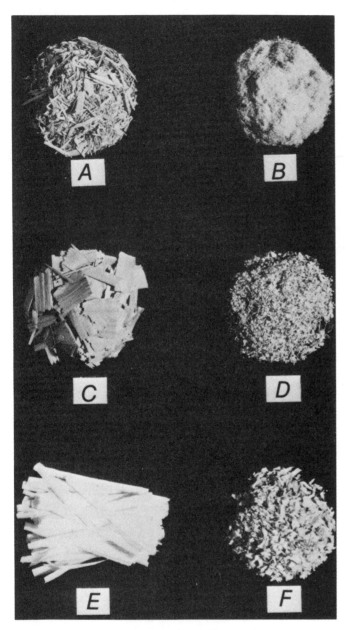

Figure 22−2—Numerous types of furnish suitable (M84 0289) for particle panel production: *A* strand-type flakes, *B* fiber bundles, *C* wafer-type flakes, *D* sawdust, *E* long flakes, and *F* planer shavings.

Although the amount of resin does not significantly affect some properties, the method of application does. The size of the resin droplets and the uniformity of their distribution on the wood particles is critical to the development of particle-to-particle bonds (fig. 22−4). Fine droplets produce a well-dispersed bond area for particles, which increases the mechanical properties of the board.

Resin Type

The bonds produced by different resin systems may be classified in two categories: those sufficiently durable for interior uses and those that are durable enough for protected exterior uses. Urea-formaldehyde resin is typically used to bind particleboard for interior uses. Phenol-formaldehyde or isocyanate resins are typical binders for protected exterior applications.

Density

All physical strength and stiffness properties may be improved by increasing the final density of the board. Bending, tensile and compressive strength, and stiffness increase linearly with density. Internal bond strength, nailholding and screwholding strength, and hardness are very sensitive to board density. Most hot-pressed particleboards possess a gradient of density levels through the thickness of the board, which was created during the pressing sequence. This density gradient is characteristically similar for most particleboards (fig. 22−5), but the differences are significant when assessing thickness swell or linear expansion properties. Board density also affects the rate of water absorption and the equilibrium moisture content (fig. 22−6).

A controlling relationship for the properties of particleboard is the ratio of board density to species density (compaction ratio). The bending strength (MOR) increases with increasing panel density but decreases with increasing species density (fig. 22−7). This relationship appears to be independent of species mix.

Particle Alignment

Alignment of the furnish is a processing variable that has tremendous influence on mechanical and dimensional properties. Typically the type of furnish that lends itself to mechanical alignment has particles longer than 3/4 inch and somewhat narrower in width. These particles may be oriented in either orthogonal direction and may be formed into several distinct layers in the mat. The aligned particles improve the mechanical and dimensional properties in the direction parallel to the orientation, but these improvements are at the expense of those same properties in the opposite direction. The strength and stiffness of panels with aligned particles is capable of exceeding 2.5 times those of panels with random orientation.

Panel shear strength of an aligned panel is less than a similar panel with random orientation. The alignment of flakes produces planes of weakness parallel to the alignment direction. Shear strength perpendicular to the plane of the board (interlaminar shear) is increased significantly due to improved bonding between similarly oriented flakes.

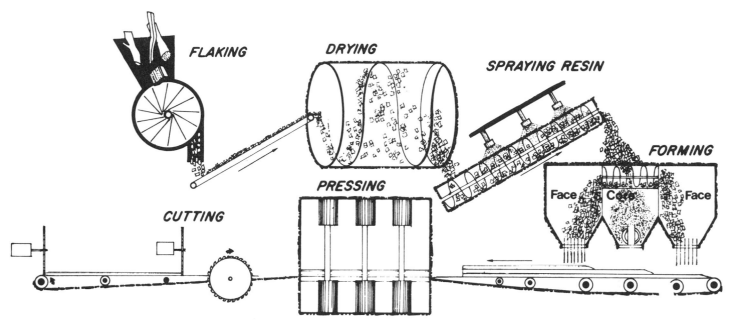

Figure 22 – 3—Process line schematic for production of mat-pressed particleboards.

(ML83 5387)

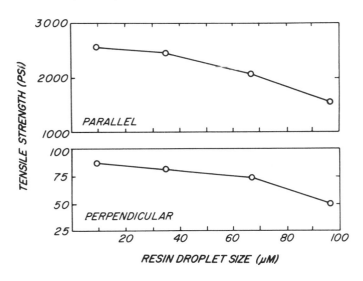

Figure 22 – 4—Influence of resin droplet size on tensile strength perpendicular and parallel to particleboard surface.

(ML84 5813)

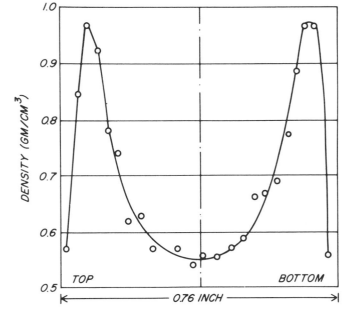

Figure 22 – 5—Vertical density gradient of a three-layer particleboard.

(ML84 5817)

Hot Pressing

Though limited theoretical analysis has been done on the physics of particleboard pressing, a great deal of information is available from empirical studies. The most obvious "signature" from any particular pressing schedule is the density gradient through the thickness of the panel (fig. 22–5). Press schedule variables for any one furnish type include rate of closing, moisture content distribution in the mat, press temperature and pressure (which is influenced by closing rate), and length of time in the press.

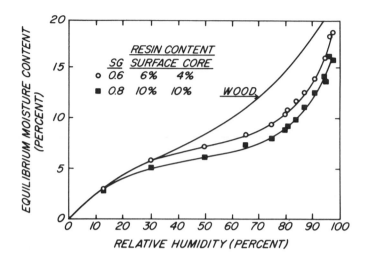

Figure 22−6—Absorption isotherms for two particleboards compared to wood, at 70 °F.   (ML84 5819)

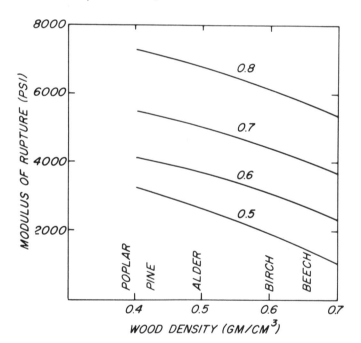

Figure 22−7—Relationship between wood density and modulus of rupture for four particleboard densities. (Use the curves with care, as a board density of 0.5 is not currently possible with wood of 0.6 density.)   (ML84 5816)

The most common manipulation of the pressing process utilizes a higher moisture content furnish on the faces of the mat. As the hot platens contact the mat, significantly greater amounts of steam are produced than during pressing of a mat with lower face moisture. This "steam shock" moves through the unpressed mat toward the centerline, heating the mat and

effectively shortening the time needed in the press to cure the resin. Another effect of the steam shock method is that the surface layer particles are plasticized and easily densified. In the particle panel product these high-density faces enhance bending properties and surface hardness and smoothness at the expense of internal bond strength and edge integrity.

Specifications for Particle Panel Products

Two approaches have been taken in providing specifications for particle panel products. The traditional approach is to set prescriptive standards for a particular product with minimum mechanical and dimensional properties. An alternate method is to specify what properties a panel must have to perform in a given end use, allowing any number of materials to qualify under the performance standard for that end use.

A majority of the particle panel products commercially available are addressed in the American National Standard, ANSI A 208.1, for mat-formed wood particleboard. This standard specifies minimum mechanical and physical properties as well as dimensional tolerances for the panel. A summary of the properties required by that specification may be found in table 22−1.

Two panel types are recognized: (a) type 1, a particle panel made with urea-formaldehyde (or equivalent) (table 22−1), and (b) type 2, made with phenol-formaldehyde (or equivalent) (table 22−2). Simply stated, type 1 panels are intended for interior use and type 2 panels are for protected exterior and sheathing uses. Three density classes exist within each panel type. High density (H) denotes a density at 5 percent moisture content which exceeds 53 pcf; medium density (M), 38 to 53 pcf; and low density (L), less than 38 pcf. All mechanical property values listed in table 22−1 are used for quality control and grade certification and must not be misconstrued as design allowable values. In-service conditions, material variability, and installation practices preclude the use of these test values for engineering design. Properties of these panels are determined by the test conditions and methods set forth in ASTM D 1037. Within the type 2 grades (table 22−2) are two grades of material that identify a specific particle type to be used in their manufacture.

Phenolic-bonded particleboards, waferboards, and OSB are finding increasing use as sheathing and combination subfloor-underlayment. Many boards are recognized for use based on performance testing rather than on prescriptive standards. These are marketed under a quality control program designed to assure that structural properties do not fall below those of panels successfully passing performance testing. The American Plywood Association has several panel product performance standards that specify the structure of these quality

Table 22 – 1—Property requirements for Type 1 mat-formed particleboard in ANSI A 208.1 (average values for sample consisting of five panels¹)

Grade,² type 1	Length and width tolerance	Thickness tolerance		Modulus of rupture	Modulus of elasticity	Internal bond	Hardness	Linear expansion, maximum average	Screwholding	
		Panel average³	Within panel⁴						Face	Edge
		Inch		Pounds per square inch			Pounds	Percent	Pounds	
1-H-1	±1/16	±0.010	±0.005	2,400	350,000	130	500	NS⁵	400	300
1-H-2	±1/16	±.010	±.005	3,000	350,000	130	1,000	NS	425	350
1-H-3	±1/16	±.010	±.005	3,400	400,000	140	1,500	NS	450	350
1-M-1	±0-1/8	±.015	±.010	1,600	250,000	60	500	0.35	NS	NS
1-M-2	±1/16	±.010	±.005	2,100	325,000	60	500	.35	225	200
1-M-3	±1/16	±.010	±.005	2,400	400,000	80	500	.35	250	225
1-L-1	±1/16	±.005 -.015	±.005	800	150,000	20	NS	.30	125	NS

¹ Except for dimensional tolerances which are individual panel values.
² Made with urea-formaldehyde resin binders or equivalent bonding systems.
³ From nominal.
⁴ Individual measurement from panel average.
⁵ NS—not specified.

Table 22 – 2—*Property requirements for Type 2 mat-formed particleboard in ANSI A 208.1 (average values for sample consisting of five panels[1])*

Grade,[3] type 2	Length and width tolerance	Thickness tolerance[2] Panel average[4]	Within panel[5]	Modulus of rupture	Modulus of elasticity	Internal bond	Hardness	Linear expansion, maximum average[6]	Screwholding Face	Edge
	- - - - - - - - - - - Inch - - - - - - - - - - -			- - - Pounds per square inch - - -			Pounds	Percent	Pounds	
2-H-1	±1/16	±0.015	±0.005	2,400	350,000	125	500	NS	400	300
2-H-2	±1/16	±.015	±.005	3,400	400,000	300	1,800	NS	450	350
2-M-1	+0 -1/8	±.015	±.010	1,800	250,000	60	500	0.35	225	160
2-M-2	+0 -1/8	±.015	±.010	2,500	450,000	60	500	.35	250	200
2-M-3	±0 -1/8	±.015	±.010	3,000	500,000	60	500	.35	NS	NS
2-MW[7]	+0 -1/8	±.015	±.010	2,500	450,000	50	500	.20	NS	NS
2-MF[8]	±0 -1/8	±.015	±.010	3,000	500,000	50	500	.20	NS	NS

[1] Except for dimensional tolerances which are individual panel values.
[2] Values shown are for sanded panels as defined by the manufacturer. Values for unsanded panels for all 2-M grades shall be ±0.030 for panel average and ±0.030 within panels.
[3] Made with phenol-formaldehyde resins or equivalent bonding system.
[4] From nominal.
[5] Individual measurement from panel average.
[6] NS—not specified.
[7] Product is made from wafers.
[8] Product is made from flakes.

control programs. This program also establishes minimum criteria for physical properties and durability.

Some standards on specific uses of particleboard are issued by trade associations such as the National Particleboard Association (NPA), building code organizations, and government agencies. A summary of these standards and the uses to which they apply is in table 22—3.

Particleboard Properties and Uses

The particle panel product that is typically made from small wood particles of mill residue is designated particleboard. While particleboards are difficult to characterize because of the infinite variations in the process variables discussed previously, the properties given in table 22—4 should pro-

Table 22—3—*Particleboard application reference table[1]*

Applications	Grade	Product reference[2]
INTERIOR		
Floor underlayment	1—M—1	ICBO, SBCC, BOCA, HUD/FHA MPS, One and Two Family Dwelling Code
Mobile home decking	Class D—2 Class D—3	NPA 1—73 HUD-Mobile Home Construction and Safety Standards
Shelving	1—M—1 1—M—2 1—M—3	
Countertops	1—M—2 1—M—3	ANSI A 161.2
Kitchen cabinets	1—M—1 1—M—2	ANSI A 161.1
Door core	1—L—1	NWMA Industry Standard Series I.S. 1—78 (wood flush doors)
Stair treads	1—M—3	HUD/FHA UM 70, ICBO RR 3390
Moldings	1—M—3	WMMP Standard WM 2—73

Applications	Grade	Product reference[2]
EXTERIOR		
Roof sheathing	2—M—W	ICBO, SBCC, BOCA, One and Two Family Dwelling Code
Wall sheathing	2—M—1	ICBO, SBCC, BOCA, One and Two Family Dwelling Code
Wall sheathing	2—M—W	ICBO, SBCC, BOCA, One and Two Family Dwelling Code
Combined subfloor underlayment	2—M—3 2—M—W	ICBO, SBCC, BOCA, One and Two Family Dwelling Code
Factory built decking	NPA 2—72	HUD/FHA UM—57a
Siding	2—M—1	ICBO, SBCC, BOCA, One and Two Family Dwelling Code
Siding	2—M—W	SBCC, BOCA

[1] Grades shown refer to ANSI A 208.1 except for mobile home decking which refers to NPA 1—73 and factory-built decking which refers to NPA 2—72.

ICBO — International Conference of Building Officials, Whittier, CA.

SBCC — Southern Building Code Congress International, Birmingham, AL.

BOCA — Building Officials and Code Administrators International, Chicago, IL.

HUD/FHA — Housing and Urban Development/Federal Housing Authority, Washington, DC.

NWMA — National Woodwork Manufacturers Association, Park Ridge, IL.

WMMP — Wood Moulding and Millwork Producers, Portland, OR.

vide an idea of the ranges possible for the mat-formed product. Again, the primary distinction between flakeboards and particleboards is in the size of the particle or flake. The properties listed do not reflect the properties attainable when using a flake furnish.

Mat-Formed Particleboard

Approximately 85 percent of interior-type mat-formed particleboard produced has traditionally been used as core stock for a wide variety of furniture and cabinet applications and floor underlayment in light-frame construction. The majority of production is between 1/2 and 2 inches thick for mat-formed boards. Low-density panels are produced in thicknesses to 1-1/2 inches for the solid core door market.

As corestock, particleboard has moved into the market formerly held by lumber-core panels and, to a limited extent, plywood. Certain grades of hardwood plywood now permit the use of particleboard as the core ply, where formerly lumber core was specified. The type of facing or finishing system applied commonly controls the construction of the particleboard produced. A panel to be overlaid with 0.0015-inch ultraviolet-cured vinyl overlay requires fine particles, or possibly fibers, on the surfaces to reduce showthrough. An overlay of veneer or high-density plastic can accommodate a coarser face particle. Balanced construction in layups using particleboard is important to minimize warping, cupping, or twisting in service.

Edges of particleboard exposed in furniture or cabinetry are frequently covered as there are coarse particles in the lower density core. Edge banding with plastic extrusions or a high-density plastic are common treatments. Filling of edge voids and subsequent finishing, or bonding solid wood to exposed edges are other options.

As floor underlayment, particleboard provides a smooth, stiff, and hard surface for coverings of carpet, resilient tile, and seamless floor coverings. Particleboard for this use is produced in 4- by 8-foot panels 1/4 to 3/4 inch thick. Specifications for these numerous uses have been written to cover particleboard floor underlayments, and manufacturers provide individual application instructions to ensure proper construction techniques. Particleboard underlayment is sold under a certified quality program where established grademarks clearly identify the use, quality, grade, and originating mill.

Other uses for particleboard require a more durable type 2 (ANSI A 208.1) adhesive in the board. Siding, combined siding-sheathing, soffit linings, ceilings for carports, porches, and other protected exterior applications are examples of uses approved by some of the model building codes. Satisfactory performance of particleboard in protected exterior environments depends not only on the manufacturing process and kind of adhesive used, but on the protection afforded by a finish. This may be a plastic film, paper overlay, or paint which is factory applied.

Manufacturers of mobile homes and factory-built conventional housing also use particleboard. Because these uses may require larger sized panels than those used for conventional construction particleboard is manufactured in sizes as great as 8 by 60 feet. With mechanical handling available in factories, large-sized panels can be positioned and attached to structural members effectively and economically. Two particleboard products have been developed to satisfy these uses. Mobile home decking is used for combined subfloor and underlayment. It is a board with a type 1 bond, but is intended to be protected from moisture when in use by a subfloor which is exposed to the exterior environment. Thus, the board is generally regarded as giving satisfactory service for mobile home use. Particleboard decking for factory-built housing is similar to mobile home decking, but it has a type 2 bond for longer life. The National Particleboard Association has established separate standards for these products. They are marketed under a certified product quality program with each product adequately identified.

Extruded Particleboard

Extruded products are typically "fluted" in that the product is not of uniform thickness over its width. This configuration is used in flush doors or overlaid with a structural panel product such as a hardboard, flakeboard, or plywood for sandwich constructions. The extruded products have distinct zones of weakness across the width of the panel as extruded; thus they are rarely used without facings of some kind glued to them. These facings control the physical and mechanical properties of the sandwich construction. This type of particleboard also has a strong tendency to swell in the lengthwise direction due to particle orientation and subsequent compression as particles are rammed through the heated die during manufacture.

Flakeboard Properties and Uses

Flakeboard is a generic subset of products included under particle panel products. Flakeboards are structural panels made from specially produced flakes, typically from relatively low-density species, such as aspen or pine, and bonded with an exterior-type water-resistant adhesive. The industry began production in the 1950's with a small plant in Sandpoint, Idaho. Subsequent plants built in the 1960's in Hudson Bay, Saskatchewan, proved to be the major developing ground for the flakeboard industry. Producing mills are now located in

Table 22–4—Physical and mechanical properties of mat-formed (platen-pressed) wood particleboard[1]

Type of particleboard	Density	Specific gravity	Modulus of elasticity (bending)	Modulus of rupture	Tensile strength		Shear strength		24-hour water adsorption	Thickness swelling, 24-hour soaking	Linear expansion[2]	Thermal conductivity[3]
					Parallel	Perpendicular	In board plane	Across board plane				
	Pcf		1,000 psi	Psi	Psi		Psi		Pct by wt	Pct	Pct	Btu·in/hr ·ft²·°F
Low density	[4]25–37	[4]0.40–0.59	[5]150–250	[5]800–1,400	—	[5]20–30	—	—	—	—	[6]0.30	0.55–0.75
Medium density	37–50	.59–.80	250–700	1,600–3,000	500–2,000	30–200	100–450	200–1,000	10–50	5–50	.2–.6	.75–1.00
High density	50–70	.80–1.12	350–1,000	2,400–7,500	1,000–5,000	125–450	200–800	—	15–40	15–40	.2–.85	1.00–1.25

[1] General round-figure values, accumulated from numerous sources; for more exact figures on a specific product, individual manufacturers should be consulted or tests made. Values are for general laboratory conditions of temperature and humidity.
[2] From measurements made on material at equilibrium from 50 to 90 percent relative humidity at room temperature.
[3] At a mean temperature of 75 °F.
[4] Lower of values is for boards as generally manufactured; lower density products with lower properties may be made.
[5] Only limited production of low-density particleboard so values presented are specification limits.
[6] Maximum permitted by specification.

Table 22–5—Ranges of physical and mechanical properties for commercially available flakeboard

Type of flakeboard	Property[1]								
	Density[2]	Water absorption	Thickness swell	Linear expansion	Modulus of rupture	Modulus of elasticity	In-plane shear strength	Hardness	Internal bond
	Pcf	Percent	Percent	Percent	Psi	1,000 psi	Psi	Lb	Psi
Waferboard	38–45	10–30	10–16	0.08–0.15	2,000–3,500	450–650	1,200–1,800	700–1,000	50–100
Oriented strand board									
Parallel to alignment	38–50	10–30	10–20	.05–.10	4,000–7,000	750–1,300	1,000–1,500	700–1,000	70–100
Perpendicular to alignment	—	—	—	.12–.30	1,500–3,500	300–500	—	—	—

[1] Not design allowables.
[2] Limited by specification (ANSI A 208.1) to 38–53 pcf.

the Great Lakes region, New England, and most Canadian provinces.

The youth of the flakeboard industry has not prevented it from making major inroads into the light-frame construction industry. Beginning in the early 1970's, interest in flakeboards increased for a variety of reasons including a strong demand for structural board products and an increase in softwood plywood prices. Utilization of the lower priced aspen stumpage in the Great Lakes and Hudson Bay regions, coupled with closer proximity to the large eastern markets than western or southern plywood mills, gave builders a low-cost alternative to plywood for sheathing uses.

Two major types of flakeboard are recognized, waferboard and OSB. Waferboard, the product produced almost exclusively from aspen wafers (wide flakes), traditionally possesses no intentional orientation of these flakes, and is bonded with an exterior-type resin. OSB is a flakeboard product that emerged in the market place during the early 1980's. Exterior-type resin is applied to wood strands (long and narrow flakes) that are formed into a mat of three to five layers. The strands in each layer are aligned in a direction 90° from the adjacent layer. Flake alignment gives the OSB bending properties (in the aligned direction) that are generally superior to those of a randomly oriented flake waferboard. Some important physical and mechanical properties for flakeboards are summarized in table 22−5. As with any particle panel product, the properties are highly dependent upon the process used to manufacture the panel; the values in tables 22−5 and 22−6 were generated from typical commercially available flakeboards.

The properties of flakeboards indicate that products are suitable for many applications now dominated by softwood plywood. Primary markets for flakeboards are wall and roof sheathing, single-layer flooring, and underlayment in light-frame construction. Several building codes have approved flakeboards for siding materials either over sheathing or nailed directly to the studs. Exterior ceilings, soffits, and interior walls or ceilings are other uses in light residential construction that have received building code approval. In nonresidential applications, flakeboard is used as sheathing/siding of farm structures, industrial packaging, crating, and pallet decks.

Other Wood-Based Particle Products

Veneer/Particle Composite Panels

Structural composite panels marketed since the mid-70's are made with face veneers bonded to a core layer made from oriented strands or random particles. This product (sometimes referred to as COM-PLY) is marketed as a plywood product and is considered to be interchangeable with construction grades of plywood (see ch. 11). Thicknesses of 3/8 to 1/2 inch are common for sheathing uses of this panel, and thicknesses up to 1-1/8 inches have been produced for single-layer flooring applications. The performance standard for structural-use panels sponsored by the American Plywood Association embodies procedures for grademarking this type of panel.

Mende-Process Particleboard

A unique process for producing thin particleboard, the Mende process, involves continuous pressing of the mat by application of heat and pressure through a large rotating cylinder. The board thickness ranges up to 1/4 inch, which permits it to be formed on this cylindrical platen and subsequently flattened. Furnish for the process may range from fibers to flakes. The panels produced in this process may be

Table 22 − 6—Effect of waferboard thickness on thermal conductivity and lateral and direct nail withdrawal

Board thickness	Property		
	Thermal conductivity	Lateral nail loads[1,2]	Withdrawal nail loads[1]
In	*Btu·in/h·ft²·°F*	- - - - - - - - - - - - - - *Lb* - - - - - - - - - - - - - -	
5/16	2.4	200 +	20 +
7/16	1.8	400 +	40 +
5/8	1.2	550 +	65 +

[1] For 6d common nails.
[2] Edge distance 1/2 inch.

used as wall paneling (with printed overlay or other surface finish), furniture backing, drawer bottoms, and case goods.

Molded Particle Products

Moldings from wood particles can be defined as parts that are formed from furnish blended with less than 25 percent binder resin and cured in dies under heat and pressure. Limited flow of the furnish during pressing restricts the kind of items that may be profitably molded. Exterior siding, door jambs, window sills, table tops, pallets, casket tops, and many other items may be molded using conventional particles. By using finer particles which approach wood flour in size, items with large relief such as toilet seats and croquet balls may be compression molded.

Cement-Bonded Particle Products

Portland cement is commercially used as a binder for a special class of wood particle panels. The wood particles typically used are called excelsior, or wood-wool, as they are long (up to 10 in) and stringy. Medium- to low-density species are reduced to excelsior, blended with cement, formed into mats, and pressed to a density of 30 to 40 pcf. A common use for this product is as roof decking due to its sound-absorbing and fire-resistive properties. Other cement-bonded particle products include building blocks and a panel made with flakes that can be used in doors, floors, load-bearing walls, partitions, concrete forms, and exterior siding.

Selected References

American National Standards Institute. Mat-formed wood particleboard. ANSI A 208.1. New York; 1979.

American Plywood Association. Performance standards and policies for structural-use panels. Tacoma, WA: APA; 1982.

American Society for Testing and Materials. Standard definitions of terms relating to wood-base fiber and particle panel materials. ASTM D 1554. Philadelphia, PA: ASTM; (see current edition).

American Society for Testing and Materials. Standard methods of evaluating the properties of wood-base fiber and particle panel materials. ASTM D 1037. Philadelphia, PA: ASTM; (see current edition).

Canadian Waferboard Association. Waferboard in residential construction. Don Mills, ON, Canada; 1982.

Geimer, R. L.; Hoover, W. L.; Hunt, M. O. Design and construction of large-scale reconstituted wood roof decking. RB 973. West Lafayette, IN: Purdue University, Agricultural Experiment Station; 1982.

Geimer, R. L. Data basic to the engineering design of a reconstituted flakeboard. In: Proceedings, 13th Washington State University International Symposium on Particleboard. Pullman, WA: Washington State University;, 1979. 105−125.

Jessome, A. P. Canadian waferboard properties. Tech. Rep. 523E. Vancouver, BC, Canada: FORINTEK Canada Corporation; 1979.

Kelly, M. W. Critical literature review of relationships between processing parameters and physical properties of particleboard. Gen. Tech. Rep. FPL−10. Madison, WI: U.S. Department of Agriculture, Forest Service, Forest Products Laboratory; 1977.

Kieser, J.; Steck, E. F. The influence of flake orientation on the MOR and MOE of strandboards. In: Proceedings, 12th Washington State University Symposium on Particleboard. Pullman, WA: Washington State University; 1978: 99−122.

National Particleboard Association. Standard for particleboard for mobile home decking. NPA 1−73. Silver Springs, MD: NPA; 1973.

National Particleboard Association. Standard for particleboard floor underlayment coated with wax-polymer type hot melt coatings. NPA 3−73. Silver Springs, MD: NPA; 1973.

National Particleboard Association. Filling particleboard. Tech. Bull. 9. Silver Springs, MD: NPA; 1972.

National Particleboard Association. Standard for particleboard decking for factory-built housing. NPA 2−72. Silver Springs, MD: NPA; 1972.

National Particleboard Association. Edge banding and veneering of particleboard. Tech. Bull. 7. Silver Springs, MD: NPA; 1971.

National Research Board. APA structural-use panels. Rep. NRB-108. Whittier, CA: Council of American Building Officials; 1982.

Nestler, F. H. M. The formaldehyde problem in wood-based products - an annotated bibliography. Gen. Tech. Rep. FPL−8. Madison, WI: U.S. Department of Agriculture, Forest Service, Forest Products Laboratory; 1977.

Snodgrass J. D. Manufacture of oriented-strand core for composite plywood. In: Proceedings, 11th Washington State University Symposium on Particleboard. Pullman, WA: Washington State University; 1977: 453−470.

Stillinger, J. R.; Wentworth, Irv W. Product, process and economics of producing structural wood-cement panels by the Bison-Werke system. In: Proceedings 11th Washington State University Symposium on Particleboard. Pullman, WA: Washington State University; 1977: 383−410.

Chapter 23

Modified Woods and Paper-Base Laminates

Modified Woods and Paper-Base Laminates*

Materials with properties substantially different than the base material are obtained by chemically treating a wood or wood-base material, compressing it under specially controlled conditions, or by combining the processes of chemical treatment and compression. Sheets of paper treated with chemicals are laminated and hot-pressed into thicker panels that have the appearance of plastic rather than paper, and they are used in special applications because of their structural properties and in items requiring hard, impervious, and decorative surfaces.

Modified woods, modified wood-base materials, and paper-base laminates are normally more expensive than wood because of the cost of the chemicals and the special processing required to produce them. Thus, their use is generally limited to special applications where the increased cost is justified by the special properties needed.

Modified Woods

Wood is treated with chemicals to increase hardness and other mechanical properties, as well as its resistance to decay, fire, and moisture. Application of water-resistant chemicals to the surface of wood, impregnation of the wood with such chemicals dissolved in volatile solvents, or bonding chemicals to the cell wall polymer reduces the rate of swelling and shrinking of the wood when in contact with water. Such treatments may also reduce the rate at which wood changes dimension because of humidity changes, even though they do not affect the final dimension changes caused by long-duration exposures. Paints, varnishes, lacquers, wood-penetrating water repellents, and plastic and metallic films retard the rate of moisture absorption, but have little effect on total dimension change if exposures are long enough.

Resin-Treated Wood (Impreg)

Permanent stabilization of the dimensions of wood is needed for certain specialty uses. This can be accomplished by depositing a bulking agent within the swollen structure of the wood fibers. The most successful bulking agents that have been commercially applied are highly water-soluble, thermosetting, phenol-formaldehyde resin-forming systems, with initially low molecular weights. No thermoplastic resins have been found that effectively stabilize the dimensions of wood.

Wood treated with a thermosetting fiber-penetrating resin and cured without compression is known as impreg. The wood (preferably green veneer to facilitate resin pickup) is soaked in the aqueous resin-forming solution or, if air dry, is impregnated with the solution under pressure until the resin content equals 25 to 35 percent of the weight of dry wood. The treated wood is allowed to stand under nondrying conditions for a day or two to permit uniform distribution of the solution throughout the wood. The resin-containing wood is dried at moderate temperatures to remove the water and then heated to higher temperatures to set the resin.

Uniform distribution of the resin has been effectively accomplished with thick wood specimens only in sapwood of readily penetrated species. Although thicker material can be treated, the process is usually applied to veneers up to 1/3 inch thick, since treating time increases rapidly with increases in thickness. Drying thick resin-treated wood may result in checking and honeycombing. For these reasons, treatments should be confined to veneer and the treated-cured veneer used to build up the desired products. Any species can be used for the veneer except the resinous pines. The stronger the original wood, the stronger will be the product.

Impreg has a number of properties differing from those of normal wood and ordinary plywood. These are given in table 23−1, together with similar generalized findings for other modified woods. Data for the strength properties of birch impreg are given in table 23−2. Information on thermal expansion properties of ovendry impreg is given in table 23−3.

The good dimensional stability of impreg has been the basis of one use where its cost was no deterrent to its acceptability. Wood dies of automobile body parts serve as the master from which the metal-forming dies are made for actual manufacture of parts. Small changes in moisture content, even with the most dimensionally stable wood, produce changes in dimension and curvature of an unmodified wood die; such changes create major problems in making the metal-forming dies where close final tolerances are required. The substitution of impreg, with its high antishrink efficiency (ASE) (table 23−4), almost entirely eliminated the problem of dimensional change during the entire period that the wood master dies were needed. Despite the tendency of the resins to dull cutting tools, patternmakers accepted the impreg readily because it machines without splitting more easily than unmodified wood.

Patterns made from impreg also are superior to unmodified wood in resisting heat when used with shell-molding techniques where temperatures as high as 400 °F were required to ure the resin in the molding sand.

Resin-Treated Compressed Wood (Compreg)

Compreg is similar to impreg except that it is compressed before the resin is cured within the wood. The resin-forming chemicals (usually phenol-formaldehyde) act as plasticizers for the wood so that it can be compressed under modest

*Revision by Donald S. Fahey, Forest Products Technologist; Roger M. Rowell, Chemist; and Theodore H. Wegner, Chemical Engineer.

Table 23 – 1—*Properties of modified woods*

Property	Impreg	Compreg	Staypak
Specific gravity	15 to 20 pct greater than normal wood	Usually 1.0 to 1.4	1.25 to 1.40
Equilibrium swelling and shrinking	1/4 to 1/3 that of normal wood	1/4 to 1/3 that of normal wood at right angle to direction of compression, greater in direction of compression but very slow to attain	Same as normal wood at right angle to compression, greater in direction of compression but very slow to attain
Springback	None	Very small when properly made	Moderate when properly made
Face checking	Practically eliminated	Practically eliminated for specific gravities below 1.3	About the same as in normal wood
Grain raising	Greatly reduced	Greatly reduced for uniform-texture woods, considerable for contrasting-grain woods	About the same as in normal wood
Surface finish	Similar to normal wood	Varnished-like appearance for specific gravities above about 1.0. Cut faces can be given this surface by sanding and buffing	Varnished-like appearance. Cut surfaces can be given this surface by sanding and buffing
Permeability to water vapor	About 1/10 that of normal wood	No data, but presumably much lower than impreg	No data, but presumably lower than impreg
Decay and termite resistance	Considerably better than normal wood	Considerably better than normal wood	Normal, but decay occurs somewhat more slowly
Acid resistance	Considerably better than normal wood	Better than impreg because of impermeability	Better than normal wood because of impermeability but not as good as compreg
Alkali resistance	Same as normal wood	Somewhat better than normal wood because of impermeability	Somewhat better than normal wood because of impermeability
Fire resistance	Same as normal wood	Same as normal wood for long exposures, somewhat better for short exposures	Same as normal wood for long exposures, somewhat better for short exposures
Heat resistance	Greatly increased	Greatly increased	No data
Electrical conductivity	1/10 that of normal wood at 30 pct RH; 1/1,000 that of normal wood at 90 pct RH	Slightly more than impreg at low relative humidity values due to entrapped water	No data

Table 23 – 1—*Properties of modified woods*—Continued

Property	Impreg	Compreg	Staypak
Heat conductivity	Slightly increased	Increased about in proportion to specific gravity increase	No data, but should increase about in proportion to specific gravity increase
Compressive strength	Increased more than proportional to specific gravity increase	Increased considerably more than proportional to specific gravity increase	Increased about in proportion to specific gravity increase parallel to grain, increased more perpendicular to grain
Tensile strength	Decreased significantly	Increased less than proportional to specific gravity increase	Increased about in proportion to specific gravity increase
Flexural strength	Increased less than proportional to specific gravity increase	Increased less than proportional to specific gravity increase parallel to grain, increased more perpendicular to grain	Increased proportional to specific gravity increase parallel to grain, increased more perpendicular to grain
Hardness	Increased considerably more than proportional to specific gravity increase	10 to 20 times that of normal wood	10 to 18 times that of normal wood
Impact strength: Toughness	About 1/2 of value for normal wood but very susceptible to the variables of manufacture	1/2 to 3/4 of value for normal wood but very susceptible to the variables of manufacture	Same to somewhat greater than normal wood
Izod	About 1/5 of value for normal wood	1/3 to 3/4 of value for normal wood	Same to somewhat greater than normal wood
Abrasion resistance (tangential)	About 1/2 of value for normal wood	Increased about in proportion to specific gravity increase	Increased about in proportion to specific gravity increase
Machinability	Cuts cleaner than normal wood, but dulls tools more	Requires metalworking tools and metalworking tool speeds	Requires metalworking tools and metalworking tool speeds
Moldability	Cannot be molded, but can be formed to single curvatures at time of assembly	Can be molded by compression and expansion molding methods	Cannot be molded
Gluability	Same as normal wood	Same as normal wood after light sanding, or, in the case of thick stock, machining surfaces plane	Same as normal wood after light sanding, or, in the case of thick stock, machining surfaces plane

Table 23 – 2—Strength properties of normal and modified laminates[1] of yellow birch and a laminated paper plastic

Property	Normal laminated wood[2]	Impreg (impregnated, uncompressed)[3]	Compreg (impregnated, highly compressed)[3]	Staypak (unimpregnated, highly compressed)[2]	Paper laminate (impregnated, highly compressed)[4]
Thickness (t) of laminate (in)	0.94	1.03	0.63	0.48	0.126 0.512
Moisture content at time of test (pct)	9.2	5.0	5.0	4.0	—
Specific gravity (based on weight and volume at test)	0.7	0.8	1.3	1.4	1.4
PARALLEL LAMINATES					
Flexure—grain parallel to span (flatwise):[5]					
Proportional limit stress (psi)	11,500	15,900	26,700	20,100	15,900
Modulus of rupture (psi)	20,400	18,800	36,300	39,400	36,600
Modulus of elasticity (1,000 psi)	2,320	2,380	3,690	4,450	3,010
Flexure—grain perpendicular to span (flatwise):[5]					
Proportional limit stress (psi)	1,000	1,300	4,200	3,200	10,500
Modulus of rupture (psi)	1,900	1,700	4,600	5,000	24,300
Modulus of elasticity (1,000 psi)	153	220	626	602	1,480
Compression parallel to grain (edgewise):[6]					
Proportional limit stress (psi)	6,400	10,200	16,400	9,700	7,200
Ultimate strength (psi)	9,500	15,400	26,100	19,100	20,900
Modulus of elasticity (1,000 psi)	2,300	2,470	3,790	4,670	3,120
Compression perpendicular to grain (edgewise):[6]					
Proportional limit stress (psi)	670	1,000	4,800	2,600	4,200
Ultimate strength (psi)	2,100	3,600	14,000	9,400	18,200
Modulus of elasticity (1,000 psi)	162	243	571	583	1,600
Compression perpendicular to grain (flatwise):[5]					
Maximum crushing strength (psi)	—	4,280	16,700	13,200	42,200
Tension parallel to grain (lengthwise):					
Ultimate strength (psi)	22,200	15,800	37,000	45,000	35,600
Modulus of elasticity (1,000 psi)	2,300	2,510	3,950	4,610	3,640

23-5

Table 23–2—*Strength properties of normal and modified laminates¹ of yellow birch and a laminated paper plastic*—Continued

Property	Normal laminated wood[2]	Impreg (impregnated, uncompressed)[3]	Compreg (impregnated, highly compressed)[3]	Staypak (unimpregnated, highly compressed)[2]	Paper laminate (impregnated, highly compressed)[4]
Tension perpendicular to grain (edgewise):					
Ultimate strength (psi)	1,400	1,400	3,200	3,300	20,000
Modulus of elasticity (1,000 psi)	166	227	622	575	1,710
Shear strength parallel to grain (edgewise):[6]					
Johnson, double shear across laminations (psi)	2,980	3,460	7,370	6,370	17,800
Cylindrical, double shear parallel to laminations (psi)	3,020	3,560	5,690	3,080	3,000
Shear modulus:					
Tension method (1,000 psi)	182	255	454	—	—
Plate shear method (FPL test) (1,000 psi)	—	—	—	385	909
Toughness (FPL test, edgewise)[6] (in–lb)	235	125	145	250	—
Do (in-lb per in of width)	250	120	230	515	—
Impact strength (Izod)—grain lengthwise:					
Flatwise (notch in face) (ft–lb per in of notch)	14.0	2.3	4.3	12.7	4.7
Edgewise (notch in face) (ft–lb per in of notch)	11.3	1.9	[7]3.2	—	0.67
Hardness:					
Rockwell, flatwise[5] (M–numbers)	—	–22	84	—	110
Load to embed 0.444-inch steel ball to 1/2 its diameter (lb)	1,600	2,400	—	—	—
Hardness modulus (H_M)[8] (psi)	5,400	9,200	41,300	43,800	35,600
Abrasion-Navy wear-test machine (flatwise),[5] wear per 1,000 revolutions (in)	0.030	0.057	0.018	0.015	0.018
Water absorption (24-hr immersion), increase in weight (pct)	43.6	13.7	2.7	4.3	2.2
Dimensional stability in thickness direction:					
Equilibrium swelling (pct)	9.9	2.8	8.0	29	—
Recovery from compression (pct)	—	0	0	4	—

Table 23–2—Strength properties of normal and modified laminates[1] of yellow birch and a laminated paper plastic—Continued

Property	Normal laminated wood[2]	Impreg (impregnated, uncompressed)[3]	Compreg (impregnated, highly compressed)[3]	Staypak (unimpregnated, highly compressed)[2]	Paper laminate (impregnated, highly compressed)[4]
CROSSBAND LAMINATES					
Flexure—face grain parallel to span (flatwise):[5]					
Proportional limit stress (psi)	6,900	8,100	14,400	11,400	12,600
Modulus of rupture (psi)	13,100	11,400	22,800	25,100	31,300
Modulus of elasticity (1,000 psi)	1,310	1,670	2,480	2,900	2,240
Compression parallel to face grain (edgewise):[6]					
Proportional limit stress (psi)	3,300	5,200	8,700	5,200	5,000
Ultimate strength (psi)	5,800	11,400	23,900	14,000	18,900
Modulus of elasticity (1,000 psi)	1,360	1,500	2,300	2,700	2,370
Tension parallel to face grain (lengthwise):					
Ultimate strength (psi)	12,300	7,900	16,500	24,500	27,200
Modulus of elasticity (1,000 psi)	1,290	1,460	2,190	2,570	2,700
Toughness (FPL test edgewise)[6] (in-lb per in of width)	105	40	115	320	—

[1] Laminates made from 17 plies of 1/16-in rotary-cut yellow birch veneer.
[2] Veneer conditioned at 80 °F and 65 percent relative humidity before assembly with phenol resin film glue.
[3] Impregnation, 25 to 30 percent of water-soluble phenol-formaldehyde resin based on the dry weight of untreated veneer.
[4] High-strength paper (0.003-in thickness) made from commercial unbleached black spruce pulp (Mitscherlich sulfite), phenol resin content 36.3 percent, based on weight of treated paper. Izod impact, abrasion, flatwise compression, and shear specimens, all on 1/2-in-thick laminate.
[5] Load applied to the surface of the original material (parallel to laminating pressure direction).
[6] Load applied to the edge of the laminations (perpendicular to laminating pressure direction).
[7] Values as high as 10.0 ft–lb per in of notch have been reported for compreg made with alcohol-soluble resins and 7.0 ft–lb with water-soluble resins.
[8] Values based on the average slope of load-penetration plots, where H_M is an expression for load per unit of spherical area of penetration of the 0.444-in steel ball expressed in pounds per square inch:

$$H_M = \frac{P}{2\pi r h} \text{ or } 0.717\,\frac{P}{h}$$

pressures (1,000 psi) to a specific gravity of 1.35. Some of its properties are similar to those of impreg, and others vary considerably (tables 23–1 and 23–2). Its advantages over impreg are its natural lustrous finish that can be developed on any cut surface by sanding with fine-grit paper and buffing, its greater strength properties, and the fact that it can be molded (tables 23–1 and 23–2). Thermal expansion coefficients of ovendry compreg, however, are also increased (table 23–3).

Compreg can be molded by: (1) Gluing blocks of resin-treated (but still uncured) wood with a phenolic glue so that the gluelines and resin within the plies are only partially set; (2) cutting to the desired length and width but two to three times the desired thickness; and (3) compressing in a split mold at about 300 °F. Only a small flash squeezeout at the parting line between the two halves of the mold need be machined off. This technique was used for molding motor-test propellers and airplane antenna masts during World War II.

A more generally satisfactory molding technique, known as expansion molding, has been developed. The method consists of rapidly precompressing dry but uncured single sheets of resin-treated veneer in a cold press after preheating the sheets to 200 to 240 °F. The heat-plasticized wood responds to compression before cooling. The heat is insufficient to cure the resin, but the subsequent cooling sets the resin temporarily. These compressed sheets are cut to the desired size, and the assembly of plies is placed in a split mold of the final desired dimensions. Because the wood was pre-compressed, the filled mold can be closed and locked. When the mold is heated, the wood is again plasticized and tends to recover its uncompressed dimensions. This exerts an internal pressure in all directions against the mold equal to about half of the original compressing pressure. On continued heating, the resin is set. After cooling, the object may be removed from the mold in finished form. Metal inserts or metal surfaces can be molded to compreg or compreg handles molded onto tools by this means. Compreg bands have been molded to the outside of turned wood cylinders without compressing the core. Compreg tubes and small airplane propellers have been molded in this way.

Past uses for compreg once related largely to aircraft; however, it is a suitable material where bolt-bearing strength

Table 23 – 3—Coefficients of linear thermal expansion per degree Celsius of wood, hydrolyzed wood, and paper products[1]

Material[2]	Specific gravity of product	Glue plus resin content[3]	Linear expansion per °C by 10^6			Cubical expansion per °C by 10^6
			Fiber or machine direction	Perpendicular to fiber or machine direction in plane of laminations	Pressing direction	
		Pct				
Yellow birch laminate	0.72	3.1	3.254	40.29	36.64	80.18
Yellow birch staypak laminate	1.30	4.7	3.406	37.88	65.34	106.63
Yellow birch impreg laminate	.86	33.2	4.648	35.11	37.05	76.81
Yellow birch compreg laminate	1.30	24.8	4.251	39.47	59.14	102.86
Do.	1.31	34.3	4.931	39.32	54.83	99.08
Sitka spruce laminate	.53	[4] 6.0	3.887	37.14	27.67	68.65
Parallel-laminated paper laminate	1.40	36.5	5.73	15.14	65.10	85.97
Crossbanded paper laminate	1.40	36.5	10.89	[5] 11.00	62.20	84.09
Molded hydrolyzed-wood plastic	1.33	25	42.69	42.69	42.69	128.07
Hydrolyzed-wood sheet laminate	1.39	18	13.49	24.68	77.41	115.58

[1] These coefficients refer to bone-dry material. Generally, air-dry material has a negative thermal coefficient of expansion, because the shrinkage resulting from the loss in moisture is greater than the normal thermal expansion.

[2] All wood laminates made from rotary-cut veneer, annual rings in plane of sheet.

[3] On basis of dry weight of product.

[4] Approximate.

[5] Calculated value.

is required, as in connector plates, because of its good specific strength (strength per unit of weight). Layers of veneer making up the compreg for such uses are often cross laminated (alternate plies at right angles to each other, as in plywood) to give nearly equal properties in all directions. It is extremely useful for aluminum drawing and forming dies, drilling jigs, and jigs for holding parts in place while welding, because of its excellent strength properties, dimensional stability, low thermal conductivity, and ease of fabrication.

Compreg has also been used in silent gears, pulleys, water-lubricated bearings, fan blades, shuttles, bobbins and picker sticks for looms, nuts and bolts, instrument bases and cases, musical instruments, electrical insulators, tool handles, and various novelties. Compreg at present finds considerable use in handles for knives and other cutlery. Both the expansion-molding techniques of forming and curing the compreg around the metal parts of the handle and attaching previously made compreg with rivets are used.

Veneer of any nonresinous species can be used for making compreg. Most properties depend upon the specific gravity to which the wood is compressed rather than the species used. Up to the present, however, compreg has been made almost exclusively from yellow birch or sugar maple.

Untreated Compressed Wood (Staypak)

Resin-treated wood in both the uncompressed (impreg) and compressed (compreg) forms is more brittle than the original wood. To meet the demand for a tougher compressed product than compreg, a compressed wood containing no resin (staypak) was developed. It will not lose its compression under swelling conditions as will ordinary compressed untreated wood. In making staypak, the compressing conditions are modified so that the lignin-cementing material between the cellulose fibers flows sufficiently to eliminate internal stresses.

Staypak is not as water resistant as compreg, but it is about twice as tough and has higher tensile and flexural strength properties, as shown in tables 23–1 and 23–2. The natural finish of staypak is almost equal to that of compreg. Under weathering conditions, however, it is definitely inferior to compreg. For outdoor use, a good synthetic resin varnish or paint finish should be applied to staypak.

Staypak can be used in the same way as compreg where extremely high water resistance is not needed. It shows promise in tool handles, forming dies, connector plates, propellers, and picker sticks and shuttles for weaving, where high impact strength is needed. As staypak is not impregnated, it can be made from solid wood as well as from veneer. Its cost is less than compreg.

A material similar to staypak was produced in Germany prior to World War II. It was a compressed solid wood with much less dimensional stability than staypak and was known as lignostone. Another similar German product was a laminated compressed wood known as lignofol.

Untreated Heated Wood (Staybwood)

Heating wood under drying conditions at higher temperatures (200 to 600° F) than those normally used in kiln drying produces a product known as staybwood that reduces the hygroscopicity and subsequent swelling and shrinking of the

Table 23–4—Comparison of wood treatments and the degree of dimensional stability achieved

Treatment	Antishrink efficiency[1]
	Pct
Simple wax dip	2–5
Wood-plastic combination	10–15
Staypak/staybwood	30–40
Impreg	65–70
Chemical modification	65–75
Polyethylene glycol	80–85
Formaldehyde	82–87
Compreg	90–95

[1] Calculated from:

$$S = \frac{V_2 - V_1}{V_1} \times 100$$

where

S = volumetric swelling coefficient,
V_2 = wood volume after humidity conditioning or wetting with water, and
V_1 = wood volume of ovendried sample before conditioning or wetting,

then

$$ASE = \frac{S_2 - S_1}{S_1} \times 100$$

where

ASE = reduction in swelling or antishrink efficiency resulting from a treatment,
S_2 = treated volumetric swelling coefficient,
S_1 = untreated volumetric swelling coefficient.

wood appreciably. The stabilization, however, is always accompanied by loss of mechanical properties. Toughness and resistance to abrasion are most seriously affected.

Under conditions that cause a reduction of 40 percent in shrinking and swelling, the toughness is reduced to less than half that of the original wood. Extensive research to minimize this loss was not successful. Because of the reduction in strength properties from heating at such high temperatures, wood that is dimensionally stabilized in this manner is not used commercially.

Polyethylene Glycol-Treated Wood (PEG)

The dimensional stabilization of wood with polyethylene glycol-1000 (PEG), also known as Carbowax, is accomplished by bulking the fibers to keep the wood in a partially swollen condition. PEG acts in the same manner as does the previously described phenolic resin. It cannot be further cured. The only reason for heating the wood after treatment is to drive off water. PEG remains water soluble. Above 60 percent relative humidity it is a strong humectant and, unless used with care and properly protected, PEG-treated wood can become sticky at high relative humidities. Because of this, PEG-treated wood is usually finished with polyurethane varnish.

Treatment with PEG is facilitated by using green wood. Here, pressure is not applied since the treatment is based on diffusion. Treating times are such that uniform uptakes of 25 to 30 percent of chemical are achieved (based on dry weight of wood). The time necessary for this uptake depends on the thickness of the wood and may require weeks. This treatment is being effectively used for walnut gunstocks for high-quality rifles. The dimensional stability of such gunstocks greatly enhances the continued accuracy of the rifles. Tabletops of high-quality furniture stay remarkably flat and dimensionally stable when made from PEG-treated wood.

Another application of this chemical is to reduce the checking of green wood during drying. For this application a high degree of polyethylene glycol penetration is not required. This method of treatment has been used to reduce checking during drying of small wood blanks or turnings.

Cracking and distortion that old, waterlogged wood undergoes when it is dried can be substantially reduced by treating the wood with polyethylene glycol. The process was used to dry 200-year-old waterlogged wood boats raised from Lake George, NY. The "Vasa," a Swedish ship that sank on its initial trial voyage in 1628, has also been treated after it was raised. There have been many applications of PEG treatment for the restoration of waterlogged wood from archeological sites.

Wood-Plastic Combination

In the modified wood products previously discussed, most of the chemical resides in cell walls; the lumens are essentially empty. If wood is vacuum impregnated with certain liquid vinyl monomers that do not swell wood, and which are later polymerized by gamma radiation or chemical catalyst-heat systems, the resulting polymer resides almost exclusively in the lumens. Methyl methacrylate is a common monomer used for a wood-plastic combination. It is converted to polymethyl methacrylate. Such wood-plastic combinations with polymer contents of 75 to 100 percent (based on the dry weight of wood) resist moisture movement through them. Moisture movement is extremely slow so that normal equilibrium swelling is reached very slowly. Wood-plastic combination materials are much stronger than untreated wood (table 23–5) and commercial application of these products is largely based on increased strength properties.

The main commercial use of this modified wood at present is as parquet flooring where it is produced in squares about 5-1/2 inches on a side from strips about 7/8 inch wide and 5/16 inch thick. It has a specific gravity of 1.0. Comparative tests with conventional wood flooring indicate wood-plastic materials resisted indentation from rolling, concentrated, and impact loads better than white oak. This is largely attributed to improved hardness, which was increased 40 percent in regular wood-plastic combination and 20 percent in the same material treated with a fire retardant. Abrasion resistance was no better than white oak; but because the finish is built in, buffing is all that is required to provide the luster. The finish is maintained even under severe traffic conditions.

Wood-plastic combinations are also being used in sporting goods, musical instruments, and novelty items.

Chemical Modification

Through chemical reactions it is possible to add an organic chemical to the hydroxyl groups on wood cell wall components. This type of treatment bulks the cell wall with a permanently bonded chemical. Many reactive chemicals have been used experimentally to chemically modify wood. For best results, chemicals used should be capable of reacting with wood hydroxyls under neutral or mildly alkaline conditions at temperatures below 248 °F. The chemical system should be simple and must be capable of swelling the wood structure to facilitate penetration. The complete molecule must react quickly with wood components to yield stable chemical bonds while the treated wood retains the desirable properties of untreated wood. Chemicals such as anhydrides, epoxides, isocyanates, acid chlorides, carboxylic acids, lactones, alkyl chlorides, and nitriles all have antishrink efficiency values of

65 to 75 percent at chemical weight gains of 20 to 30 percent.

Reaction of these chemicals with wood yields a modified wood with increased dimensional stability and improved resistance to termites, decay, and marine organisms. Mechanical properties of chemically modified wood are somewhat reduced (10 to 15 percent) as compared to untreated wood.

The reaction of formaldehyde with wood hydroxyl groups is an interesting variation of chemical modification. At weight gains as low as 2 percent, formaldehyde-treated wood is not attacked by wood-destroying fungi. An antishrink efficiency of 47 percent is achieved at a weight gain of 3.1 percent, 55 percent at 4.1, 60 percent at 5.5, and 90 percent at 7.

The mechanical properties of formaldehyde-treated wood are all reduced from those of untreated wood. A definite embrittlement is observed, toughness and abrasion resistance are greatly reduced, crushing strength and bending strengths are reduced about 20 percent, and impact bending strength is reduced up to 50 percent.

Paper-Base Plastic Laminates

Commercially, paper-base plastic laminates are of two types—industrial and decorative. The total annual production is equally divided between the two types. They are made by superimposing layers of paper that have been impregnated with a resinous binder and curing the assembly under heat and pressure.

Industrial Laminates

Industrial laminates are produced to perform specific functions requiring materials with predetermined balances of mechanical, electrical, and chemical properties. The most common use of such laminates is for electrical insulation. The paper reinforcements used in the laminates are of kraft pulp, alpha pulp, cotton linters, or blends of these. Kraft paper emphasizes mechanical strength and dielectric strength perpendicular to laminations. Alpha paper is used for its electric and electronic properties, machinability, and dimensional stability. The cotton linter paper combines greater strength than alpha paper with excellent moisture resistance.

Phenolic resins are the most suitable resins for impregnating the paper from the standpoint of high water resistance, low swelling and shrinking, and high strength properties (except for impact). Phenolics are also lower in cost than other resins that give comparable properties. Water-soluble resins of the type used for impreg impart the highest water resistance and compressive strength properties to the product, but they make the product brittle (low impact strength). Alcohol-soluble phenolic resins produce a considerably

Table 23 − 5—*Strength properties of wood-plastic combination*[1]

Strength property	Unit	Untreated[2]	Treated[2]
STATIC BENDING			
Modulus of elasticity	10^6 psi	1.356	1.691
Fiber stress at proportional limit	psi	6,387	11,582
Modulus of rupture	psi	10,649	18,944
Work to proportional limit	in−lb/in^3	1.66	4.22
Work to maximum load	in−lb/in^3	10.06	17.81
COMPRESSION PARALLEL TO GRAIN			
Modulus of elasticity	10^6 psi	1.113	1.650
Fiber stress at proportional limit	psi	4,295	7,543
Maximum crushing strength	psi	6,505	9,864
Work to proportional limit	in−lb/in^3	11.28	21.41
Toughness	in−lb/in^3	41.8	62.6

[1] Methyl methacrylate impregnated basswood.
[2] Moisture content—7.2 percent.

tougher product, but the resins fail to penetrate the fibers as well as water-soluble resins and thus impart less water resistance and dimensional stability to the product. In practice, alcohol-soluble phenolic resins are generally used.

Paper-base plastic laminates inherit their final properties from the paper from which they are made. High-strength papers yield higher strength plastic laminates than do low-strength papers. Papers with definite directional properties result in plastic laminates with definite directional properties unless they are cross-laminated (alternate sheets oriented with the machine direction at 90° to each other).

Improving the paper used has helped develop paper-base laminates suitable for structural use. Pulping under milder conditions and operating the paper machines to give optimum orientation of the fibers in one direction, together with the desired absorbency, contribute markedly to improvements in strength.

Strength and some other properties of a paper plastic laminate are shown in table 23−2. The National Electric Manufacturers Association L1−1 specification has further information on industrial laminates. As paper is considerably less expensive than glass fabric or other woven fabric mats and can be molded at considerably lower pressures, the paper-base laminates generally have an appreciable price advantage over fabric laminates. Some fabric laminates, however, give superior electrical properties and the highest impact properties. Glass fabric laminates can be molded to greater double curvatures than paper laminates.

During World War II, a high-strength paper plastic known as papreg was used for molding nonstructural and semi-structural airplane parts, such as gunner's seats and turrets, ammunition boxes, wing tabs, and the surfaces of cargo aircraft flooring and catwalks. Papreg was tried to a limited extent for the skin surface of airplane structural parts, such as wing tips. One major objection to its use for such parts is that it is more brittle than aluminum and requires special fittings. Papreg has been used to some extent for heavy-duty truck floors and industrial processing trays for nonedible materials. Because it can be molded at low pressures and is made from thin papers, it is advantageous for use where accurate control of panel thickness is required.

Decorative Laminates

Although decorative laminates are made by the same process as industrial laminates, they are used for very different purposes and bear little outward resemblance to each other. They are used as facings for doors and walls, and for tops of counters, tables, desks, and other furniture.

These decorative laminates are usually composed of a combination of phenolic- and melamine-impregnated sheets of paper. The phenolic-impregnated sheets are brown because of the impregnating resins and comprise most of the built-up thickness of the laminate. The phenolic sheets are overlaid with paper impregnated with melamine resin. One sheet of the overlay is usually a relatively thick one of high opacity and has the color or design printed on it. Then one or more tissue-thin sheets, which become transparent after the resin is cured, are overlaid on the printed sheet to protect it in service. The thin sheets generally contain more melamine resin than the printed sheet, providing stain and abrasion resistance as well as resistance to cigarette burns, boiling water, and common household solvents.

The resin-impregnated sheets of paper are hot-pressed, cured, and then bonded to a wood-base core, usually plywood, hardboard, or particleboard. The thin transparent (when cured) papers impregnated with melamine resin can be used alone as a covering for decorative veneers in furniture to provide a permanent finish. In this use the impregnated sheet is bonded to the wood surface in hot presses at the same time the resin is cured. The heat and stain resistance and the strength of this kind of film make it a superior finish.

The overall thickness of a laminate may obviously be varied by the number of sheets of kraft-phenolic used in the core assembly. Some years ago the 1/16-inch thickness was used with little exception because of its very high impact strength and resistance to substrate showthrough. More recently the 1/32-inch thickness has been popular on vertical surfaces such as walls, cabinet doors, and vertical furniture faces. This results in better economy, since the greater strength of the heavier laminate is not necessary. As applications have proliferated, a whole series of thicknesses have been offered from about 20 to 62 mils, or even up to 150 mils when self-supporting types are needed. The laminate may have decorative faces on both sides if desired, especially in the heavier thicknesses.

The phenolic sheets may also contain special postforming-type phenolic resins or extensible papers that make it possible to postform the laminate. By heating to 325 °F for a short time, the structure can undergo simple bending to a radius of 1/2 inch readily and of 3/16 to 1/4 inch with careful control. Rolled furniture edges, decorative moldings, curved counter tops, shower enclosures, and many other applications are served by this technique. Finally, the core composition may be modified to yield fire-retardant, low-smoking laminates to comply with fire codes. These high-pressure decorative laminates are covered by the National Electrical Manufacturers Association Specification LD3.

Paper will absorb or give off moisture, depending upon conditions of exposure. This moisture change causes paper to shrink and swell, usually more across the machine direction than along it. Likewise, the laminated paper plastics shrink and swell, although at a much slower rate. Cross-laminating minimizes the amount of this shrinking and swelling. In many uses in furniture where laminates are bonded to cores, these changes in dimension due to moisture changes with the change of seasons are different than those of the core material. To balance the construction, a paper plastic with similar properties may be glued to the opposite face of the core to prevent bowing or cupping from the moisture changes.

Lignin-Filled Laminates

The cost of phenolic resins at one time resulted in considerable effort to find impregnating and bonding agents that were less expensive and yet readily available. Lignin-filled laminates made with lignin recovered from the spent liquor of the soda pulping process have been produced as a result of this search. Lignin is precipitated from solution within the pulp or added in a pre-precipitated form before the paper is made. The lignin-filled sheets of paper can be laminated without the addition of other resins, but their water resistance is considerably enhanced when some phenolic resin is applied to the paper in a second operation. The water resistance can also be improved by impregnating only the surface sheet with phenolic resin. It is also possible to introduce lignin, together with phenolic resin, into untreated paper sheets.

The lignin-filled laminates are always dark brown or black. They have better toughness than phenolic laminates. In most other strength properties they are comparable or lower. In spite of the fact that lignin is somewhat thermoplastic, the loss in strength on heating to 200 °F is proportionately no more than for phenolic laminates.

Reduction in costs of phenolic resins has virtually eliminated the lignin-filled laminates from American commerce. They have a number of potential applications, however, where a cheaper laminate with less critical properties than phenolic laminates can be used.

Paper-Face Overlays

Paper has found considerable use as an overlay material for veneer or plywood. Overlays can be classified into three different types according to their use—masking, structural, and decorative. Masking overlays are used to cover minor defects in plywood, such as face checks and patches, minimize grain raising, and provide a more uniform paintable surface, thus making possible the use of lower grade veneer. Paper for this purpose need not be of high strength, as the overlays need not add strength to the product. For adequate masking a single surface sheet with a thickness of 0.012 to 0.030 inch is desirable. Paper impregnated with phenolic resins at 17 to 25 percent of the weight of the paper gives the

best all-around product. Higher resin contents make the product too costly and tend to make the overlay more transparent. Appreciably lower resin contents give a product with low scratch and abrasion resistance, especially when the panels are wet or exposed to high relative humidities.

The paper faces can be applied at the same time that the veneer is assembled into plywood in a hot press. Thermal stresses that might result in checking are not set up if the machine direction of the paper overlays is at right angles to the grain direction of the face plies of the plywood.

The masking paper-base overlays or vulcanized fiber sheets have been used for such applications as wood house siding that is to be painted. These overlays mask defects in the wood, prevent bleedthrough of resins and extractives in the wood, and provide a better substrate for paint. The paper-base overlays improve the across-the-board stability from changes in dimension due to changes in moisture content.

The structural overlay, also known as the high-density overlay, contains no less than 45 percent thermosetting resin, generally phenolic. It consists of one or more plies of paper similar to that used in the industrial laminates described previously. The resin-impregnated papers can be bonded directly to the surface of a wood substrate during cure of the sheet, thus requiring only a single pressing operation.

The decorative-type overlay has been described in the section on decorative laminates.

Selected References

Erickson, E.C.O. Mechanical properties of laminated modified woods. FPL Rep. 1639. Madison, WI: U.S. Department of Agriculture, Forest Service, Forest Products Laboratory; 1947.

Forest Products Laboratory. Physical and mechanical properties of lignin-filled laminated paper plastic. FPL Rep. 1579. Madison, WI: U.S. Department of Agriculture, Forest Service, Forest Products Laboratory; rev. 1962.

Heebink, B. G. Importance of balanced construction in plastic-faced wood panels. Res. Note FPL−021. Madison, WI: U.S. Department of Agriculture, Forest Service, Forest Products Laboratory; 1963.

Heebink, B. G.; Haskell, H. H. Effect of heat and humidity on the properties of high-pressure laminates. Madison, WI: Forest Products Journal. 12(11): 542−548; 1962.

Langwig, J. E.; Meyer, J. A.; Davidson, R. W. Influence of polymer impregnation on mechanical properties of basswood. Madison, WI: Forest Products Journal. 18(7): 33−36; 1968.

Meyer, J. A. Treatment of wood-polymer systems using catalyst-heat techniques. Madison, WI: Forest Products Journal. 15(9): 362−364; 1965.

Meyer, J. A.; Loos, W. E. Treating southern pine wood for modification of properties. Madison, WI: Forest Products journal. 19(12): 32−38; 1969.

Mitchell, H. L. How PEG helps the hobbyist who works with wood. FPL Unnumbered publication. Madison, WI: U.S. Department of Agriculture, Forest Service, Forest Products Laboratory. n.d.

National Electrical Manufacturers Association. Standard specification for industrial laminated thermosetting products. Designation L1−1. Washington, DC: NEMA; (see current edition).

National Electrical Manufacturers Association. Standard specification for high-pressure decorative laminates. Designation LD–3. Washington, DC: NEMA; (see current edition).

Rowell, R. M. Chemical modification of wood: advantages and disadvantages. American Wood-Preservers Association Proceedings. 71: 41–51; 1975.

Rowell, R. M., ed. The chemistry of solid wood. American Chemical Society, Advances in Chemistry Series No. 207; 1984.

Seborg, R. M.; Inverarity, R. B. Preservation of old, waterlogged wood by treatment with polyethylene glycol. Science. 136(3516): 649–650; 1962.

Seborg, R. M.; Millett, M. A.; Stamm, A. J. Heat-stabilized compressed wood (staypak). Mechanical Engineering. 67(1): 25–31; 1945.

Seborg, R. M.; Vallier, A. E. Application of impreg for patterns and die models. Madison, WI: Forest Products Journal 4(5): 305–312; 1954.

Seidl, R. J. Paper and plastic overlays for veneer and plywood. Madison, WI: Forest Products Research Society Proceedings. 1: 23–32; 1947.

Stamm, A. J. Effect of polyethylene glycol on dimensional stability of wood. Madison, WI: Forest Products Journal. 9(10): 375–381; 1959.

Stamm, A. J. Wood and cellulose science. New York: Ronald Press; 1964.

Stamm, A. J.; Seborg, R. M. Forest Products Laboratory resin-treated laminated, compressed wood (compreg). FPL Rep. 1381. Madison, WI: U.S. Department of Agriculture, Forest Service, Forest Products Laboratory; 1951.

Stamm A. J.; Seborg, R. M. Forest Products Laboratory resin-treated wood (impreg). FPL Rep. 1380. Madison WI: U.S. Department of Agriculture, Forest Service, Forest Products Laboratory; rev. 1962.

Weatherwax, R. C.; Stamm, A. J. Electrical resistivity of resin-treated wood (impreg and compreg), hydrolyzed-wood sheet (hydroxylin), and laminated resin-treated paper (papreg). FPL Rep. 1385. Madison, WI: U.S. Department of Agriculture, Forest Service, Forest Products Laboratory; 1945.

Weatherwax, R. C.; Stamm, A. J. The coefficients of thermal expansion of wood and wood products. Transactions of American Society of Mechanical Engineering 69(44): 421–432; 1946.

Glossary

Adherend. A body that is held to another body by an adhesive.

Adhesion. The state in which two surfaces are held together by interfacial forces which may consist of valence forces or interlocking action or both.

Adhesive. A substance capable of holding materials together by surface attachment. It is a general term and includes cements, mucilage, and paste, as well as glue.

   Assembly Adhesive—An adhesive that can be used for bonding parts together, such as in the manufacture of a boat, airplane, furniture, and the like.

   Cold-Setting Adhesive—An adhesive that sets at temperatures below 20 °C (68 °F).

   Construction Adhesive—Any adhesive used to assemble primary building materials into components during building construction—most commonly applied to elastomer-based mastic-type adhesives.

   Contact Adhesive—An adhesive that is apparently dry to the touch and which will adhere to itself instantaneously upon contact; also called contact bond adhesive or dry bond adhesive.

   Gap-Filling Adhesive—Adhesive suitable for use where the surfaces to be joined may not be in close or continuous contact owing either to the impossibility of applying adequate pressure or to slight inaccuracies in matching mating surfaces.

   Hot-Melt Adhesive—An ahesive that is applied in a molten state and forms a bond on cooling to a solid state.

   Hot-Setting Adhesive—An adhesive that requires a temperature at or above 100 °C (212 °F) to set it.

   Room-Temperature Setting Adhesive—An adhesive that sets in the temperature range of 20 to 30 °C (68 to 86 °F), in accordance with the limits for Standard Room Temperature specified in the Standard Methods of Conditioning Plastics and Electrical Insulating Materials for Testing (ASTM Designation: D 618).

   Solvent Adhesive—An adhesive having a volatile organic liquid as a vehicle. Note: This term excludes water-based adhesives.

   Structural Adhesive—A bonding agent used for transferring required loads between adherends exposed to service environments typical for the structure involved.

Air-Dried. (See **Seasoning.**)

Allowable Property. The value of a property normally published for design use. Allowable properties are identified with grade descriptions and standards, reflect the orthotropic structure of wood, and anticipate certain end uses.

Allowable Stress. (See **Allowable Property.**)

American Lumber Standards. American lumber standards embody provisions for softwood lumber dealing with recognized classifications, nomenclature, basic grades, sizes, description, measurements, tally, shipping provisions, grademarking, and inspection of lumber. The primary purpose of these standards is to serve as a guide in the preparation or revision of the grading rules of the various lumber manufacturers' associations. A purchaser must, however, make use of association rules because the basic standards are not in themselves commercial rules.

Anisotropic. Exhibiting different properties when measured along different axes. In general, fibrous materials such as wood are anisotropic.

Annual Growth Ring. The layer of wood growth put on a tree during a single growing season. In the temperate zone the annual growth rings of many species (e.g., oaks and pines) are readily distinguished because of differences in the cells formed during the early and late parts of the season. In some temperate zone species (black gum and sweetgum) and many tropical species, annual growth rings are not easily recognized.

Assembly Joint. (See **Joint.**)

Assembly Time. (See **Time, Assembly.**)

Balanced Construction. A construction such that the forces induced by uniformly distributed changes in moisture content will not cause warping. Symmetrical construction of plywood in which the grain direction of each ply is perpendicular to that of adjacent plies is balanced construction.

Bark Pocket. An opening between annual growth rings that contains bark. Bark pockets appear as dark streaks on radial surfaces and as rounded areas on tangential surfaces.

Bastard Sawn. Lumber (primarily hardwoods) in which the annual rings make angles of 30° to 60° with the surface of the piece.

Beam. A structural member supporting a load applied transversely to it.

Bending, Steam. The process of forming curved wood members by steaming or boiling the wood and bending it to a form.

Bent Wood. (See **Bending, Steam.**)

Bird Peck. A small hole or patch of distorted grain resulting from birds pecking through the growing cells in the tree. In shape, bird peck usually resembles a carpet tack with the point towards the bark; bird peck is usually accompanied by discoloration extending for considerable distance along the grain and to a much lesser extent across the grain.

Birdseye. Small localized areas in wood with the fibers indented and otherwise contorted to form few to many small circular or elliptical figures remotely resembling birds' eyes on the tangential surface. Sometimes found in sugar maple and used for decorative purposes; rare in other hardwood species.

Blister. An elevation of the surface of an adherend, somewhat resembling in shape a blister on the human skin; its boundaries may be indefinitely outlined, and it may have burst and become flattened. Note: A blister may be caused by insufficient adhesive; inadequate curing time, temperature or pressure; or trapped air, water, or solvent vapor.

Bloom. Crystals formed on the surface of treated wood by exudation and evaporation of the solvent in preservative solutions.

Blow. In plywood and particleboard especially, the development of steam pockets during hot pressing of the panel, resulting in an internal separation or rupture when pressure is released, sometimes with an audible report.

Blue Stain. (See **Stain.**)

Board. (See **Lumber.**)

Board Foot. A unit of measurement of lumber represented by a board 1 foot long, 12 inches wide, and 1 inch thick or its cubic equivalent. In practice, the board foot calculation for lumber 1 inch or more in thickness is based on its nominal thickness and width and the actual length. Lumber with a nominal thickness of less than 1 inch is calculated as 1 inch.

Bole. The main stem of a tree of substantial diameter—roughly, capable of yielding sawtimber, veneer logs, or large poles. Seedlings, saplings, and small-diameter trees have stems, not boles.

Bolt. (1) A short section of a tree trunk; (2) in veneer production, a short log of a length suitable for peeling in a lathe.

Bond (noun). The union of materials by adhesives.

Bond (verb). To unite materials by means of an adhesive.

Bondability. Term indicating ease or difficulty in bonding a material with adhesive.

Bond Failure. Rupture of adhesive bond.

Bondline. The layer of adhesive that attaches two adherends.

Bondline Slip. Movement within and parallel to the bondline during shear.

Bond Strength. The unit load applied in tension, compression, flexure, peel impact, cleavage, or shear, required to break an adhesive assembly with failure occurring in or near the plane of the bond.

Bow. The distortion of lumber in which there is a deviation, in a direction perpendicular to the flat face, from a straight line from end-to-end of the piece.

Box Beam. A built-up beam with solid wood flanges and plywood or wood-base panel product webs.

Boxed Heart. The term used when the pith falls entirely within the four faces of a piece of wood anywhere in its length. Also called boxed pith.

Brashness. A condition that causes some pieces of wood to be relatively low in shock resistance for the species and, when broken in bending, to fail abruptly without splintering at comparatively small deflections.

Breaking Radius. The limiting radius of curvature to which wood or plywood can be bent without breaking.

Bright. Free from discoloration.

Broad-Leaved Trees. (See **Hardwoods**.)

Brown Rot. In wood, any decay in which the attack concentrates on the cellulose and associated carbohydrates rather than on the lignin, producing a light to dark brown friable residue—hence loosely termed ''dry rot.'' An advanced stage where the wood splits along rectangular planes, in shrinking, is termed ''cubical rot.''

Brown Stain. (See **Stain**.)

Built-Up Timbers. An assembly made by joining layers of lumber together with mechanical fastenings so that the grain of all laminations is essentially parallel.

Burl. (1) A hard, woody outgrowth on a tree, more or less rounded in form, usually resulting from the entwined growth of a cluster of adventitious buds. Such burls are the source of the highly figured burl veneers used for purely ornamental purposes. (2) In lumber or veneer, a localized severe distortion of the grain generally rounded in outline, usually resulting from overgrowth of dead branch stubs, varying from 1/2 inch to several inches in diameter; frequently includes one or more clusters of several small contiguous conical protuberances, each usually having a core or pith but no appreciable amount of end grain (in tangential view) surrounding it.

Butt Joint. (See **Joint**.)

Buttress. A ridge of wood developed in the angle between a lateral root and the butt of a tree, which may extend up the stem to a considerable height.

Cambium. A thin layer of tissue between the bark and wood that repeatedly subdivides to form new wood and bark cells.

Cant. A log that has been slabbed on one or more sides. Ordinarily, cants are intended for resawing at right angles to their widest sawn face. The term is loosely used. (See **Flitch**.)

Casehardening. A condition of stress and set in dry lumber characterized by compressive stress in the outer layers and tensile stress in the center or core.

Catalyst. A substance, usually present in small amounts relative to the reactants, that modifies the rate of a chemical reaction (such as the cure of an adhesive) without being consumed in the process.

Cell. A general term for the anatomical units of plant tissue, including wood fibers, vessel members, and other elements of diverse structure and function.

Cellulose. The carbohydrate that is the principal constituent of wood and forms the framework of the wood cells.

Check. A lengthwise separation of the wood that usually extends across the rings of annual growth and commonly results from stresses set up in wood during seasoning.

Chemical Brown Stain. (See **Stain**.)

Chipboard. A paperboard used for many purposes that may or may not have specifications for strength, color, or other characteristics. It is normally made from paper stock with a relatively low density in the thickness of 0.006 inch and up.

Close Grained. (See **Grain**.)

Coarse Grain. (See **Grain**.)

Cohesion. The state in which the particles of a single substance are held together by primary or secondary valence forces. As used in the adhesive field, the state in which the particles of adhesive (or adherend) are held together internally.

Cold Pressing. A bonding operation in which an assembly is subjected to pressure without the application of heat.

Cold-Press Plywood. (See **Plywood**.)

Collapse. The flattening of single cells or rows of cells in heartwood during the drying or pressure treatment of wood. Often characterized by a caved-in or corrugated appearance of the wood surface.

Compartment Kiln. (See **Kiln**.)

Composite Panel. A veneer-faced panel with a reconstituted wood core. The flakeboard core may be random or have alignment in the direction 90° from the grain direction of the veneer faces.

Compound Curvature. Wood bent to a compound curvature has curved surfaces, no element of which is a straight line.

Compreg. Wood in which the cell walls have been impregnated with synthetic resin and compressed to give it reduced swelling and shrinking characteristics and increased density and strength properties.

Compression Failure. Deformation of the wood fibers resulting from excessive compression along the grain either in direct end compression or in bending. It may develop in standing trees due to bending by wind or snow or to internal longitudinal stresses developed in growth, or it may result from stresses imposed after the tree is cut. In surfaced lumber, compression failures may appear as fine wrinkles across the face of the piece.

Compression Wood. Abnormal wood formed on the lower side of branches and inclined trunks of softwood trees. Compression wood is identified by its relatively wide annual rings (usually eccentric when viewed on cross section of branch or trunk), relatively large amount of summerwood, sometimes more than 50 percent of the width of the annual rings in which it occurs, and its lack of demarcation between earlywood and latewood in the same annual rings. Compression wood shrinks excessively lengthwise, as compared with normal wood.

Conditioning (pre and post). The exposure of a material to the influence of a prescribed atmosphere for a stipulated period of time or until a stipulated relation is reached between material and atmosphere.

Conifer. (See **Softwoods**.)

Connector, Timber. Metal rings, plates, or grids which are embedded in the wood of adjacent members, as at the bolted points of a truss, to increase the strength of the joint.

Consistency. That property of a liquid adhesive by virtue of which it tends to resist deformation. Note: Consistency is not a fundamental property but is comprised of rheological properties such as viscosity, plasticity, and other phenomena.

Construction Adhesive. (See **Adhesive**.)

Contact Angle. The angle (θ) between a substrate plane and the free surface of a liquid droplet at the line of contact with substrate.

Cooperage. Containers consisting of two round heads and a body composed of staves held together with hoops, such as barrels and kegs.

>**Slack Cooperage**—Cooperage used as containers for dry, semidry or solid products. The staves are usually not closely fitted and are held together with beaded steel, wire, or wood hoops.

>**Tight Cooperage**—Cooperage used as containers for liquids, semisolids, and heavy solids. Staves are well fitted and held tightly with cooperage-grade steel hoops.

Copolymer. Substance obtained when two or more types of monomers polymerize.

Corbel. A projection from the face of a wall or column supporting a weight.

Core Stock. A solid or discontinuous center ply used in panel-type glued structures (such as furniture panels and solid or hollowcore doors).

Crook. The distortion of lumber in which there is a deviation, in a direction perpendicular to the edge, from a straight line from end to-end of the piece.

Crossband. To place the grain of layers of wood at right angles in order to minimize shrinking and swelling; also, in plywood of three or more plies, a layer of veneer whose grain direction is at right angles to that of the face plies.

Cross Break. A separation of the wood cells across the grain. Such breaks may be due to internal stress resulting from unequal longitudinal shrinkage or to external forces.

Cross Grain. (See **Grain**.)

Cross-Link. An atom or group connecting adjacent molecules in a complex molecular structure.

Cup. A distortion of a board in which there is a deviation flatwise from a straight line across the width of the board.

Cure. To change the properties of an adhesive by chemical reaction (which may be condensation, polymerization, or vulcanization) and thereby develop maximum strength. Generally accomplished by the action of heat or a catalyst, with or without pressure.

Curing Agent. (See **Hardener**.)

Curing Temperature. (See **Temperature, Curing**.)

Curing Time. (See **Time, Curing**.)

Curly Grain. (See **Grain**.)

Curtain Coating. Applying liquid adhesive to adherend by passing the adherend under a thin curtain of liquid falling by gravity or pressure.

Cut Stock. A term for softwood stock comparable to dimension parts in hardwoods. (See **Dimension Parts**.)

Cuttings. In hardwoods, portions of a board or plant having the quality required by a specific grade or for a particular use. Obtained from a board by crosscutting or ripping.

Decay. The decomposition of wood substance by fungi.

> **Advanced** (or **Typical Decay**)—The older stage of decay in which the destruction is readily recognized because the wood has become punky, soft and spongy, stringy, ringshaked, pitted, or crumbly. Decided discoloration or bleaching of the rotted wood is often apparent.

> **Incipient Decay**—The early stage of decay that has not proceeded far enough to soften or otherwise perceptibly impair the hardness of the wood. It is usually accompanied by a slight discoloration or bleaching of the wood.

Delamination. The separation of layers in laminated wood or plywood because of failure of the adhesive, either within the adhesive itself or at the interface between the adhesive and the adherend.

Delignification. Removal of part or all of the lignin from wood by chemical treatment.

Density. As usually applied to wood of normal cellular form, density is the mass of wood substance enclosed within the boundary surfaces of a wood-plus-voids complex having unit volume. It is variously expressed as pounds per cubic foot, kilograms per cubic meter, or grams per cubic centimeter at a specified moisture content.

Density Rules. A procedure for segregating wood according to density, based on percentage of latewood and number of growth rings per inch of radius.

Dew Point. The temperature at which a vapor begins to deposit as a liquid. Applies especially to water in the atmosphere.

Diagonal Grain. (See **Grain**.)

Diffuse-Porous Wood. Certain hardwoods in which the pores tend to be uniform in size and distribution throughout each annual ring or to decrease in size slightly and gradually toward the outer border of the ring.

Dimension. (See **Lumber**.)

Dimension Parts. A term largely superseded by the term "hardwood dimension lumber." It is hardwood stock processed to a point where the maximum waste is left at the mill, and the maximum utility is delivered to the user. It is stock of specified thickness, width, and length, or multiples thereof. According to specification it may be solid or glued up, rough or surfaced, semifabricated or completely fabricated.

Dimensional Stabilization. Special treatment of wood to reduce the swelling and shrinking that is caused by changes in its moisture content with changes in relative humidity.

Dote. "Dote," "doze," and "rot" are synonymous with "decay" and are any form of decay that may be evident as either a discoloration or a softening of the wood.

Double Spread. (See **Spread**.)

Dry-Bulb Temperature. The temperature of air as indicated by a standard thermometer. (See **Psychrometer**.)

Dry Kiln. (See **Kiln**.)

Dry Rot. A term loosely applied to any dry, crumbly rot but especially to that which, when in an advanced stage, permits the wood to be crushed easily to a dry powder. The term is actually a misnomer for any decay, since all fungi require considerable moisture for growth.

Dry Wall. Interior covering material, such as gypsum board, hardboard, or plywood, which is applied in large sheets or panels.

Durability. A general term for permanence or resistance to deterioration. Frequently used to refer to the degree of resistance of a species of wood to attack by wood-destroying fungi under conditions that favor such attack. In this connection the term "decay resistance" is more specific. As applied to bondlines, the life expectancy of the structural qualities of the adhesive under the anticipated service conditions of the structure.

Earlywood. The portion of the annual growth ring that is formed during the early part of the growing season. It is usually less dense and weaker mechanically than latewood.

Edge Grain. (See **Grain**.)

Edge Joint. (See **Joint**.)

Elastomer. A macromolecular material which, at room temperature, is deformed by application of a relatively low force and is capable of recovering substantially in size and shape after removal of the force.

Empty-Cell Process. Any process for impregnating wood with preservatives or chemicals in which air, imprisoned in the wood under pressure, expands when pressure is released to drive out part of the injected preservative or chemical. The distinguishing characteristic of the empty-cell process is that no vacuum is drawn before applying the preservative. The aim is to obtain good preservative distribution in the wood and leave the cell cavities only partially filled.

Encased Knot. (See **Knot**.)

End Grain. (See **Grain**.)

End Joint. (See **Joint**.)

Equilibrium Moisture Content. The moisture content at which wood neither gains nor loses moisture when surrounded by air at a given relative humidity and temperature.

Excelsior. (See **Wood Wool**.)

Extender. A substance, generally having some adhesive action, added to an adhesive to reduce the amount of the primary binder required per unit area.

Exterior Plywood. (See **Plywood**.)

Extractive. Substances in wood, not an integral part of the cellular structure, that can be removed by solution in hot or cold water, ether, benzene, or other solvents that do not react chemically with wood components.

Extrusion Spreading. A method of adhesive application in which adhesive is forced through small openings in the spreader head.

Factory And Shop Lumber. (See **Lumber**.)

Failure, Adherend. Rupture of an adhesive joint, such that the separation appears to be within the adherend.

Failure, Adhesive. Rupture of an adhesive joint, such that the plane of separation appears to be at the adhesive-adherend interface.

Failure, Cohesive. Rupture of an adhesive joint, such that the separation appears to be within the adhesive.

Feed Rate. The distance that the stock being processed moves during a given interval of time or operational cycle.

Fiber, Wood. A wood cell comparatively long (1/25 or less to 1/3 in), narrow, tapering, and closed at both ends.

Fiberboard. A broad generic term inclusive of sheet materials of widely varying densities manufactured of refined or partially refined wood (or other vegetable) fibers. Bonding agents and other materials may be added to increase strength, resistance to moisture, fire, or decay, or to improve some other property. (See **Medium-Density Fiberboard.**)

Fiber Saturation Point. The stage in the drying or wetting of wood at which the cell walls are saturated and the cell cavities free from water. It applies to an individual cell or group of cells, not to whole boards. It is usually taken as approximately 30 percent moisture content, based on ovendry weight.

Fibril. A threadlike component of cell walls, visible under a light microscope.

Fiddleback. (See **Grain.**)

Figure. The pattern produced in a wood surface by annual growth rings, rays, knots, deviations from regular grain such as interlocked and wavy grain, and irregular coloration.

Filler. In woodworking, any substance used to fill the holes and irregularities in planed or sanded surfaces to decrease the porosity of the surface before applying finish coatings. As applied to adhesives a relatively nonadhesive substance added to an adhesive to improve its working properties, strength, or other qualities.

Fine Grain. (See **Grain.**)

Finger Joint. (See **Joint.**)

Finish (Finishing). Wood products such as doors, stairs, and other fine work required to complete a building, especially the interior. Also, coatings of paint, varnish, lacquer, wax, etc., applied to wood surfaces to protect and enhance their durability or appearance.

Fire Endurance. A measure of the time during which a material or assembly continues to exhibit fire resistance under specified conditions of test and performance.

Fire Resistance. The property of a material or assembly to withstand fire or give protection from it. As applied to elements of buildings, it is characterized by the ability to confine a fire or to continue to perform a given structural function, or both.

Fire Retardant. A chemical or preparation of chemicals used to reduce flammability or to retard spread of a fire over the surface.

Flake. A small flat wood particle of predetermined dimensions, uniform thickness, with fiber direction essentially in the plane of the flake; in overall character resembling a small piece of veneer. Produced by special equipment for use in the manufacture of flakeboard.

Flakeboard. (See **Particleboard.**)

Flame Spread. The propagation of a flame away from the source of ignition across the surface of a liquid or a solid, or through the volume of a gaseous mixture.

Flat Grain. (See **Grain.**)

Flat Sawn. (See **Grain.**)

Flecks. (See **Rays, Wood.**)

Flitch. A portion of a log sawn on two or more faces—commonly on opposite faces leaving two waney edges. When intended for resawing into lumber, it is resawn parallel to its original wide faces. Or, it may be sliced or sawn into veneer, in which case the resulting sheets of veneer laid together in the sequence of cutting are called a flitch. The term is loosely used. (See **Cant.**)

Framing. Lumber used for the structural member of a building, such as studs and joists.

Full-Cell Process. Any process for impregnating wood with preservatives or chemicals in which a vacuum is drawn to remove air from the wood before admitting the preservative. This favors heavy adsorption and retention of preservative in the treated portions.

Furnish. The wood material which has been reduced for incorporation into wood-based fiber or particle panel products.

Gap-Filling Adhesive. (See **Adhesive.**)

Gelatinous Fibers. Modified fibers that are associated with tension wood in hardwoods.

Girder. A large or principal beam used to support concentrated loads at isolated points along its length.

Gluability. (See **Bondability.**)

Glue. Originally, a hard gelatin obtained from hides, tendons, cartilage, bones, etc., of animals. Also, an adhesive prepared from this substance by heating with water. Through general use the term is now synonymous with the term "Adhesive."

Glue Laminating. Production of structural or nonstructural wood members by bonding two or more layers of wood together with adhesive.

Glueline. (See **Bondline.**)

Grade. The designation of the quality of a manufactured piece of wood or of logs.

Grain. The direction, size, arrangement, appearance, or quality of the fibers in wood or lumber. To have a specific meaning the term must be qualified.

Close-Grained Wood—Wood with narrow, inconspicuous annual rings. The term is sometimes used to designate wood having small and closely spaced pores, but in this sense the term "fine textured" is more often used.

Coarse-Grained Wood—Wood with wide conspicuous annual rings in which there is considerable difference between springwood and summerwood. The term is sometimes used to designate wood with large pores, such as oak, ash, chestnut, and walnut, but in this sense the term "coarse textured" is more often used.

Cross-Grained Wood—Wood in which the fibers deviate from a line parallel to the sides of the piece. Cross grain may be either diagonal or spiral grain or a combination of the two.

Curly-Grained Wood—Wood in which the fibers are distorted so that they have a curled appearance, as in "birdseye" wood. The areas showing curly grain may vary up to several inches in diameter.

Diagonal-Grained Wood—Wood in which the annual rings are at an angle with the axis of a piece as a result of sawing at an angle with the bark of the tree or log. A form of cross-grain.

Edge-Grained Lumber—Lumber that has been sawed so that the wide surfaces extend approximately at right angles to the annual growth rings. Lumber is considered edge grained when the rings form an angle of 45° to 90° with the wide surface of the piece.

End-Grained Wood—The grain as seen on a cut made at a right angle to the direction of the fibers (e.g., on a cross section of a tree).

Fiddleback-Grained Wood—Figure produced by a type of fine wavy grain found, for example, in species of maple; such wood being traditionally used for the backs of violins.

Fine-Grained Wood—(See **Grain.**)

Flat-Grained Wood—Lumber that has been sawed parallel to the pith and approximately tangent to the growth rings. Lumber is considered flat grained when the annual growth rings make an angle of less than 45° with the surface of the piece.

Interlocked-Grained Wood—Grain in which the fibers put on for several years may slope in a right-handed direction, and then for a number of years the slope reverses to a left-handed direction, and later changes back to a right-handed pitch, and so on. Such wood is exceedingly difficult to split radially, though tangentially it may split fairly easily.

Open-Grained Wood—Common classification for woods with large pores, such as oak, ash, chestnut, and walnut. Also known as "coarse textured."

Plainsawed Lumber—Another term for flat-grained lumber.

Quartersawed Lumber—Another term for edge-grained lumber.

Side-Grained Wood—Another term for flat-grained lumber.

Slash-Grained Wood—Another term for flat-grained lumber.

Spiral-Grained Wood—Wood in which the fibers take a spiral course about the trunk of a tree instead of the normal vertical course. The spiral may extend in a right-handed or left-handed direction around the tree trunk. Spiral grain is a form of cross grain.

Straight-Grained Wood—Wood in which the fibers run parallel to the axis of a piece.

Vertical-Grained Lumber—Another term for edge-grained lumber.

Wavy-Grained Wood—Wood in which the fibers collectively take the form of waves or undulations.

Green. Freshly sawed or undried wood. Wood that has become completely wet after immersion in water would not be considered green, but may be said to be in the ''green condition.''

Growth Ring. (See **Annual Growth Ring**.)

Gum. A comprehensive term for nonvolatile viscous plant exudates, which either dissolve or swell up in contact with water. Many substances referred to as gums such as pine and spruce gum are actually oleoresins.

Hardboard. A generic term for a panel manufactured primarily from interfelted ligno-cellulosic fibers (usually wood), consolidated under heat and pressure in a hot press to a density of 31 pounds per cubic foot or greater, and to which other materials may have been added during manufacture to improve certain properties.

Hardener (noun). A substance or mixture of substances added to an adhesive to promote or control the curing reaction by taking part in it. The term is also used to designate a substance added to control the degree of hardness of the cured film. (See **Catalyst**.)

Hardness. A property of wood that enables it to resist indentation.

Hardwoods. Generally one of the botanical groups of trees that have broad leaves in contrast to the conifers or softwoods. The term has no reference to the actual hardness of the wood.

Heart Rot. Any rot characteristically confined to the heartwood. It generally originates in the living tree.

Heartwood. The wood extending from the pith to the sapwood, the cells of which no longer participate in the life processes of the tree. Heartwood may contain phenolic compounds, gums, resins, and other materials that usually make it darker and more decay resistant than sapwood.

Hemicellulose. A celluloselike material (in wood) that is easily decomposable as by dilute acid, yielding several different simple sugars.

Hertz. A unit of frequency equal to one cycle per second.

High Frequency Curing. (See **Radiofrequency Curing**.)

Hollow-Core Construction. A panel construction with faces of plywood, hardboard, or similar material bonded to a framed-core assembly of wood lattice, paperboard rings, or the like, which support the facing at spaced intervals.

Honeycomb Core. A sandwich core material constructed of thin sheet materials or ribbons formed to honeycomblike configurations.

Honeycombing. Checks, often not visible at the surface, that occur in the interior of a piece of wood, usually along the wood rays.

Horizontally Laminated Timber. (See **Laminated Timbers**.)

Hot-Setting Adhesive. (See **Adhesive**.)

Impreg. Wood in which the cell walls have been impregnated with synthetic resin so as to reduce materially its swelling and shrinking. Impreg is not compressed.

Increment Borer. An augerlike instrument with a hollow bit and an extractor, used to extract thin radial cylinders of wood from trees to determine age and growth rate. Also used in wood preservation to determine the depth of penetration of a preservative.

Insulating Board. (See **Structural Insulating Board**.)

Intergrown Knot. (See **Knot**.)

Interlocked-Grained Wood. (See **Grain**.)

Internal Stresses. Stresses that exist within an adhesive joint even in the absence of applied external forces.

Interphase. In wood bonding, a region of finite thickness as a gradient between the bulk adherend and bulk adhesive in which the adhesive penetrates and alters the adherend's properties and in which the presence of the adherend influences the chemical and/or physical properties of the adhesive.

Intumesce. To expand with heat to provide a low-density film; used in reference to certain fire-retardant coatings.

Isotropic. Exhibiting the same properties in all directions.

Joint. The junction of two pieces of wood or veneer.

Adhesive Joint—The location at which two adherends are held together with a layer of adhesive.

Assembly joint—Joints between variously shaped parts or subassemblies such as in wood furniture (as opposed to joints in plywood and laminates that are all quite similar).

Butt Joint—An end joint formed by abutting the squared ends of two pieces.

Edge Joint—A joint made by bonding two pieces of wood together edge to edge, commonly by gluing. The joints may be made by gluing two squared edges as in a plain edge joint or by using machined joints of various kinds, such as tongued-and-grooved joints.

End Joint—A joint made by bonding two pieces of wood together end to end, commonly by finger or scarf joint.

Finger Joint—An end joint made up of several meshing wedges or fingers of wood bonded together with an adhesive. Fingers are sloped and may be cut parallel to either the wide or narrow face of the piece.

Lap Joint—A joint made by placing one member partly over another and bonding the overlapped portions.

Scarf Joint—An end joint formed by joining with adhesive the ends of two pieces that have been tapered or beveled to form sloping plane surfaces, usually to a featheredge, and with the same slope of the plane with respect to the length in both pieces. In some cases, a step or hook may be machined into the scarf to facilitate alinement of the two ends, in which case the plane is discontinuous and the joint is known as a stepped or hooked scarf joint.

Starved Joint—A glue joint that is poorly bonded because an insufficient quantity of adhesive remained in the joint.

Sunken Joint—Depression in wood surface at a joint (usually an edge joint) caused by surfacing material too soon after bonding. (Inadequate time was allowed for moisture added with the adhesive to diffuse away from the joint.)

Joint Efficiency or Factor. The strength of a joint expressed as a percentage of the strength of clear straight-grained material.

Joist. One of a series of parallel beams used to support floor and ceiling loads and supported in turn by larger beams, girders, or bearing walls.

Kiln. A chamber having controlled air-flow, temperature, and relative humidity for drying lumber, veneer, and other wood products.

Compartment Kiln—A kiln in which the total charge of lumber is dried as a single unit. It is designed so that, at any given time, the temperature and relative humidity are essentially uniform throughout the kiln. The temperature is increased as drying progresses, and the relative humidity is adjusted to the needs of the lumber.

Progressive Kiln—A kiln in which the total charge of lumber is not dried as a single unit but as several units, such as kiln truckloads, that move progressively through the kiln. The kiln is designed so that the temperature is lower and the relative humidity higher at the end where the lumber enters than at the discharge end.

Kiln Dried. (See **Seasoning**.)

Knot. That portion of a branch or limb that has been surrounded by subsequent growth of the stem. The shape of the knot as it appears on a cut surface depends on the angle of the cut relative to the long axis of the knot.

Encased Knot—A knot whose rings of annual growth are not intergrown with those of the surrounding wood.

Intergrown Knot—A knot whose rings of annual growth are completely intergrown with those of the surrounding wood.

Loose Knot—A knot that is not held firmly in place by growth or position and that cannot be relied upon to remain in place.

Pin Knot—A knot that is not more than 1/2 inch in diameter.

Sound Knot—A knot that is solid across its face, at least as hard as the surrounding wood, and shows no indication of decay.

Spike Knot—A knot cut approximately parallel to its long axis so that the exposed section is definitely elongated.

Laminate. A product made by bonding together two or more layers (laminations) of material or materials.

Laminate, Paper-Base. A multilayered panel made by compressing sheets of resin-impregnated paper together into a coherent solid mass.

Laminated Timbers. An assembly made by bonding layers of veneer or lumber with an adhesive so that the grain of all laminations is essentially parallel. (See **Built-Up Timbers**.)

Horizontally Laminated Timbers—Laminated timbers designed to resist bending loads applied perpendicular to the wide faces of the laminations.

Vertically Laminated Timbers—Laminated timbers designed to resist bending loads applied parallel to the wide faces of the laminations.

Laminated Veneer Lumber (LVL). A structural lumber manufactured from veneers laminated into a panel with the grain of all veneer running parallel to each other. The resulting panel is normally manufactured in 3/4- to 1-1/2-inch thicknesses and ripped to common lumber widths of 1-1/2 to 11-1/2 inches, or wider.

Lap Joint. (See **Joint**.)

Latewood. The portion of the annual growth ring that is formed after the earlywood formation has ceased. It is usually denser and stronger mechanically than earlywood.

Latex Paint. A paint containing pigments and a stable water suspension of synthetic resins (produced by emulsion polymerization) that forms an opaque film through coalescence of the resin during water evaporation and subsequent curing.

Layup. The process of loosely assembling the adhesive-coated components of a unit, particularly a panel, to be pressed or clamped.

Lbs/MSGL. Abbreviation for rate of adhesive application in pounds of adhesive per thousand square feet of single glueline (bondline). (See **Spread**.) When both faces of an adherend are spread as in some plywood manufacturing processes, the total weight of adhesive applied may be expressed as Lbs/MDGL (pounds per 1,000 square feet double glueline).

Lignin. The second most abundant constituent of wood, located principally in the secondary wall and the middle lamella, which is the thin cementing layer between wood cells. Chemically it is an irregular polymer of substituted propylphenol groups, and thus no simple chemical formula can be written for it.

Longitudinal. Generally, parallel to the direction of the wood fibers.

Loose Knot. (See **Knot**.)

Lumber. The product of the saw and planing mill not further manufactured than by sawing, resawing, passing lengthwise through a standard planing machine, crosscutting to length, and matching.

Boards—Lumber that is nominally less than 2 inches thick and 2 or more inches wide. Boards less than 6 inches wide are sometimes called strips.

Dimension—Lumber with a nominal thickness of from 2 up to but not including 5 inches and a nominal width of 2 inches or more.

Dressed Size—The dimensions of lumber after being surfaced with a planing machine. The dressed size is usually 1/2 to 3/4 inch less than the nominal or rough size. A 2- by 4-inch stud, for example, actually measures about 1-1/2 by 3-1/2 inches.

Factory and Shop Lumber—Lumber intended to be cut up for use in further manufacture. It is graded on the basis of the percentage of the area that will produce a limited number of cuttings of a specified minimum size and quality.

Matched Lumber—Lumber that is edge dressed and shaped to make a close tongued-and-grooved joint at the edges or ends when laid edge to edge or end to end.

Nominal Size—As applied to timber or lumber, the size by which it is known and sold in the market; often differs from the actual size. (See **Lumber**.)

Patterned Lumber—Lumber that is shaped to a pattern or to a molded form in addition to being dressed, matched, or shiplapped, or any combination of these workings.

Rough Lumber—Lumber that has not been dressed (surfaced) but which has been sawed, edged, and trimmed.

Shiplapped Lumber—Lumber that is edge dressed to make a lapped joint.

Shipping-Dry Lumber—Lumber that is partially dried to prevent stain and mold in transit.

Shop Lumber—(See **Lumber**.)

Side Lumber—A board from the outer portion of the log—ordinarily one produced when squaring off a log for a tie or timber.

Structural Lumber—Lumber that is intended for use where allowable properties are required. The grading of structural lumber is based on the strength or stiffness of the piece as related to anticipated uses.

Surfaced Lumber—Lumber that is dressed by running it through a planer.

Timbers—Lumber that is nominally 5 or more inches in least dimension. Timbers may be used as beams, stringers, posts, caps, sills, girders, purlins, etc.

Yard Lumber—A little-used term for lumber of all sizes and patterns that is intended for general building purposes having no design property requirements.

Lumen. In wood anatomy, the cell cavity.

Manufacturing Defects. Includes all defects or blemishes that are produced in manufacturing, such as chipped grain, loosened grain, raised grain, torn grain, skips in dressing, hit and miss (series of surfaced areas with skips between them), variation in sawing, miscut lumber, machine burn, machine gouge, mismatching, and insufficient tongue or groove.

Mastic. A material with adhesive properties, usually used in relatively thick sections, that can be readily applied by extrusion, trowel, or spatula. (See **Adhesive**.)

Matched Lumber. (See **Lumber**.)

Medium-Density Fiberboard. A panel product manufactured from ligno-cellulosic fibers combined with a synthetic resin or other suitable binder. The panels are manufactured to a density of 31 pcf (0.50 specific gravity) to 55 pcf (0.88 specific gravity) by the application of heat and pressure by a process in which the interfiber bond is substantially created by the added binder. Other materials may have been added during manufacturing to improve certain properties.

Millwork. Planed and patterned lumber for finish work in buildings, including items such as sash, doors, cornices, panelwork, and other items of interior or exterior trim. Does not include flooring, ceiling, or siding.

Mineral Streak. An olive to greenish-black or brown discoloration of undetermined cause in hardwoods.

Modified Wood. Wood processed by chemical treatment, compression, or other means (with or without heat) to impart properties quite different from those of the original wood.

Moisture Content. The amount of water contained in the wood, usually expressed as a percentage of the weight of the ovendry wood.

Moulded Plywood. (See **Plywood**.)

Moulding. A wood strip having a curved or projecting surface, used for decorative purposes.

Monomer. A relatively simple molecular compound which can react at more than one site to form a polymer.

Mortise. A slot cut into a board, plank, or timber, usually edgewise, to receive the tenon of another board, plank, or timber to form a joint.

Naval Stores. A term applied to the oils, resins, tars, and pitches derived from oleoresin contained in, exuded by, or extracted from trees, chiefly species of pines (genus *Pinus*). Historically, these were important items in the stores of wood sailing vessels.

Nominal-Size Lumber. (See **Lumber**.)

Oil Paint. A paint containing a suspension of pigments in an organic solvent and a drying oil, modified drying oil, or synthetic polymer that forms an opaque film through a combination of solvent evaporation and curing of the oil or polymer.

Old Growth. Timber in or from a mature, naturally established forest. When the trees have grown during most if not all of their individual lives in active competition with their companions for sunlight and moisture, this timber is usually straight and relatively free of knots.

Oleoresin. A solution of resin in an essential oil that occurs in or exudes from many plants, especially softwoods. The oleoresin from pine is a solution of pine resin (rosin) in turpentine.

Open Assembly Time. (See **Time, Assembly**.)

Open Grain. (See **Grain**.)

Orthotropic. Having unique and independent properties in three mutually orthogonal (perpendicular) planes of symmetry. A special case of anisotropy.

Ovendry Wood. Wood dried to a relatively constant weight in a ventilated oven at 102 to 105 °C.

Overlay. A thin layer of paper, plastic, film, metal foil, or other material bonded to one or both faces of panel products or to lumber to provide a protective or decorative face or a base for painting.

Paint. Any pigmented liquid, liquifiable, or mastic composition designed for application to a substrate in a thin layer that converts to an opaque solid film after application.

Pallet. A low wood or metal platform on which material can be stacked to facilitate mechanical handling, moving, and storage.

Paperboard. The distinction between paper and paperboard is not sharp, but broadly speaking, the thicker (over 0.012 in), heavier, and more rigid grades of paper are called paperboard.

Papreg. Any of various paper products made by impregnating sheets of specially manufactured high-strength paper with synthetic resin and laminating the sheets to form a dense, moisture-resistant product.

Parenchyma. Short cells having simple pits and functioning primarily in the metabolism and storage of plant food materials. They remain alive longer than the tracheids, fibers, and vessel segments, sometimes for many years. Two kinds of parenchyma cells are recognized—those in vertical strands, known more specifically as axial parenchyma, and those in horizontal series in the rays, known as ray parenchyma.

Particleboard. A generic term for a material manufactured from wood particles or other ligno-cellulosic material and a synthetic resin or other suitable binder.

  Extruded Particleboard—A particleboard made by ramming binder-coated particles into a heated die, which subsequently cures the binder and forms a rigid mass as the material is moved through the die.

  Flakeboard—A particle panel product composed of flakes.

  Mat-Formed Particleboard—A particleboard in which the particles (being previously coated with the binding agent) are formed into a mat having substantially the same length and width as the finished panel. This mat is then duly pressed in a heated flat-platen press to cure the binding agent.

  Mende-Process Board—A particleboard made in a continuous ribbon from wood particles with thermosetting resins used to bond the particles. Thickness ranges from 1/32 to 1/4 inch.

  Multilayer Particleboard—A type of construction in which the wood particles are made or classified into different sizes and placed into the preprocessed panel configuration to produce a panel with specific properties. Panels which are destined for primarily nonstructural uses requiring smooth faces are configured with small particles on the outside and coarser particles on the interior (core). Panels designed for structural application may have flakes aligned in orthogonal directions in various layers which mimic the structure of plywood. Three- and five-layer constructions are most common.

  Oriented Strand Board—A type of particle panel product composed of strand-type flakes which are purposefully alined in directions which make a panel stronger, stiffer, and with improved dimensional properties in the alinement directions than a panel with random flake orientation.

  Waferboard—A particle panel product made of wafer-type flakes. Usually manufactured to possess equal properties in all directions parallel to the plane of the panel.

Particles. The aggregate component of particleboard manufactured by mechanical means from wood. These include all small subdivisions of wood such as chips, curls, flakes, sawdust, shavings, slivers, strands, wafers, wood flour, and wood wool.

Peck. Pockets or areas of disintegrated wood caused by advanced stages of localized decay in the living tree. It is usually associated with cypress and incense-cedar. There is no further development of peck once the lumber is seasoned.

Peel. To convert a log into veneer by rotary cutting.

Phloem. The tissues of the inner bark, characterized by the presence of sieve tubes and serving for the transport of elaborated foodstuffs.

Pile. A long, heavy timber, round or square, that is driven deep into the ground to provide a secure foundation for structures built on soft, wet, or submerged sites; e.g., landing stages, bridge abutments.

Pin-Knot. (See **Knot**.)

Pitch Pocket. An opening extending parallel to the annual growth rings and containing, or that has contained, pitch, either solid or liquid.

Pitch Streaks. A well-defined accumulation of pitch in a more or less regular streak in the wood of certain conifers.

Pith. The small, soft core occurring near the center of a tree trunk, branch, twig, or log.

Pith Fleck. A narrow streak, resembling pith on the surface of a piece; usually brownish, up to several inches in length; results from burrowing of larvae in the growing tissues of the tree.

Plainsawed. (See **Grain**.)

Planing Mill Products. Products worked to pattern, such as flooring, ceiling, and siding.

Plank. A broad board, usually more than 1 inch thick, laid with its wide dimension horizontal and used as a bearing surface.

Plasticizing Wood. Softening wood by hot water, steam, or chemical treatment to increase its moldability.

Plywood. A glued wood panel made up of relatively thin layers of veneer with the grain of adjacent layers at right angles, or of veneer in combination with a core of lumber or of reconstituted wood. (See **Composite Panel**.) The usual constructions have an odd number of layers.

  Cold-Pressed Plywood—Refers to interior-type plywood manufactured in a press without external applications of heat.

  Exterior Plywood—A general term for plywood bonded with a type of adhesive that by systematic tests and service records has proved highly resistant to weather; micro-organisms; cold, hot, and boiling water; steam; and dry heat.

Interior Plywood—A general term for plywood manufactured for indoor use or in construction subjected to only temporary moisture. The adhesive used may be interior, intermediate, or exterior.

Marine Plywood—Plywood panels manufactured with the same glueline durability requirements as other exterior-type panels but with more restrictive veneer quality requirements.

Molded Plywood—Plywood that is glued to the desired shape either between curved forms or more commonly by fluid pressure applied with flexible bags or blankets (bag molding) or other means.

Postformed Plywood—The product formed when flat plywood is reshaped into a curved configuration by steaming or plasticizing agents.

Pocket Rot. Advanced decay that appears in the form of a hole or pocket, usually surrounded by apparently sound wood.

Polymer. A compound formed by the reaction of simple molecules having functional groups which permit their combination to proceed to high molecular weights under suitable conditions. Polymers may be formed by polymerization (addition polymer) or polycondensation (condensation polymer). When two or more different monomers are involved, the product is called a copolymer.

Polymerization. A chemical reaction in which the molecules of a monomer are linked together to form large molecules whose molecular weight is a multiple of that of the original substance. When two or more different monomers are involved, the process is called copolymerization.

Pore. (See **Vessels.**)

Porous Woods. Hardwoods having vessels or pores large enough to be seen readily without magnification.

Postformed Plywood. (See **Plywood.**)

Post Cure (noun). A treatment (normally involving heat) applied to an adhesive assembly following the initial cure, to complete cure, or to modify specific properties.

Post Cure (verb). To expose an adhesive assembly to an additional cure, following the initial cure, to complete cure or to modify specific properties.

Pot Life. (See **Working Life.**)

Precure. Condition of too much cure, set, or solvent loss of the adhesive before pressure is applied, resulting in inadequate flow, transfer, and bonding.

Preservative. Any substance that, for a reasonable length of time, is effective in preventing the development and action of wood-rotting fungi, borers of various kinds, and harmful insects that deteriorate wood.

Pressure Process. Any process of treating wood in a closed container whereby the preservative or fire retardant is forced into the wood under pressures greater than 1 atmosphere. Pressure is generally preceded or followed by vacuum, as in the vacuum-pressure and empty-cell processes respectively; or they may alternate, as in the full-cell and alternating-pressure processes.

Progressive Kiln. (See **Kiln.**)

Psychrometer. An instrument for measuring the amount of water vapor in the atmosphere. It has both a dry-bulb and wet-bulb thermometer. The bulb of the wet-bulb thermometer is kept moistened and is, therefore, cooled by evaporation to a temperature lower than that shown by the dry-bulb thermometer. Because evaporation is greater in dry air, the difference between the two thermometer readings will be greater when the air is dry than when it is moist.

Qualification Test (noun). A series of tests conducted in advance of use in regular manufacturing, to determine conformance of materials, or material systems, to the requirements of published specifications. Note: Generally, qualification under a specification requires a conformance to all tests in the specification; alternately it may be limited to conformance to a specific type or class, or both, under the specification.

Quartersawed. (See **Grain.**)

Radial. Coincident with a radius from the axis of the tree or log to the circumference. A radial section is a lengthwise section in a plane that passes through the centerline of the tree trunk.

Radiofrequency (RF) Curing. Curing of bondlines by the application of radiofrequency energy. (Sometimes called high-frequency curing.)

Rafter. One of a series of structural members of a roof designed to support roof loads. The rafters of a flat roof are sometimes called roof joists.

Raised Grain. A roughened condition of the surface of dressed lumber in which the hard latewood is raised above the softer earlywood but not torn loose from it.

Rays, Wood. Strips of cells extending radially within a tree and varying in height from a few cells in some species to 4 or more inches in oak. The rays serve primarily to store food and transport it horizontally in the tree. On quartersawed oak, the rays form a conspicuous figure, sometimes referred to as flecks.

Reaction Wood. Wood with more or less distinctive anatomical characters, formed typically in parts of leaning or crooked stems and in branches. In hardwoods this consists of tension wood and in softwoods of compression wood.

Relative Humidity. Ratio of the amount of water vapor present in the air to that which the air would hold at saturation at the same temperature. It is usually considered on the basis of the weight of the vapor but, for accuracy, should be considered on the basis of vapor pressures.

Resilience. The property whereby a strained body gives up its stored energy on the removal of the deforming force.

Resin. Inflammable, water-soluble, vegetable substances secreted by certain plants or trees, and characterizing the wood of many coniferous species. The term is also applied to synthetic organic products related to the natural resins.

Resin Ducts. Intercellular passages that contain and transmit resinous materials. On a cut surface, they are usually inconspicuous. They may extend vertically parallel to the axis of the tree or at right angles to the axis and parallel to the rays.

Retention by Assay. The determination of preservative retention in a specific zone of treated wood by extraction or analysis of specified samples.

Rheology. The study of the deformation and flow of matter.

Ring Failure. A separation of the wood during seasoning, occurring along the grain and parallel to the growth rings. (See SHAKE.)

Ring-Porous Woods. A group of hardwoods in which the pores are comparatively large at the beginning of each annual ring and decrease in size more or less abruptly toward the outer portion of the ring, thus forming a distinct inner zone of pores, known as the earlywood, and an outer zone with smaller pores, known as the latewood.

Ring Shake. (See **Shake.**)

Rip. To cut lengthwise, parallel to the grain.

Roll Spreading. Application of a film of a liquid material to a surface by means of rollers.

Room-Temperature-Setting Adhesive. (See **Adhesive.**)

Rot. (See **Decay.**)

Rotary-Cut Veneer. (See **Veneer.**)

Rough Lumber. (See **Lumber.**)

Sandwich Construction. (See **Structural Sandwich Construction.**)

Sap Stain. (See **Stain.**)

Sapwood. The wood of pale color near the outside of the log. Under most conditions the sapwood is more susceptible to decay than heartwood.

Sash. A frame structure, normally glazed (e.g., a window), that is hung or fixed in a frame set in an opening.

Sawed Veneer. (See **Veneer.**)

Saw Kerf. (1) Grooves or notches made in cutting with a saw; (2) that portion of a log, timber, or other piece of wood removed by the saw in parting the material into two pieces.

Scarf Joint. (See **Joint.**)

Schedule, Kiln Drying. A prescribed series of dry- and wet-bulb temperatures and air velocities used in drying a kiln charge of lumber or other wood products.

Seasoning. Removing moisture from green wood to improve its serviceability.

    Air-Dried.—Dried by exposure to air in a yard or shed, without artificial heat.

    Kiln-Dried.—Dried in a kiln with the use of artificial heat.

Second Growth. Timber that has grown after the removal, whether by cutting, fire, wind, or other agency, of all or a large part of the previous stand.

Semitransparent Stain. A suspension of pigments in either a drying oil/organic solvent mixure or a water/polymer emulsion, designed to color and protect wood surfaces by penetration without forming a surface film, and without hiding wood grain.

Set. A permanent or semipermanent deformation. In reference to adhesives, to convert an adhesive into a fixed or hardened state by chemical or physical action, such as condensation, polymerization, oxidation, vulcanization, gelation, hydration, or evaporation of volatile constituents.

Shake. A separation along the grain, the greater part of which occurs between the rings of annual growth. Usually considered to have occurred in the standing tree or during felling.

Shakes. In construction, shakes are a type of shingle usually hand cleft from a bolt and used for roofing or weatherboarding.

Shaving. A small wood particle of indefinite dimensions developed incidental to certain woodworking operations involving rotary cutterheads usually turning in the direction of the grain. This cutting action produces a thin chip of varying thickness, usually feathered along at least one edge and thick at another and generally curled.

Shear. A condition of stress or strain where parallel planes slide relative to one another.

Sheathing. The structural covering, usually of boards, building fiberboards, or plywood, placed over exterior studding or rafters of a structure.

Shelf Life. (See **Storage Life.**)

Shiplapped Lumber. (See **Lumber.**)

Shipping-Dry Lumber. (See **Lumber.**)

Shop Lumber. (See **Lumber.**)

Side-Grained. (See **Grain.**)

Side Lumber. (See **Lumber.**)

Siding. The finish covering of the outside wall of a frame building, whether made of horizontal weatherboards, vertical boards with battens, shingles, or other material.

Slash-Grained. (See **Grain.**)

Sliced Veneer. (See **Veneer.**)

Soft Rot. A special type of decay developing under very wet conditions (as in cooling towers and boat timbers) in the outer wood layers, caused by cellulose-destroying microfungi that attack the secondary cell walls and not the intercellular layer.

Softwoods. Generally, one of the botanical groups of trees that in most cases have needlelike or scalelike leaves, the conifers, also the wood produced by such trees. The term has no reference to the actual hardness of the wood.

Solid Color Stains (Opaque Stains). A suspension of pigments in either a drying oil/organic solvent mixture or a water/polymer emulsion designed to color and protect a wood surface by forming a film. Solid color stains are similar to paints in application techniques and in performance.

Solids Content. The percentage of weight of the nonvolatile matter in an adhesive.

Solvent Adhesive. (See **Adhesive.**)

Sound Knot. (See **Knot.**)

Specific Gravity. As applied to wood, the ratio of the ovendry weight of a sample to the weight of a volume of water equal to the volume of the sample at a specified moisture content (green, air-dry, or ovendry).

Spike Knot. (See **Knot.**)

Spiral Grain. (See **Grain.**)

Spread. The quantity of adhesive per unit joint area applied to an adherend, usually expressed in pounds of adhesive per thousand square feet of joint area. (See **Lbs/MSGL.**)

    Single spread—Refers to application of adhesive to only one adherend of a joint.

    Double spread—Refers to application of adhesive to both adherends of a joint.

Springwood. (See **Earlywood.**)

Squeezeout. Bead of adhesive squeezed out of a joint when pressure is applied.

Stain. A discoloration in wood that may be caused by such diverse agencies as micro-organisms, metal, or chemicals. The term also applies to materials used to impart color to wood.

    Blue Stain—A bluish or grayish discoloration of the sapwood caused by the growth of certain dark-colored fungi on the surface and in the interior of the wood; made possible by the same conditions that favor the growth of other fungi.

    Brown Stain—A rich brown to deep chocolate-brown discoloration of the sapwood of some pines caused by a fungus that acts much like the blue-stain fungi.

    Chemical Brown Stain—A chemical discoloration of wood, which sometimes occurs during the air drying or kiln drying of several species, apparently caused by the concentration and modification of extractives.

    Sap Stain—(See **Stain.**)

    Sticker Stain—A brown or blue stain that develops in seasoning lumber where it has been in contact with the stickers.

Starved Joint. (See **Joint.**)

Static Bending. Bending under a constant or slowly applied load; flexure.

Staypak. Wood that is compressed in its natural state (that is, without resin or other chemical treatment) under controlled conditions of moisture, temperature, and pressure that practically eliminate springback or recovery from compression. The product has increased density and strength characteristics.

Stickers. Strips or boards used to separate the layers of lumber in a pile and thus improve air circulation.

Sticker Stain. (See **Stain.**)

Storage Life. The period of time during which a packaged adhesive can be stored under specific temperature conditions and remain suitable for use. Sometimes called shelf life.

Straight Grained. (See **Grain.**)

Strength. (1) The ability of a member to sustain stress without failure. (2) In a specific mode of test, the maximum stress sustained by a member loaded to failure.

Strength Ratio. The hypothetical ratio of the strength of a structural member to that which it would have if it contained no strength-reducing characteristics (knots, slope-of-grain, shake, etc).

Stress-Wave Timing. A method of measuring the apparent stiffness of a material by measuring the speed of an induced compression stress as it propagates through the material.

Stressed-Skin Construction. A construction in which panels are separated from one another by a central partition of spaced strips with the whole assembly bonded so that it acts as a unit when loaded.

Stringer. A timber or other support for cross members in floors or ceilings. In stairs, the support on which the stair treads rest.

Structural Insulating Board. A generic term for a homogeneous panel made from lignocellulosic fibers (usually wood or cane) characterized by an integral bond produced by interfelting of the fibers, to which other materials may have been added during manufacture to improve certain properties, but which has not been consolidated under heat and pressure as a separate stage in manufacture, said board having a density of less than 31 pcf (specific gravity 0.50) but having a density of more than 10 pcf (specific gravity 0.16).

Structural Lumber. (See **Lumber.**)

Structural Sandwich Construction. A layered construction comprising a combination of relatively high-strength facing materials intimately bonded to and acting integrally with a low-density core material.

Structural Timbers. Pieces of wood of relatively large size, the strength or stiffness of which is the controlling element in their selection and use. Examples of structural timbers are trestle timbers (stringers, caps, posts, sills, bracing, bridge ties, guardrails); car timbers (car framing, including upper framing, car sills); framing for building (posts, sills, girders); ship timber (ship timbers, ship decking); and crossarms for poles.

Stud. One of a series of slender wood structural members used as supporting elements in walls and partitions.

Substrate. A material upon the surface of which an adhesive-containing substance is spread for any purpose, such as bonding or coating. A broader term than adherend. (See **Adherend.**)

Summerwood. (See **Latewood.**)

Surfaced Lumber. (See **Lumber.**)

Symmetrical Construction. Plywood panels in which the plies on one side of a center ply or core are essentially equal in thickness, grain direction, properties, and arrangement to those on the other side of the core.

Tack. The property of an adhesive that enables it to form a bond of measurable strength immediately after adhesive and adherend are brought into contact under low pressure.

Tangential. Strictly, coincident with a tangent at the circumference of a tree or log, or parallel to such a tangent. In practice, however, it often means roughly coincident with a growth ring. A tangential section is a longitudinal section through a tree or limb perpendicular to a radius. Flat-grained lumber is sawed tangentially.

Temperature, Curing. The temperature to which an adhesive or an assembly is subjected to cure the adhesive. Note: The temperature attained by the adhesive in the process of curing (adhesive curing temperature) may differ from the temperature of the atmosphere surrounding the assembly (assembly curing temperature).

Temperature, Setting. (See **Temperature, Curing.**)

Tenon. A projecting member left by cutting away the wood around it for insertion into a mortise to make a joint.

Tension Wood. Abnormal wood found in leaning trees of some hardwood species and characterized by the presence of gelatinous fibers and excessive longitudinal shrinkage. Tension wood fibers hold together tenaciously, so that sawed surfaces usually have projecting fibers, and planed surfaces often are torn or have raised grain. Tension wood may cause warping.

Texture. A term often used interchangeably with grain. Sometimes used to combine the concepts of density and degree of contrast between earlywood and latewood. In this handbook, texture refers to the finer structure of the wood (see **Grain**) rather than the annual rings.

Thermoplastic Glues And Resins. Glues and resins that are capable of being repeatedly softened by heat and hardened by cooling.

Thermosetting Glues And Resins. Glues and resins that are cured with heat but do not soften when subsequently subjected to high temperatures.

Timbers, Round. Timbers used in the original round form, such as poles, piling, posts, and mine timbers.

Timber, Standing. Timber still on the stump.

Timbers. (See **Lumber.**)

Time, Assembly. The time interval between the spreading of the adhesive on the adherend and the application of pressure or heat, or both, to the assembly. Note: For assemblies involving multiple layers or parts, the assembly time begins with the spreading of the adhesive on the first adherend.

Open Assembly Time—The time interval between the spreading of the adhesive on the adherend and the completion of assembly of the parts for bonding.

Closed Assembly Time—The time interval between completion of assembly of the parts for bonding and the application of pressure or heat, or both, to the assembly.

Time, Curing. The period of time during which an assembly is subjected to heat or pressure, or both, to cure the adhesive.

Time, Setting. (See **Time, Curing.**)

Toughness. A quality of wood which permits the material to absorb a relatively large amount of energy, to withstand repeated shocks, and to undergo considerable deformation before breaking.

Tracheid. The elongated cells that constitute the greater part of the structure of the softwoods (frequently referred to as fibers). Also present in some hardwoods.

Transfer. In wood bonding, the sharing of adhesive between a spread and an unspread surface when the two adherends are brought into contact.

Transverse. Directions in wood at right angles to the wood fibers. Includes radial and tangential directions. A transverse section is a section through a tree or timber at right angles to the pith.

Treenail. A wooden pin, peg, or spike used chiefly for fastening planking and ceiling to a framework.

Trim. The finish materials in a building, such as moldings, applied around openings (window trim, door trim) or at the floor and ceiling of rooms (baseboard, cornice, and other moldings).

Truss. An assembly of members, such as beams, bars, rods, and the like, so combined as to form a rigid framework. All members are interconnected to form triangles.

Twist. A distortion caused by the turning or winding of the edges of a board so that the four corners of any face are no longer in the same plane.

Tyloses. Masses of parenchyma cells appearing somewhat like froth in the pores of some hardwoods, notably the white oaks and black locust. Tyloses are formed by the extension of the cell wall of the living cells surrounding vessels of hardwood.

Ultrasonics. (See **Stress-Wave Timing.**)

Vapor Retarder. A material with a high resistance to vapor movement, such as foil, plastic film, or specially coated paper, that is used in combination with insulation to control condensation.

Veneer. A thin layer or sheet of wood.

Rotary-Cut Veneer—Veneer cut in a lathe which rotates a log or bolt, chucked in the center, against a knife.

Sawed Veneer—Veneer produced by sawing.

Sliced Veneer—Veneer that is sliced off a log, bolt, or flitch with a knife.

Vertical Grain. (See **Grain.**)

Vertically Laminated Timbers. (See **Laminated Timbers.**)

Vessels. Wood cells of comparatively large diameter that have open ends and are set one above the other to form continuous tubes. The openings of the vessels on the surface of a piece of wood are usually referred to as pores.

Virgin Growth. The growth of mature trees in the original forests.

Viscoelasticity. The ability of a material to simultaneously exhibit viscous and elastic responses to deformation.

Viscosity. The ratio of the shear stress existing between laminae of moving fluid and the rate of shear between these laminae.

Wane. Bark or lack of wood from any cause on edge or corner of a piece except for eased edges.

Warp. Any variation from a true or plane surface. Warp includes bow, crook, cup, and twist, or any combination thereof.

Water Repellent. A liquid that penetrates wood which, after drying, materially retards changes in moisture content and in dimensions without adversely altering the desirable properties of wood.

Water-Repellent Preservative. A water repellent that contains a preservative which, after application to wood and drying, accomplishes the dual purpose of imparting resistance to attack by fungi or insects and also retards changes in moisture content.

Weathering. The mechanical or chemical disintegration and discoloration of the surface of wood caused by exposure to light, the action of dust and sand carried by winds, and the alternate shrinking and swelling of the surface fibers with the continual variation in moisture content brought by changes in the weather. Weathering does not include decay.

Wet-Bulb Temperature. The temperature indicated by the wet-bulb thermometer of a psychrometer.

Wettability. A condition of a surface that determines how fast a liquid will wet and spread on the surface or if it will be repelled and not spread on the surface.

Wetting. The process in which a liquid spontaneously adheres to and spreads on a solid surface.

White-Rot. In wood, any decay or rot attacking both the cellulose and the lignin, producing a generally whitish residue that may be spongy or stringy rot, or occur as pocket rot.

Wood Failure. The rupturing of wood fibers in strength tests of bonded joints usually expressed as the percentage of the total area involved which shows such failure. (See **Failure, Adherend.**)

Wood Flour. Wood reduced to finely divided particles approximately those of cereal flours in size, appearance, and texture, and passing a 40−100 mesh screen.

Wood Substance. The solid material of which wood is composed. It usually refers to the extractive-free solid substance of which the cell walls are composed, but this is not always true. There is no wide variation in chemical composition or specific gravity between the wood substance of various species, the characteristic differences of species being largely due to differences in extractives, and variations in relative amounts of cell walls and cell cavities.

Wood Wool. Long, curly, slender strands of wood used as an aggregate component for some particleboards.

Workability. The degree of ease and smoothness of cut obtainable with hand or machine tools.

Working Life. The period of time during which an adhesive, after mixing with catalyst, solvent, or other compounding ingredients, remains suitable for use. Also called pot life.

Working Properties. The properties of an adhesive that affect or dictate the manner of application to the adherends to be bonded and the assembly of the joint before pressure application, i.e. viscosity, pot life, assembly time, setting time, etc.

Xylem. The portion of the tree trunk, branches, and roots that lies between the pith and the cambium.

Yard Lumber. (See **Lumber.**)

Index

Metric Conversion Table

To convert from	To	Multiply by
British Thermal Unit (Btu) (international table)	joule (J)	[1]1.055 056 E+03
Btu (international table) · in/(h·ft²·°F) (k, thermal conductivity)	W/(m·K)	1.442 279 E−01
Btu (international table)/(h·ft²·°F) (k, thermal conductance)	W/(m²·K)	5.678 263 E+00
day (mean solar)	second (s)	8.640 000 E+04
degree (angular)	radian (rad)	1.745 329 E−02
degree Celsius (°C)	kelvin (K)	[K = °C + 273.15]
degree Fahrenheit (°F)	kelvin (K)	[K = (°F + 459.67)/1.8]
degree Fahrenheit (°F)	Celsius (°C)	[°C = (°F − 32)/1.8]
°F·h·ft²/Btu (international table) (R, thermal resistance)	K·m²/W	1.761 102 E−01
°F·h·ft²/(Btu (international table)·in) (R, thermal resistivity)	K·m/W	6.933 471 E+00
electron volt (X10⁶) (MEV)	joule (J)	1.602 19 E−13
foot (ft)	meter (m)	3.048 E−01
foot−pound−force (ft−lbf)	joule (J)	1.355 818 E+00
foot² (ft²)	meter² (m²)	9.290 304 E−02
gallon (U.S. liquid)	meter³ (m³)	3.785 412 E−03
hour (mean solar)	second (s)	3.60 E+03
inch (in)	meter (m)	2.540 E−02
inch² (in²)	meter² (m²)	6.451 600 E−04
inch−pound−force (in-lbf)	joule (J)	1.129 848 E−01
mile (international)	meter (m)	1.609 344 E+03
mile/hour	meter/second (m/s)	4.470 400 E−01
minute (mean solar)	second (s)	6.0 E+01
month (mean)	second (s)	2.628 E+06
pound−force	Newton (N)	4.448 222 E+00
pound−force/foot² (psf)	pascal (Pa)	4.788 026 E+01
pound−force/inch² (psi)	pascal (Pa)	6.894 757 E+03
pound−mass/foot³ (pcf)	kilogram/meter³ (kg/m³)	1.601 846 E+01
rad(X10⁶) (megarad)	gray (Gy)	1.0 E+04
year (sidereal)	second (s)	3.155 815 E+07

[1] The letter E (for exponent), the plus or minus symbol, and the two digits which follow indicate the power of 10 by which the number preceding the E must be multiplied to obtain the correct conversion factor. For example:

$1.055056\ E + 03 = 1.055056 \times 10^3 = 1055.056$; $1.442279\ E − 01 = 1.442279 \times 10^{-1} = 0.1442279$.

Reference: American Society for Testing and Materials, "Standard for Metric Practice," E 380−79 (approved 12/19/79), Philadelphia, PA: ASTM.